Amateur Radio Operating Manual

FIFTH EDITION

Editor: Ray Eckersley, G4FTJ

Contributors to this edition:
John Bazley, G3HCT
Geoff Bond, G4GJB
Geoff Brown, G4ICD
T C Bryant, GW3SB
Dave Coomber, G8UYZ
Mike Dixon, G3PFR
George Dobbs, G3RJV
Geoff Dover, G4AFJ
Michael Fadil, G4CCA
Don Field, G3XTT
Bill Hall, G4FRN
John Hall, G3KVA
Dennis Kitchen, G0FCL
Dave Lawley, G4BUO
Chris Lorek, G4HCL
Robin Page-Jones, G3JWI
Dave Pick, G3YXM
Phillip Smith, GW1XBG
Steve Telenius-Lowe, G4JVG
Steve Thompson, G8GSQ
Robert Vickers, G3ORI
Mike Wooding, G6IQM

Radio Society of Great Britain

Published by the Radio Society of Great Britain, Cranborne Road, Potters Bar, Herts EN6 3JE.

First published 1979
Fifth edition 2000

ISBN 1 872309 63 1

Cover design: Tyga, Potters Bar.
Illustrations: Derek Cole and Bob Ryan (Radio Society of Great Britain), and Ray Eckersley.
Typography: Ray Eckersley, Seven Stars Publishing, Marlow.
Production: Mike Dennison.

Disclaimer

Printed in Great Britain by Whitstable Litho.

There is a web page for this book at www.rsgb.org/books/extra/opman.htm

Contents

Preface iv

1   The Amateur Service 1

2   Setting up a station 7

3   Operating practices and procedures 18

4   DX 33

5   Contests 87

6   Mobile and portable operation 114

7   Amateur satellite and space communications 129

8   Data communications 139

9   Image techniques 156

10 Special-event stations 165

Appendices

1   Continental and regional maps 173

2   International callsign series holders 179

3   Callsign list 203

4   DXCC entities 208

5   Amateur Service frequency allocations 216

6   Foreign-language phone contacts 219

7   Band plans 224

8   Beacons 233

9   Repeaters 246

10 Packet radio mailboxes 252

Index 256

Preface

Welcome to the fifth edition of the *Amateur Radio Operating Manual*. The aim of this edition, as with the previous ones, is to present a detailed guide to good operating practice on the amateur bands, combined with a comprehensive set of operating aids. It is not intended to be a technical book and where necessary it refers the reader to appropriate sources of information.

In the interval since the last edition was published, some five years ago, the Internet has become the premier source of up-to-date amateur radio information worldwide. This edition reflects this revolution, with many references to web pages – even to its own page at the RSGB web site! It also includes for the first time advice on organising a DXpedition, which will be of great interest to those contemplating operation abroad.

The editor is greatly indebted to the many amateurs who have enthusiastically contributed in some way, large or small, to this book over the last 20 years. Several chapters contain revised versions of material supplied by various contributors to previous editions, and due acknowledgement is made to this fact.

Chapter 5 includes material taken from the *VHF Contesting Guide* written by the VHF Contests Committee and published by the RSGB.

Appendix 6 is based on the memory sheets devised by the late L Craven, G4EQI, for which the following provided translations and assistance: W Baker, G3HDQ, J Gaudron, F9OJ, G G Goodfield, GW4CNL, and H Hahn, DJ7TF.

The *RSGB Prefix Guide* by John Forward, G3HTA, was used as a reference for callsign data in various parts of the book, and is recommended for those who require more detailed information on this topic.

Grateful thanks are due to several amateurs for helpful comments on the previous edition, in particular Jack Barker, M0ABP, the late Etienne Heritier, HB9DX, and R E Parkes, G3REP.

Ray Eckersley, G4FTJ

1 The Amateur Service

AMATEUR radio is an officially recognised radio service, the *Amateur Service*, bound and protected by both national and international legislation. It also has its own codes of operating practice and procedures which set it apart from other radio services. Although many of these are unofficial (though nevertheless established as good practice), many others are recommended and standardised by national and international societies of radio amateurs.

In this first chapter a brief look is taken at this formal background to amateur radio as it affects operating matters, in order to give some perspective to the more detailed practical discussions in later chapters.

INTERNATIONAL TELECOMMUNICATION UNION

The ITU is the intergovernmental agency responsible for the co-ordination, standardisation and planning of world telecommunications. This United Nations organisation, founded in 1865, unites the telecommunication administrations of over 150 member countries. Its headquarters are located in Switzerland [1].

ITU regulations cover the activities of all telecommunication services, including amateur radio, and are published by the general secretariat of the ITU [2]. These regulations define international telecommunication law, and therefore are the cornerstone of amateur licence conditions in all countries.

From time to time, the ITU organises conferences attended by delegates from national administrations where aspects of the current regulations are reviewed to take into account developments in communications techniques. Major conferences used to be held every 20 years or so, and the last of this series took place in 1992 (the 1992 World Administrative Radio Conference). At this conference it was realised that such conferences in future were going to have to take place more frequently because technology was changing so rapidly and more countries were requiring access to sections of the spectrum. It was also realised that they would have to be more specialised, either in subject or range of frequencies under consideration. It was therefore decided to hold conferences every two years in future.

It is important that amateurs understand the basis of the service that they use.

Definitions

Within the ITU Radio Regulations the Amateur Service is defined as:

A radiocommunication service for the purpose of self-training, intercommunication and technical investigations carried out by amateurs, that is, by duly authorised persons interested in radio technique solely with a personal aim and without pecuniary interest.

It should be noted that the ITU considers amateur-satellite activities as a separate *Amateur-Satellite Service* defined as:

A radiocommunication service using space stations on earth satellites for the same purposes as those of the Amateur Service.

The section of the regulations dealing specifically with amateur radio is Article 32, and this is shown in the panel overleaf.

Callsigns

It is a requirement of the regulations that all radio amateurs identify themselves by transmitting an identification code, known as the *callsign*, at short intervals during their transmissions.

The initial characters of the callsign denote the country to which the amateur station belongs and is operating from. The ITU allocates the telecommunication administration of each country blocks of letters and numbers in order to form callsigns, not just for amateur stations but for all radio stations in that country. The ITU allocation blocks are of three types:

(a) letter-letter-letter, eg LAA–LNZ;
(b) digit-letter-letter, eg 2AA–2ZZ; and
(c) letter-digit-letter, eg H4A–H4Z.

Amateur callsigns are normally made up of the first *two* characters from the allocation, followed by a *single digit* and then a *group of not more than three letters*. However, if a country has the whole of a letter-letter-letter block, eg WAA–WZZ, then it is also entitled to form callsigns consisting of the first letter in front of the digit (in this case W).

For example, the UK has been allocated GAA–GZZ, MAA–MZZ and 2AA–2ZZ. This means that UK amateur callsigns can have GA, GB, GC etc as the first two characters. (As the UK has the whole GAA–GZZ block, it is also entitled to use the single letter G.) Similarly, it can form callsigns starting with MA, MB, MC etc and 2A, 2B, 2C etc.

The first two characters of the allocation are usually unique to a country but the ITU occasionally allocates 'half-series': eg 3DA–3DM to Swaziland and 3DN–3DZ to Fiji. In practice, Swaziland callsigns commence with 3DA and Fiji amateur radio callsigns with 3D2.

Certain limitations are imposed by the ITU to ensure that callsigns issued cannot be confused with internationally agreed distress signals such as SOS (eg G4SOS was not issued) or the international Q-code (QAA–QZZ was not allocated by the ITU and also, for example, G4QAA was not issued).

Further information on the allocation of callsigns is given in the section later on dealing with the work of the national telecommunications administration.

Designation of emissions

There are very many ways in which a radio signal can be modulated so as to convey information. In order to define the types of emission a special code has been internationally agreed and implemented by the ITU.

The code is in two parts, the first specifying the *necessary bandwidth* and the second the *classification of emission*.

ARTICLE 32

I. Amateur Service

1. Radio communications between amateur stations of different countries shall be forbidden if the administration of one of the countries concerned has notified that it objects to such radio communications.

2. (1) When transmissions between amateur stations of different countries are permitted, they shall be made in plain language and shall be limited to messages of a technical nature relating to tests and to remarks of a personal character for which, by reason of their unimportance, recourse to the public telecommunications service is not justified.

(1A) It is absolutely forbidden for amateur stations to be used for transmitting international communications on behalf of third parties.

(2) The preceding provisions may be modified by special arrangements between the administrations of the countries concerned.

3. (1) Any person seeking a licence to operate the apparatus of an amateur station shall have proved that he is able to send correctly by hand and to receive correctly by ear, texts in Morse code signals. The administrations concerned may, however, waive this requirement in the case of stations making use exclusively of frequencies above 30MHz.

(2) Administrations shall take such measures as they judge necessary to verify the operational and technical qualifications of any person operating the apparatus of an amateur station.

4. The maximum power of amateur stations shall be fixed by the administrations concerned, having regard to the technical qualifications of the operators and to the conditions under which these stations are to operate.

5. (1) All the general rules of the Convention and of these Regulations shall apply to amateur stations. In particular, the emitted frequency shall be as stable and as free from spurious emissions as the state of technical development for such stations permits.

(2) During the course of their transmissions, amateur stations shall transmit their callsign at short intervals.

II. Amateur-Satellite Service

6. The provisions of Section I of this Article shall apply equally, as appropriate, to the Amateur-Satellite Service.

7. Space stations in the Amateur-Satellite Service operating in bands shared with other services shall be fitted with appropriate devices for controlling emissions in the event that harmful interference is reported in accordance with the procedure laid down in Article 22. Administrations authorising such space stations shall inform the IFRB and shall ensure that sufficient earth command stations are established before launch to guarantee that any harmful interference which might be reported can be terminated by the authorising administration.

The necessary bandwidth is specified as follows: between 0.001 and 999Hz in hertz (H); between 10 and 999kHz in kilohertz (K); between 1.00 and 999MHz in megahertz (M); between 100 and 999GHz in gigahertz (G). For example, 400Hz would be '400H', 2.4kHz as '2K40' and 12.5kHz as '12K5'.

The classification is specified by three symbols. The first denotes the type of modulation of the main carrier, the second the nature of the modulating signal(s) and the third the type of information to be transmitted. Table 1 gives the symbols and their meanings. Some examples relevant to amateur radio are also shown.

Note that for the purposes of the UK licence, modulation used only for short periods and for incidental purposes, such as

Table 1. Classification of emissions

FIRST SYMBOL – Type of modulation of main carrier

1. Emission of unmodulated carrier: N.
2. Emission in which the main carrier is amplitude modulated including cases where sub-carriers are angle modulated. Double sideband: A. Single sideband, full carrier: H. Single sideband, reduced or variable carrier: R. Single sideband, suppressed carrier: J. Independent sideband: B. Vestigial sideband: C.
3. Emission in which the main carrier is angle modulated. Frequency modulation: F. Phase modulation: G.
4. Emission in which the main carrier is amplitude or angle modulated either simultaneously or in a pre-arranged sequence: D.
5. Emission of pulses. Unmodulated sequence of pulses: P. A sequence of pulses (a) modulated in amplitude: K, (b) modulated in width/duration: L, (c) modulated in position/phase: M, (d) in which the carrier is angle modulated during the period of the pulse: Q, (e) which is a combination of the foregoing or is produced by other means: V.
6. Cases not covered above, in which an emission consists of the main carrier modulated, either simultaneously or in a pre-established sequence, in a combination of two or more of the following modes – amplitude, angle, pulse: W.
7. Cases not otherwise covered: X.

Note: Emissions where the main carrier is directly modulated by a signal which has been coded into quantised form (eg pulse code modulation) should be designated by A, H, R, J, B, C, F, or G as appropriate.

SECOND SYMBOL – Nature of signal(s) modulating main carrier

1. No modulating signal: 0.
2. A single channel containing quantised or digital information without the use of a modulating subcarrier (excluding time-division multiplex): 1.
3. A single channel containing quantised or digital information with the use of a modulating subcarrier (excluding time-division multiplex): 2.
4. A single channel containing analogue information: 3.
5. Two or more channels containing quantised or digital information: 7.
6. Two or more channels containing analogue information: 8.
7. Composite system with one or more channels containing quantised or digital information, together with one or more channels containing analogue information: 9.
8. Cases not otherwise covered: X.

THIRD SYMBOL – Type of information to be transmitted

1. No information transmitted: N.
2. Telegraphy – for aural reception: A
3. Telegraphy – for automatic reception: B.
4. Facsimile: C.
5. Data transmission, telemetry, telecommand: D.
6. Telephony (including sound broadcasting): E.
7. Television (video): F.
8. Combination of the above: W.
9. Cases not otherwise covered: X.

Note: In this context the word 'information' does not include information of a constant, unvarying nature such as provided by standard frequency emissions, continuous wave and pulse radars etc.

EXAMPLES OF THE USE OF EMISSION CODES

Telephony (speech)

Single sideband, suppressed carrier (SSB)	J3E
Frequency modulation (FM)	F3E
Phase modulation (PM)	G3E
Amplitude modulation	A3E

Morse code

Hand sent, on/off keying of carrier	A1A
Hand-sent, on/off keying of the audio tone (FM transmitter)	F2A

RTTY/AMTOR

Direct frequency shift keying of carrier	F1B
Frequency shift keyed audio tone (FM transmitter)	F2B
Frequency shift keyed audio tone (SSB transmitter)	J2B

Packet/Data

Direct frequency shift keying of carrier	F1D
Frequency shift keyed audio tone (FM transmitter)	F2D
Frequency shift keyed audio tone (SSB transmitter)	J2D

Television

Vestigial sideband (AM transmitter)	C3F
Slow-scan TV (SSB transmitter)	J2F

Facsimile

Frequency shift keyed audio tone (SSB transmitter)	J2C

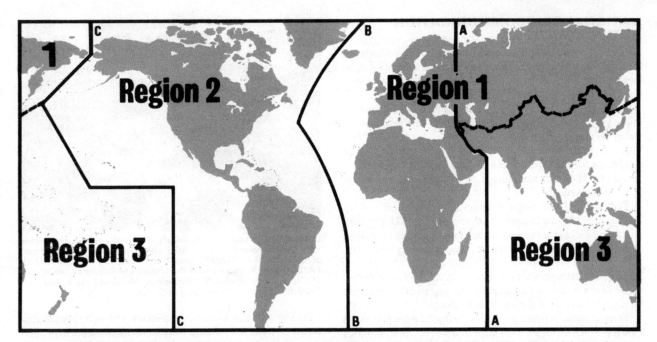

Fig 1. The three ITU regions. Region 1 includes the area limited on the east by line A (lines A, B and C are defined below) and on the west by line B, excluding any of the territory of Iran which lies between these limits. It also includes that part of the territory of Turkey and the former USSR lying outside of these limits, the territory of the Mongolian People's Republic, and the area to the north of the former USSR which lies between lines A and C.

Region 2 includes the area limited on the east by line B and on the west by line C.

Region 3 includes the area limited on the east by line C and on the west by line A, except the territories of the Mongolian People's Republic, Turkey, the former USSR, and the area to the north of the former USSR. It also includes that part of the territory of Iran lying outside of those limits. Lines A, B and C are defined as follows:

Line A extends from the North Pole along meridian 40°E to parallel 40°N, thence by great-circle arc to the intersection of meridian 60°E and the Tropic of Cancer; thence along the meridian 60°E to the South Pole.

Line B extends from the North Pole along meridian 10°W to its intersection with parallel 72°N, thence by great-circle arc to the intersection of meridian 20°W and parallel 10°S; thence along meridian 20°W to the South Pole.

Line C extends from the North Pole by great-circle arc to the intersection of parallel 65° 30'N with the international boundary in Behring Strait, thence by great-circle arc to the intersection of meridian 165°E and parallel 50°N; thence by great-circle arc to the intersection of meridian 170°W and parallel 10°N, thence along parallel 10°N to its intersection with meridian 120°W; thence along meridian 120°W to the South Pole

identification or calling, may be ignored when calculating the emission designator. Double sideband emissions with reduced or suppressed carrier are included in the first character A.

Frequency allocations

One of the most important tasks of the ITU is to allocate parts of the radio spectrum to various radio services in such a way that the spectrum is utilised as efficiently as possible without mutual interference. The claims on the available frequency space are many and are continually changing.

The range of frequencies covered by the present ITU regulations is 9kHz–400GHz. For the allocation of these frequencies, and for various administrative reasons, the world has been divided into three regions (see Fig 1).

Region 1 comprises Europe and Africa, and including Mongolia, Turkey and the whole of the CIS. Region 2 comprises North and South America, while Region 3 comprises Asia, Australasia and Oceania.

The more formal ITU definitions of these areas are given in Fig 1. Note that these three ITU regions should not be confused with the ITU broadcasting zones, which are used for scoring purposes in some amateur radio contests (see Appendices 1 and 3).

Sometimes the ITU finds it necessary to allocate part of the radio spectrum to two or more radio services. Where this is done it is always specified which if any of the services has priority over the others, by placing the service in one of three categories: *primary, permitted* or *secondary*.

Permitted and primary services have equal status except that in the preparation of ITU frequency plans the primary service has prior choice of frequencies.

Services allocated a band on a secondary basis must not cause harmful interference to stations within that allocation having primary or permitted status, and cannot claim protection from harmful interference caused by the latter. They can, however, claim protection from interference generated by stations in the same service or other secondary services.

What this means is that primary and permitted services have the 'right of way' over secondary services. When operating in a shared allocation in which the Amateur Service is a secondary service, the amateur operator must take care not to cause interference to primary users. If these start transmitting on the same frequency, even while an amateur contact is already in progress, the amateurs must change frequency and leave the channel clear.

Appendix 5 shows all amateur allocations between 9kHz and 275GHz in the three ITU regions, the status of these allocations and the nature of services (if any) sharing the allocations. The information is shown in generalised form, and may not apply to particular countries within each ITU region. Most variations from these allocations are however given in Chapter 4 on a band-by-band basis.

NATIONAL TELECOMMUNICATION ADMINISTRATION

This is the government department in each country which has the responsibility of controlling all its telecommunication services and the issue of licences. This same body usually represents that government at the ITU conferences.

In the UK this function is undertaken by the Radiocommunications Agency (RA) of the Department of Trade and Industry (DTI).

Although based on the ITU regulations, the amateur licence conditions for each country can be quite different: some countries do not allow amateur radio at all, while in others its operation is restricted.

The national telecommunication administration assigns callsigns to individual amateur stations using its ITU-allocated character blocks as previously noted. It may choose to use these to denote the geographical area or licence category of the station. Similarly the single digit following the ITU characters is sometimes used to denote licence category or geographical area, while in other cases it has no particular significance and may be regarded as part of the serial letter group.

The serial letters following the single digit are the part of the callsign which identifies individual stations. In a few countries the first or last letter of this group indicates geographical area or licence category. The serial letters are usually assigned in strict alphabetical order, but in some smaller countries it is possible to request letters to correspond with the licensee's initials.

The part of the callsign which precedes the serial letters is known as the *prefix*. If the single digit is simply part of the serial allocation it is not included in the prefix, though it is customary in amateur usage not to split two consecutive figures, eg C31 (not C3, which is more correct) is often given as the Andorran prefix.

Notes on systems are given for each country where known in Appendix 2.

Sometimes the administration will temporarily allocate special callsign prefixes outside its normal series to stations commemorating some national event or celebration, and this results in some strange callsigns being heard on the air from time to time which may be difficult to identify. The important thing to note is that *any* prefix so used must conform with the ITU block allocation(s) for the country concerned, and the list of these in Appendix 3 should be consulted in cases of doubt.

It is also possible for a callsign to have a *suffix*, which is always separated from the rest of the callsign by a solidus (/). A suffix usually indicates temporary operation away from the registered address of the station. In the UK, for example, '/M' denotes a mobile station and '/P' a temporary location. Another use for a suffix is to denote a licence issued to a foreign amateur (see below).

Operation abroad

Most countries allow foreign amateurs to operate within their boundaries, but the conditions are set by the national administration and vary widely. In some cases, the administration acts unilaterally and permits amateurs from certain countries to apply for licences while in others there must exist a formal reciprocal agreement with the amateur's country.

A welcome development is the increasing number of countries worldwide which permit temporary operation according to Recommendation T/R-61-01 of the European Conference of Postal and Telecommunication Administrations (CEPT), which is a group of national telecommunication administrations from over 40 European countries. Once an

IARU meetings take place regularly all over the world, and liaison with the telecommunication administrations is an important part of its work. Here IARU Region 1 chairman Lou Van de Nadort, PA0LOU (left) is shown with the Minister of Telecommunications of Oman, H E Ahmed Bin Suwaidan Al-Balushi, at a Region 1 Executive Committee meeting in Oman

administration has confirmed that its amateur radio licence conforms to the CEPT minimum standard, its amateurs may operate portable or mobile within the other participating countries with the minimum of formality. The text of the Recommendation is shown in the panel. It should be noted that not all member administrations have implemented this Recommendation but over 20 have done so and the list is growing all the time. In addition, some non-CEPT countries such as Canada, Israel, New Zealand, Peru and the USA are prepared to authorise operation on similar lines.

Note: the CEPT Radio Amateur Licence has been incorporated into the UK licence, and is not a separate document.

The temporary callsigns issued to foreign amateurs vary. Usually the foreign country prefix is added to the amateur's own call before the call (but sometimes after it in non-CEPT countries) and separated by a solidus (/). For example, G4FTJ may become OE/G4FTJ in Austria but G4FTJ/OA in Peru. In some smaller countries, or in other cases for extended operating periods, a separate callsign may be issued. Since CEPT operation is mobile or portable, the host country will require a /M or /P suffix to be added.

Further information on operating abroad is given in Chapter 6 and Appendix 2.

INTERNATIONAL AMATEUR RADIO UNION

The Amateur Service and Amateur-Satellite Service have their own international organisation which has the specific objective of promoting, protecting and advancing these services within the framework of regulations established by the International Telecommunication Union. It also provides support to member societies in the pursuit of these objectives at the national level, with special reference to the following:

(a) representation of the interests of amateur radio at and between conferences and meetings of international telecommunications organisations;

(b) encouragement of agreements between national amateur radio societies on matters of common interest;

(c) enhancement of amateur radio as a means of technical self-training for young people;

(d) promotion of technical and scientific investigations in the field of radio communication;

RECOMMENDATION T/R 61-01 (NICE 1985)

Concerning the CEPT Radio Amateur Licence

The European Conference of Posts and Telecommunications Administrations,

Whereas:

(a) the Amateur Service is a telecommunications service governed by the Radio Regulations and by the national regulations of the countries of the CEPT;

(b) administration are responsible, in accordance with Article 32 Para 3 of the Radio Regulations, for taking such measures as they judge necessary to verify the operational and technical qualifications of radio amateurs, and for ensuring that they undergo a Morse telegraphy test, permitting them to operate on frequencies below 30MHz;

(c) the issue and administration of temporary authorisations based on existing bilateral agreements involves a considerable increase in work for CEPT member administrations;

(d) technical developments and the growing standardisation of amateur radio equipment in the CEPT member countries mean a reduction of the risks of harmful interference;

(e) certain CEPT member administration have concluded or are drawing up agreements to simplify the current procedure for the issue of temporary licences;

Noting, however:

(f) that this Recommendation bears no relation to the import and export of amateur radio equipment, which is subject only to the relevant customs regulations,

Recommends:

that the CEPT member administration recognise the principle of a 'CEPT radio amateur licence' under the conditions specified in Appendix 1, on which the member administration will levy neither duties nor taxes, only the Administration issuing the licence being entitled to do so.

Appendix I – General conditions for allocation of a CEPT licence

1. General provisions relating to the 'CEPT amateur radio licence'

The 'CEPT amateur radio licence' will take a form similar to a national licence or a special document issued by the same authority, and will be drafted in the national language, and in German, English and French; it will be valid for non-residents only, for the duration of their temporary stay in CEPT member countries having adopted the Recommendation, and within the limit of validity of the national licence. Radio amateurs holding a temporary licence may not benefit from the provisions of the Recommendation.

The minimum requirements for a CEPT licence will be:

- a declaration according to which the holder is authorised to utilise his amateur radio station in accordance with this Recommendation in countries where the latter applies;
- the name and address of the holder;
- the callsign;
- the CEPT licence class;
- the validity;
- the issuing authority.

It is recommended that a list be added of the administrations applying the Recommendation.

2. Licence classes

Each of the CEPT classes described below will only be considered as equivalent to a national class in cases where conditions of utilisation in another country are not considerably broader than those in the country where the licence was issued. Equivalents are given in columns 3 and 4 of the Table in Appendix II *[not shown in this book]*.

Class 1 – This class permits utilisation of all frequencies allocated to the Amateur Service and authorised in the country where the station is to be operated. It will be open only to those amateurs who have proved their competence with Morse code to their own administration.

Class 2 – This class limits utilisation of stations to frequency allocations above 144MHz *[Note: revised in 1992 to 30MHz]*, authorised for the Amateur Service in the country where the station is to be installed.

3. Conditions of utilisation

3.1. The holder shall present his CEPT radio amateur licence on request to the supervisory authorities in the country visited.

3.2. Authorisation is granted for utilisation of a portable or mobile station only. A portable station shall, for the purposes of this Recommendation, include any station using mains electricity at a temporary location, eg a hotel.

3.3. Authorisation is also granted for utilisation of the station of a radio amateur holding a licence in the host country.

3.4. The holder shall observe the provisions of the Radio Regulations, this Recommendation and the regulations in force in the country visited; he shall, furthermore, respect any restriction placed on him concerning local conditions of a technical nature regarding the public authorities.

(e) promotion of amateur radio as a means of providing relief in the event of natural disasters;

(f) encouragement of international goodwill and friendship;

(g) support of member societies in developing amateur radio as a valuable national resource, particularly in developing countries; and

(h) development of amateur radio in those countries not represented by member societies.

The International Amateur Radio Union was founded in 1925 and its International Secretariat is located at ARRL HQ in Newington, Connecticut, USA; it has separate organisations in each of the ITU regions.

Membership is restricted to national societies just as the members of the ITU are national administrations, not radio services or individuals. There are nearly 150 national societies which are members of the IARU. The largest region is Region 1 which contains more than 80 societies located in Europe,

Africa, the Middle East and the Commonwealth of Independent States.

In a similar way to that in which the ITU regulates and co-ordinates frequency allocations as a whole, the IARU regulates and co-ordinates amateur activities within individual amateur bands. This is on a voluntary basis within the UK but mandatory in some other countries. Whether mandatory or voluntary, they should always be obeyed because they have resulted from long debate and discussion within the world's national amateur radio societies. Band plans are discussed in Chapter 3, and the UK plans are detailed in Appendix 7.

The IARU also agrees and makes recommendations concerning specialised activities such as meteor scatter and direction finding.

Another valuable service, the IARU Monitoring System (IARUMS), monitors unauthorised transmissions by other services within amateur allocations. Volunteers located throughout the world record the characteristics of the signal and if possible identify the country of origin. These reports are collected and collated, and the administrations notified that they are causing problems. This can be very effective. More details of the RSGB Intruder Watch, which forms part of the IARU Monitoring System, are given in the *RSGB Yearbook* [4].

NATIONAL SOCIETY

The fourth arbiter of operating standards and practice is the national society of the country concerned. Recommendations may come from the specialist committees within each society whose recommendations are considered and debated at regional conferences. Within the UK these areas are represented by the RSGB Licensing Advisory Committee, Microwave Committee, VHF Committee and the HF Committee.

The Amateur Radio Observation Service is an advisory and reporting service of the RSGB which is intended to assist radio amateurs and others who may be affected by problems which occur within the amateur bands or which develop on other frequencies as a result of amateur transmissions. The Service investigates reports of licence infringements, or instances of poor operating practice which might bring the Amateur Service into disrepute.

After investigation and where there is evidence of deliberate malpractice or malicious abuse of Amateur Radio facilities, a formal report may be made to the appropriate authorities. This report will contain sufficient detail and evidence to enable further investigations to be made and the authorities may take such action as is appropriate. However, AROS prefers to settle problems – great or small – within the Amateur Service. Problems arising are referred to the authorities as a last resort. For more details, see the *RSGB Yearbook* [4].

FURTHER INFORMATION

[1] International Telecommunication Union, Place des Nations, CH-1211 Geneva 20, Switzerland. Web: www.itu.ch.
[2] *Radio Regulations,* ITU, Geneva, 1994.
[3] International Amateur Radio Union, Box 310 905, Newington, Connecticut 06131-0905, USA. Web: www.iaru.org.
[4] *RSGB Yearbook*, ed M Dennison, G3XDV, RSGB (published annually).

2 Setting up a station

MOST radio amateurs begin as listeners, perhaps using a simple home-constructed receiver or a factory-made general-coverage receiver. As interest in the hobby grows, various items of equipment are then added to the basic set-up – perhaps a pair of headphones to assist concentration on weak signals, an audio filter for CW reception or an antenna tuning unit. Fairly soon it becomes inconvenient to set this system up on the kitchen table for each listening session, and a permanent place in the home for it (known as the *shack)* is required, often to the consternation of the rest of the family!

Later on, when the coveted licence arrives, a fresh flurry of building, buying and borrowing takes place, and another amateur transmitting station is born, usually to be completely transformed just a few months later.

This chapter is intended to assist with some of the decisions involved in this process, possibly at risk of making them sound too complex and important. This is of course not so – a little common sense as an antidote to an overdose of enthusiasm works wonders!

CHOOSING EQUIPMENT

When you're setting up a station, or adding to your existing equipment, you should ideally think of your station as a system. Even if you're just looking for a low-cost 70cm handheld to use when out and about, remember that you could also use this from home, either connected to a rooftop antenna from your shack or, under UK licensing conditions, even from your living room so as to remotely use a high-power transceive system in your loft or outdoor shack. Many amateurs already have a PC in their home, so think about the possibilities of combining this with your station to add to its capabilities, ie adding modes such as SSTV and data with no extra hardware, rather than buying an 'all-singing, all-dancing' transceiver with capabilities you might not initially need.

If you're starting out for the first time, then take a step back and consider where your interests lie. To try to become equipped for every possible operating mode on every allocated band is still a dream for virtually every amateur. First of all, are your interests primarily VHF/UHF or HF? Or maybe both? There are a number of combined HF, 6m, 2m and 70cm transceivers available in a compact size that can offer you a taste of all these bands. Alternatively, if you're aspiring to a licence at some time in the future then maybe a wide-coverage receiver,

possibly an all-mode (ie SSB, CW and FM etc) scanner typically covering 100kHz to 2000MHz, could be the thing for you to let you decide what you're interested in.

Many prospective amateurs put off buying a transceiver until they get the magical pass slip in their hands, and then rush off and buy the most attractive new transceiver they see at their local dealer or advertised in the amateur press. But do consider previously used equipment – you can often get a good bargain which could give you virtually the same performance and features at a fraction of the price. You can the use this until you 'find your feet', and upgrade to a higher-specification (and higher-price) transceiver should you feel the need, or indeed buy further equipment instead for different bands and operating modes.

Try before you buy

Why not spend a half or even a full day at your local amateur radio dealer and actually try the equipment out on the air. Get their opinion on different types of equipment. Many established dealers have demonstration facilities with equipment linked to antennas for you to try, and some will even keep you refreshed with cups of coffee throughout your visit!

You could be tempted to buy a small dual-band handheld as a 'main' base station for 2m/70cm FM. But as soon as you connect a rooftop antenna you might well find it continually suffers from out-of-band interference, preventing you having

Your local dealer is a good place to try different sets out before you buy *(photo courtesy Waters and Stanton)*

A second-hand multi-mode transceiver like this FT290 allows you to explore various VHF modes

contacts. Micro-miniature HF equipment is often ideal for mobile use and similarly useful from home with a simple antenna. But it won't offer you the performance of a top-flight base transceiver to squeeze that weak DX signal out of the QRM.

Most, if not all, equipment manufacturers publish comprehensive leaflets on their equipment, which naturally reflect their equipment in the best possible light. The technical specifications are often bewildering to many amateurs, and may not be in a form where you can compare 'like with like' between manufacturers. Monthly magazines, such as *RadCom* and *Radio Today* each published by the RSGB, and *QST* published by the ARRL in the USA, feature equipment reviews with full technical measurements, each magazine using their own methods of similar measurements. From these you can compare the technical performance between sets, as well as gaining a useful and independent insight to the good and bad points of each.

Buying new

Once you've decided upon a 'short list', or even a specific receiver or transceiver, your first port of call could very well be the local dealer you tried the equipment at. They're likely to be able to offer you a good back-up, probably also with an on-site service facility. Although the reliability of equipment gets better each year with each new model, it's comforting to know you won't be kept waiting while your precious rig gets sent away for repair to another service centre by the dealer. Mail order can be an attractive proposition price-wise, but remember that you'll be dealing with someone on the end of a phone rather than face to face – the latter usually gets any potential future problems sorted out rather more quickly. Beware of 'box shifters' with no back-up facilities. If you're buying from abroad, then check whether your country's manufacturer's (ie Yaesu, Icom, Kenwood etc) authorised distributor of your chosen set can offer an in-country back-up in case of any warranty repairs, otherwise you may face hefty overseas carriage charges and delays. Check also that the equipment parameters are suitable for use in your country, for example US mains-operated equipment may be set for 110V and not 230V as for European Union use. Also 2m and 70cm band limits, repeater splits and channel spacing can vary between countries, especially so between the USA and the EU.

Second-hand

Many dealers have a well-stocked second-hand shelf, usually from equipment 'traded in' which they usually offer with a guarantee of a few months, and amateur radio magazines carry classified readers ads offering equipment from private individuals. The golden rule when buying from private ads is make sure you buy from someone you trust. If at all possible visit the seller to try the equipment on air, and take an experienced amateur with you for a 'second opinion'. Some new transceivers offer

digital signal processing as an enhancement, either as an audio-based DSP, or IF based, the latter being infinitely preferable. Audio DSP can easily be added on as an external accessory to an earlier second-hand radio. You can even build your own DSP unit from a kit, or buy a new or second-hand commercial unit. Many second-hand transceivers may also have been fitted with additional optional IF filters which would otherwise add substantially to the cost of a new transceiver.

Conclusion

Step back and see what you really want, rather than 'everything'. Examine the options, new or second-hand, and where to buy. Read independent reviews and try the equipment out before buying. Then, relax and have fun on the air with the knowledge that you haven't rushed out and made a hasty and possibly expensive decision.

CHOOSING ANTENNAS

The choice of an antenna usually comes down to one making the best use of a particular location but the beginner is unlikely to choose the right one if the elementary theory of antennas is not understood, and it also pays to be conversant with the basic details of all the popular antennas before making a choice. Information on both these aspects is given in references [3–5]. The beginner is advised to start by building simple wire antennas such as dipoles which should work first time.

If a beam antenna is required, then it is probably best to buy a commercial one which should work right away. Construction of beam antennas is definitely not for the newcomer. Although such antennas have the useful property of increasing both effective radiated power and received signals in the desired direction, it should not be thought that it is impossible to work DX without them, particularly on HF. A gain of 5dB, typical of many HF beam antennas, corresponds to only 1–2 S-points and under favourable conditions may only mean the difference between a strong signal and an extremely strong signal at the other end. In general a lower-performance antenna may have the effect of decreasing the amount of time that DX can be worked rather than prohibiting it altogether, and this may be perfectly acceptable, particularly near the peak of the sunspot cycle on HF, or in summer when sporadic-E gives frequent 'lifts' on VHF.

Many other factors have to be taken in account in choosing an antenna for installation near the home, such as space, whether a mast or tower can be erected, the attitude of neighbours, whether planning permission may be required etc. Some further notes on siting antennas are given later on.

The newcomer should be careful not to underestimate the importance of the correct choice of feeder, particularly on VHF and UHF. Losses in poor-quality coaxial cable can be unacceptably high at these frequencies, and many VHF/UHF enthusiasts find it worthwhile to spend quite large sums of money on their feeder system. The characteristics of typical coaxial cables are given in references [3–5], and should be consulted before purchase. Be sure to also check with the dealer that the cable is suitable for the purpose for which it is intended, and is of the correct impedance for the system.

The antenna and its feeder system is often undervalued by amateurs, almost as if it were an optional extra, but attention to detail in this area is however one of the most cost-effective ways of improving station performance.

Some further notes on selecting HF antennas for optimum DX performance are given in Chapter 4.

CHOOSING A SITE FOR THE SHACK

In any household, there is often only one reasonable possibility open to the amateur in respect of siting the shack. In many cases this will be a spare room or the garden shed, already overflowing with domestic bric-à-brac. In others it will be the only empty corner of the living room or even the cupboard underneath the stairs.

In other cases a choice may be made between several possibilities, and this section will discuss briefly the various factors involved, as follows:

- Accessibility and privacy
- Acoustics and acoustic noise
- Comfort
- Distance from antenna
- Power
- Space

Accessibility and privacy

Accessibility is used here in its negative sense. It is the duty of the operator to ensure that the station is inaccessible to any other person who might inadvertently injure themself or deliberately operate the station. If young children are present in the household a separate room or shed is desirable, with a door which can be locked when the station is not in use. If this is not possible some way of discreetly disabling the power to the station will be necessary.

Privacy is often another requirement. The garden shed may have windows through which the neighbours and their children can peer, and the noises coming from such a shack might attract the attention of the curious. For some amateurs this may not matter, for others it could almost amount to harassment.

Acoustics and acoustic noise

If phone operation is contemplated, the site should be reasonably free from room echo and external noise. Such extraneous noises will be made much worse if some form of speech processing is used. If such effects are present, it should be ascertained whether it will be possible to remove them. Heavy curtains, acoustic tiles on walls or ceilings, carpets and soft furniture may be required to deaden room echo, while it may be necessary to fit double glazing to lessen external noise if a busy road, railway or airport is nearby. Phone operation also means that the station will generate considerable acoustic noise itself, particularly when the operator is 'shouting for DX'! It is possible that this may be audible to neighbours, or disturb other members of the household.

Comfort

Comfort in the shack allows an operator to concentrate and enjoy the hobby. It is foolish to spend hundreds or maybe thousands of pounds on equipment and then to sit shivering in the cold with hands too numb to operate a key properly.

A site should be chosen which can be kept between about 19–24°C (65–75°F) summer or winter, at least during use. Many lofts heat up on a summer's day to well over 40°C (104°F) even in the UK, yet can be unbearably cold in the winter. Apart from operator discomfort, temperature extremes are not good for equipment. Oscillators can malfunction in the cold, and high-power operation in very hot conditions might lead to damage to transmitters unless cooling is very efficient.

Unheated sites are also likely to be damp, and again this is bad for both the operator and equipment. Damp chairs are not only uncomfortable but are liable to be damaged by mildew.

Condensation can form inside equipment and cause malfunctions which may be difficult to trace.

Other environmental factors such as ventilation and lighting may also be important in some cases and should also be considered.

Distance from antenna

This is not normally a problem on bands below 432MHz as good-quality low-loss feeders are readily available. However, if VHF and UHF operation is the main interest, the site-to-antenna distance may become important, especially if the chosen site is on the opposite side of the house to the antenna. It should be checked that there will be no problems running feeders from the station to the antenna. Some large-diameter feeders are quite stiff and cannot in any case be bent round sharp corners due to risk of bruising. A hole to take the feeder may be needed in a window pane or window frame. Feeders should not be run through an open window, except on a temporary basis, to avoid any security risk.

Power

A correctly wired mains power supply with an effective earth should be available. Lighting circuits should not be used. As a temporary measure, low-power mobile equipment can be run off an old car battery, but the cost of a new battery is likely to be just as high as that of a proper mains-operated supply, which is far more convenient.

When arranging the power supply, consideration should be given to future expansion of the station and the additional demands this will make on the supply.

Space

Adequate space should also be available not only for the present equipment but future expansion. It is highly undesirable to set up a station site with considerable trouble only to have to move when an interest develops which requires additional equipment. In general, completely home-constructed stations can require up to three times the space of 'off-the shelf' shacks. Not only is home-constructed equipment often somewhat larger than factory made items, but a separate workbench is desirable. A test facility area may also be required to accommodate the oscilloscope, signal generators and other equipment.

SITING THE ANTENNA

For maximum efficiency, the antenna should be sited as high as possible, away from the ground, buildings and other obstructions. The advantages of even a hill-top site will not be realised unless this is done. For this purpose distances and heights should be thought of in terms of wavelengths rather than metres or feet, and the various publications describing particular antennas often give useful information on their performance at varying heights. A rule-of-thumb employed by many amateurs is a gain of 6dB for each doubling of the antenna height, but this can be somewhat optimistic, particularly in urban conditions.

Transmitting antennas should be kept away from metalwork, such as iron drainpipes. This can influence the directional properties of the antenna, and can also lead to radiation of harmonics of the transmitted signal if the metalwork contains joints capable of acting as metal-oxide diodes (the 'rusty bolt effect').

TV sets radiate timebase interference on harmonics of 15kHz in the HF bands, and antennas intended for HF reception should be kept as far as possible from such equipment. Transmitting antennas should also be located as far as possible from TV antennas and their feeders to avoid electromagnetic compatibility problems (see later). TV feeders usually run vertically, and are therefore particularly prone to RF pick-up from vertical transmitting antennas.

Unfortunately TV equipment is not the only category of equipment susceptible to high RF field strengths. Hi-fi equipment, radios, electronic organs and solid-state switching circuits can all be adversely affected. Similarly other domestic equipment can also radiate interference. Electric motors, thermostats, and fluorescent or neon lighting are frequent culprits.

If the antenna is fixed in position and thereby unable to rotate, it may be of advantage to position it so that equipment susceptible to RF (or generating interference) is in a position of minimum antenna gain. This is particularly important where high transmitter powers or high-gain antennas are in use. The safety and social aspects of the antenna site should not be forgotten. Antennas such as dipoles have dangerous RF voltages at their ends when high-power transmissions are being made, and it should not be possible for children or animals to reach up and touch them.

The aesthetic aspect is more a function of the type of antenna used than its location, but it may sometimes be better to have an antenna located lower than optimum to avoid antagonising neighbours or drawing attention to the presence of a radio station.

For more information on the practical aspects of siting antennas, see references [3–5, 9].

PLANNING PERMISSION FOR AMATEUR RADIO MASTS AND ANTENNAS

Town and Country Planning legislation is aimed at securing 'the proper use of land' and, as part of this, to safeguard amenities, natural and otherwise, enjoyed by local inhabitants. In the UK, permission is been required before undertaking 'development' which is defined as "carrying out of building, engineering, mining or other operations in, on, over or under land, or the making of any material change in the use of any buildings or other land".

Such applications are considered by Local Planning Authorities (usually the District Council in the case of amateur radio masts or antennas) and will entail an assessment of the pros and cons, balancing the issues involved and making a final judgement upon the merits of the proposal. The Secretary of State for the Environment and his opposite numbers for Wales, Scotland and Northern Ireland have appellate roles over these decisions and can also issue guidance on procedures. Councils also may publish guidance policies, either in the form of 'Local Plans' relating to all or part of their areas or on specific topics. Look out for them.

The overall definition of 'development' set out in the first paragraph is subject to some exclusions set out in Section 55 of the Town and Country Planning Act 1990 (as amended). Two are relevant to radio amateurs. Firstly the carrying out for the maintenance, improvement or other alteration of any building of works which (i) affect only the interior of the building or (ii) do not materially affect the external appearance of the building are not 'development'. Secondly, the use of any buildings or other land within the curtilage (usually the garden) of a dwelling for any purpose incidental to the enjoyment of the dwelling as such is not 'development' and therefore does not require planning permission. Amateur radio, as a hobby, falls within this – thus the 'transmitting' component of the hobby cannot

be challenged in the way that, say, a PMR base station might be.

In addition to the exclusions set out above, the Act provides for General Development Orders (GDOs), to be prepared which in effect represent automatic permission for things which may not warrant developers or local authorities having to go to the trouble and cost of preparing and processing applications of little moment to the community or where the 'benefits', in a national context, will override any possible costs. Such GDO permissions can be qualified in a variety of ways not practicable in general legislation. This Secondary Statutory Regulation system has a more simple Parliamentary procedure and thus can be revised more easily. The provisions which allow the extensions to a dwelling by 10 or 15% of its cubic capacity, subject to not coming in front of the forward-most part, nor exceeding the maximum height of the roof of the original dwelling, for many years allowed substantial dormer windows to be built on the front and/or back faces of dwelling roofs and so provided a very cheap way of enlarging dwellings. However, in 1988 this provision was changed to preclude such dormers on roofs facing a public highway and thus reducing some unsightly alterations to street scenes.

Satellite antennas on houses have also caused much criticism, more so on blocks of flats (where in fact the GDO permissions do not apply), but also because the 90cm size constraint produced difficulties in those northern parts of the British Isles where the satellite footprints were less strong. Adverse amenity impacts from the numerous dishes in street schemes and the adverse picture effects were dealt with in July 1991 by changes which varied the size according to the location in the country, allowing smaller (45cm) dishes on the chimney and 70cm elsewhere on the building but retained the 90cm below the roof top in the South West and the area North of the Humber.

There are also two other parts of the GDO which relate specifically to telecommunications; one is the Telecom Code Systems (BT and Mercury) and the other deals with other commercial microwave systems. Neither is relevant to radio amateurs as, apart from these limitations to commercial systems, they specifically exclude siting on dwelling houses.

The 'exclusions', set out in the third paragraph above, together with the automatic permissions referred to in the fourth paragraph may well meet the needs of many amateurs. Remember that neighbours and others may not regard a mast and antenna as a thing of beauty!

In other cases it will be necessary to make an application for permission. A study of the brochure *Planning Permission – Advice to Members,* which is available free to members from RSGB Headquarters, is strongly recommended. Think carefully what you really *need* – not what you would like because a local amateur has received permission for a pair of 17m telescopic masts with a variety of HF, VHF and UHF antennas. Don't forget that his house is set in a pocket of woodland and that your corner bungalow on an estate will pose a very different set of 'costs' when your application is being considered. Do make sure that your application sets out fully why you *need* whatever it is that you show on the application forms and plans. Think about the siting and design of your proposal. Slimline (box-section or tubular) masts are more acceptable than square or triangular lattice masts although they won't carry so high wind loads at such heights. Some antennas are of better *visual* design than others – they don't have so many excrescences or breaks in their general lines which draw attention to them in an irritating way. Look at your proposal as if through a neighbour's eyes and show them where and why you want it. If you want a high mast to get antennas up to reduce EMC problems already existing, can you get help to have a report prepared to illustrate how the new set-up will reduce close-in field strengths? Don't try to mislead with antenna details. The Planning Authority is fully entitled to require details and a failure to supply them may result in permission subject to a condition which requires approval of detail drawings of the antenna before the mast is erected! At least one inspector endorsed this course of action; others have dismissed appeals because of the absence of clear information about antennas.

Make sure the mast, any guys and the antenna elements are *entirely* within your own land and do not trespass onto next door's airspace; if you do encroach, you must get his permission in writing and refer to this in supporting material, saying why the space is needed. Remember that an antenna which cannot be accommodated on your own land invites the comment "It's too big for this site".

At GW4YKL a large piece of wood over a cabinet makes a good 'table top' – the wooden shelf has several vertical supporting pieces to take the weight of the equipment which can be considerable. Note the convenient power sockets

Make good plans and fair copies of any explanatory material and show them to your neighbours; the Council has to consult your neighbours and it is better if you get to them first and explain things. Ask the Case Officer in the Planning Department if he or his chief has any query. If so, offer to come into the office and explain things.

Some amateurs believe in erecting things first and hoping that nobody will notice them! If you are lucky, you may get away with this for four years after which the Council cannot serve an enforcement notice. It often happens, however, that a new neighbour arrives after three years and can cause problems. In any event, if you have altered the antenna at all within the four years, you may well have put the mast and all at risk. This is currently a grey area and the enforcement law is being steadily tightened. The four-year rule now only applies to building or engineering works and changes of use can now be challenged for longer periods from the change taking place.

One final point. You will know, if you have got the planning permission brochure mentioned above, that the RSGB has a panel of advisers who may be able to help you, perhaps to avoid problems or to overcome them by negotiation with the authority or to appeal to the Secretary of State if you have been refused permission.

STATION LAYOUT

Assuming a suitable site for the shack has been found the next question is how the equipment shall be arranged in the most convenient manner. Obviously this depends upon the type of operation envisaged, and the quantity of equipment to be accommodated, but there are some basic rules common to most situations.

In general, the equipment should be arranged so that all controls in frequent use are within reach of the comfortably seated operator, without any need to bend very far forward. Similarly, it should be possible to read meters and dials accurately without straining to one side or peering.

Units which generate appreciable heat should not be stacked

on top of one another, and care should be taken that any ventilation holes in the cabinets are unobstructed. Solid-state transceivers may have cooling fins on the rear of the cabinet, which should have unobstructed air flow. No heating radiator should be located in the vicinity.

Sufficient clearance behind the equipment should be allowed for the connecting leads, especially if large-diameter coaxial cable is in use. Adequate lengths of cable should be used to connect equipment together, so as to permit the sliding forward of any unit for inspection or adjustment while it remains connected to the rest of the station. Any strain in connecting leads *will* lead to failure sooner or later (Murphy's law).

The transceiver is the most important item of electronic equipment in an amateur radio station, and it is recommended

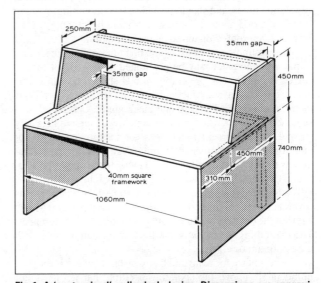

Fig 1. A 'customised' radio desk design. Dimensions are approximate, and assume the station equipment is fairly bulky. For example, the 450mm separation of the shelf and desk could be reduced if compact modern equipment was in use. Note the 35mm clearance between the rear of the desk shelf and wall to allow for connecting cables. If required, multiple power sockets may be installed under the table on one of the side pieces

Siting a well-equipped station in a corner is a natural way to achieve an L-shaped layout so that controls on all the gear are to hand. This fine example belonged to G3HTA some years ago

that the rest of the operating area be built around it. It should be placed on a large firm table which allows plenty of space in front for a logbook, scrap pad and any other material, and so positioned that its tuning knob can be turned easily with the elbow resting lightly on the table.

Opinion varies as to whether the transceiver should be placed to the left or right of the operator or dead centre. Possibly one could get used to it placed in any of these positions. If, however, the set is placed to the left, the right-handed operator can tune the set with the left hand, leaving the right free to write the log and operate the key or microphone. It may be found convenient to tilt the transceiver slightly so that its tuning dial is more easily visible, or so that its controls are more easily operated. Some transceiver manufacturers have extensions already fitted to the front pair of rubber feet for this purpose. If there is still insufficient tilt, packing can be inserted under the front rubber feet of the set. Such packing should not obstruct ventilation holes, and should provide a firm mounting. A scrap length of wood about 150mm wide should be suitable, with a strip of foam rubber on top to deaden vibration.

Other equipment such as an antenna tuning unit, linear amplifier and computer may be positioned either side of the receiver as convenient, preferably in a U-shaped layout as this ensures all displays are facing the operator. The power/SWR meter, monitor oscilloscope and clock are best arranged on a shelf above the operating position, but low enough to be easily viewed. All equipment which does not need to be adjusted or monitored while routinely operating, such as a low-pass filter or frequency meter, does not really belong to the operating area, and can be positioned elsewhere in the shack to save space.

A suggested line-up on the table in front of the electronics is (from left to right) microphone, logbook, scrap pad and Morse key, but the operator will no doubt have individual preferences. The main thing to ensure is that these items can be used with the minimum of effort.

The best place for maps and lists constantly in use is the wall immediately above the operating position, but if this is not possible then the material can be laid out on the table and a thin sheet of glass placed on top to keep it clean and in position.

Last, but not least, a comfortable chair with adequate back support should be chosen, for this will allow the operator to concentrate properly on the job without any fears of aches and pains when a long listening session is over.

SAFETY PRECAUTIONS

Safety is of paramount importance and every precaution should be taken to ensure that the equipment is perfectly safe, not only for the operator himself but also for the other members of the household or visitors. Double-pole switches should be used for all AC supply circuits, and interconnected switches should be fitted to home-constructed equipment so that no part of it can have high voltage applied to it until the valve heaters and low-power stages have been switched on. This precaution may not only save the life of the operator, but also protects the transmitter against damage.

Linear amplifiers using valves should have a microswitch fitted so that the EHT is switched off when the cabinet is opened. Where interconnecting plugs and sockets are used for high voltages, ensure that the female connector is connected to the supply unit, and the male to the unit to be supplied.

It should be possible to turn off power to the entire station by operating one master switch, located in a very prominent position, and all members of the household should know that in the event of an emergency this must be switched off before anything is touched.

The antenna may require the provision of a lightning conductor in the immediate vicinity, or the use of lightning arresters. The most satisfactory type of arrester is the gas-filled type fitted in an adequately insulated mounting which includes a parallel air spark gap. Arresters for coaxial cable are also available. Lightning protection should always be located outside the building, providing as direct a path to earth as possible – "lightning doesn't go round corners!" Great care should be exercised before touching feeders which have been disconnected before a thunderstorm as they may hold a dangerous charge for a considerable time after the storm. Further information on lightning protection is given in references [3].

An often-overlooked precaution concerns the current-carrying capability of the mains supply to the station. An amateur station fully equipped with ancillary apparatus can draw quite a heavy current from the supply, and when assembling the equipment it is important to calculate the current that will be drawn when everything is in use and to check that the house wiring will carry this amount. If there is any doubt new wiring should be installed.

It is most important that every amateur should develop a strict code of safety discipline for use when handling his radio equipment. It should be the rule never to work on equipment which is plugged into the AC supply if this can possibly be avoided. However, there are occasions when this is unavoidable and under these circumstances the following precautions should be followed:

1. Keep one hand in a pocket.
2. Remove metal bracelets or watch straps.
3. Never wear headphones.
4. Be certain that no part of the body is touching an object

SAFETY RECOMMENDATIONS FOR THE AMATEUR RADIO STATION

1. All equipment should be controlled by one master switch, the position of which should be well known to others in the house or club.
1. All equipment should be properly connected to a good and permanent earth (but see box on PME later in this chapter and Note A).
2. Wiring should be adequately insulated, especially where voltages greater than 500V are used. Terminals should be suitably protected.
3. Transformers operating at more than 100V RMS should be fitted with an earthed screen between the primary and secondary windings or have them in separate slots in the bobbin.
4. Capacitors of more than 0.01µF capacitance operating in power packs, modulators, etc (other than for RF bypass or coupling) should have a bleeder resistor connected directly across their terminals. The value of the bleeder resistor should be low enough to ensure rapid discharge. A value of 1/C megohms (where C is in microfarads) is recommended. The use of earthed probe leads for discharging capacitors in case the bleeder resistor is defective is also recommended. (Note B). Low-leakage capacitors, such as paper and oil-filled types, should be stored with their terminals short-circuited to prevent static charging.
5. Indicator lamps should be installed showing that the equipment is live. These should be clearly visible at the operating and test position. Faulty indicator lamps should be replaced immediately. Gas-filled (neon) lamps and LEDs are more reliable than filament types.
6. Double-pole switches should be used for breaking mains circuits on equipment. Fuses of correct rating should be connected to the equipment side of each switch in the live lead only. (Note C.) Always switch off before changing a fuse. The use of AC/DC equipment should be avoided.
7. In metal-enclosed equipment install primary circuit breakers, such as micro-switches, which operate when the door or lid is opened. Check their operation frequently.
8. Test prods and test lamps should be of the insulated pattern.
9. A rubber mat should be used when the equipment is installed on a floor that is likely to become damp.
10. Switch off before making any adjustments. If adjustments must be made while the equipment is live, use one hand only and keep the other in your pocket. Never attempt two-handed work without switching off first. Use good-quality insulated tools for adjustments.

11. Do not wear headphones while making internal adjustments on live equipment.
12. Ensure that the metal cases of microphones, Morse keys etc are properly connected to the chassis.
13. Do not use meters with metal zero-adjusting screws in high-voltage circuits. Beware of live shafts projecting through panels, particularly when metal grub screws are used in control knobs.
14. Antennas should not, under any circumstances, be connected to the mains or other HT source. Where feeders are connected through a capacitor, which may have HT on the other side, a low resistance DC path to earth should be provided (RF choke).
15. Antennas must be designed with due allowance for wind loading. For this, guidance from the antenna manufacturer is necessary and British Standard (BS) CP3 Chapter 5 for guyed masts and BS 8100 for self-supporting masts should be consulted [7].
16. Certain chemicals occur in electronic devices which are harmful. Notable amongst these are the *polychlorinated biphenyls* (PCBs) which have been used in the past to fill transformers and high-voltage capacitors and *beryllium oxide* (BeO) which is used as an insulator *inside* the case of some high-power semiconductors. In the case of PCBs, the names to look out for on capacitors are: ARACLOR, PYROCHLOR, PYRANOL, ASBESTOL, NO-FLAMOL, SAF-T-KUL and others. If one of these is present in a device, it must be disposed of carefully. The local Health and Safety Authority will advise. In the case of beryllium oxide, the simple rule is DON'T OPEN ANY DEVICE THAT *MAY* CONTAIN IT.

Note A. – Owing to the common use of plastic water main and sections of plastic pipe in effecting repairs, it is no longer safe to assume that a mains water pipe is effectively connected to earth. Steps must be taken, therefore, to ensure that the earth connection is of sufficiently low resistance to provide safety in the event of a fault. Checks should be made whenever repairs are made to the mains water system in the building.

Note B. – A 'wandering earth lead' or an 'insulated earthed probe lead' is an insulated lead permanently connected via a high-power 1kΩ resistor or a 15W 250V lamp at one end to the chassis of the equipment; at the other end a suitable length of bare wire with an insulated handle is provided for touch contacting the high-potential terminals to be discharged.

Note C. – Where necessary, surge-proof fuses can be used.

which is earthed and use a rubber or similar non-conductive covering over concrete floors.

5. Use insulated tools.

Before working on equipment of any kind, plugged into the mains or not, it is vital to make sure that all filter capacitors are fully discharged – these are capable of retaining what could be a lethal charge for a considerable time. Use an insulated screwdriver to short each capacitor in turn.

The vast majority of shocks sustained from electrical apparatus are derived from the 230V mains line lead. Every year there are 100 or more deaths in the UK due to electrocution, mostly as a result of accidental contact with mains voltage. There is evidence to suggest that because of the different physiological effects, those who receive shocks from voltages of more than 1000V have a better chance of survival than those subjected to severe medium-voltage shocks. Voltages as low as 32V have been known to cause death – as the jingle says: "It's volts that jolts but mils that kills".

The danger of electrocution is increased where the victim's

skin resistance is lowered by dampness or perspiration, or where he grips an extensive area of 'live' metal while in good contact with earth. It is against this second possibility that particular care is needed in amateur stations.

A particular hazard is equipment which has a chassis connected to the mains which is being used under conditions for which it was not intended. Some old British TV sets and domestic broadcast receivers fall into this category; not only the 'AC/DC' sets but also – and this is not always appreciated – a large proportion of 'AC only' models. If there is any such equipment in your station, the exposed parts should be checked with a neon screwdriver to make certain that they are not 'live' with the on/off switch in either position. Remember that a single-pole switch in the neutral lead will leave the chassis 'live' even with the set apparently turned off. After such checks have been carried out non-reversible plugs should be fitted to AC supply leads.

It is wise to check all three-pin supply sockets in the house to see whether they have been correctly wired; all too often

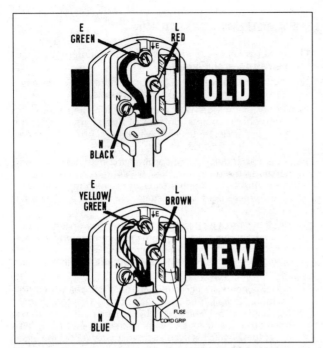

Fig 2. The correct wiring for three-pin plugs in the UK. To test that a socket is correctly wired, a lamp should light when connected between 'L' and 'N' or 'L' and 'E', but not when connected between 'N' and 'E'. A neon bulb will glow when touched against 'L'

this is not the case. A three-pin plug with the 'earth' contact at the top should have the 'neutral' contact on the bottom left, and the live 'line' contact on the bottom right – these directions apply when looking at the back of the plug for wiring purposes (see Fig 2). Correct colour coding of leads in the UK has been for many years: 'live', brown ; 'neutral', blue ; 'earth', yellow and green.

It is very important to note that this coding may not apply to the wiring on some imported equipment, and the manufacturer's instructions should be very carefully studied before plugging into the supply. The use of modern fused plugs is recommended.

An even greater hazard, because it is seldom anticipated, can arise under fault conditions on equipment fitted with a double wound (ie 'isolating') transformer of the type so often used in amateur equipment. It is by no means unusual or unknown for the primary winding to short circuit to the screen between the primary and secondary, the core, or to one of the secondary windings, so that the chassis of the equipment becomes 'live'. Such equipment will often continue to operate quite normally and can thus represent a very real danger over a considerable period. The best safeguard against this danger is to ensure that the screen between the primary and the other windings, the core and the chassis are all effectively earthed.

The earth connection must be of very low resistance otherwise the supply fuses may not blow. These fuses should be of the minimum practicable rating – it is no use having a 50Ω resistance to earth and a 10A fuse – if this should be the case the size of the electricity bills may be surprising, but the hazard is likely to remain undetected!

Another source of danger is the electric tool which has developed a fault and which has a 'live' casing. This can happen, for example, with soldering irons and electric drills. A very careful check should be kept on the leads to all such tools, and any 'tingles' felt when they are in use must be investigated immediately.

Many amateurs fit extra power sockets in their stations and the control arrangements may call for quite a lot of semi-permanent AC wiring and switching. In the UK these should always conform with the high standards laid down in the IEE Wiring Regulations. These are rather formidable reading for the non-professional but a number of books giving sound advice on modern wiring practice, based on the IEE recommendations, have been published and can often be obtained from local libraries. Advice can also be obtained from the offices of local electricity boards. In most countries overseas, similar regulations exist and operators in these areas are recommended to obtain copies or seek the advice of the supply authorities.

Another problem is the use of protective multiple earthing (PME) in some UK houses (see panel later in this chapter).

Finally, taking the worst possible event into consideration the operator and members of his household are advised to familiarise themselves with the procedures for the treatment of electric shock.

It is often not realised that low-voltage, high-current equipment can also have dangerous aspects. Some recent solid-state transceivers have power supply requirements of some 20–30A at 12V. It is vital that suitable cable is used to connect any external power supply and that all electrical connections are of low resistance, otherwise there could be a fire risk. There is also a considerable personal danger through hot or molten metal if the user inadvertently short-circuits such a power supply – the resultant current through, say, a wedding ring could cause severe burns.

It is a wise precaution to have a fire extinguisher of the type suitable for use on electrical equipment in the shack. The best type is that which directs a stream of carbon dioxide gas on to the burning area; the powder and Halon types may be used but are liable to cause further damage to electrical equipment with which they come into contact.

SECURITY

Many amateur stations contain easily portable equipment worth thousands of pounds. Sometimes this is located in a shack remote from the house, possibly on view to the passer-by. It is not surprising that an amateur station can attract attention of the wrong sort, and some elementary precautions should always be taken.

Make a note of the *model number, serial number* and *distinguishing features* of each item of equipment and keep this somewhere in the house (*not* in the shack) where it is unlikely to be seen by an intruder. Make sure the shack is *secure* by checking (and using) the locks on windows and doors. Consider fitting a *burglar alarm*, especially if a separate shack in the garden is in use.

If in doubt on these matters, seek the advice of the crime prevention officer at the local police station. When going on holiday, let the police know.

Equipment left in cars is at particular risk. Never leave hand-portable transceivers (or any other valuables, for that matter) lying around on the seats of cars. Keep them with you or, if this is not convenient, locked up in the boot. Remove mobile transceivers when the car is left parked outside the house. Make sure the car is fitted with an alarm.

If the worst comes to the worst, make it difficult for the thief. Give the details of the theft and the equipment to the local radio club(s) and the RSGB (or appropriate national society) as well as the police and insurers. The local radio shops should also be warned, particularly any dealing in second-hand equipment.

INSURANCE

While every amateur will try to construct and maintain his station so that it is completely safe for himself and any others who may visit it, there is always the possibility that an accident may occur. Owing to a component failure a visitor may receive an electric shock, or an antenna or mast may fall and injure someone or damage property. Such an occurrence can result in a legal action, and these days it can result in the award of very substantial damages against the person held to be responsible for the accident. This risk can, and should, be insured against, either by an extension to the existing Householder's Comprehensive policy or by taking out a separate public liability policy if only fire insurance is held. The annual premium for this will only be quite a small amount and readers cannot be too strongly urged to consult their insurance advisers over this matter.

The other insurance risk is of course theft. Again, the existing household insurance should be checked to ensure that the equipment is covered and the cover should be increased if necessary. Some years ago the RSGB introduced a special insurance scheme for its members which fully covers the above liabilities and risks.

TRANSMITTER CHECKS AND ADJUSTMENTS

The UK licence stipulates that transmitters should be tested from time to time to ensure that the licence requirements concerning non-interference to other telecommunications can be met; details of such tests must be recorded in the station logbook.

Adverse comments made on the air concerning modulation quality (including that of CW signals) should be treated with concern, and the fault or incorrect adjustment traced before further transmissions are made. Most amateurs tend to be uncritical of other stations' signals out of politeness, and therefore when comments are made it is often the case that something is quite seriously wrong.

Most transceivers have built-in calibrators or digital frequency readouts. As with any other frequency measuring equipment, these should be checked periodically.

If available, the manufacturer's handbook should be carefully consulted for details of adjustments and tuning up procedure, and the operator should be fully conversant with these *before* putting the transmitter on the air.

The standard handbooks (in particular [8]) should be consulted for test methods and circuits. Most test gear required is either inexpensively obtainable on the second-hand market or easy to construct at home. Its regular use is not only the mark of a responsible operator, but also one who is sure that his or her transmitter is really giving peak performance in those DX pile-ups!

EMC – DEALING WITH INTERFERENCE

One of the most troublesome problems in amateur radio is interference caused by the fundamental transmission getting into all types of electronic equipment. The term 'breakthrough' is normally used to describe this phenomenon, emphasising the fact that it is really a shortcoming on the part of the equipment being interfered with, and not a transmitter fault.

Good radio housekeeping

The main object of good radio housekeeping is to minimise breakthrough, by making sure that as little as possible of the precious RF energy finds its way into neighbouring TV, videos, telephones, and the multitude of electronic gadgets which are part and parcel of the modern home.

It could be argued that the immunity of the domestic equipment is inadequate, but this does not absolve the amateur from the responsibility of keeping his RF under reasonable control. Many of the features which contribute to minimising breakthrough also help in reducing received interference, so that the virtue of good neighbourliness has the bonus of better all-round station performance.

Antennas

By far the most important factor in preventing both breakthrough and received interference problems is the antenna and its siting. The aim is to site the antenna as high as you can, and as far as possible from your own house and from neighbouring houses (see Fig 3). If there is any choice to be made in this regard give your neighbours the benefit of the increased distance – it is usually much easier to deal with any problems in your own home. It is a sad fact that many amateurs are persuaded by social pressures into using low, poorly sited, antennas only to find that breakthrough problems sour the local relations far more than fears of obtrusive antennas would have done.

HF antennas

The question of which antenna to use is a perennial topic and the last thing that anyone would want to do is to discourage experimentation, but there is no doubt that certain types of antenna are more likely to cause breakthrough than others. It is simply a question of horses for courses; what you can get away with in a large garden, or on HF field day, may well be unsuitable for a confined city location. Where EMC is of prime importance, the antenna system should be:

(a) *Horizontally polarised.* TV down leads and other household wiring tend to look like an earthed vertical antenna so far as HF is concerned, and are more susceptible to vertically polarised radiation.

(b) *Balanced.* This avoids out-of-balance currents in feeders giving rise to radiation which has a large vertically polarised component. Generally, end-fed antennas are unsatisfactory from the EMC point of view and are best kept for portable and low-power operation. Where a balanced antenna is fed with coaxial feeder, then a balun must be used.

(c) *Compact.* So that neither end comes close to the house and consequently to TV down leads and mains wiring. Antennas to be careful with are the extended types such as the W3DZZ trap dipole or the G5RV, because almost inevitably, in restricted situations, one end is close to the house.

On frequencies of 14MHz upwards it is not too difficult to arrange an antenna fulfilling these requirements, even in quite a small garden. A half-wave dipole or small beam up as high as possible and 15m or more from the house is the sort of thing to aim for.

At lower frequencies compromise becomes inevitable, and at 3.5MHz most of us have no choice but to have one end of the antenna near the house, or to go for a loaded vertical antenna which can be mounted further away. A small loop antenna is another possibility, but in general any antenna which is very small compared to a wavelength will have a narrow bandwidth and a relatively low efficiency. Many stations use a

G5RV or W3DZZ trap dipole for the lower frequencies but have separate dipoles (or a beam) for the higher frequencies, sited as far down the garden as possible.

VHF antennas

The main problem with VHF is that large beams can cause very high field strengths. For instance 100W fed to an isotropic transmitting antenna in free space would give a field strength of about 3.6V/m at a distance of 15m.

The same transmitter into a beam with a gain of 20dB would give a field strength, in the direction of the beam, of 36V/m the

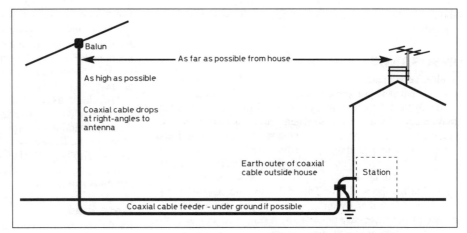

Fig 3. Good radio housekeeping – site your antenna and feeder system well away from the house

same distance away. Again, it comes down to the fact that if you want to run high power to a high-gain beam, the antenna must be kept as far from neighbouring houses as possible, and of course as high as practical.

Operation in adverse situations

The obvious question arises as to what to do if your garden is small or non-existent, or domestic conditions make a simple wire tuned against ground the only possibility.

First of all, and most important, don't get discouraged – many amateurs operate very well from amazingly unpromising locations. It is really a question of cutting your coat according to your cloth. If there is no choice but to have antennas very close to the house, or even in the loft, then it will almost certainly be necessary to restrict the transmitted power. It is worth remembering that it is good radio operating practice not to use more power than is required for satisfactory communication. In many cases relations with neighbours could be significantly improved by observance of this simple rule.

Not all modes are equally 'EMC friendly', and it is worth looking at some of the more frequently used modes from this point of view.

SSB This is one of the least EMC friendly modes, particularly where audio breakthrough is concerned.

FM This is a very EMC friendly mode, mainly because in most cases the susceptible equipment sees only a constant carrier turned on and off every minute or so.

CW This is the old faithful for those with breakthrough problems because it has two very big advantages. First, providing the keying waveform is well shaped with rise and fall times of about 10ms or so, the rectified carrier is not such a problem to audio equipment as SSB. Secondly it is a very 'power efficient' mode, so that it is possible to use much lower power for a given contact. There is no doubt that CW has a big advantage over voice communication when it comes to the simple criterion of 'getting through' to the other station, ignoring the rate of information exchange (and also the social aspects of the contact). Figures up to 20dB are sometimes quoted but such high figures depend, to some extent, on the fact that, under poor conditions, CW messages are often very simple, with important parts repeated several times.

Data Generally the data modes used by amateurs are based on frequency shift keying (FSK), and should be EMC

friendly. All data systems involve the carrier being keyed on and off – when going from receive to transmit, and vice versa – and consideration should be given the carrier rise and fall times, just as in CW. Some of the new data modes which can now be generated on a home computer rival CW in their ability to 'get through' with minimum power. These open up new horizons to amateurs living in difficult locations.

Earths

From the EMC point of view, the purpose of an earth is to provide a low impedance path for RF currents which would otherwise find their way into household wiring, and hence into susceptible electronic equipment in the vicinity. As shown in Fig 4, the RF earth is effectively in parallel with the mains earth path. Good EMC practice dictates that any earth currents should be reduced to a minimum by making sure that antennas are balanced as well as possible. An inductively coupled ATU can be used to improve the isolation between the antenna/RF earth system and the mains earth. The impedance of the mains earth path can be increased by winding the mains lead supplying the transceiver and its ancillaries onto ferrite cores to form a common-mode choke.

WARNING
Protective Multiple Earthing (PME)

Some houses, particularly those built or wired since the middle 'seventies, are wired on what is known as the *PME system*. In this system the earth conductor of the consumer's installation is bonded to the neutral close to where the supply enters the premises, and there is no separate earth conductor going back to the sub-station.

With a PME system a small voltage may exist between the consumer's earth conductor, and any metal work connected to it, and the true earth (the earth out in the garden). Under certain very rare supply system faults this voltage could rise to a dangerous level. Because of this supply companies advise certain precautions relating to the bonding of metal work inside the house, and also to the connection of external earths.

WHERE A HOUSE IS WIRED ON THE PME SYSTEM DO NOT CONNECT ANY EXTERNAL (ie radio) EARTHS TO APPARATUS INSIDE THE HOUSE unless suitable precautions are taken.

A free leaflet EMC 07 *Protective Multiple Earthing* is available on request from RSGB.

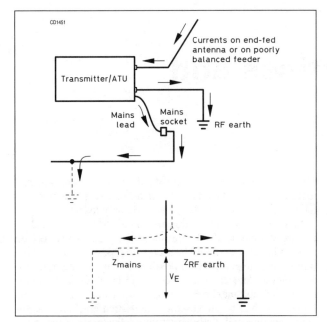

Fig 4. Earth current divides between RF earth and mains. The current down each path will depend on the impedances. The transmitter earth terminal will be at V_E relative to 'true' earth potential

Antennas which use the earth as part of the radiating system – that is, antennas tuned against earth – should be avoided since these inevitably involve large RF currents flowing in the earth system. If this type of antenna must be used, arrange for it to be fed through coaxial cable so that the earth, or better some form of counterpoise, can be arranged at some distance from the house.

The minimum requirement for an RF earth is several copper pipes 1.5m long or more, driven into the ground at least 1m apart and connected together by thick cable. The connection to the station should be as short as possible using thick cable or flat copper strip/braid.

Where the shack is installed in an upstairs room, the provision of a satisfactory RF earth is a difficult problem, and sometimes it may found that connecting an RF earth makes interference problems worse. In such cases it is probably best to avoid the need for an RF earth by using a well-balanced antenna system – but don't forget to provide lightning protection.

Harmonics

Harmonics are much less of a problem than formerly, partly due to the closing down of VHF TV in the UK and partly to the much greater awareness by home-brewers and commercial manufacturers alike of the importance of good design and construction. Not withstanding this, if there is any doubt about the harmonic performance of a transceiver, then a low-pass filter should be used. Care should be taken where harmonics can fall into broadcast radio or TV bands, in particular:

1. The harmonics of some HF bands fall into the VHF broadcast band, 88–108MHz, as does the second harmonic of 50MHz.
2. The fourth harmonic from the 144MHz band could cause problems on TV channels 34 and 35 and the fifth on channels 52 and 53.
3. The second harmonic of 18MHz falls into the IF band of TV receivers.

Interference to amateur reception

While breakthrough causes the most problems so far as the relationship between an amateur and his (or her) neighbours is concerned, the most serious long-term threat to the future of amateur radio is the pollution of the spectrum by the increasing number of interference sources. Interference generated by motors and similar devices is still with us, but by far the most serious problem is the broad-band noise generated by digital electronic equipment.

The EC EMC Standards put limits on permissible emissions from all types of electrical equipment but these are intended to protect relatively large signal services such as broadcasting. They are not stringent enough to prevent interference to amateur reception. Fortunately most products are nowhere near the maximum limit for emissions, or may be so only on specific frequencies, so in practice the situation is not as bad as it appears at first sight.

Techniques for the distribution of high-speed data signals over telephone lines, and even over the electricity mains, are being pushed forward as an economical way of distribution of data for the Internet. This raises entirely different problems. Systems operating at relatively low frequencies are unlikely to give problems unless there is a system faulty, but the demand for higher data speeds will mean pressure to move to frequencies in the HF band. Inevitably the signals will be broad-band and there is a risk that the competitive pressures will force providers to push emissions to the limit to minimise costs. The RSGB, in co-operation with other national societies, is lobbying for tight limits on permissible emissions from such systems, to ensure that the spectrum is protected for the use of small-signal services such as amateur radio.

Dealing with interference to reception is too big a question to be addressed in these notes. Further information can be found in reference [9]. The RSGB EMC Committee also publishes a leaflet EMC 04 *Interference to Amateur Reception*. This is available free of charge from the RSGB. Up to date information on EMC issues can be found in the 'EMC Column' in *RadCom*. Information on leaflets and also of filters available from RSGB can be found in the *RSGB Yearbook*, and on the EMC Committee web site, accessible via the RSGB site www.rsgb.org.uk.

FURTHER INFORMATION

[1] *The Buyer's Guide to Amateur Radio*, Angus MacKenzie, G3OSS, RSGB, 1986 (out of print).

[2] *Your First Amateur Station,* Colin Redwood, G6MXL, RSGB, 1997.

[3] *Radio Communication Handbook,* 7th edn, ed Dick Biddulph, M0CGN, RSGB, 1999.

[4] *VHF/UHF Handbook*, ed Dick Biddulph, G8DPS, RSGB, 1997.

[5] *Backyard Antennas*, Peter Dodd, G3LDO, RSGB, 2000.

[6] 'Wind loading', D J Reynolds, G3ZPF, *Radio Communication* 1988, pp252–255, 340–344 and 629.

[7] 'Technical Topics', Pat Hawker, G3VA, *Radio Communication* 1984, p295.

[8] *Test Equipment for the Radio Amateur*, 3rd edn, C V Smith, G4FZH, RSGB, 1995.

[9] *The RSGB Guide to EMC,* Robin Page-Jones, G3JWI, RSGB, 1998.

3 Operating practices and procedures

HOWEVER much time and effort is spent on setting up a station, its success or otherwise depends on the skill and behaviour of the operator. Unfortunately there is no mandatory period for acquiring listening experience or a practical test of operating technique prior to gaining an amateur transmitting licence, and it is possible to tune around the bands at almost any time and hear dreadful examples of incompetent operating. These are usually more the result of lack of experience and knowledge than a deliberate attempt to be a nuisance.

Operating techniques and procedures vary from band to band and many regular HF band users might find themselves in difficulty when using a VHF repeater for the first time. Each set of techniques is appropriate for the particular circumstances, and the only way to become familiar with them is to spend time listening to the good operators, noting how and why their methods succeed where those of the poor operators do not.

It is quite important that the newcomer should be discriminating in this respect, and not simply mimic whatever he or she has heard previously when going on the air for the first time. To put it more bluntly – copy the good operators, not the bad ones!

Whatever band, mode or type of operating one is undertaking, there are three fundamental things to remember. The first is that *courtesy costs very little* and is very often amply rewarded by bringing out the best in others. The second is that the aim of each radio contact should be *100% effective communication*. The good operator is never satisfied with anything less, and adjusts his or her technique accordingly. The third is that the 'private' conversation with the other station is *actually in public*. Never discuss emotive subjects such as politics, race or religion over the air to avoid any possibility of giving offence to others. Never give any information over the air which might be of assistance to the criminally inclined, such as the time the house is unoccupied, dates of holidays and valuables in the house.

Preparing to transmit

Older transmitters and receivers with a free-running VFO can exhibit appreciable drift within the first 15 minutes or so of operation and should therefore be switched on and allowed a reasonable period to warm up before they are actually used, particularly if a narrow-band mode such as CW is going to be employed. There is nothing more irritating than 'following' a CW station down the band, and of course it may drift onto another station's frequency and be lost altogether. Part of this warm-up period can usefully be employed by searching the band for interesting stations. Each time one is heard, its callsign and frequency can be noted on the scrap pad. In this way a short list of frequencies to be monitored is built up, which will be useful during the operating session. This process can be automated if a modern transceiver with memories is used.

Only those tuning adjustments which directly involve the antenna system should be carried out with the latter connected to the transmitter. All others should be made with the transmitter connected to a suitable screened dummy load.

In a typical system the transmitter should first be tuned up into the dummy load. The transmitter may then be connected to the ATU. Power should be decreased if possible at this stage (not by detuning!) to avoid possible damage to the transmitter's output devices. The ATU should be adjusted for minimum SWR and maximum forward power in the normal way. Power can then be increased to the desired level and the adjustments repeated. Only very minor adjustment of the transmitter output stage should be necessary to ensure optimum working conditions at this point.

If the transmitter has grid or preselector circuits which require peaking every 50kHz or so, a most useful accessory is a screened coaxial switch which will allow the dummy load to be quickly switched in for tune-up purposes.

It is most important that any tune-up signal radiated by the antenna should *not* cause interference to stations already using the band, and a careful check of the chosen frequency should first be made. In particular, any temptation to tune up on the frequency of a DX station before calling should be very firmly resisted for obvious reasons.

When near a band-edge, great care must be taken to ensure that the transmission does not accidentally occur outside the amateur band. Allowance should be made for drift, calibration accuracy and sidebands, particularly those of FM signals (NBFM carriers should be at least 10kHz within the limits of the band in use).

By common agreement in IARU Region 1, each amateur band has been divided into sections which cater for SSB or CW operation, and also more specialised modes such as RTTY and SSTV. These *band plans* are given in Appendix 7. Transmissions in a particular mode should only be made in the appropriate section of the band.

Establishing contact

There are two accepted ways of establishing contact with another amateur station. The first method is to put out a general call or broadcast to all stations, known as *a CQ call,* and to hope that another station responds. The second is to call another specific station by prior arrangement (known as *keeping a sked*) or after the other station has just finished a contact or made a CQ call.

Calling a specific station which has just finished a contact or made a CQ call has the advantage that one knows who is likely to reply, and the signal strength of that reply. This is important on the HF bands because likely fading can be taken into account to determine whether a contact can be sustained. Another factor is that any rare or distant (DX) station on the band is probably already having a contact and would not reply to a CQ call.

When calling a specific station it is good practice to keep calls short and to use the callsign of the station called once or twice only, followed by one's own repeated several times and pronounced or keyed carefully and clearly (particularly when signal strengths are not good or in the presence of interference). The calling procedure should be repeated as required until a reply is obtained or it is clear that someone else is in contact. VOX operation (or break-in on CW) is very useful, because if it is properly adjusted it will be possible to listen between words and know what is happening at all times.

This basic procedure may be summarised as follows:

1. Listen – to find out just what is happening on the frequency.
2. Be patient – wait until any other station already in contact has finished.
3. Make calls short – give the other station's callsign once or at the most twice, but your own two or three times.
4. Listen carefully – between words if possible.
5. Be ready to stop transmitting if the station being called replies to someone else.
6. If unlucky, be patient! Wait for another chance and call again.

If activity on the band is low and a reasonable amount of power is available, a CQ call may be useful. When this is transmitted just as 'CQ' it can be answered by any other station. If the call is 'CQ DX' this limits replies to calls from distant (DX) stations. The definition of DX varies from band to band, and also on conditions (see Chapter 4). Make sure this is understood before making (or answering) a CQ DX call. If stations which are not DX call they should be politely ignored, and it should be a practice never to call anyone else who has made a CQ DX call unless one is DX to that station.

If it is desired to make contact with a particular country a directional CQ call can be made, such as 'CQ VK', which means that only replies from Australia will be answered.

Before making a CQ call it is important to find a frequency which appears unoccupied by any other station. This may not be easy, particularly in crowded band conditions. Listen carefully – perhaps a DX station is on the frequency which is difficult to hear. If a beam antenna is in use, rotate it to make sure. If after a reasonable time the frequency still seems clear, ask if the frequency is in use before transmitting the CQ call (On CW this can be done by transmitting 'QRL?' or just '?'). If two or three CQ calls produce no answer it may be that interference is present on that frequency for distant stations and a new frequency should be sought.

Each CQ call should be short with a break to listen for replies. It may then be repeated as often as required. Long calls without breaks denote poor operating technique – interference may be unwittingly caused to stations which were already on the frequency but which the caller did not hear in the initial check, and moreover stations intending to reply to the call may become impatient and move on to another frequency.

CQ calls on VHF should include details of the location so that those stations using a beam antenna can work out the rough bearing and swing their antenna round before answering the call.

Summarising, the important points in making a CQ call are:

1. Find a clear frequency – check that this is so.
2. Keep calls short as possible and listen frequently.
3. Do not answer directional calls unless in the category of station being called.

Conducting the contact

If contact has been established on a special calling channel, the frequency should be vacated as soon as possible, and the contact completed elsewhere.

After the usual greetings it is customary to exchange details of signal strength, name, location and information on the equipment being used, the latter often consisting only of the model number of the transceiver, the antenna and, if not immediately obvious, the power in use. Often on the VHF bands the height above sea level (ASL) of the location is also given, while HF operators usually mention the local weather and radio conditions.

A contact limited to such information exchanged (quite common on the HF bands) is often known, somewhat disparagingly, as a 'rubber stamp' contact. Yet many overseas amateurs deserve great credit for learning enough of a foreign language (such as English) to enable them to do this. They may not be familiar with other words and expressions, certainly not colloquialisms, and if a contact goes beyond the basic details with a foreign-language amateur it is always wise to use the simplest words possible (including international Q-codes on telephony). A further point is that in poor conditions it may be necessary to keep the whole contact short in case fading or interference occur. The good operator takes both these factors into account when expanding on a basic contact.

On VHF contacts tend to be much less formal due to the usual absence of language difficulty, and 'rubber stamp' contacts are rare except under good conditions when DX contacts are possible.

On both HF and VHF it is good operating practice to use the minimum amount of RF power output consistent with 100% copy at the other station. Any reports of better than S9 received, eg "S9 + 20dB", indicate too much power is in use and this should be reduced if possible to avoid interference to other band users.

Concluding the contact

It is customary in a final transmission for an operator to express some gratitude for the contact, and to convey best wishes to the other person, with often a hope that another contact may be made at a later date.

After concluding the contact, both stations will listen carefully for callers, and what happens next depends on which station was using the frequency or channel prior to the contact. This station has some claim to the frequency, and it is usual and polite for the other station to move off elsewhere after the contact. However, this convention may be modified to suit the circumstances. Good operators move off a frequency where they have been fortunate enough to be called by a rare or DX station, to enable others to contact it. No band is so crowded that it is not possible to find another frequency.

CW OPERATING

The amateur who never learns or uses the Morse code (Table 2) is missing a very large part of the enjoyment available from holding a transmitting licence. These days there are many aids to learning Morse, including books [1], tapes [2], computer programs and even a CD-ROM [3]. Though some effort is involved, it is not as difficult to do this as many seem to fear.

To say that it is easier to copy Morse signals than telephony under adverse conditions can be something of an understatement. Actually for many amateurs CW provides the *only* possible method of working DX stations. So superior is CW to even RF-clipped SSB as a DX mode on the HF bands, that it has been estimated that the difference is roughly equivalent to at least a 20dB power gain.

Another advantage of CW is that it is possible to communicate and exchange information with another person who does not know one word of one's own language by using standard abbreviations and codes.

A further advantage is that CW takes up much less frequency space in bands which are severely overcrowded, and it usually seems to be easier to find another station to contact without interference, especially if a good CW filter is in use.

Why with all these advantages is not CW even more popular? It must be admitted that it takes much longer to really feel at home on CW than on phone, and many operators do not feel the effort is worth it once they have passed the elementary Morse test to qualify for a licence. The high powers and beam antennas many use on the HF bands mean that is possible to work worldwide using phone, and thus the DX advantage of CW is to some degree eroded. Last but not least, many operators are able to exchange signal reports, name and location at quite high speed using CW, but their ability falls off quickly when the conversation becomes less stereotyped and the available codes and abbreviations no longer apply. This does not apply to phone operation, provided of course that both stations are fully readable and the operators speak the same language fluently.

Sending CW

The best CW operator is the one who is easy to read, and who does not impress by sending faster than he or she is competent to. The correct sending speed depends on individual circumstances, and it is sensible to reduce it when conditions are poor, signals are weak or the other operator is a novice. It is good practice to adjust one's speed to match that of the person with whom one is in contact, and not to be reluctant to slow down if it is clear that he or she is not copying fully.

The maximum sending speed should never be faster than the maximum receiving speed. In particular, newcomers to CW operating are advised not to answer a high-speed CQ call at the same speed if they cannot receive at that speed! The correct procedure is to answer at normal speed and request that the other operator slows down. This may be done by use of the QRS signal (see later).

There is nothing like practice for those who wish to become CW operators, and the aim of those using a straight key should be to produce Morse which sounds exactly like that being sent by an automatic Morse generator. There is one, and only one, proper way to form Morse code with character length and spacing correctly timed according to international recommendations (Table 1).

However, operators are not machines, and the natural rhythm of sending (known as the 'fist') does vary slightly from person to person – the experienced operator can sometimes recognise the person sending the code! Some operators deliberately send the first dah of each character longer than the others, giving rise to a syncopation of the normal Morse rhythm. This 'swing' may make for a distinctive 'fist' but can make copying more difficult for the beginner.

A steady rhythm is important and failure to achieve this may be due to use of an inferior key or not holding the key properly. Some operators use fast character speeds with relatively slow overall speeds. It is difficult to keep the inter-character spacing uniform in such circumstances and the result may be an unsteady rhythm.

It is most important to monitor all CW transmissions as this

Table 1. The Morse code and sound equivalents

Alphabet and numerals

A	di-dah	S	di-di-dit
B	dah-di-di-dit	T	dah
C	dah-di-dah-dit	U	di-di-dah
D	dah-di-dit	V	di-di-di-dah
E	dit	W	di-dah-dah
F	di-di-dah-dit	X	dah-di-di-dah
G	dah-dah-dit	Y	dah-di-dah-dah
H	di-di-di-dit	Z	dah-dah-di-dit
I	di-dit	1	di-dah-dah-dah-dah
J	di-dah-dah-dah	2	di-di-dah-dah-dah
K	dah-di-dah	3	di-di-di-dah-dah
L	di-dah-di-dit	4	di-di-di-di-dah
M	dah-dah	5	di-di-di-di-di
N	dah-dit	6	dah-di-di-di-dit
O	dah-dah-dah	7	dah-dah-di-di-dit
P	di-dah-dah-dit	8	dah-dah-dah-di-dit
Q	dah-dah-di-dah	9	dah-dah-dah-dah-dit
R	di-dah-dit	0	dah-dah-dah-dah-dah

Accented letters

à, á, â	di-dah-dah-di-dah	ê	dah-di-di-dah-dit
ä	di-dah-di-dah	ñ	dah-dah-di-dah-dah
ç	dah-di-dah-di-dit	ö, ó, ô	dah-dah-dah-dit
ch	dah-dah-dah-dah	ü, û	di-di-dah-dah
è, é	di-di-dah-di-dit		

Abbreviated numerals

1	di-dah	6	dah-di-di-di-dit
2	di-di-dah	7	dah-di-di-dit
3	di-di-di-dah	8	dah-di-dit
4	di-di-di-di-dah	9	dah-dit
5	di-di-di-di-dit	0	daaah (long dash)

Punctuation

Full stop (.)	di-dah-di-dah-di-dah
Comma (,)	dah-dah-di-di-dah-dah
Colon (:)	dah-dah-dah-di-di-dit
Question mark (?)	di-di-dah-dah-di-dit
Apostrophe (')	di-dah-dah-dah-dah-dit
Hyphen or dash (-)	dah-di-di-di-di-dah
Fraction bar or solidus (/)	dah-di-di-dah-dit
Brackets – open [(]	dah-di-dah-dah-dit
– close [)]	dah-di-dah-dah-di-dah
Double hyphen (=)	dah-di-di-di-dah
Quotation marks (")	di-dah-di-di-dah-dit
Error	di-di-di-di-di-di-di-dit

Spacing and length of signals
1. A dash is equal to three dots.
2. The space between the signals which form the same letter is equal to one dot.
3. The space between two letters is equal to three dots.
4. The space between two words is equal to seven dots.

is the only way to tell whether they are being sent properly, and a tape recording of these should show up any defects in the sender's 'fist'.

Complete and properly adjusted full break-in facilities are a great asset because they enable received signals to be heard between transmitted characters. This means that any interference appearing on the frequency can be detected instantly, and it also becomes possible to conduct a two-way dialogue similar to that possible on phone using VOX. However, don't get the impression that full break-in is essential – slick operators do manage snappy break-in QSOs without it.

CW abbreviations, codes and procedure signals

Amateur CW operating procedures are in the most part derived from commercial telegraphy procedures and, while not so strict, retain two valuable features which assist fast and efficient communication.

The first is the extensive use of codes and abbreviations to cut down transmission time, for if brevity is the soul of wit it is also the soul of good CW operating! The two commonest codes

Table 2. International Q-code (extract)

QRA	What is the name of your station? The name of my station is . . .	QSO	Can you communicate with . . . direct (or by relay)? I can communicate with . . . direct (or by relay through . . .)
QRB	How far approximately are you from my station? The approximate distance between our stations is . . . kilometres.	QSP	Will you relay to . . .? I will relay to . . .
		QSR	Shall I repeat the call on the calling frequency? Repeat your call on the calling frequency; did not hear you (or have interference).
QRG	Will you tell me my exact frequency (or that of . . .)? Your exact frequency (or that of . . .) is . . . kHz (or MHz).	QSS	What working frequency will you use? I will use the working frequency . . . kHz (Normally only the last three figures of the frequency need be given).
QRH	Does my frequency vary? Your frequency varies.		
QRI	How is the tone of my transmission? The tone of your transmission is . . . (amateur T1-T9).		
QRK	What is the intelligibility of my signals (or those of . . .)? The intelligibility of your signals (or those of . . .) is . . . (amateur R1-R5).	QSU	Shall I send or reply on this frequency (or on . . . kHz (or MHz)) (with emissions of class . . .)? Send or reply on this frequency (or on . . . kHz (or MHz)) (with emissions of class . . .).
QRL	Are you busy? I am busy (or I am busy with . . .). Please do not interfere.	QSV	Shall I send a series of V's on this frequency (or . . . kHz (or MHz))? Send a series of V's on this frequency (or . . . kHz (or MHz)).
QRM	Are you being interfered with? I am being interfered with.		
QRN	Are you troubled by static? I am troubled by static.	QSW	Will you send on this frequency (or on . . . kHz (or . . . MHz)) (with emissions of class . . .)? I am going to send on this frequency (or on . . . kHz (or MHz)) (with emissions of class . . .).
QRO	Shall I increase transmitter power? Increase transmitter power.		
QRP	Shall I decrease transmitter power? Decrease transmitter power.	QSX	Will you listen to . . . (callsign(s)) on . . . kHz (or MHz)? I am listening to . . . (callsign(s)) on . . . kHz (or MHz).
QRQ	Shall I send faster? Send faster (. . . words per minute).	QSY	Shall I change to transmission on another frequency? Change to transmission on another frequency (or on . . . kHz (or MHz)).
QRR	Are you ready for automatic operation? I am ready for automatic operation. Send at . . . words per minute.		
QRS	Shall I send more slowly? Send more slowly (. . . words per minute).	QSZ	Shall I send each word or group more than once? Send each word or group twice (or . . . times).
QRT	Shall I stop sending? Stop sending.	QTC	How many messages have you to send? I have . . . messages for you (or for . . .).
QRU	Have you anything for me? I have nothing for you.		
QRV	Are you ready? I am ready.	QTH	What is your position in latitude and longitude (or according to any other indication)? My position is . . . latitude . . . longitude (or according to any other indication).
QRW	Shall I inform . . . that you are calling him on . . . kHz (or MHz)? Please inform . . . that I am calling him on . . . kHz (or MHz).		
QRX	When will you call me again? I will call you again at . . . hours (on . . . kHz (or MHz)).	QTQ	Can you communicate with my station by means of the International Code of Signals? I am going to communicate with your station by means of the International Code of Signals.
QRY	What is my turn? (Relates to communication). Your turn is Number . . . (or according to any other indication). (Relates to communication.)		
		QTR	What is the correct time? The correct time is . . . hours.
QRZ	Who is calling me? You are being called by . . . (on . . . kHz (or MHz)).	QTS	Will you send your callsign for tuning purposes or so that your frequency can be measured now (or at . . . hours) on . . . kHz (or MHz)? I will send my callsign for tuning purposes or so that my frequency may be measured now (or at . . . hours) on . . . kHz (or MHz).
QSA	What is the strength of my signals (or those of . . .)? The strength of your signals (or those of . . .) is . . . (amateur S1-S9).		
QSB	Are my signals fading? Your signals are fading.	QTV	Shall I stand guard for you on the frequency of . . . kHz (or MHz) (from . . . to . . . hours)? Stand guard for me on the frequency of . . . kHz (or MHz) (from . . . to . . . hours).
QSD	Is my keying defective? Your keying is defective.		
QSI	I have been unable to break in on your transmission or Will you inform . . . (callsign) that I have been unable to break in on his transmission (on . . . kHz (or MHz)).	QTX	Will you keep your station open for further communication with me until further notice (or until . . . hours)? I will keep my station open for further communication with you until further notice (or until . . . hours)
QSK	Can you hear me between your signals and if so can I break in on your transmission? I can hear you between my signals, break in on my transmission.	QUA	Have you news of . . . (callsign)? Here is news of . . . (callsign).
QSL	Can you acknowledge receipt? I am acknowledging receipt.	QUM	May I resume normal working? Normal working may be resumed.
QSN	Did you hear me (or . . . (callsign)) on . . . kHz (or MHz)? I did hear you (or . . . (callsign)) on . . . kHz (or MHz).		

are the Q-code for information and procedure signals, and the RST code for signal reports. The second feature is the use of a special procedure signal at the end of each transmission to indicate to the other station (and any other listeners) that the transmission is over, the status of that transmission, and other procedural information. Table 2 gives the official meanings of Q-codes of possible value in the amateur service. When a 'Q' group is followed by a question mark an answer is required, and where appropriate this should be followed by the addition of a number according to the following classification: 1 – very slight; 2 – slight; 3 – moderate; 4 – severe; 5 – extreme. For example 'QRM?' means "is there any interference?" and 'QRM5' means "there is extreme interference".

It should be noted that the very value of the Q-code is that it is an internationally agreed code, and care should be taken to use it correctly if misunderstandings are not to arise. For example, 'QRZ?' is not an alternative to 'CQ'!

Some Q-codes have also come to be used as abbreviated nouns or verbs in amateur parlance, eg 'QSO' means 'contact' in the sentence "Thank you for the QSO". See Table 3.

The RST code (Table 4), due originally to W2BSR, gives readability (R1–R5), signal strength (S1–S9) and tone (T1–T9). The report is usually given as one three-digit combination, eg

Table 3. Informal use of the most common Q-codes

QRG	Frequency.
QRM	Interference from other stations.
QRN	Interference from atmospheric noise or from nearby electrical apparatus.
QRO	High-power.
QRP	Low-power.
QRT	Close(d) down.
QRV	Ready.
QRX	Stand by.
QSB	Fading.
QSL	Verification card; confirm contact.
QSO	Radio contact.
QSY	Change frequency.
QTC	Message.
QTH	Location.
QTR	Time.

Note also: QSLL (card sent in receipt of yours) and QTHR (address is correct in the current RSGB Yearbook)

Table 4. RST code

Readability

R1 Unreadable.
R2 Barely readable, occasional words distinguishable.
R3 Readable with considerable difficulty.
R4 Readable with practically no difficulty.
R5 Perfectly readable.

Signal strength

S1 Faint, signals barely perceptible.
S2 Very weak signals.
S3 Weak signals.
S4 Fair signals.
S5 Fairly good signals.
S6 Good signals.
S7 Moderately strong signals.
S8 Strong signals.
S9 Extremely strong signals.

Tone

T1 Extremely rough hissing note.
T2 Very rough AC note, no trace of musicality.
T3 Rough, low-pitched AC note, slightly musical.
T4 Rather rough AC note, moderately musical.
T5 Musically modulated note.
T6 Modulated note, slight trace of whistle.
T7 Near DC note, smooth ripple.
T8 Good DC note, just a trace of ripple.
T9 Purest DC note.

'579' denotes "Fully readable signals, moderately strong with pure DC note."

The report can be qualified further by adding letters: 'X' denotes tone as steady as if crystal control was in use and 'C' denotes 'chirp'. However, the use of such qualifiers these days is rare, probably due to the widespread use of modern factory-made transmitters.

One important point about giving signal reports using this code is that they should be as honest as possible if they are to have any value at all. "A fully readable signal" means just that, and 'R5' should not be given if the signal is being drowned in noise and interference – it is not an insult to the other station to send 'R3' or as appropriate. Such action also warns the other station to slow down or take other steps to improve readability. Nowadays nearly every receiver has an S-meter which gives some relative indication of signal strength.

This meter is more often than not *useless* for giving signal reports. On many receivers it is possible for a CW signal (and often even a phone signal) to be fully readable and yet not lift the S-meter needle off the zero position. This does *not* mean the report should be 'S0'! In this case the meter should be ignored and an aural estimate made of the signal strength using the RST definitions. The appropriate S-number may then be chosen.

In addition to the Q-code and the RST code, a great many abbreviations are in use, such as 'tnx' or 'tks' for "thanks" (Table 5). There is, however, no need for the newcomer to try to start using all these at once. It is perhaps better to use the most common ones and leave the rest until code proficiency increases.

Last but certainly not least are the procedure signals sent at the end of each message (Table 5). A bar over letters in what follows denotes these are sent as one character.

Table 5. CW abbreviations and procedure signals

AA	All after . . . (used after a question mark or RPT to request a repetition)	BD	bad	NR	number
		BLV	believe	OM	old man
		B4	before	OP	operator
AB	All before . . . (see AA)	CK	check	OT	old-timer
$\overline{AR}$	End of transmission	CLD	called	PA	power amplifier
$\overline{AS}$	Wait a moment	CNT	cannot	PP	push-pull
BK	Signal used to interrupt a transmission in progress	CNDX	conditions	PSE	please
		CPSE	counterpoise	PWR	power
BN	All between . . . and . . . (see AA)	CRD	card	RCVR	receiver
		CUD	could	RPRT	report
$\overline{BT}$	Signal to mark the separation between different parts of the same transmission	CUAGN	see you again	RX	receiver
		CUL	see you later	SA	say
		DR	dear	SED	said
CFM	Confirm (or I confirm)	DX	long distance	SIGS	signals
CL	I am closing my station	ELBUG	electronic key	SKED	schedule (prearranged transmission)
CQ	General call to all stations	ENUF	enough		
DE	"from . . ." (precedes the callsign of the station transmitting)	ES	and	SN	soon
		FB	fine business	SRI	sorry
K	Invitation to transmit	FER	for	STN	station
$\overline{KA}$	Starting signal	FONE	telephony	SUM	some
$\overline{KN}$	Invitation to transmit (named station only)	FQ	frequency	SWL	short-wave listener
		FREQ	frequency	TDA	today
NW	Now	GA	good afternoon (or go ahead)	TKS	thanks
OK	We agree (or It is correct)	GB	goodbye	TMW	tomorrow
PSE	Please	GD	good day	TNX	thanks
R	Received	GE	good evening	TRX	transceiver
RPT	Repeat (or I repeat or Repeat . . .)	GLD	glad	TT	that
$\overline{SK}$	End of work	GM	good morning	TX	transmitter
TFC	Traffic	GN	good night	U	you
TU	Thank you	GND	ground (earth)	UR	your
WA	Word after . . . (see AA)	GUD	good	VY	very
WB	Word before . . . (see AA)	HI	laughter	WID	with
WX	Weather report (or Weather report follows)	HPE	hope	WKD	worked
		HR	here	WKG	working
HRD	heard	HV	have	WL	will
HV	have	HVY	heavy	WUD	would
		HW	how	XYL	wife
Informal amateur CW abbreviations		INPT	input	YDA	yesterday
ABT	about	LID	poor operator	YF	wife
ADR	address	LOC	locator	YL	young lady
AGN	again	LSN	listen	55	best success
ANI	any	MNI	many	73	best regards
ANT	antenna	MSG	message	88	love and kisses
BCNU	be seeing you	ND	nothing doing		

'K' means "invitation to transmit". It is sent right at the end of a transmission in all cases where a reply or call is desired. '$\overline{KN}$' is more specific. It means "invitation to transmit, named station only". It is therefore sent when replies are desired from the specific station given only. It may be sent after each transmission when other stations joining the contact ('breakers') are not welcome. It is also of use to DX or rare stations to send it at the end of each transmission to make it quite clear they were not making a CQ call, in case someone just hears their callsign and a 'K'.

'$\overline{AR}$' means "end of message". It is usually sent after the callsigns when replying to a CQ call or before them when contact has been established.

'$\overline{SK}$' (often written '$\overline{VA}$') means "end of work". It is sent immediately after the callsigns in each operator's final transmission. It signals to listening stations that they are now free to call.

'CL' means "closing down" and is sent right at the end of a transmission if that station is closing down and is not listening for any other calls. These signals are usually sent in association with callsigns but there are two procedure signals which are not.

'BK' means "break" and is used where a quick response to a question is desired without the inconvenience of repeating callsigns. 'BK' is sent after the question, before and after the answer, and before normal transmission resumes.

As noted earlier, if full break-in facilities are available the whole contact can be conducted as a two-way dialogue and in this case 'BK' is superfluous. It may however be used after a CQ call to denote full break-in facilities are available, for example "CQ CQ CQ DE G4ZZZ BK". The Q-code combination 'QSK' may alternatively be used in this case instead of 'BK'. The other such procedure signal is '$\overline{AS}$' which means "stand by for a moment".

There is some slight worldwide variance in the order of CW procedure signals. The ARRL recommends that '$\overline{SK}$' be sent *before* the callsigns in the final transmission, and also that the 'K' is not necessary after '$\overline{AR}$' at the end of a call to a specific station before contact has been established (some East European stations do not use '$\overline{AR}$' at all in this case, sending 'PSE K' instead). In practice these points are unlikely to lead to any difficulty and are only mentioned here so that the reader is aware of the variations.

There is occasionally a couple of spaced dots sent after the final transmission which are usually echoed by the other station. This is just a final goodbye and has no special significance.

A typical CW contact

All these codes and abbreviations may seem rather daunting to the newcomer – it is difficult enough to learn the Morse code without having a whole new set of codes superimposed. Yet after a few CW contacts have been heard and understood the usage begins to become clear and the value of the codes and abbreviations can be appreciated. To assist this learning process, an imaginary CW contact and its literal equivalent is reproduced below as an example.

CQ CQ CQ CQ DE G4ZZZ G4ZZZ K

"General call to all stations from G4ZZZ. Over"

This is known as a 'four by two' CQ call, meaning that 'CQ' is sent four times and the callsign twice. A much longer call is not recommended, as noted earlier.

G4ZZZ G4ZZZ DE WD9ZZZ WD9ZZZ $\overline{AR}$

"G4ZZZ from WD9ZZZ. Over"

Note that the call is short, only 'two by two'. Quite often it is possible to hear a station still calling when the station called has replied to another station. Note also that the USA station did not send 'K'.

WD9ZZZ DE G4ZZZ GA OM ES MNI TNX FER CALL = UR RST 57 = NAME JOHN ES QTH LONDON = SO HW CPY? $\overline{AR}$ WD9ZZZ DE G4ZZZ K

"WD9ZZZ from G4ZZZ. Good afternoon old man and many thanks for the call. Your signals are fully readable and moderately strong, with pure DC note. My name is John and my location is London. So how do you copy me? WD9ZZZ from G4ZZZ. Over"

Each station has the other's callsign correct, so there is now no need to give callsigns more than once.

G4ZZZ DE WD9ZZZ R FB JOHN ES GM OM = UR RST 559 = QTH SR [error] SPRINGFIELD, ILL = NAME IS ED = SO HW? $\overline{AR}$ G4ZZZ DE WD9ZZZ K

"G4ZZZ from WD9ZZZ. Roger. Fine business John and good morning old man. Your signals are fully readable, fairly good strength, and pure DC note. My location is Springfield, Illinois. My name is Ed. So how do you copy me? G4ZZZ from WD9ZZZ. Over."

'R' denotes *all* received correct and should not be sent if there was any part of the message which was not copied or understood. Some operators send it before the callsigns. Note the error made in giving the location and correction; '[error]' means the eight-dot error signal was sent.

WD9ZZZ DE G4ZZZ SRI OM QRM5 = PSE RPT UR NAME?? BK

"WD9ZZZ from G4ZZZ. Sorry old man, extreme interference here. Please repeat your name. Break."

BK NAME IS ED ED ED BK

"My name is Ed. Break."

BK R R TNX ED = QRM GONE = RIG IS HOMEBREW WID 75W INPT = ANT IS DIPOLE = MNI TNX FER QSO ES CUAGN = 73 ES GB $\overline{AR}$ WD9ZZZ DE G4ZZZ $\overline{SK}$

"Roger. Thanks (for the repeat) Ed. The interference has gone now. My rig is homebrew with an input power of 75W. My antenna is a dipole. Many thanks for the contact and I hope to see you again some time. Best wishes to you and goodbye now. WD9ZZZ from G4ZZZ. Over."

G4ZZZ DE WD9ZZZ R UR RIG T9X ES FB = RIG HR IS TS520 = ANT IS 2EL QUAD = WL QSL VIA BURO = SO 73 ES GUD DX = GB $\overline{SK}$ G4ZZZ DE WD9ZZZ

"G4ZZZ from WD9ZZZ. Roger. Your rig sounds crystal controlled, and very nice. The rig here is a TS520. The antenna is a two-element quad. I will send my QSL card via the bureau (and I hope to receive one in return). So best wishes and good DX. Goodbye. G4ZZZ from WD9ZZZ."

With both stations having sent '$\overline{SK}$', the contact is at an end and either station may be called by a third station for another contact. The reader will appreciate that this is a basic contact with only the essential details exchanged between the stations. Nevertheless this type of contact can still give much pleasure. There is of course nothing to prevent much more information being transmitted in such favourable circumstances, although the individual transmissions should be kept reasonably short. Strictly speaking, such details as name, location and signal report should be sent only once as shown here, and if the other operator does not copy the details he or she should ask for a repeat. However most amateurs find it easier to repeat these basic details once, rather than risking time-wasting queries, especially if conditions are poor. The established CW operator may also like to bear this point in mind when contacting stations obviously new to CW operating.

CW operation in a foreign language

It is sometimes claimed that CW operation consists entirely of code groups and thus amateurs who do not possess a common language can communicate with each other. This is only partly true. A typical reply by a Russian amateur to a call would be "GM OM TNX FR QSO UR RST 589 QTH MOSCOW OP VLAD HW?" Apart from the two Q-code groups, every word is an English abbreviation.

The Russian amateur does not necessarily see it like this – unless he is a linguist, he will probably have a list of phrases with the Russian equivalents and he will treat these phrases as code groups and not as abbreviations of English words.

Given a similar list of phrases in other languages, there is nothing to prevent a monoglot English speaker from having contacts in those languages. No knowledge of the language concerned is necessary, and on CW there is no problem of pronunciation or understanding the accent of the foreign amateur.

For generations, English speakers have expected everyone else to learn English in order to communicate with them but this attitude is not universally popular! Amateurs, with a minimum of effort, can make a real contribution to international goodwill by greeting fellow amateurs in their mother tongues. A list of 24 phrases (see next page), which may be combined in various ways, will cover the requirements of most simple contacts and enable a QSL card to be made out in the appropriate language.

Although this author can only speak two, he regularly enjoys CW contacts in 14 different languages. By noting phrases used by other amateurs and written on incoming QSL cards, a stock of idiomatic phrases, in addition to the basic ones, will soon be acquired. It is even possible to adapt phrases which have been collected when travelling abroad or half-remembered from one's schooldays!

There is always the danger that the other amateur may assume that one's knowledge of his language is greater than it really is. A combination of phrases 17 and 18 will meet this situation without giving offence.

The words in roman type are for general use in almost all contacts. Those in *italic* type are for occasional use or for use on QSL cards. It is desirable to write the month in words on QSL cards to avoid the confusion which may arise from different conventions when figures only are used. After each list will be found information regarding accented letters for easy reference. It should, however, be noted that in French and Spanish working, the accents are often omitted and in German 'Ä' is sometimes sent as 'AE', 'Ö' as 'OE' and 'Ü' as 'UE'.

In English, the use of endearments between persons of the same sex would be regarded as unusual, but these are quite common in some languages – *lieber* Hans, *cher* Marcel, *amigo* Juan etc. To avoid complications, feminine forms have not been given and so care should be taken when working YLs!

Phrase 8 can be easily adapted; if the rig is an internationally

Table 6. 'Goodbye' in 15 languages

Danish:	FARVEL
Dutch:	TOT WERKENS
English:	GB (CHEERS)
Esperanto:	ĜIS
Finnish:	HEI
French:	AU REVOIR
German:	AWDH
Italian:	CIAO
Norwegian:	HEI
Polish:	CZESC
Portuguese:	ATÉ BREVE
Russian:	DSW
Spanish:	ADIOS (HLV)
Swedish:	HEJ
Welsh:	POB HWYL

Ĝ = dah-dah-di-dah-dit   É = di-di-dah-di-dit

known one, the type number can be inserted in place of '100 WATTS'. As to antennas, those known by the inventor's callsign present no difficulty and the expressions '3 EL', 'YAGI' and 'QUAD' are used in all languages.

It is suggested that German should be the first language to be attempted because it is already used extensively in CW operation and the procedure and abbreviations are well known. In addition to Austria and Germany, the language is used in a large part of Switzerland. All these countries are easy to contact on a number of bands and it is much better for one's first contacts in a foreign language to be with nearby countries when readability is usually good.

French is not so easily abbreviated but it is used over the air quite extensively. The language is still spoken in parts of Africa and it is the first language in Quebec Province (VE2). French-speaking Canadians particularly value contacts in their own language.

Spanish is used throughout South America (except in Brazil) and it is for that reason that it has been selected as one of the languages for inclusion here.

It will be noticed that in some cases, the English abbreviations are retained (or given as alternatives) when the equivalent wording in the foreign language is rather long. In contacts between (say) two Italian amateurs, one will often hear a number of English expressions.

In recent years, it has become popular to say 'Goodbye' in the appropriate language, and a list is given in Table 6.

The effort involved in using a foreign language is well repaid by the thanks which will be expressed in almost every contact, and the final courtesy should be to include a few remarks in the language on the outgoing QSL card.

It is necessary to remember that, in some countries, language is an explosive political matter. This usually occurs in a country where only a very small minority of the people speak a particular language. In these circumstances, it is wise to make sure that use of the language concerned will be welcome before using it over the air.

TELEPHONY TECHNIQUES AND PROCEDURES

Most amateur radio traffic is now carried out using telephony, especially on VHF and above. Though this mode does not require the knowledge of codes and abbreviations, correct operation is more difficult than it may appear at first sight, as is only too apparent after a listen to any amateur band.

Part of the problem is that many operators will have acquired some bad habits in their pronunciation, intonation and phraseology even before entering amateur radio. To these are then

CW PHRASES IN ENGLISH, FRENCH, GERMAN AND SPANISH

ENGLISH
1. GM/GA/GE/GN OM
2. TNX FR CALL (ES FIRST QSO)
3. *Nice to meet u agn*
4. *We have met B4*
5. UR RST . . . QTH . . . NAME . . .
6. HW?
7. TNX . . . FR (FB) REPORT
8. HR RIG [100 WATTS] ANT [DIPOLE]
9. WX (1) FINE (2) CLEAR (3) CLOUDY (4) RAINY (5) WINDY (6) FOGGY (7) WARM (8) COLD (9) SNOW TEMP . . . C
10. TNX . . . FR INFO ON UR RIG (ES WX)
11. *Only partly OK*
12. *All OK except ur QTH*
13. *Some/much/too much QSB, QRM (etc)*
14. *Pse rpt my report/ur name/ur QTH*
15. *Pse QSY up/down . . . kHz*
16. *Band vy noisy*
17. *Sri but I do not understand completely*
18. *I speak only a little English*
19. QSL OK VIA BURO
20. *I have recd ur crd*
21. *Have u recd my crd?*
22. *I am QRU nw*
23. *Best 73 es DX*
24. TNX . . . FR QSO NW QRU 73 DX ES GB

January	April	July	October
February	May	August	November
March	June	September	December

FRENCH
1. BJR/—/BSR/BN MON VIEUX
2. MCI BCP POUR VOTRE APPEL (ET POUR PREMIER QSO)
3. *Je suis enchanté de vous rencontrer de nouveau*
4. *Nous avons déjà fait QSO*
5. VOTRE RST . . . QTH . . . NOM . . .
6. HW? (*Quel est mon contrôle?*)
7. MCI BCP . . . POUR RPRT (*contrôle*) (*Bien aimable*)
8. ICI RIG (*appareil*) [100 WATTS] ANT [DIPOLE]
9. WX (1) MERVEILLEUX (2) CLAIR (3) NUAGEUX (4) PLUVIEUX (5) IL Y A DU VENT (6) IL Y A BROUILLARD (7) CHAUD (8) FROID (9) IL NEIGE TEMP . . . C
10. MCI BCP . . . POUR L'INFO SUR VOTRE RIG (*appareil*) (ET WX)
11. *OK seulement en partie*
12. *Complètement OK excepté votre QTH*
13. *Peu/bcp/trop de QSB, de QRM (etc)*
14. *Repetez mon rprt (contrôle)/votre nom/votre QTH svp*
15. *QSY plus haut/plus bas . . . kHz svp*
16. *Il y a bcp de bruit sur la bande*
17. *Je regrette bcp mais je ne vous ai pas complètement compris*
18. *Je parle un peu le français*
19. QSL OK PAR BUREAU
20. *J'ai reçu votre carte mci bcp*
21. *Avez vous reçu ma carte?*
22. *Je suis maintenant QRU*
23. *Mes meilleurs amitiés et bonne chance pour le DX*
24. MCI BCP . . . POUR QSO ICI QRU 73 DX ET AU REVOIR

janvier	avril	juillet	octobre
février	mai	août	novembre
mars	juin	septembre	décembre

À = di-dah-dah-di-dah
Ô = dah-dah-dah-dit
Ê = dah-di-di-dah-dit
È and É = di-di-dah-di-dit
Ç = dah-di-dah-di-dit
Û = di-di-dah-dah

GERMAN
1. GM/GT/GA/GN LBR FRD
2. VLN DK . . . FR DEN ANRUF (UND ERSTES QSO)
3. *Es freut mich sehr Sie wiederzutreffen*
4. *Wir haben uns schon mal getroffen*
5. IHR RST . . . QTH . . . NAME . . .
6. WIE?
7. VLN DK FR DEN (NETTEN) RPRT
8. RIG (*Gerät*) HR IST [100 WATT] UND ANT IST [EIN DIPOL]
9. DAS WX IST (1) SCHÖN (2) KLAR (3) BEWÖLKT (4) REGNERISCH (5) WINDIG (6) NEBELIG (7) WARM (8) KALT (9) ES SCHNEIT TEMP . . . C
10. DKE . . . FR BRT ÜBER IHR RIG (*Gerät*) (UND WX)
11. *Nur teilweise OK*
12. *Alles OK ausser Ihrem QTH*
13. *Etwas/vl/zu vl QSB, QRM (etc)*
14. *Bte wiederholen Sie meinen Rprt/Ihren Namen/Ihr QTH*
15. *Bte QSY/höher/tiefer . . . kHz*
16. *Auf dem Bande ist vl Lärm*
17. *Leider habe ich nicht alles verstanden*
18. *Ich spreche nur wenig Deutsch*
19. QSL OK VIA BÜRO
20. *Ich habe Ihre Karte bekommen*
21. *Haben Sie meine Karte bekommen?*
22. *Ich bin nun QRU*
23. *Die besten Grüsse und DX*
24. DKE SEHR . . . FR QSO NUN QRU 73 GUTES DX UND AWDH (*auf wiederhören*)

Januar	April	Juli	Oktober
Februar	Mai	August	November
März	Juni	September	Dezember

Ä = di-dah-di-dah
Ü = di-di-dah-dah
Ö = dah-dah-dah-dit

SPANISH
1. GM/GA/GE/GN OM
2. MUCHAS GRACIAS POR TU LLAMADA (Y POR ESTE PRIMER QSO)
3. *Tengo un gran placer en encontrarle*
4. *Nos hemos encontrado ya antes*
5. TU RST . . . QTH . . . NOMBRE . . .
6. HW?
7. MUY TNX (*muchas gracias*) . . . POR RPRT
8. MI TX TIENE [100 VATIOS] MI ANT ES [UN DIPOLO]
9. EL WX AQUÍ ES (1) MUY BUENO (2) CLARO (3) NUBLADO (4) LLUVOSO (5) VENTOSO (6) NEBULOSO (7) ACALORADO (8) FRIO (9) ESTA NEVANDO TEMP . . . C
10. TNX FR INFO
11. *OK sólo en parte*
12. *Completamente OK con excepción de tu QTH*
13. *Poco/mucho/demasiado QSB, QRM (etc)*
14. *Pse repetirme mi RST/tu nombre/tu QTH*
15. *Pse QSY más alto/más bajo cerca . . . kHz*
16. *La banda está muy turbulenta*
17. *Sri pero no he comprendido completamente*
18. *Hablo solamente un poco español*
19. QSL OK VIA BURO
20. *He recibido tu QSL tnx*
21. *Has recibido mi QSL?*
22. *Tengo ahora QRU*
23. *73 y buenos DX*
24. MUCHAS GRACIAS . . . POR QSO QRU 73 DX Y ADIOS

Enero	Abril	Julio	Octobre
Febrero	Mayo	Agosto	Novembre
Marzo	Junio	Septembre	Dicembre

Á = di-dah-dah-di-dah
Ó = dah-dah-dah-dit
Ñ = dah-dah-di-dah-dah

added a whole new set of clichés and mannerisms derived from listening to bad operators. Some of these can be extremely difficult to remove once learnt, even if a conscious effort is made.

Microphone technique

Unless there is very little external noise and room echo, it is best to hold the microphone fairly near the mouth, between 70 and 140mm away. Some microphones are unduly sensitive to letters like 'S' and 'P', and better audio quality may then be obtained by speaking *across* the microphone.

The audio quality should be tested by running the transmitter into a dummy load and monitoring the output on a receiver with headphones. In this way the characteristics of the microphone and the optimum speaking distance will be apparent. If a separate receiver is not available, the opinion of a local amateur should be sought on the air. Speech processing is often a mixed blessing. If too much is used, the modulation quality suffers greatly and the PA transistors or valves could overheat due to the increased duty cycle if SSB is used. Tests should be made to determine the maximum level which can effectively be used, and this noted or marked on the control. Be ready to turn it down if it is not really required during a contact.

Conversation

It is important to speak clearly and not too quickly, not just when talking to someone who does not fully understand the language, but at all times as this is excellent practice.

Plain language should be used when you are having a conversation with the other person – radio clichés should be kept to a minimum. In particular, avoid the use of 'we' when 'I' is meant and 'handle' when 'name' is meant. Of course, use of phrases like "That's a roger" has to be a matter of personal taste but don't feel you have to talk in slang like this to have an amateur radio contact.

The use of CW abbreviations (including 'HI') should also normally be avoided and the Q-code should really only be used when there is a language difficulty or where it has become accepted practice, eg "QRZ?" or "Please QSL via the bureau."

Phonetic alphabets should only be employed when they are necessary to clarify the spelling of a word or callsign. If they are to be effective the listener should know them, and it follows that only the internationally recommended phonetic alphabet should be used (see Table 7). It is confusing to the listener to use other phonetic alphabets and especially to generate new letter words. An exception is when it is found that a letter is being consistently miscopied. Although the recommended alphabet was selected to avoid ambiguity, it can happen in very poor reception conditions that, for example, 'Oscar' sounds like 'Alpha'. In this case, the use of an alternative like 'Ontario' might be the solution.

Unlike CW operation, it is very easy to forget that the conversation is not taking place down a telephone line. The listening station cannot interject a query if something is not understood and cannot give an answer until the transmitting station has finished. The result, especially on VHF, is often a long monologue in which the listening station has to take notes of all the points raised and questions asked if a useful reply is to be given. This should not be necessary if these points are dealt with one at a time.

Voice-operated changeover (VOX) operation should be used where possible as it enables a more normal two-way conversation to be carried on, which tends to avoid long monologues. However, it must be remembered that relays are supposed to be changing over frequently and prolonged "aaahs" spoken to

Table 7. International phonetic alphabet

Letter	Word	Pronounced
A	Alfa	*AL* FAH
B	Bravo	*BRAH* VOH
C	Charlie	*CHAR* LEE or *SHAR* LEE
D	Delta	*DELL* TAH
E	Echo	*ECK* OH
F	Foxtrot	*FOKS* TROT
G	Golf	GOLF
H	Hotel	HOH *TELL*
I	India	*IN* DEE AH
J	Juliett	*JEW* LEE *ETT*
K	Kilo	*KEY* LOH
L	Lima	*LEE* MAH
M	Mike	MIKE
N	November	NO *VEM* BER
O	Oscar	*OSS* CAH
P	Papa	PAH *PAH*
Q	Quebec	KEH *BECK*
R	Romeo	*ROW* ME OH
S	Sierra	SEE *AIR* RA
T	Tango	*TANG* GO
U	Uniform	*YOU* NEE FORM or *OO* NEE FORM
V	Victor	*VIK* TAH
W	Whiskey	*WISS* KEY
X	X-Ray	*ECKS* RAY
Y	Yankee	*YANG* KEY
Z	Zulu	*ZOO* LOO

prevent this happening are bad practice. Remember to give callsigns as often as required by the licensing authority in this type of contact. If VOX is not available, nimble use of the push-to-talk (PTT) switch is an acceptable substitute, but of course this requires more effort.

Both PTT and VOX operation are *simplex* systems, ie communication can take place in only one direction at any given time. The capability of instant and spontaneous response typical of an ordinary telephone conversation is only possible if *duplex* operation is used, ie simultaneous communication in both directions. This requires two communication channels which must be well spaced to avoid the transmitter desensitising the receiver, they may be located in the same band (*in-band* duplex) or in two separate bands (*cross-band* duplex).

Generally in-band duplex is not practical on the narrow HF phone bands due to receiver desensitisation and not too popular even at VHF and UHF because separate transmitters and receivers for those frequencies are unusual items these days, and many amateurs cannot afford two transceivers for the same band! Most duplex work therefore takes place cross-band (usually 144–432MHz).

If harmonically related bands are chosen, the receiver frequency needs to be well clear of any transmitter harmonics. Care should also be taken to comply with the relevant band plan(s) which may be difficult with in-band duplex. It is best to avoid the usual simplex channels for what will appear to others as a one-sided conversation. In any case, it is wise to check that the transmit channel is clear at, say, five-minute intervals. As with VOX operation, remember to give callsigns as frequently as required by the licence authority, in this case also stating the other station's transmit frequency. Generally headphones should be used at least at one end of a duplex contact to avoid 'howl-round' but a carefully adjusted loudspeaker output does allow the other operator to hear how his transmissions sound and may even give a more natural quality to the contact; after all, this is what duplex operation 'is all about'!

Procedure

As noted earlier, when calling a specific station it is good practice to keep calls short and to use the callsign of the station called once or twice only, followed by one's own callsign pronounced carefully and clearly at least twice using a phonetic alphabet, for example:

"WD9ZZZ. This is Golf Four Zulu Zulu Zulu calling, and Golf Four Zulu Zulu Zulu standing by."

Emphasis should be placed on the caller's own callsign and not on that of the station called. If there is no response, the caller's callsign may be repeated once more after a brief listen. Some stations send very long calls because operators think that when everybody else has finished calling they will still be heard. The reader is advised to forget this 'trick of the trade', partly because it rarely seems to work(!), and partly because it causes severe interference to the station called when it goes back to another station. As in CW operation CQ calls should also be kept short and repeated as often as desired. An example would be:

"CQ, CQ, CQ, CQ. This is Golf Four Zulu Zulu Zulu calling, Golf Four Zulu Zulu Zulu calling CQ and standing by."

There is no need to say what band is being used, and certainly no need to add "for any possible calls, dah-di-dah!" or "K someone please" etc!

When replying to a call both callsigns should be given clearly, usually in the phonetic alphabet, so that the calling station can check its callsign has been received correctly. From then on it is not necessary to use the phonetic alphabet for callsigns until the final transmissions. An example would be:

"Whiskey Delta Nine Zulu Zulu Zulu. This is Golf Four Zulu Zulu Zulu."

Once contact is established it is only necessary to give one's callsign at the intervals required by the licensing authority. A normal two-way conversation can thus be enjoyed, without the need for continual identification. If necessary the words 'break' or 'over' may be added at the end of a transmission to signal a reply from the other station. In good conditions this will not normally be found necessary. When FM is in use it is self-evident when the other station has stopped transmitting and is listening, because the carrier drops.

The situation is more complex where three or more stations are involved, and it is a good idea to give one's own callsign briefly before each transmission, for example:

"From G4ZZZ . . ."

At the end of the transmission the callsign should again be given together with an indication of whose turn to speak it is next, for example:

". . . WD9ZZZ to transmit. G4ZZZ in the group."

It is not necessary to run through a list of who is in the group, who has just signed off (and who may possibly be listening) after each transmission, although it may be useful for one person in the group to do this occasionally.

Signal reports on telephony are usually given using the RST code although of course in this case an indication of tone is not required. The report is given as a single two-digit number, in a similar fashion to the three-digit CW RST code.

When the time comes to end the contact, *end it*. Thank the other operator (once) for the pleasure of the contact and say goodbye. This is all that is required. Unless the operator is a good friend there is no need to start sending best wishes to everyone in the household including the family dog! Nor is this the time to start digging up extra comments on the contact which will require a 'final final' from the other station to answer – there

may be other stations patiently waiting to call. It is recommended that both callsigns be given in the final transmission using the phonetic alphabet so that listening stations can check that they have them correct before calling, for example:

". . . This is Golf Four Zulu Zulu Zulu signing clear with Whiskey Delta Nine Zulu Zulu Zulu and Golf Four Zulu Zulu Zulu is now standing by for a call."

Note that some indication to listening stations is useful to indicate what is planned next. Such an indication is also appropriate if an immediate change to another frequency is intended, for example:

". . . This is Golf Four Zulu Zulu Zulu signing clear with Golf Two X-Ray Yankee Zulu Mobile. Golf Four Zulu Zulu Zulu now monitoring V40 for a call."

Further notes on operating procedures for more specialised modes and activities will be found in the relevant chapters.

Phone operation in a foreign language

The ability to conduct a straightforward contact in, say, French or Spanish is most useful for the DXer. Unlike the method of CW operation described earlier this does require an elementary knowledge of the language, but it can still be broadly based on a few stock phrases committed to memory such as those given in Appendix 6. The best way of learning them is to listen to contacts in the target language, recording them on tape if possible.

If the two speakers are fluent a good deal of the conversation will not be understood but this is not important. The main thing is to discover how contacts are initiated, how details like name, QTH and signal report are exchanged, and how contacts are completed. If a few other useful phrases can be acquired at the same time, so much the better but be sure to learn the appropriate responses.

The process should be completed before any attempt is made to use this new-found knowledge, otherwise the results could be a little embarrassing. However, most foreign amateurs do appreciate the effort being made and assist where possible, so perfection is certainly not required. Many of the comments made in the foreign CW section are just as relevant to phone operation and should be noted.

THAT FIRST CONTACT

No matter how much has been read about amateur radio operating nor how many hours have been spent listening to other amateurs on the air, the thrill of the first contact one makes is usually an unforgettable experience. It can, however, also be an unnerving one if too much is attempted too soon!

It is recommended that the first few contacts should be with local stations, for three reasons, First, the high signal strength will allow any defects in the transmitter to show up at the other station and it will be able to give a useful report on audio quality, splatter, key clicks, drift etc, as a final check that nothing is wrong. Second, the high received signal strength should allow concentration on the contact itself, without any need to worry about fading, noise or interference. Third, it is much better to make any mistakes when going on the air for the first time in a non-competitive situation, preferably with a friend operating the other station.

Working through a repeater (Chapter 6) is not recommended until the operator has made a few simplex contacts and feels 'in control' of the standard procedures. Similarly someone who has previously held a VHF telephony-only licence is advised to have a few SSB contacts on HF before attempting to use the key.

One common problem is 'microphone shyness', which often results in new operators suddenly feeling lost for words in the middle of a transmission. This can be particularly noticeable if the other operator has made a long transmission and a long reply is felt to be necessary. Of course this is not the case, and each contact should be regarded as a *conversation* rather than an exchange of 'speeches'. Fortunately microphone shyness tends to disappear as soon as the operator has made a few successful contacts and it is often a case of needing to limit each transmission!

NET OPERATION

As experience is gained an operator will undoubtedly wish to participate in one of the many nets which operate regularly throughout the British Isles. These nets can involve three or four (or 30 or 40 stations) in a multiway link-up. A tidy size for a net is, however, no more than 10 or a dozen operators; even with this number a time limit needs to be set on the length of overs. In a net of 10, it will take 10 minutes before the first operator's turn comes round again even if overs are set at only one minute each.

Such 'round tables' are preferably confined to local bands such as 3.5MHz for inter-UK working and 1.8MHz and 144/433MHz for local working. It is positively anti-social to occupy intercontinental bands for such activity. Some regular nets favour the use of a net control station (popularly, a *master of ceremonies*). In fact this procedure can become insufferably long-winded and a better substitute for it is 'callbook order'. This system utilises the very simple alphanumerical sequence used in the *RSGB Yearbook*. In other words, every station with a G0 + 3 callsign transmits first, then come the G1 + 3, and so on. Any two-letter callsign may be slotted in at the relevant position when his or her turn to transmit comes around. In practice the 'batting order' of such a net could look like this: G3OUF, G4FTJ, G5UM, G6BBC, G8LM, G8IBA. Every operator thus knows his or her place.

Note the special provisions for callsign identification for net operating in the UK licence conditions.

EMERGENCY OPERATION IN THE UK

UK radio amateurs are advised against getting involved in requests for assistance from foreign countries. Requests for assistance or distress calls from ships or aircraft must *immediately* be passed to HM Coastguards or the police. Under no circumstances should the radio amateur 'go it alone'.

In the UK radio amateurs are permitted to pass third-party traffic on behalf of four user services (the British Red Cross Society, the St John Ambulance Brigade, the County Emergency Planning Officers or any police force in the UK) during an emergency or properly constituted exercise. Amateurs interested in such emergency working are urged to join Raynet, the radio amateurs' emergency network. There are at present over 150 groups and over 3000 members of Raynet, all trained and prepared to react against a known plan to provide communications under emergency conditions. Clearly it is desirable that emergency communicators be trained in message handling and procedure since this type of working is very different from that described elsewhere in this chapter.

Further details of Raynet, membership of which is open to both listeners and licensed amateurs, may be obtained from the RSGB's emergency communications officer (at the address given in the *RSGB Yearbook*) or from the Raynet web site at www.sgi.leeds.ac.uk/raynet.

LOGKEEPING

Radio amateurs in most countries are obliged by their licensing administration to keep a log of their transmissions so that if interference to other services is experienced it can be ascertained whether or not the amateur was transmitting at the time. For some amateurs the keeping of a log appears to be an inescapable chore and an official intrusion into their privacy, but in fact the log is a most useful part of the station.

Because it is a history of the amateur operator's activity over the years it is deserving of more attention than simply the minimum called for by the licensing authority. The log may act as a permanent record of band conditions and propagation, station performance, and the other stations details. In addition, in conjunction with a simple indexing system, it may be used to assist the operator in gaining various operating awards.

Fig 1 shows a sample of a standard logbook correctly filled in according to current UK licence regulations. No log entry need be made if the station is only used for a period of reception. Only the frequency *band* (not the exact frequency) is required, although the keen operator may find it useful to record the exact frequency (especially of DX or rare stations). The 'Mode' column should be filled in with the international emission code (see Chapter 1), and not 'CW' or 'FM'. Calls to specific stations which went unanswered should be logged, as should both successful and unsuccessful CQ calls. 'No contact' may be written in the 'Remarks' column to differentiate any unsuccessful attempts.

Any further information added to the log naturally depends on the operator's interests. The casual operator may perhaps record only the names and locations of stations worked for the first time. Those interested in propagation may find it useful to record details of the equipment and antennas at the other stations so that their ERP can be roughly ascertained. Award-chasers may note when a new station, country, prefix, locator square etc is worked for the first time. QSL information can likewise be recorded, and a 'D' or 'B' written in the 'QSL' columns to indicate direct' or 'via bureau'.

There is, however, a limit to how much information can be entered in a standard printed log. No matter how carefully this is designed it has to cater for all interests and is bound to be a compromise. If found inadequate there is nothing to prevent the operator from designing one more suited to his or her needs, provided of course that it complies with the licence regulations.

Another possibility is to use a standard log in conjunction with a scrap pad/notebook. A small reporter's notebook is ideal for this purpose because it can be folded back on itself to take minimal space on the shack table. The date may be entered at the top of each page for future reference, and then rough notes made from each contact. This will be found useful for slow CW contacts or the long monologues often heard on VHF FM.

Logkeeping using a computer

The UK licence allows you to maintain a log on your personal computer, and this has many benefits, particularly for contests. Typical non-contest features include: major awards progress indication, distance and bearing indication, instant indexing, printing of records, and access to databases (such as callbook

AMATEUR RADIO STATION LOG

DATE	TIME (UTC) start	finish	FREQUENCY (MHz)	MODE	POWER (dBW)	STATION called/worked	REPORT sent	received	QSL sent	rcvd	REMARKS
2 Nov '88	0800	0810	3	J3E	20	GM5ABC	59+10	59+5			Bert
''	0811	0820	145	F3E	16	G7XYZ	57	56			Terry first G7
''	0825	0830	14	J3E	20	CQ					No reply
''	1725	1735	145	F2D	16	GB7XYZ					Local packet mailbox
''	1740		Station closed down								
4 Nov '88	1030	/P	from 73 Antenna Lane, Squelch-on-Sea								
''	1031	1036	50	J3E	10	G1ØXYZ	55	56			Jim, Bridgetown
''	1036	1045	50	J3E	10	G7XYL	58	58			Anne, Nr Squelch-on-Sea. QRM
''	1205	1215	433	F3E	13	G2XYZ	46	47			
''	1220		Station closed down								
5 Nov '88	0945	/P	from 73 Antenna Lane, Squelch-on-Sea								
''	0950	1005	144	J3E	16	GB2GUY	56	56	✓		Catherine Fawkesville
''	1010	1015	144	A1A	16	GD5ZZZ	542	541	✓		QSB! QSL via WF9XYZ
''	1526	1530	144	A1A	16	G7CW	579	589			Good keying!
''	1535		Station closed down and dismantled								
7 Nov '88	1810	1902	435	C3F	10	G7ZZZ	P3	P3			Ted, first ATV contact!
''	1930	1945	21	J2B	16	VK2ABC	559	569	✓		RTTY, Sid at Bandedge
''	1946	2005	21	J2B	16	ZL3ZZZ	569	559	✓		1st ZL on RTTY
''	2010		Test for TVI/Harmonic Radiation - Nothing noted								
''	2020		Station closed down								
8 Nov '88	1735	1737	7	J3E	20	CQ					
	1738	1805	7	J3E	20	G1Ø2ZZZ	58	58			Nobby - chatted about G5RV ant
	1930	1945	51	F3E	10	GØ5IX	55	55			Allen, wanted W19B ref.
	1950		Station closed down.								
NOTES											

Fig 1. Typical logbook entries

directories). Some sophisticated software is available commercially but there are also some simpler programs available as shareware.

QSL CARDS

In the early days of amateur radio, contacts were infrequent and long-distance (DX) contacts were often records. Operators were keen to obtain written confirmation but writing and exchanging letters became rather too repetitive and onerous. Someone had the bright idea of printing the callsign and all the station details on a postcard, leaving spaces for details of the contact, weather report, interference etc to be filled in. Thus many contacts could be confirmed with the minimum amount of effort. These cards became known as *QSL cards* from the international Q-code meaning "I acknowledge receipt" (see p22).

The pioneering days on many amateur bands are long past, but the exchanging of QSL cards is still one of the most popular activities within amateur radio. The cards serve as a reminder of interesting contacts or a rare country worked or heard and are often required as confirmation for one of the many operating awards available (see Chapter 4).

To lessen the heavy cost of posting each card individually to other stations, most national societies operate what is

called a *QSL bureau.* This is a 'clearing house' which accepts cards in from members, sorts them into countries, and mails them in batches to bureaux in those countries which in turn sort them into batches for individual stations.

The great advantage is that all posting is done in bulk, a considerable saving to everyone concerned. Furthermore it is only necessary to send cards to one well-known address in one's own country, which is considerably more convenient than taking foreign addresses down over the air, with the possibility of errors.

Nevertheless many amateurs still occasionally send cards

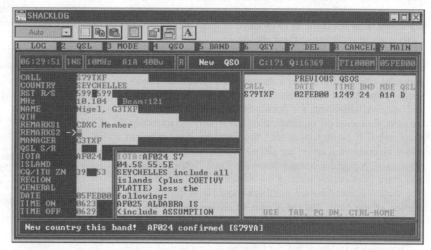

Computer logging programs do much more than just record contacts. Here Shacklog is keeping track of countries worked, giving beam headings, distances and more

With the Compliments

of

2. U. V.

IN ACKNOWLEDGEMENT OF

SIGNALS

W. E. F. CORSHAM,
104, HARLESDEN GARDENS,
HARLESDEN. N.W. 10.   *Date* JAN 1922

The first QSL card? Way back in 1922, printed cards were already starting to be used to confirm contacts

direct, even if they are members of national societies. Sometimes the urge to receive confirmation of contact with a rare or DX station is too great to wait the few months cards take to work their way through the QSL bureaux system. In fact, some DX stations only reply to direct QSLs. Furthermore, a few countries do not have a QSL bureau and all cards must be sent direct in these cases.

Designing cards

There are various ways in which QSL cards can be produced. Some cards are designed by the operator and then home-made or printed professionally, while others are selected from a range of standard designs produced by a firm specialising in the printing of QSL cards. Some large firms provide QSL cards to members of their staff who are radio amateurs, and such cards may also be provided to special-event stations.

The card should not exceed the standard postcard size (140 by 80mm), especially if a QSL bureau is to be used. Try and use the recommended size of 140 by 80mm. Cards any larger or smaller than that are difficult to sort and are likely to be lost or damaged.

Cards should be printed on ordinary postcard board, not on thick heavy cardboard which makes them difficult to handle and more expensive in the post. The use of unorthodox materials is *not* advised. Over the years, the RSGB QSL Bureau has handled cards printed on cork, wafer-thin sheets of wood, plastic and silk. The worst were those printed on thin aluminium or copper sheet which could cause painful cuts to the unsuspecting sorter. It is not difficult to produce original and attractive cards without resorting to these means.

The object of sending a QSL card is primarily to confirm a contact but it is also a means of conveying brief information about the contact, the sender's station and its location. The essential printed data should therefore include the sender s name and address with space for details of the contact (date, time, frequency, method of communication and report) and the equipment used. It is important that the wording does confirm two-way communication; the phrase 'Confirming contact' is often used. A popular layout is shown in Fig 2. Some operators add extra information such as the station's locator (see p64). The space for the addressee's callsign should be in the left-hand portion of the card, and the sender's callsign should feature prominently on the front of the card either as part of the design or as an overprint in a contrasting colour.

The artistic design of the card can be the subject of much ingenuity. Undoubtedly the more original and attractive it is,

the more interest it will create for the recipient. This is perhaps a subject best left to the reader's own imagination but a few suggestions for the fronts of printed QSL cards are as follows:

1. A colour picture postcard of the locality overprinted with the callsign.
2. A black-and-white photograph of the station which shows the callsign prominently, either as part of the photo or overprinted in a contrasting colour.
3. A humorous cartoon or message of radio interest.
4. A sketch map of the country, with some indication of the whereabouts of the station.

It is not generally necessary to use more than two colours (or a colour-tinted card and one other colour) when printing cards. Each additional colour will significantly increase the cost of the card and may add little to the overall impact. When deciding how many cards to print, think ahead and estimate how many are going to be required over a reasonable period, say, two years. It is much more economical to have one large printing than several smaller ones, particularly if any original design is being used.

The design of listener cards should follow the same basic principles as those for transmitting stations in respect of size, material and information, except that it should be clear that they are reception reports. The registered listener 'ARS' or 'BRS' number should be prominent in the case of RSGB members. The card should also have spaces for details of radio conditions and other countries audible at the time, and also the callsign of the station which the recipient was heard working or calling, a typical layout being shown in Fig 3.

If you don't require many cards and have access to a computer with a good-quality printer that can cope with thin card, you may be able to produce all the cards you need that way, especially if you can use a drawing software package to design an interesting layout or include a photo image. If the printer is capable of colour, so much the better, but don't overdo it.

Sending cards

It is usual to send cards for three good reasons: because a return card is wanted; to acknowledge the receipt of a card; or because a card has been requested during the contact.

The sending of a card for every contact is both costly and wasteful, resulting in large numbers of unclaimed cards having to be destroyed. Many amateurs, particularly in the rarer countries, do not send QSL cards unless they receive one first from the other station.

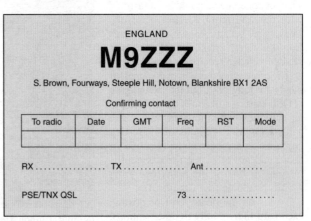

ENGLAND

M9ZZZ

S. Brown, Fourways, Steeple Hill, Notown, Blankshire BX1 2AS

Confirming contact

To radio	Date	GMT	Freq	RST	Mode

RX .... TX .... Ant ....

PSE/TNX QSL   73 ....

Fig 2. Transmitting-station basic QSL card layout

SCOTLAND

RS90000

J. McNab, 21 Salburgh Close, Blankirk, Lowlands LO1 2XX

To radio	Date	GMT	Clg/Wkg	Freq	RST	Mode

CONDX . . . .   RX . . . .   Ant . . . .

PSE QSL   73 . . . .

Fig 3. Receiving-station basic QSL card layout

If a card is sent direct to an amateur with the object of receiving one direct in return, sufficient International Reply Coupons should be enclosed to cover the cost of the return postage. These coupons are available at main post offices, although IRCs have assumed the status of a 'currency' among DXers and can be purchased more economically from amateurs with a surplus who advertise from time to time in the DX news-sheets. It is also common practice, particularly among American amateurs, to enclose dollar bills rather than IRCs. This carries a risk of theft and, worse, citizens of some countries can be imprisoned for being in possession of foreign currency. This method of payment is therefore not recommended. If unused stamps from the country concerned can be obtained, these can be used as payment for return postage, as an alternative to IRCs.

The addresses of United Kingdom amateurs can be found in reference [4] and the addresses of amateurs worldwide in [5].

Quite often rare stations will give their QSL mailing address over the air, but not necessarily in every contact if they are working a string of stations rapidly. It must be appreciated that such a station has to spend a quite considerable amount of time and money in preparing and sending QSL cards because practically every station worked will want a card. With this in mind, some DX stations appoint their own QSL manager to cope with the workload. Cards should then be sent *direct* to the manager, and marked 'via . . .'.

Much useful QSL information is given in *RadCom, Radio Today* and other amateur radio magazines, especially the DX news-sheets (see Chapter 4). The Internet is also becoming an important source of QSL news.

A self-addressed envelope should always be enclosed when QSLing rare DX stations or their managers direct. This not only saves them stationery costs but also a great deal of time and effort copying out your name and address from your card. The rare station's name and address should be added to the reverse of the envelope prefixed by 'From . . .' for the same reason. Some DXers even send out their own special QSL card which merely has to be endorsed and returned by the DX station, saving further expense.

It is difficult to give hard and fast rules on how many IRCs to enclose. Most rare DX stations or their managers make sure they are not out-of-pocket on QSL costs, and may subtract the price of the card from the total value of any enclosed coupons before deciding what balance is available for return postage. Normally, however, two IRCs should be sufficient for most DX stations. A single coupon will only pay the postage for airmail return and leave nothing for other expenses.

Avoid sending cards through QSL bureaux with IRCs attached,

A selection of QSL cards showing different styles – perhaps one will inspire you!

as these have a habit of disappearing on route. For the same reason, use normal stamps on direct QSLs, rather than special commemorative issues.

When making out the card, be careful not to make any amendment, or to cross out anything written, or alter the callsign. Some societies which issue awards automatically reject 'altered cards', even if the alteration was made by the sender.

Many amateur transmitting stations will send their QSL cards

Cards being sorted at the RSGB QSL Bureau

THE RSGB QSL BUREAU

The purpose of the Bureau is to facilitate the exchange of QSL cards between RSGB members and other radio amateurs. Most national radio societies operate such a bureau, some making an extra charge for this service. The RSGB provides it as a free service, though only members may make use of the outgoing service.

Use of the bureau is the cheapest way of sending cards in these days of high postal charges.

Sending cards through the Bureau

All cards for distribution should be sent to the RSGB QSL Bureau at Headquarters. There is no limit to the number of cards which may be sent at any one time.

When the cards arrive at the bureau, those destined for abroad are sorted into countries and despatched in bulk to the appropriate overseas QSL bureaux, most of which are operated by members societies of the International Amateur Radio Union.

Cards for stations within the UK are sorted into callsign groups, each of which is in the charge of a volunteer sub-manager. It is this person's task to associate the cards sent to the sub-bureau from the main QSL Bureau with the envelopes which are on file.

Collecting cards from the bureau

Supply your sub-manager with stamped self-addressed envelopes of a suitable size and strong material – 140 × 80mm is ideal. Use only 1st or 2nd class postage stamps. They are valid indefinitely once purchased whereas the ones bearing a monetary value only last for as long as the postage rate stays at that level. To use stamps other than 1st or 2nd class causes immense problems for QSL Bureau sub-managers.

Print your callsign or RS number in the top left hand corner of each envelope.

Envelopes should be numbered and 'Last envelope' marked on one so that it is known when a fresh batch is needed.

Envelopes are not normally returned until full weight has been reached for the postage paid; those wishing to receive cards at more frequent intervals should mark their envelopes 'Wait 6' etc.

An up-to-date list of names and addresses of sub-managers is available from RSGB HQ on request. Changes to the list are broadcast on GB2RS and in the QSL column in *RadCom*.

General notes

1. Licensed UK amateurs who are non-members of the RSGB may send stamped addressed envelopes to their sub-manager for collection of their cards, but they *may not* send cards for distribution.
2. Cards for amateurs who have neglected to send envelopes are retained for three months, after which the cards are destroyed. Amateurs who do not wish to collect cards should notify the QSL Bureau accordingly.
3. Amateurs who operate from a part of the United Kingdom which has a different prefix should deposit envelopes with the appropriate sub-manager for the different prefix. For example a G7 station who operates temporarily from Wales and who wishes to receive cards should leave envelopes with the GW7 submanager.
4. Overseas members of RSGB in countries where there is no QSL service operated by the IARU member society for that country may send their cards to the RSGB QSL Bureau for distribution.
5. Overseas members who are not members of the RSGB may send cards addressed to UK stations only, direct to the RSGB QSL Bureau.
6. The facilities of the RSGB QSL Bureau are available both to transmitting and receiving members of the Society. Listeners are reminded, however, that their reports should contain sufficient information to be of genuine value to the transmitting amateurs concerned. Reception reports relating to short-wave broadcasting stations unfortunately cannot be accepted.

All QSL cards and correspondence relating to the RSGB QSL Bureau should be sent to the QSL Bureau at RSGB Headquarters.

Adhesive address labels are available free of charge on receipt of an SAE.

Envelopes for the collection of cards and correspondence concerning incoming cards should be sent to the appropriate sub-manager.

Sending cards through the bureau

Choose QSL cards which do not exceed normal postcard size, viz 5.5in × 3.5in (140 × 80 mm). Large cards invariably have to be folded, while small ones and those of a thin nature are difficult to handle.

Print the addressee's callsign on both sides of the cards.

Separate cards destined for within the UK from foreign-going ones. Sort all cards alphabetically by prefix. Sort USA cards into call areas regardless of prefix. When a QSL Manager is involved, sort under *his* callsign. Please pack the cards so that they are all the same way up, and do not space cards with markers or similar. Pack cards adequately, so that they do not get damaged in transit, and with the correct postage to reach RSGB, and send them to:

RSGB QSL Bureau
PO Box 1773
POTTERS BAR
Herts.
EN6 3EP

in exchange for listeners' cards, and these can be used to gain the various listener operating awards (see Chapter 4). However, this is likely only if the listener report is of interest or of value. For example, a station in the USA working long strings of UK stations on 14MHz is unlikely to find a UK listener report helpful. He *knows* he can work the UK! It is therefore advisable to send reports to some of the weaker signals or to the people who do not seem to be making many contacts.

To be useful to the recipient, a report should preferably contain observations made at several different times, a comparison with other signals heard at the same time, together with useful comments on quality, depth of modulation etc, although there may not be very much room on a card for all this. Some listeners go to the extreme of sending separate QSL cards each time they hear the same station – this is a waste of time and money as one card is quite sufficient.

One last tip: if you need confirmation badly for an operating award, why not say so on the card? It might just make the difference!

FURTHER INFORMATION

[1] *The Morse Code for Radio Amateurs*, George Benbow, G3HB, RSGB, 1994.

[2] *Your Introduction to Morse Code* tapes, ARRL.

[3] *Instant Morse* CD-ROM, RSGB.

[4] *RSGB Yearbook* (published annually). Also available on CD-ROM as the *CallSeeker* call book program with the information directory pages in Adobe Acrobat format.

[5] *Radio Amateur Callbook*, Radio Amateur Callbook Inc (published annually on CD-ROM).

4 DX

ALTHOUGH one of the oldest aspects of amateur radio, 'working DX' remains as popular a pastime today as ever. Basically 'DX' means 'long distance' but of course this is relative and depends on the band, mode and even the equipment in use. In fact DX has more to do with 'difficulty' than with distance in absolute terms, and herein lies its fascination, for of course the more effort is put into making a contact, the greater the satisfaction that will be the reward. Perhaps this point is worth emphasising in these days of so-called 'push-button DX' and kilowatt linears.

This chapter deals first with the HF bands, then with the VHF/UHF region and microwaves. These arbitrary divisions nevertheless reflect the fact that many, perhaps most, operators choose to use one set of these bands in preference to the other, although the object of DXing remains much the same whatever frequency is in use.

DX ON THE HF BANDS

Generally speaking, on the bands below 30MHz, signals emanating from one's own continent would not qualify as 'DX'. However, those using very low power and simple equipment might consider that this is not so. There are different degrees of 'DX'ness' and quite obviously the distance between transmitter and listener is the most important factor here. Another is the band being used – it is no great achievement to hear Japanese signals on 14, 21 and 28MHz under the right circumstances, it is more difficult on 7 and 3.5MHz, but very difficult on 1.8MHz. There is also a rarity factor – it is much more probable that, given suitable propagation conditions, signals from the southern USA will be heard than from northern Mexico. The reason is of course that there are very few XE2 amateurs, but many thousands north of the border.

LISTENING FOR DX AS AN SWL

This is one of the most interesting aspects of amateur radio, and attracts a very large number of those who never aspire to a transmitting licence but prefer to listen to what is happening on our bands. The number of enthusiastic shortwave listeners probably exceeds that of licensed amateurs, and it is these very same individuals who often make the best operators when they do obtain a licence.

Keen short-wave listeners are especially cared for by the International Short Wave League [1] and the International Listeners Association [2]. The former organisation produces an excellent monthly magazine which contains a great deal of information on both amateur and broadcast DX listening. The League also has many transmitting members.

Most of the tips given below are just as applicable for listeners as transmitting amateurs.

'RARE' COUNTRIES

What is a 'rare' country? An idea may be gained by reference to the following lists.

1. **Countries where legitimate amateur radio activity is infrequent:**

Afghanistan (YA)	Libya (5A)
Bangladesh (S2)	North Korea (P5)
Bhutan (A5)	Yemen (7O)

2. **Countries where activity is usually by an operator at a weather station, military base or scientific research base:**

Crozet Is (FT–W)	S Georgia (VP8)
Kerguelen Is (FT–X)	S Orkney (VP8)
Amsterdam & St Paul	S Shetland (VP8)
Is (FT–Z)	Auckland I & Campbell Is
Glorioso Is (FR/G)	(ZL9)
Tromelin Is (FR/T)	Kermadec Is (ZL8)
Kure Is (KH7K)	Marion Is (ZS8)
Wake Is (KH9)	Jan Mayen (JX)
Midway Is (KH4)	N Cooks Is (ZK1)
Willis Is (VK9)	Guantanamo Bay (KG4)
Macquarie Is (VK0)	

3. **Countries only activated by special expeditions:**

San Felix Is (CE0X)	Heard Is (VK0)
Clipperton Is (FO)	S Sandwich Is (VP8)
Juan de Nova Is (FR/J)	Andaman Is (VU)
Malpelo Is (HK0)	Laccadive Is (VU)
Navassa Is (KP1)	Revilla Gigedo Is (XF4)
Desecheo Is (KP5)	Aves Is (YV0)
Kingman Reef (KH5K)	Spratly Is (9M0)
Palmyra Is (KH5)	Annobon Is (3C0)
Market Reef (OJ0)	Bouvet Is (3Y)
Malyj Vysotskij Is (R1M)	Peter 1st Is (3Y)
Sovereign Military	Conway Reef (3D2)
Order of Malta (1A0)	Rotuma (3D2)
Sable Is (CY0)	Banaba (T33)
St Paul Is (CY9)	Wallis and Futuna Is (FW)
Mellish Reef (VK9)	

GETTING IN SHAPE FOR DXING

Most rare DX stations will have many stations calling them so DXing is very competitive. How can you can ensure that there's a good chance of the DX station coming back to *your* call? Three key elements of your station need to work together to achieve this goal – the antenna, the radio and *you*.

The antenna

Some years ago the RSGB conducted a survey of leading UK DXers, covering antennas, equipment and operating practice, and this, together with other information, was published in *Radio Communication* [3]. Some of the information in this series of articles is summarised here but they are well worth

You and your station had better be in good shape if you want to get exotic QSL cards like this one from VK9YG/AX9YG, operated by Steve Telenius-Lowe, G4JVG, in 1988

consulting if you are serious about working DX. The survey found that most of the respondents use triband Yagi beams on towers of about 60ft for 14, 21 and 28MHz and a variety of wire antennas or verticals for the LF bands. Of course, other antennas also put out good signals, and the quad and delta loop antennas have their devotees. But the important point is that 'off-the-shelf' commercial triband beams will do the job until you find something better.

For working DX it is clear that a good front-to-back ratio is more important than forward gain – often the limiting factor in copying a weak signal is interference coming from the opposite direction, and if this can be lessened there is a much better chance of making contact.

Height is also very important. An increase in antenna height will improve the low-angle radiation, and it is generally reckoned that a doubling of antenna height from, say, 30ft to 60ft will result in something like a 6dB better signal at a DX location! A word of warning: do make sure your tower and rotator are up to it – get professional advice from your suppliers, don't guess. If even a small tower blows down, the results could be catastrophic.

Don't despair if you don't have room for a tower and a big beam. Working DX on the LF bands using a vertical can be a good bet. These antennas need very little space, and will be a lot more competitive on 7MHz than 14MHz. Another possibility is to build a wire antenna for working in certain directions. There is a lot of fun to be had in experimenting to find an antenna to suit your situation. Many designs for antennas are available in the standard handbooks covering all likely situations at your station [4–12].

Remember, if you have a poor antenna you can still work DX but you'll maybe have to wait a bit longer and be more cunning like QRP operators!

The transceiver

Modern radios will provide some or all of the following features, which are highly desirable for working DX. Older ones, particularly those of the valve era, may lack them in spite of having generally good performance.

- Twin VFOs or a VFO with memories. They are almost essential for working DXpeditions as these usually work split frequency (see later).
- Total control over the shape and width of the IF passband for SSB reception. This can be achieved by variable bandwidth tuning (VBT), slope tuning, and IF shift.

- A 500Hz filter (or, ideally, a 500Hz and a 200Hz filter) for CW reception. A good audio filter can improve the performance of a 500Hz filter.
- A notch filter for removing particular sources of interference to either CW or SSB within the IF passband. This should preferably operate at the IF rather than at AF.
- An RF attenuator for use on the LF bands such as 7MHz where cross-modulation is often a problem.
- A receiver noise blanker, ideally with variable time constants.
- The ability to work properly with a linear amplifier.
- A facility to connect a separate receiver antenna.

Of course, these features will count for little if the performance of the receiver is poor. Generally modern radios offer a high level of performance but, if you are in the market for another one, you should understand the factors which affect receiver performance so you can make the best use of reviews in magazines. These are outside the scope of this book and you should refer to references [13–15] for further information.

If you have an older radio, there is often a great deal that can be done to improve its performance, perhaps by fitting an optional internal CW filter and adding external units: an RF attenuator for LF band operation; an RF amplifier for, say, 28MHz; a speech processor; an AF filter etc. These are often simple to build; see reference [4] for details.

Linear amplifier

This is a definite asset for working competitive DX, especially if you are limited to a modest antenna or operation at evenings and weekends. The bad news is that it will make any EMC problems you have a lot worse. So you must make absolutely sure you have a clean signal into it and also do your EMC 'housekeeping'. See Chapter 3 for more information.

The cost of a decibel

In *HF Antennas for All Locations* [6] Les Moxon, G6XN, asks "How much is a decibel worth?" and this is an important question for all DXers. Every improvement in performance, even 1dB, is worth having in a pile-up, but some decibels cost a lot more than others. Make sure you have got all the cheap ones before spending money on the expensive ones!

THE DX OPERATOR

So much for the equipment – what about you? The most important requirement for a budding DXer is spare time. The RSGB survey respondents were firm on that point. If you aren't in the shack you aren't going to work anything, however wonderful your station might be. Roger Balister, G3KMA, took the theme further with the rather awkward advice: "Don't go away on holiday – ever!" And Bill Ricalton, G4ADD, likewise advised cancelling any holiday that clashes with a new DX-pedition.

To make the most of the odd spare minute an anonymous respondent recommends locating the equipment as close as possible to the living areas of the house, and the survey heard from several others who have a rig next to the bed for those early morning skeds – CW of course, so as not to wake the XYL . . .

The mention of CW is important because top DXers use all modes. If you ignore all CW activity your progress up the DX tables will be slowed. But don't be put off by any thought that vast competence is required. To be honest, all you need in many cases is the ability to read the DX callsign, recognise your own and send 599! For large split-frequency pile-ups you will have to develop the ability to recognise other calls in the QRM (as discussed later) but this will come with practice.

Background knowledge

A good knowledge of subjects such as geography, current affairs and foreign languages can also be of great help to the serious DX chaser. Information on time-zone differences, public holidays and normal working hours in different countries can also be very helpful. For example, French public holidays are observed in all overseas French provinces so that any of the overseas F prefixes are more likely to be heard on such a day. Remember also that in Islamic countries Friday is the day of prayer and equates with our Sunday. Fridays are thus good days for looking out for some of the rarer countries in North Africa, the Middle East and certain parts of Asia.

Amateur radio often gives the keen listener an interest in other countries. One essential is naturally a good atlas – not the schoolbook kind but instead one of those large (and expensive) volumes. Unfortunately the publishers of these often see fit to include much extraneous material on such subjects as the galaxy, aerial navigation or trade routes etc. Choose one in which the money has been spent instead on the maps and gazetteer, for this is what an atlas is all about!

Complementing the atlas may be one or more wall maps. These are available in great-circle or Mercator projections. Great-circle maps are centred on one particular location, usually London in the case of UK maps. They give the true bearing and distance of any other place in the world relative to that location, and are thus more useful to the operator using a beam antenna. See Fig 1.

The more familiar Mercator projection map is more generally available at lower cost than the great-circle type, and is often more detailed and colourful. Both types are available from RSGB HQ.

As time goes on, you will need to know not just where countries and towns are, but something of the internal administrative divisions in many countries. This is because a large proportion of countries vary their amateur radio prefixes according to the region of the country (see Chapter 1). A knowledge of these divisions and their corresponding amateur radio prefixes is most useful for the DXer, and DXpeditions often choose uninhabited areas of a country so that they can use a rare prefix.

The reader will already no doubt be familiar with the UK countries, the Australian and USA states, and the Canadian territories and provinces. Many other countries, however, have divisions which affect the prefix, two examples being the Australian states and the 24 Swedish läns (provinces). Full details of such callsign systems are given in Appendix 2.

A callsign list by ITU allocation is given in Appendix 3. A

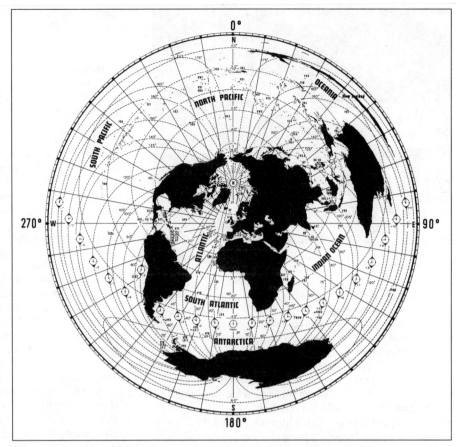

Fig 1. Example of a great-circle map, available from the RSGB in a large size suitable for wall mounting

more detailed list, which is updated regularly, is available in the *RSGB Prefix Guide* compiled by John Forward, G3HTA, and published by the RSGB. This contains a lot of other useful information and is a 'must' for any serious DXer.

LISTENING TIPS

You need to tune the bands to find out what is happening but what are you listening for? How do you separate the unusual from the commonplace?

- Check around the standard DX frequencies: 5 and 25kHz up from the band edge on CW; 3795, 14,195, 21,295, 28,495 and 28,595kHz on SSB. Don't forget the IOTA frequencies (14,260, 21,260 and 28,460kHz) as many islands are also DXCC countries. Fred, G3NSY, has all of these frequencies programmed into the memories of his transceiver to simplify the search.

- DX stations often sound 'different': weak, fluttery, or with some wobble on the transmission as a result of local power supply problems. They are often heard working pile-up-style, giving quick reports to get as many QSOs in the log as they can.

- Get to know the names of active DX operators. For example Cav is probably V63JC in Micronesia, and Tom, VP6TC, on Pitcairn Island, Antoine might have been 3D2AG on Rotuma; and Martti is probably OH2BH who could be almost anywhere rare – or soon will be.

- Watch out for an unusual accent. An American voice when the band is not open to the States indicates something different, perhaps Hawaii or Guam.

A one-man DXpedition: Nao, NX1L, operating FO0AKI on Rurutu Island in June 1994

- An operator talking about his QSL manager is normally somewhere interesting.
- Listen to the snippets of conversation as you tune across the band. You should be able to tell in seconds if it is a normal ragchew or an exotic DX station stopping for a chat. Obvious giveaways include remarks about postal delays, supply ships, missionaries, coral atolls, etc. In the run-up to a major expedition you will also hear all sorts of rumours, some of which may have a useful basis in fact!
- Check out the pile-ups. You will generally find the DX station on the same frequency but underneath all the callers, or about 5kHz lower, or on a standard DX frequency.

THE INFORMATION TRADE-OFF

Those of us with families, jobs, and other commitments can fortunately make use of other sources of information besides hours spent listening on the bands. If you know exactly when the stations you need will be active, you only have to fire up your rig at the right moment, work them, and get back to watching TV. In the survey Peter Wallis, G3YJI, and many others listed "good information" as the principal requirement for DXing success.

So how do you find out what's happening?

- Get on the DX PacketCluster network, a wonderful development which enables all active DXers to share each other's ears. If a DX station is heard on the band, an alert (called a 'spot') will be fed into the system and will be seen by all users. PacketCluster also offers other facilities, such as being able list the last 10 spots on 14MHz, or the last five spots related to stations with a particular prefix. Also available are WWV propagation data, QSL information, together with the facility to calculate beam headings, propagation predictions and other data of value to HF DXers.
- Use the RSGB DXNS Voicebank system, a computerised voice messaging system operated by British Telecom. Access is at local call rates from all over the UK, and there are two separate numbers: 01426 925240 plays back tips recorded by DXers over the last three days and 01426 910240 is available for inputting DX tips (up to one minute).
- Get on the Internet. There is an e-mail 'reflector' system which covers HF DX and HF contests. Via World Wide Web (or the Telnet/Finger protocols), you can also log on to or

read remote PacketCluster systems, download DX bulletins, receive ARRL and RSGB news bulletins, access news and other discussion groups, and acquire masses of solar/propagation information.

Both free and paid-for DX bulletins are available from the Internet. The *Ohio-Pennsylvania DX Bulletin* is sent weekly to registered subscribers (send an e-mail to opdx-request@nshore.org with the word 'subscribe' in the text field). The *425 DX News*, published by the Italian 425 DX Club, is also sent weekly to subscribers (send a e-mail to 425server@425dxn.org with the phrase 'subscribe 425eng' for the English-language version). Both these bulletins are also available via the web pages of these organisations.

Many DXpeditions set up their own web pages, covering topics such as schedules, operator CVs, predicted propagation and much more. It is also increasingly common for the DXpeditions to update these pages during the DXpedition itself, with changes to operating schedules, photographs, and log search facilities.

Other web pages of use to DXers offer QSL manager look-up, propagation data, IOTA information, award details, and much more. The DX Summit web page (http://oh2aq. kolumbus.com/dxs/) gives access to recent DX spots collected from PacketCluster systems around the world, and is therefore especially useful for real-time information about DX activity.

- Make sure your friends know which countries, islands, zones, or oblasts you still need and get them to call you on the phone if they hear one of them. This can generally be set up as a group activity, perhaps based around the most enthusiastic members of your local club.
- Ask a local. In the survey Bob Whelan, G3PJT, commented that "If you need Kure Island then ask the KH6s, who will probably know of any planned activity".

WORKING DX STATIONS

Vital to those who wish to make a success of DX working is patience and politeness. The rarest stations are often running simple equipment and using low power. Their antennas may be poor and they may be a very long way away. They may not speak the caller's language very well, and they may not have any particular desire to talk to yet another European! This means that it is helpful to listen to a few contacts before trying to call, and thereby get to know the way in which calls are accepted. Causing interference by calling at the wrong time or on the wrong frequency very rapidly loses friends – on the other hand there is a great deal of satisfaction to be gained from getting a reply from one's first call made at the right time and in the right place. DXing is indeed remarkably like fishing!

When and where are the correct times and places to call? Techniques are decided by two possibilities:

1. Is the station working callers on his own frequency?
2. Is he listening elsewhere?

In condition 1 (which often arises if the DX station was just looking for a contact and things got out of hand!), careful listening is more than ever important because ill-timed calls will cause interference. If it is found that 'tail-ender' calls are being accepted it is then permissible to drop one's callsign in just as the station currently in contact is signing off. However, this should never be done unless it is absolutely clear that that is the end of the contact, and that the DX station has the callsign of the person he is talking to logged correctly. Tail-ending is more likely to be successful if only a few are waiting on frequency and it often becomes a sign for a 'free-for-all' to develop. Note that when using SSB, calls must be made on a frequency exactly zero-beat with that of the station being called. In the case of CW calls there may be some advantage to be gained from making calls a few hundred hertz away from zero-beat. When operating on CW some stations seem to follow a pattern, perhaps taking callers from either side of their frequency alternately, or gradually tuning outwards on one side or other. Listening will establish that this is happening and may save a lot of wasted calls.

If condition 2 applies things are totally different and calling so as to cause interference on the DX station's frequency is to be avoided at all costs. The more experienced resident DX stations, and nearly all expeditionary type stations, follow this course as it is impossible to cope with many hundreds of callers within a few hertz of each other. Such stations make a practice of announcing where they will listen and often specify a section of the band over which they will tune, eg they may be transmitting on 14,195kHz and listening over the range 14,200–14,210kHz. Under these circumstances the user of a transceiver without a second VFO is at a severe disadvantage, and the serious DX operator should bear this in mind when setting up his or her station.

Once again listening is all important, first of all to establish where to transmit, and also to check that the frequencies mentioned are in fact those being tuned. It is not unusual to discover that stations calling outside these are being worked! Once again there may be a recognisable tuning pattern and it may be possible to hear stations in contact and to work out the best place to place a call.

Perhaps the DX station is tuning slowly from the bottom of the range to the top, perhaps sticking to a single frequency inside the announced range, or even listening somewhere completely different. A few DXpeditioners make a habit of listening below their transmit frequencies when they've announced "Listening up" and vice versa. It is also worth checking whether the listening frequency changes after each QSO or whether the DX operator pauses in one place until everybody discovers that frequency and the rising QRM forces him to tune elsewhere.

Another reason to listen before calling is that DXpeditions frequently limit the pile-up by asking only for stations with a particular number or letter in their callsign. To take advantage of propagation they may even ask for particular regions or countries only. For example "European 1s only" or "North American 5s call now". Failure to observe these restrictions may result in your call being added to a blacklist of people who find it difficult to get a QSO or a QSL card from that expedition.

Once the DX tuning pattern has been established, opinions differ on the best technique. In the RSGB survey, G4ADD recommended finding the quietest in-range frequency and sticking to it, while others like Pat Gowen, G3IOR, suggested anticipating the tuning direction and always trying to be next in line. Chris Eyles, G3SJH, reminded us that an instruction to the pile-up to "Spread out" is often a cue to call at the extreme ends of the announced listening range. The famous DXpeditioner SM0AGD likes to program up to 10 discrete listening frequencies into his transceiver and, randomly or predictably, to hop between them.

Frequencies are usually publicised in advance or announced on the air. If you're trying to be next in line for the DX, you first have to find the frequency of the station he's currently working. Then make your call either on that frequency or a bit further in the direction that the DX station seems to be tuning.

You may have only two or three seconds to find the other station in the QRM of the whole pile-up, and you probably won't have propagation to all callers, so even highly experienced DXers may only track down 25% of the stations being worked in a big pile-up. There is no time to stop on each signal; you have to identify calls as the dial is turning and develop a mental 'map' of those present. With time and experience you develop a sixth sense about your competitors' habits and can go immediately to their frequency if you hear the DX pick up one of their calls. When you get everything right and the DX station comes straight back to your call the thrill is indescribable!

Yet another reason for listening first is that it is always advisable to find out what kind of contacts the DX station is making before calling. Is he making short snappy ones? If so it is not sensible to try to tell him all about the weather and full details of the equipment being used. The best idea is to give information similar to that which is being sent, and if the contact is a CW one this should be sent at the same speed (or slower if conditions are poor) as that at which the distant station is operating. It is quicker and more effective to send a report once at 13wpm than twice at 25wpm! When using phone, ITU phonetics should be used and pronounced clearly and slowly.

Calls themselves should be short. The DX station knows his own callsign and does not want to hear this being repeated many times but it is best to give one's own callsign more than once. The full use of VOX or break-in enables close watch to be kept on the time that the station being called is transmitting, and further calling can be avoided until he is listening again.

It is all too often possible to hear blind calls being made by those who quite clearly cannot hear the DX station, and also to hear a group of those waiting to work him talking on the frequency where he is expected to transmit (or is or has been transmitting). Both practices are to be deplored. Either is capable of causing severe interference to those located in other parts of the world.

Having made contact it is important to make sure that one's callsign has been logged correctly – if there is any doubt it is quite acceptable to continue to call and to make a second contact. However, during the usual DXpeditionary activity one contact only should be the rule. This may be extended to one per mode or one per band if the opportunity arises. It is important to log the time of the contact accurately, and quite a good idea to note the callsigns of others who make contact just before and just after – this may be helpful to the DX station in locating a log entry at a later date.

In conclusion, it is possible to list a few general rules which help when trying to make interesting DX contacts. They are as follows:

1. Listen carefully.
2. Be aware of what is happening.
3. If in any doubt – do not call.

By following these rules a few contacts will be missed because ruder and less well-behaved callers will sometimes succeed in crushing opposition. However, in the world of DXing these individuals are well known and receive no credit for their behaviour from good operators.

More pile-up tips

What more is there to do? Well, in the survey G4ADD advised paying lots of attention to your audio quality so that your voice stands out from the pack and others suggest you make sure your voice and your call are well known to DXpeditioners. Faced with a mass of signals the familiar voice or call is much more likely to be recognised by a tired DXpeditioner. Language skills can help on the occasions when listening frequencies are announced in a foreign language – but some techniques defeat even the best prepared. For example, BT's Directory Enquiry service had a sudden surge of calls when a DXpeditioner announced he was listening on the frequency given by the last three digits of his phone number! In addition to their publicised frequencies, most DXpeditions have unannounced frequencies for contact with members of their home clubs; identifying these by careful tuning can be an easy way to avoid the main pile-up. You can usually tell when they are in use because the DXpedition will suddenly, and quickly, work a run of calls from the same country or town.

A controversial pile-up point is whether you should spell out your whole callsign (in this author's case "Golf Three Zulu Alpha Yankee") or just repeat the last two letters "Alpha Yankee, Alpha Yankee". This author's view is that the full call is better. If the DX station hears you clearly he won't want to waste time establishing your full callsign (especially as you may not be in the clear on the second go round). If he is only going to catch a fragment then "Golf Three Zulu" is probably as useful to him as "Alpha Yankee", and when the last two letters are repeated he may come back to "Yankee Alpha Yankee" and get hopelessly confused.

Before leaving the subject of pile-ups a point must be stressed again: never, ever transmit on the DX station's frequency in a split-frequency pile-up. Don't ask "Where's he listening?". And don't ask "Who's his QSL manager?" while other people are still trying to work him. In most cases it's the failure to listen that leads to unnecessary interference. However, it's also possible to call on the DX station's frequency by cancelling or inverting the split settings on the transceiver – so please take care!

Perhaps total immersion in a wild 20m weekend pile-up is not the best way to start your DXing career. So it's worth remembering that weekdays are more relaxed on all bands, and the 10, 18 and 24MHz bands are still oases of tranquillity where a barefoot transceiver and a simple dipole can crack most of the embryo pile-ups that occur.

LISTS AND NETS

An increasingly common practice among users of SSB transceivers is the formation of lists of those who wish to contact a particular station. The 'master of ceremonies' then passes the list of callsigns to the DX station who proceeds to call in each station in turn. This almost inevitably degenerates into bad behaviour and chaos caused by those who arrive on the scene after the list has been made and who cannot therefore get a chance to call, and lack the manners to play the game the way the DX station is asking.

There are some circumstances when a station can only be contacted by this method – for instance he may only have a transceiver without dual VFOs, or he may speak a language not widely understood and may need help with translation.

How does one get on a list? Yet again the first and most important thing to do is to listen carefully. If the operation is being well conducted the operator in charge will keep those listening informed and will frequently say when a further list will be taken and on what frequency. Calling on the DX station's frequency to get onto a list should not be attempted – it could result in your call being added to another type of list altogether – a blacklist! Having registered your call it is important to return to the original frequency and to remain silent but

---

LISTS AND NETS – A CODE OF PRACTICE

1. The 'master of ceremonies' (MC), when taking the list, should endeavour to ensure a fair and even representation from all those countries calling to participate.
2. It is not desirable to take a list for use at some future date. In the case of poor propagation, however, a running list may be held over and continued when possible.
3. It is desirable to establish with the DX station beforehand how much time he or she has available, or how many stations can be worked in the time available.
4. A valid QSO requires some minimum two-way exchange of information. As stations are usually addressed by callsign this information has already been imparted to the DX station; nevertheless the MC should seek to avoid passing the whole callsign if possible. Convention has established that the exchange need only be a correctly received RS report by both parties. It is therefore the responsibility of the MC at all times to ensure that this is accomplished fairly, accurately and without assistance. While repeats are in order, if necessary, verification of partly received reports is not. Should a relay or a guess be suspected by the MC, the transmitting station should be instructed to make a second attempt with a changed report. The MC should not flinch from giving "negative QSO" when not satisfied with the exchange.
5. It is acceptable practice for the MC to nominate another station to monitor and assist with the procedure in difficult circumstances due to interference or linking for example.
6. If conditions fail the MC should terminate the operation rather than allow a 'free for all' under the guise of the list.
7. It is very important that the MC gives information out at regular intervals, relating to new lists, QSL managers, length of current list etc. This will be of great assistance to waiting stations not on the list, and minimise breaking and interference.

---

listen carefully until called in to transmit. When this happens a bare minimum of information should be given – this may even be limited to a signal report if that is the normal routine of the particular operation. Many others will be waiting and to try to prolong such a contact will not impress them.

In the RSGB survey the respondents' attitudes to lists and nets varied from grudging acceptance to outright hostility – nobody had anything positive to say about them! G3PJT: "Time wasted in a list cannot be regained". Tom Austin, G3RCA: "They are a waste of time where you spend hours listening to 'No call thanks' all night".

This author's personal view is that they have a place for the less rare DX and can sometimes help the novice DXer who would not be strong enough to crack a pile-up. The notion that net QSOs are 'always less valid' than freestyle QSOs is strongly rejected; quite often a lot more information is passed between stations in a net than in a pile-up. Only too often one hears the comment after a pile-up battle "I think he came back to me but I couldn't really tell through the QRM so I'll send off a QSL and see if I get a card back" – so much for the sanctity of pile-up QSOs! In the end, it's always between yourself and your own conscience.

Of course, DX stations are entitled to operate as they wish and some really do prefer the more relaxed atmosphere of a DX net to a full-scale pile-up. In rare cases there may even be good reasons for the list technique. For example, if the DX station's only source of power is a small battery it makes sense to minimise the amount of transmitting by allowing the net controller to organise the callers and put them through one at a time.

If you use DX nets and lists then you may be interested in a regular publication by Dieter Konrad, OE2DYL, which gives the meeting times and frequencies of all such operations world-wide. Write to him at Rosengasse 1, A-5020, Salzburg, Austria, for the latest price. And one final tip: if you do use lists, never miss a chance to get on one! You can always duck out if it doesn't seem worth the wait. You never know what other DX may join the net, even while a list is running for something you already have confirmed.

HF BAND CHARACTERISTICS

A major part of the fascination of amateur radio as a hobby lies in the tremendous variety of communication paths provided by the HF amateur bands. This enables them to cater for a wide range of interests, from regular schedules across town or across the world to the search for contact with rare and exotic DX stations and a variety of competitive activities and contests. In all these aspects, satisfaction and competitive success benefit from a thorough knowledge of the possibilities presented by the individual bands at different times of day and year.

Amateurs can and do learn from their own experience and from others, either through contact or by listening, and a comprehensive account would need to draw on the experience of all users – so someone's favourite 'wrinkle' will certainly be missing from what follows. There are, however, some general rules underlying the whole process of HF communication and its variability. These are outlined and the characteristics of the individual bands are described.

The whole business of HF band conditions is moderately complex and it is not the purpose of this book to deal in detail with the theoretical aspects, which are well covered in the *Radio Communication Handbook* [20] and in much professional literature. One cannot, however, discuss the subject without touching on the main technical considerations and a general understanding of these can be both interesting and helpful to the amateur communicator.

The technical background

The ionosphere, on which we depend for all HF contacts beyond the ground-wave zone, is itself dependent on the intensity of a variety of wave and particle emissions from the Sun and it is the variation of these, both as a whole and in their individual intensities, that causes the considerable variations in propagation on the HF bands. The principal variations, affecting all those bands in their different ways, are:

Diurnal: variations within the day due to the varying altitude of the Sun in relation to a particular path as the Earth rotates.

Seasonal: variations from month to month, again due to alterations in mean solar altitude and the duration of daylight.

Solar: changes from day to day in the intensity and mix of the solar radiations responsible for ionisation in the upper atmosphere. These changes can be considerable over quite short periods of time as well as showing longer-term trends, of which the 11-year cycle is the best known.

Geomagnetic: variations in ionospheric behaviour caused by alterations in the intensity and shape of the Earth's magnetic field. The magnetic variations are themselves caused by certain solar emissions.

To complete the picture it must be said that variation as used above refers to two distinct features of the ionosphere:

(a) its ability to reflect HF waves of a particular frequency at a given time and angle of incidence;

(b) the absorption (attenuation) suffered by the waves passing through the lower regions of the atmosphere on the way to and from the reflecting regions.

Both these vary but not by any means always in step. Both reduce as frequency is increased, the absorption decreasing roughly with the square of frequency. In general, communication is possible over a given path at a given time if the frequency is low enough to be reflected and high enough not to be attenuated below the noise level at the receiver.

The practical implications of the above for the amateur user of the bands from 1.8 to 28MHz are the purpose of the rest of this section but even this requires an understanding of a few basic concepts and terms.

For any path at any time there is a maximum frequency above which the signal is not reflected to the receiving station, however powerful the transmitter and however effective the transmitting and receiving antennas ('scatter' communication is dealt with later). Since, for reasons already mentioned, that frequency will vary from day to day, most statements on ionospheric behaviour must be of a statistical nature. A common measure for both recording and predicting is the monthly median maximum usable frequency (the median is the value likely to be exceeded on 15 days in the month and not reached on 15, the maximum variation above or below the median being from 30% to as much as 50%.) This value relates to a given path at a given time and takes account of all reflection points. If, however, the monthly median MUF is determined for a single-hop path, that MUF can be regarded as a measure of the statistical behaviour of the ionosphere at the point of reflection for any path of the same length reflected at that point, irrespective of direction.

It is therefore possible to produce worldwide maps of monthly median MUF distribution at various times in the day. Figs 1–5 show such distributions on a great circle map for five times of day in October at a time of fairly high solar activity (smoothed sunspot number about 100). Such a plot is not a forecast but the consolidation of a very large number of observations. The contours shown are for the F2 layer and for a single-hop path length of 4000km which is the practical limit for most single-hop F2 communications. Hops of about 5000km can sometimes be achieved where antenna and site favour very low angle paths on the higher frequencies*.

October was chosen because it is representative of peak winter DX conditions and is the month of several major contests. To keep the plots simple only the contours for the 14, 21 and 28MHz amateur bands are shown, the frequency being printed on the side of the contour where the 4000km MUF is above the contour value. On the 0600 GMT map only, the 40MHz contour is lightly sketched in and the 'high' of over 50MHz is over the South China Sea at this time. There are no 3.5 or 7MHz contours for the simple reason that nowhere is the 4000km median MUF below 7MHz. For those to whom this form of presentation is new, the contours for hops of less than 4000km lie to the 'up-hill' side of those shown.

A smoothed sunspot number of 100 was chosen as the value likely to be relevant when the maps were originally drawn and enables the potentialities of the highest HF bands to be fully illustrated. The solar cycle is now in its declining phase and MUFs as high as those shown are becoming less and less likely.

---

* Plots of this type but on Mercator charts and with E-layer contours included are contained in OT/TRER 13, Vols 1–4 published by US Dept of Commerce, Office of Telecommunications, Boulder, Colorado 80302, USA. Contours are provided at 2h intervals for each month and for smoothed sunspot numbers of 10, 110 and 160.

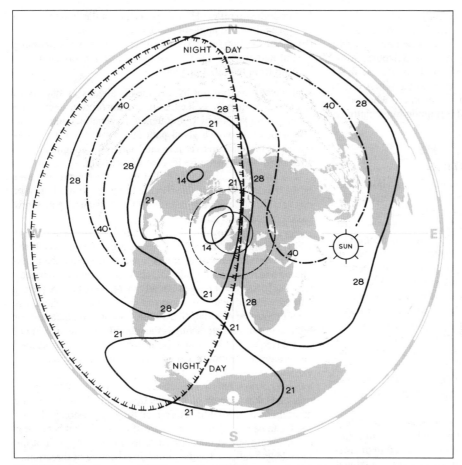

Fig 2. Global distribution of 4000km median MUF contours for 14, 21 and 28MHz, and day/night zones at 0600 GMT in October (sunspot number = 100)

28MHz signals can cross the 'mountain' when those on 21MHz are at best weak and those on 14MHz and below are attenuated below the noise level. On average, higher solar activity means higher absorption but the latter can vary considerably from day to day without a corresponding change in MUFs, and this variation is often responsible for the difference between 'good' and 'bad' conditions as seen by the amateur, the effect being particularly noticeable on the lower frequencies. The dramatic effect of this absorption can be seen from the following example. For a daytime path on 21MHz for which the absorption per hop is 6dB the corresponding figure for 7MHz is around 40dB and for 3.5MHz is around 100dB.

Band characteristics in a typical month

Against this background it is possible to see what to expect on an undisturbed day in October with fairly high solar activity. Referring to Fig 2 and looking at the bands in turn:

3.5MHz. A station in the UK is 'firing into darkness' over a sector from approximately north through 270° to 180°, and the zone of minimum absorption lies somewhat to the west of this line. This is near the peak time for contact with North America including the west coast, Central and South America, the western 'bulge' of Africa and the South Atlantic Is. If absorption is exceptionally low, contacts may be possible with the Pacific islands from Fiji to the west (seen from the UK).

The shadow line is slowly rotating clockwise and by 0625 (in the middle of the month) will pass through London and, being a great circle, will lie along the line 015°/295°, intersecting northern New Zealand. During the next 30min or so the absorption on the southerly 'long path' to New Zealand passes through a short minimum during which UK-New Zealand contact is possible. The peak time gets later as the month progresses and also depends on the location of both UK and New Zealand stations, since the times of sunrise and sunset vary and so does the direction of the great circle path.

There is no skip zone on 3.5MHz at this time and in easterly directions the absorption rises steeply so that the band is rapidly assuming its daytime role as essentially a local band.

7MHz. As regards DX much the same considerations apply as for 3.5MHz except that the lower ionospheric absorption means stronger signals and makes possible contacts that would at best be marginal on the lower band. The long path opening to New Zealand starts at about local dawn and can last an hour or more with stations from East Australia appearing somewhat later and overlapping the New Zealanders. Conditions with the Americas are at a peak but will remain good longer than on 3.5MHz. The skip zone is shrinking to the east so European interference is rising.

14MHz. The significant feature is the 'pool' in which the

To complete the picture each map shows the position of the Sun and the sunrise/sunset line. The solid circle round London marks the 2000km normal limit for a single-hop reflection and the dotted circle the normal limit for single-hop communication. Maps of this type conveniently illustrate all the main characteristics of HF amateur-band communication. In fact the main factors at work can be seen by studying the map for 0600 GMT (Fig 2).

In order to understand what is happening there is a crucial question to be answered. If the 4000km MUF is well above 7MHz in all directions why cannot contacts be made on that band and on 3.5MHz with the whole world at this time? The same question applies to 14MHz except for the sector where the 2000km line crosses a 'pool' in which the MUF is below 14MHz. The answer is that such communication is theoretically possible, given sufficient radiated power. The obstacle to be overcome is ionospheric absorption. This is relatively low on the night side of the day/night line, the lowest values being about an hour inside the area of darkness since the Sun rises earlier and sets later in the relevant parts of the ionosphere than it does at ground level. The area of daylight in Fig 2 is best seen as a 'mountain' of absorption, roughly sinusoidal in section, with absorption rising steeply after dawn and falling steeply towards dark with a relatively flat peak under the Sun.

In short, for any frequency below the MUF, long-distance communication is always best (and with amateur EIRPs often only possible) when the path lies across the lower slopes of the absorption 'mountain' and/or through areas of darkness. The higher the frequency, the smaller is the absorption so that

4000km median MUF is below 14MHz. On below-average days this grows larger and on a very low day will join with that over North America and fill much of the area bounded by the median 21MHz contour. Conversely, on above-average days it shrinks. It must also be remembered that the 'pool' is larger for hops of less than 4000km.

On an average day, therefore, there is a take-off for 14MHz from north through east and south to about 240°. In the sector from about 030° to 150° absorption is too high for more than single-hop contacts and the single-hop skip is rapidly shortening, particularly to the south-east.

Thus the path to Cape Town, which may have been open most of the night, is about to close. West Africa, the South Atlantic and South America are all good. The MUF is above 14MHz on the long path to New Zealand and East Australia and the dip in absorption which causes the LF openings will result in excellent signals on 14MHz during the next hour to those areas the islands of the West Pacific and NE Asia (Japan and the Far Eastern Russia.)

At the same time a low-absorp-

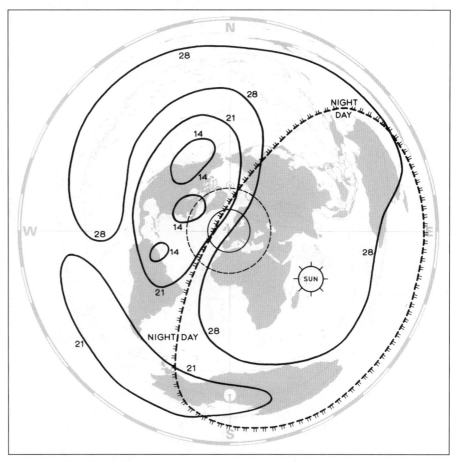

Fig 3. Fig 2 data at 0800 GMT

tion path is just opening to the north, giving access to Alaska and the Central Pacific, and this path will improve as the contours rotate clockwise (compare Fig 3 for 0800), peaking at around 0700.

Finally, on better days the 14MHz 'pool' will be small enough for there to be paths to at least parts of North America.

21MHz. At this time a 'wave' of rising MUF is sweeping in from the east. The 21MHz contour intersects the 2000km take-off circle at about 035° and 180° and contacts are possible within this sector with the shortest skip to the east, in which direction stations well within the 4000km circle will be workable. At the same time absorption is high in the direction of the Sun and on most days long-distance contacts (more than two hops) will only be good along the lower slopes of the 'mountain', ie to NE Asia and East Africa.

On a day with exceptionally high MUF and low absorption, long-path contacts with the Central Pacific should be possible at about this time – even the west coast of the USA is a possibility. On above-average days, long-path contact with ZL1/2 should open soon after 0600.

28MHz. On average or above-average days the band is open to the east (about 060°–150°), the first stations to appear being about 4000km away. High absorption towards the Sun may keep signals from long distances weak or inaudible but contact with East Africa and the Indian Ocean should be good and the short path to Australia and intermediate areas is opening, though at this time it tends to be better for stations in East and South Europe.

Finally it will be seen that on exceptional days there is a possibility of long-path contact with the Central Pacific and

the California area, particularly if helped by sporadic-E (see below) reflection south of Australia.

The above analysis shows that there is a great deal to be said about just one time period in one month. It is outside the scope of this book to extend this detailed analysis to every hour of every month in each of the 11 years of a typical solar cycle! Having established a basis, however, it becomes easier to follow the general trends.

Dealing first with the diurnal variations, let us move forward to 0800 GMT (Fig 3). Note that the whole contour pattern has rotated clockwise by about 30°. In fact some changes in shape do occur due to the influence of the geomagnetic field on the position of the contours at different geomagnetic longitudes, but these are incorporated in the charts.

The position at 0800 GMT is:

3.5MHz. Most of the path to North America is still in darkness but attenuation is rising at the UK end of the path. Contact with the east coast and with Central America will soon fade but at this time UK stations have the advantage that interference from more easterly stations in Europe has faded at the DX end. For the next eight hours the band will be essentially local in character.

7MHz. The path to North and Central America may well last another hour or more and, on good days, South America and Australia may still be workable after 0800. The skip zone will reduce to zero at about 0800 and from then on the band will become a daytime local band. Absorption will increase until noon, with the effective range of communication with

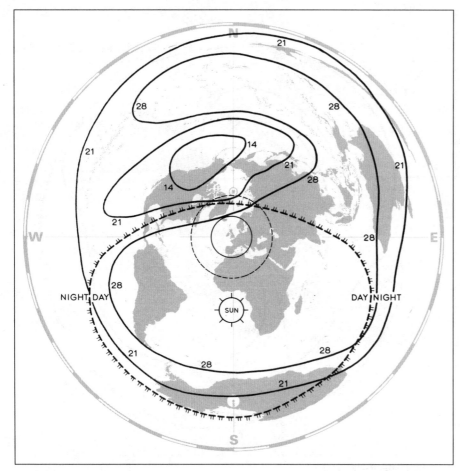

Fig 4. Fig 2 data at 1200 GMT

September than in other months. The band is now good to West Africa and the South Atlantic, and is opening to South America, the first to appear usually being Brazil. It is sunset in Tokyo and from now on the signal strengths of East Asian stations will improve as attenuation falls, the most northerly peaking first.

In East Africa noon is approaching and signals from this area may become weak or unreadable for a while, to appear again later.

28MHz. From the run of the contours it can be seen that the band is about to open to Japan and beyond in the north-east direction, is fully open to most of Asia and all of Africa and is about to open to Brazil.

This is also the peak time for contacts with East Australia, though the path may have opened soon after 0600 and may remain until after 1200, the closure being mainly determined by falling MUF at the far end and with stations in the north and west of the continent lasting longest. MUFs well above average are needed for 28MHz contacts with ZL1/2 which is most likely between 0700 and 0900. ZL3/4 is less critical, the path being similar to that to East Australia.

European stations steadily decreasing as signals from the more distant fall below the noise level.

14MHz. This is the peak time for the long-path opening to Australia, eastern areas peaking before western ones, and the path can extend 'round the back' as far as India. Note that the path to ZL3/4 is similar to that to Australia. For ZL1/2, however, the attenuation on the long path increases from 0800 onwards, while that on the short path decreases, the 'turn round' usually occurring between 0900 and 1000 and sometimes including a 'both ways' period. There is still a path to the north to parts of the Central Pacific and good contact with West Africa, South Atlantic and Central and South America but these will in turn succumb to rising attenuation during the next two hours or so. On a day with average MUFs or above there will be contact with at least parts of North America though reliable conditions may not develop until after 0900.

21MHz. There is now a take-off from the north clockwise to the south-west and the skip is short enough to the south-east to bring in the more distant Europeans.

On many days there is a low-absorption 'round the world' path at this time on the approximate line 030°/210°. This gives good communication with, for example, ZL1/2, UA0 and Japan over either short or long path, with the latter normally providing the stronger signals. Over the next two hours the long path extends to ZL3/4 and Australia, and sometimes to East Asia, eg Hong Kong, while the short path continues to ZL1/2.

On exceptional days contact is possible over the North Pole with the Central Pacific (eg Fiji) somewhere between 0800 and 1000, but this path is more likely to be good in March and

The position at 1200 GMT (Fig 4) is:

14MHz. On average and above days there is still a path to the north to Alaska, the Central Pacific and ZL1/2, and the band is opening to Japan and the West Pacific islands (which bear north-east from the UK). With most of the path now in darkness short-path signals from ZL3/4 and Australia will begin to show but will be better later in the afternoon. The same applies to East and SE Asia. Note that on days of low attenuation the long path may still be open to West Australia and SE Asia at this time, but not for long after 1200.

At this time the skip is short enough for contact with much of Europe and interference is a limiting factor on DX operation – but a good time for multipliers or bonuses in a contest. To the west, Central American stations have passed their morning peak and will soon fade but the band is well open to North America with the west coast just opening on an average day.

21MHz. Except on the best days the path to ZL1/2 will have closed and Japan is about to follow. Conditions to Australia and ZL3/4 are approaching their peak and, during the afternoon, stations in East Asia south of 040° will peak before the paths close due to falling MUFs, the northernmost peaking first.

Skip is now short enough for contact with the more distant Europeans. DX signals from the south will be weak or unreadable due to high attenuation and this is beginning to affect the path to South America. The band is now well open to Central America and the east coast of the USA and Canada.

28MHz. On the short path to Australia the MUF is falling at the DX end. On an average day the multi-hop take-off from the UK is limited to paths south of 040°. The Australia path will close in the next hour or so, VK6 lasting longest. At the same time the paths to Hong Kong, SE Asia, India and the Indian Ocean are good.

Skip is short enough for contacts of 1500km or less, a good time for European multipliers in a contest. On days of high absorption African stations may be weak but will soon improve. Communication with South America is well established and is open to Central America and the east coast of the USA and Canada. (If active, VO and some VE1 stations may show soon after 1000 on good days.)

With regard to the position at 1600 GMT (Fig 5), sunset is just after 1700 in the middle of the month and the approach of the twilight zone is a significant feature, while inside the area of daylight attenuation is low along the line 150°/330°. From 1600 the two LF bands must again be considered from a DX point of view.

3.5 and 7MHz. As attenuation falls, ranges in the sector 000° to 090° will increase. At sunset the day/night line lies along the great circle 170°/350° and on days of low absorption there is a chance to contact the Central Pacific at that time. The paths to New Zealand, the West

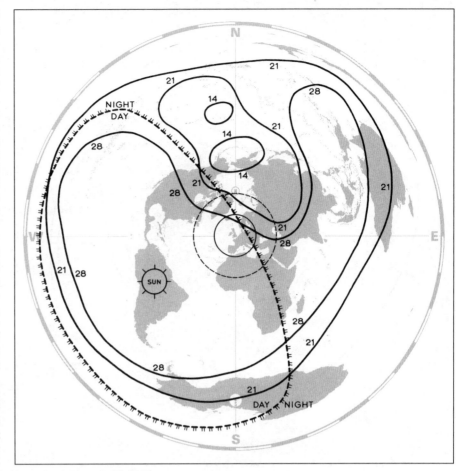

Fig 5. Fig 2 data at 1600 GMT

Pacific islands and East Asia are all in darkness but are at their best as the Sun rises at the DX end – 1800 to 1900 for New Zealand and progressively later for more westerly areas. There is a dip in the absorption on the path to East Africa during the period 1700–1800.

14MHz. In the area of darkness the only obstacle is the small pool to the north, particularly on below-average days. Communication is good to all of Asia and beyond, and to Australia and ZL3/4. On above-average days the band is open to the Central Pacific and ZL1/2, if not at 1600, certainly by 1800.

Falling attenuation means that communication is now again good with the Indian Ocean and East Africa. Note also that the long path to the western USA is mostly in darkness and there is often a good opening between about 1500 and 1600 with signals much stronger than on the short path. This path crosses the Pacific and some islands en route may also be worked in this manner at this time. (Note that in mid-December the twilight zone at sunset lies along the great circle between London and San Francisco, and LF band contacts are sometimes possible.)

High attenuation still keeps 14MHz signals weak or unreadable to West Africa and South and Central America, and may even affect contact with southern areas of the USA at this time. On more northerly paths absorption is relatively low and the band is opening to Hawaii and other Pacific islands bearing north to north-west from the UK.

21MHz. On an average day the sector from 340° through north to about 070° is only suitable for single-hop contacts up to about 4000km. This means that the band is still open to parts

of Australia and to SE Asia, the Indian Ocean, Africa, the South Atlantic and the Americas, the opening having reached the West Coast around 1400. The 21MHz contour will be optimum for the Hawaii area at or soon after 1800, and may reach due north and the Central Pacific on exceptional days.

28MHz. On a few days the path to SE Asia and parts of Australia may still be open but will soon close. Good contacts are possible over the whole arc from east through south to northwest where on all but the worst days the west coast of the USA will have opened up. On below-average days this opening may be rather brief but under exceptional conditions (28MHz contour near median contour for 21MHz) the opening to the West Coast may last several hours and extend to the Hawaii area.

The position at 2000 GMT (Fig 6) is:

3.5MHz. This is the time of the short opening to East Australia, the more westerly stations peaking later as dawn moves across the continent, reaching VK6 by 2200. Depending on absorption levels (and interference) contact is possible with all of Asia and Africa during the next two hours and with west of 90°E until midnight. The band will soon open to Brazil and Canada.

7MHz. Since it is absorption, not MUF, which controls the openings, the same pattern applies as for 3.5MHz except that openings will start earlier and last longer, and signals will be generally stronger. On average or above-average days there is still no skip zone at this time and interference is a major obstacle to DX working.

14MHz. At this time it is possible to contact all continents

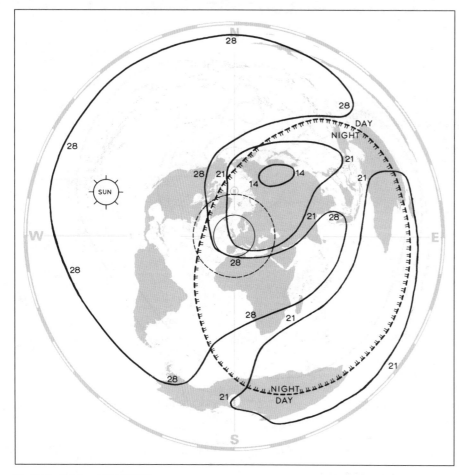

Fig 6. Fig 2 data at 2000 GMT

amateurs, the range typically covered will be only up to about 400km. As with the other bands, DX propagation is by ionospheric reflection from the E and F layers and, as the critical frequency for the E layer is almost always higher than 2MHz, there is rarely a silent zone on this band.

Dealing with one month in detail seemed to be the best way of bringing out the underlying factors but naturally the amateur is interested in the whole year.

Seasonal variations and grey-line propagation

Even with unchanging solar activity F2 layer MUFs are much higher in local winter than in the summer, being highest in the Northern Hemisphere in January/February and October/November with something of a dip in December. The situation is not all that different in March and September when thc distribution of MUFs in the two hemispheres is similar. To this extent the pattern of band behaviour described for October is broadly representative of the winter months as a whole.

Account must of course be taken of the distribution of daylight and

with ease. On average or above-average days the band is still open to at least southern New Zealand, Australia, the Far East, the rest of Asia and to Africa. Falling attenuation opens the path to South America, and North American signals will reach peak levels during the next few hours. Long hauls into the Pacific in the sector 330° to 010° are still possible but attenuation is rising rapidly as most of the path is now in daylight.

21MHz. The sector from north to south-east is now useless except on a few days. Communication is good to Africa, the South Atlantic and the Americas and on a good day the Hawaii area may still be accessible.

On days of high MUF and low absorption the long path to New Zealand and East Australia can produce good contacts at about this time.

28MHz. On an average day the band will be just about closing in all directions (28MHz contour passing out of 2000km circle) except for stations with very good low-angle take-off. On exceptional days it can still be open to most of the Americas and to the south, and under these conditions 28MHz contact on the long path to ZL3/4 and East Australia is possible.

The preceding sections show the main diurnal features of the amateur bands from 3.5 to 28MHz in October. The remarks made about the 3.5MHz band apply also in general to the 1.8MHz band, except that D layer absorption is higher and therefore the conditions for making DX contacts are even more difficult to fulfil. They can only be made when the path lies in darkness – propagation during daylight hours is mainly by ground wave only, and with the low powers permitted to UK

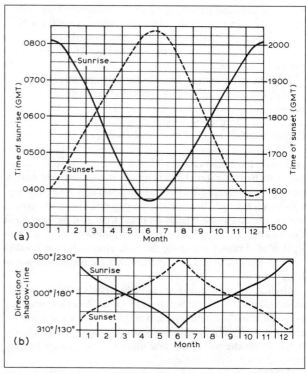

Fig 7. (a) Time of sunrise and sunset at London. (b) Approximate great-circle direction of shadow-line at local sunrise and sunset in UK

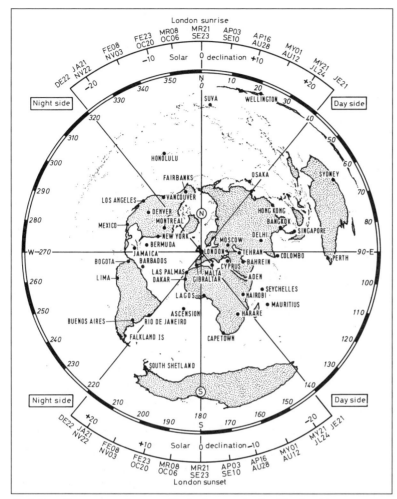

Fig 8. Grey-line propagation map *(Radio Communication Handbook)*

is accurate within a few degrees for any location in the UK at local sunrise and sunset.

Fig 8 shows this information incorporated into a great-circle map by Ray Flavell, G3LTP [20].

In this connection it should be remembered that the standard great-circle map centred on London, while applying with sufficient accuracy to the whole of the UK in most cases, can be misleading with paths to the antipodes. While the bearing of ZL4 from London is 060°/240°, it is 030°/210° from Belfast. The corresponding values for ZL1/2 are 010°/190° and 350°/170° respectively. While these differences may seem small they mean a completely different path for the two locations and should imply, for example, that the best time for GI/ZL1 contacts on 3.5 MHz is at local dawn in April to August, though this author has no practical experience of this.

The importance of the twilight zone can be seen from the following example. At sunset in mid-April and mid-August the sunset line lies along the great circle path from the UK to Ascension Is and absorption goes through a minimum about an hour later. During this period contact between these two places has been made successfully on 21, 14, 7 and 3.5MHz, with a return to 21MHz to complete the contact.

However, as G3LTP points out in reference [20], it must be remembered that it is not the ground-based shadow that determines the state of the ionosphere because the Earth's shadow is shaped like a cone; the area of darkness at F2 heights is appreciably less than that at ground level. Sunrise comes earlier than on the ground, sunset later. In fact, in mid-summer the F2 layer is in sunlight for all 24 hours of the day over the whole of the UK. So the ionospheric grey line (as opposed to the ground-based grey line) cannot be considered as a great circle and, therefore, cannot be represented by the straight line on the map. This should not stop you from trying your luck, however! But there is no point in trying to calculate the true outline of the Earth's shadow, because your signals are going to take the great-circle route no matter what you come up with on your computer. You may as well make all your plans using ground-based data, because it is easy to come by, and make up for its likely deficiencies by being generous with your timing. At the sort of frequency you are using, the beamwidth of the antenna will be wide enough to take care of direction.

To summarise, DX can be worked throughout the year on the LF bands although interference and noise level may make life difficult in the summer months. Basically the rules are simple. Most or all of the path must be in darkness and the best times, particularly for long hauls, are when much or all of the path lies just inside the area of darkness, a situation which occurs if it is near sunrise or sunset at both ends of the path. The latter condition cannot of course always be met, in which case there are two optimum times – around sunset at the western end of the path and near sunrise at the eastern end. With the shorter paths, eg to North America, the path is open between these times but with long hauls, eg to the Far East and Australasia, the openings can only be expected at these times.

The characteristics of the higher bands during summer are mainly determined by four factors:

darkness which varies continuously throughout the period. In general shorter days mean shorter openings on 14, 21 and 28MHz and longer periods during which 7 and 3.5MHz are potentially available for DX.

In the case of 3.5 and 7MHz the position of the twilight zone is also important as well as the distribution of light and dark.

Computer software is available to calculate the times and bearings when this occurs. An example is DX Edge by Xantec (available from RSGB HQ). However, there is also a method of determining this involving a great circle map, an aid which no keen DX operator should be without. Figs 7(a) and 7(b) show the variation in the time of sunrise and sunset at London throughout the year together with the great circle bearing along which the dawn/dusk line lies at those times. Applying this data to a great-circle map will show which areas are in darkness and which lie along the zone of lowest absorption which is just inside the area of darkness. Remember that the dawn/dusk line is only a straight line on a great-circle map at sunrise and sunset. Its position at other times may be roughly extrapolated by reference to Figs 1–5.

Sunrise and sunset times are different elsewhere in the UK – typically 10min later in Birmingham and as much as 40min later in Glasgow and Belfast in summer – and the differences are not the same for sunrise and sunset. To make a table for your own location, borrow a copy of Whitaker's Almanac from the local library. The bearing of the dawn/dusk line in Fig 7(b)

1. Daytime MUFs are considerably lower than in winter but the difference between day and night is much less marked apart from a drop before dawn.
2. Higher solar altitude means higher average attenuation than in winter and the longer days mean that paths in the Northern Hemisphere are in daylight for much longer periods.
3. The lower E layer (with a maximum single-hop range of 2000km) is more likely to play a part in DX communications than in winter ('sporadic-E' is dealt with later).
4. Stations in the Southern Hemisphere are experiencing winter conditions so that in communications with them the MUFs at the remote ends are more often a critical factor than in our winter.

The effects of this on the three higher bands are roughly as follows:

14MHz. DX communications are mainly conditioned by absorption. Between about 0800 and 1400 only shorter-range signals are strong enough for contact – on days of exceptionally low absorption this may include contacts with the eastern USA. The band is best at night. Particular features of summer are a good long path to the west coast of North America in the early mornings (at dawn in mid-June the twilight zone lies along the UK-California path) and this path also covers parts of the Pacific. There is also a possibility of long-path contact with Australasia in the hours before and sometimes after midnight but this is more reliable when solar activity is relatively low.

21MHz. The band is open to the south of the east-west line on most days except perhaps for a brief interruption before local dawn. Particularly on paths crossing the Equator, attenuation may make signals weak or unreadable during the period when the path passes near to the sub-solar point.

The 21MHz contour is usually not far enough to the north to provide short-path contact on more northerly paths, eg to Japan and the west coast of the USA, but the summer months provide conditions for long-path communications to these areas to the USA west coast in the early mornings and to Australasia, Japan and New Zealand both in the early morning and in the period from 2000 to midnight.

28MHz. In midsummer the median 4000km 28MHz contour lies about 2000km south of London. This means that stations in the southern UK are best placed for making DX contacts. Only on days well above average will reliable contacts be possible except to stations in the sector from roughly 120° to 240°. Contacts with North America and NE Asia are most unlikely except for a remote possibility of long-path contacts with California soon after dawn. On good days there can, however, be a good long path to ZL3/4 and eastern Australia around 2100–2200.

Sporadic-E conditions (see below) are most frequent in summer and as well as bringing in signals in the range 500–2000km may help with DX communication by providing a first hop of up to 2000km into an area of higher F2 MUF both on 28 and 21MHz.

The above discussion is concentrated on midsummer. In practice, band characteristics change gradually so that conditions in April and August show characteristics midway between extreme summer and winter.

The whole discussion has also assumed higher mean solar activity than is likely during the next few years, though there could be some days when MUFs approach those described. The best available guide to when the various bands are likely to be open to various parts of the world is to be found in the HF propagation predictions which appear monthly in *RadCom*. It should be noted, however, that brief openings due to short-lived minima in ionospheric absorption, such as the regular early morning openings to New Zealand on 3.5MHz, do not show up in this presentation.

It must be remembered that even amateurs must sleep and most must work so that their times of activity may not always correspond with the times of best communication. The charts show when the bands should be open but necessarily include times when, except perhaps during major contests, the DX stations are not likely to be very active.

The 10, 18 and 24MHz bands

Each of these bands is located in terms of frequency approximately midway between two of the amateur bands already discussed and, as might be expected, their propagation characteristics are intermediate between those of their neighbours. One way of looking at this is to sketch in the 10, 18 and 24MHz contours in Figs 1–5. Those for 10MHz lie inside those given for 14MHz. The contours for 18 and 24MHz lie approximately midway between those for 14, 21 and 28 MHz and illustrate the main propagation features of the two bands.

The 10MHz band

During sunspot maxima this band is not usually 'MUF-limited' except for a short period around dawn in midwinter, but as solar activity declines there will be an increasing tendency for the MUF on more northerly paths to fall below 10MHz during darkness. Apart from this, absorption will be the main factor affecting DX working which will generally be best when much of the path is in darkness or, if not, is reasonably close to the twilight zone. DX openings will therefore follow the same pattern as those on 7MHz but, because absorption is lower, they will start earlier and end later than on the lower band, and peak signals will be stronger. During daylight short-range communication within Europe will be possible by both E and F layers with zero-skip conditions around noon on better days in winter.

The 18 and 24MHz bands

While there will be detailed differences with different levels of solar and geomagnetic activity the propagation characteristics of these two bands show marked similarities to those of their immediate neighbours, sometimes behaving more like the next lower band and sometimes resembling the next higher. As the solar cycle declines it will be found that 24MHz will remain open for DX on days when 28MHz is no longer useful, while the 1.8MHz band should prove good for daytime DX working even during solar minimum when the 18MHz median MUF contour will occupy the approximate position shown for the 28MHz contour in Figs 1–5.

Anomalous propagation, disturbances and blackouts

There are various ways in which band characteristics depart from the general pattern and the effect of those of most concern to the amateur are briefly described below.

Sporadic-E (Es)

Patches of high ionisation can appear without warning and disappear as suddenly. In the UK area they are most likely in the daytime and from March to September, and can result in communication over ranges of 500–2000km at frequencies very much above the F2 MUF. For the HF operator their effect is most noticeable on 21 and 28MHz where, as already mentioned, they not only give good contact with Europe but may 'help' a DX signal into a region of higher F2 MUF. Es elsewhere in the world can also help the DX operator. It is common throughout the year in daytime in equatorial regions and mainly at night in the auroral zones. The so-called 'M' reflection, involving a signal on its way down from the F2 region being 'bounced' back off the *top* of an Es cloud, can help your signal cross an area of low F2 MUF and has low attenuation because the signal still only passes twice through the absorbing regions. Es is also discussed in the VHF DX section (see p70).

Extended ('chordal') hops

When transmitting in a direction in which there is a 'valley' in the MUF contours (eg to the south-west in Fig 3) the F2 ionisation gradient, though not steep enough to return your signal to earth, may bend it sufficiently for it to follow the curvature of the Earth until it encounters rising MUF on the far side of the 'valley' when it will be deflected downwards. Such hops have low attenuation and can be very much longer than 4000km.

Ground scatter

At certain times of the day and in certain directions the path attenuation may be low enough for quite strong signals to be scattered from the point of ground reflection both sideways and back along the transmitting path. This is useful to the amateur in three ways:

1. If a beam antenna is used to determine the direction from which the back scatter from other UK or Continental stations is arriving, this is evidence of a good DX path in that direction.
2. It can be used to contact stations who are in the skip zone if both beam towards the scatter source.
3. When there is no normal path to, say, the USA it may still be possible to communicate if both stations beam towards West Africa or South America.

An interesting example of this can be seen in the evenings on 14 and sometimes 21MHz when there is no direct path to Australasia but a good path to South Africa. Good communication is then possible with the Australasian and UK stations both beaming on Cape Town. Whether the scatter occurs in Africa or in the South Polar Region is an open question. In fact both the long and short paths to stations near the antipode sometimes show significant deviations from the great circle route from 'over the pole' before the opening is fully developed to north of (LP) or south of (SP) the true direction as the path closes. Side scatter, ionospheric tilts and polar Es may all contribute to these effects.

Skip focussing

When operating well below the MUF the various vertical rays from and to your antenna follow different vertical paths with different hop lengths When, however, the operating frequency is close to the MUF for the path, these rays tend to converge and this *skip focussing* can provide significant signal enhancement, equivalent to the difference between a typical dipole and a good beam antenna (6–9dB).

This accounts for the signal peak which often occurs soon after an MUF-limited path has opened and shortly before it closes, and is the best time for the operator with low power (see p57).

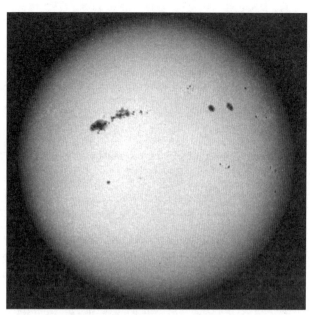

Sunspots like these have a major influence on the Earth's ionosphere and hence radio propagation worldwide

Disturbances and blackouts

Events taking place in the Sun can cause major and minor upsets to HF communication in basically two ways. As already mentioned, a rise in certain emissions can cause daytime absorption to rise to high levels. On such occasions paths in a generally westerly direction which are normally open at a given time may show signals for a short time after opening, but signals will then fade to return after dark if the path MUF is still high enough.

In extreme cases, usually associated with solar flares, the absorption rises suddenly to a high value and all daylight paths are 'blacked out'. Such a *sudden ionospheric disturbance* (SID) may last minutes or hours. Sometimes solar noise is high on 21 and 28MHz before the blackout and MUFs may be abnormally high both before and after the disturbance.

Particle radiation from flares or through *coronal holes* affects the intensity and shape of the Earth's magnetic field and therefore the shape of the MUF contours. These magnetic disturbances or 'storms' generally result in lower MUFs and higher absorption in high latitudes, particularly in the auroral zones. Under such conditions paths in northerly directions, notably to North America, may be badly affected while conditions to the south may even be improved. Disturbances resulting from persistent solar anomalies are predictable since they recur at 27-day intervals (see below and also p71).

Solar variations

The various ionising radiations from the Sun are not constant in their effect. Since the sources of radiation are not evenly distributed over the Sun's surface there is usually a 27-day cycle of variation due to the Sun's rotation relative to the Earth.

The daily count of sunspots has been the traditional measure of solar activity, a more modern and somewhat more objective one being the power received on the Earth at 2800MHz (solar flux). On average the two vary together and are nearly linearly related. For practical purposes the relation:

$$R = 1.1 \ (\text{S.F.} - 60) \quad (R = \text{sunspot number})$$

gives an answer sufficiently accurate for amateur purposes, enabling the WWV 18 minutes-after-the-hour broadcast of the daily solar flux value (SFU) to be converted to an approximate sunspot number if so desired (see below).

The most well-known aspect of solar variation is the approximately 11-year cycle of rise and fall in the 12-month mean of the daily sunspot numbers and solar flux – 'approximately' since Cycles 15–19 were all nearer 10 years while a recent one lasted nearly 12 years. The average rise time is about four years.

Since the cause of this cyclic behaviour is not yet fully understood (there is even evidence that it is not always present, the last absence being in the 75 years from 1650AD), prediction of future trends is based on extrapolation from the past. For the amateur communicator it is important to remember that professional propagation predictions are based on forecasts of solar activity and that these are based on heavily smoothed data. While the ionosphere does not respond instantly to solar variations its smoothing effect spreads over days rather than months, and generally the higher the mean solar activity, the greater the short-term fluctuations. For example, during the peak months from November 1957 to March 1958 the highest daily sunspot number was 342 and the lowest 90. Since forecasts have to be prepared well ahead of events one should always be on the look out for conditions which are outside the limits predicted.

There are also from time to time shorter periodicities within the 11-year cycle. On its down slope one recent cycle had distinct subsidiary peaks at intervals of rather more than two years [21], and over the years 1975–77 there was a fairly regular rise and fall with a period of approximately 120 days.

Professional forecasts smooth out these variations and, since they perforce are based on data which is many months old by the time the prediction appears in print, they can even get 180° out of phase with them. The amateur, on the other hand, would like to know what band conditions will be like the next day, or for the following weekend's contest. As with weather forecasting there can be no certainty but a surprising amount can be done with relatively little effort.

Do-it-yourself prediction

The 27-day (28 in the early stages of a cycle) solar rotation is a key factor in short-term prediction. Some amateurs keep a record of their own assessment of band conditions on this basis and know that similar conditions, both good and bad, can often be predicted 27 days ahead. The criteria to be used depend on individual interests. A useful one is the duration and penetration of the North American opening because there is plenty of activity and the path is a good indicator of general conditions.

Not everyone has the time to make the necessary observations and an alternative or supplementary approach is to make use of the WWV broadcasts at 18 minutes past each hour. These give the most recent value of 2800MHz flux which is recorded at 1700 GMT each day, ie WWV broadcasts from 1818 GMT usually contain the current day's value. The broadcast also gives a daily geomagnetic index (A-index) and mentions events such as solar flares and gives a 24-hour projection of solar and geomagnetic trends. Reliable copy of WWV needs a good antenna, preferably with a good null to the east to reduce interference from other standard transmissions and intruders, but if the flux and 'A' values can be copied on most days and the gaps filled by a friendly North American contact, a really up-to-date picture of solar activity and trends is available [22, 23]. If you have difficulty in receiving WWV, remember that propagation data is available on the packet network, Internet (World Wide Web) and the GB2RS news bulletins.

Fig 9 plots the solar flux as published by NOAA Boulder,

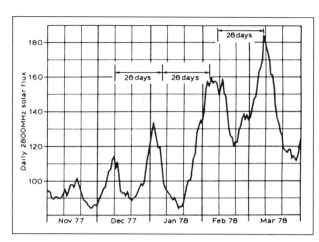

Fig 9. Daily 2800MHz solar flux from November 1977 to March 1978

over the period from November 1977 to March 1978. It will be seen that the excellent high-band conditions in early February and March were easily predictable. A plot of daily sunspot numbers is very similar in shape so that if WWV cannot be received with sufficient reliability the numbers can be obtained from the Sunspot Index Data Centre, Brussels for a small fee. They arrive in monthly batches in the first week of the following month, providing a somewhat less timely but still valuable base for short-term prediction. The A-index is a measure of worldwide geomagnetic activity. In itself, a low value (15 or less) means a stable magnetic field and stable radio conditions, usually with low absorption. A low A-index with a high flux value means excellent conditions, particularly on the high bands (it was around 10 during the early February and March flux peaks in Fig 9). For good LF band conditions look out for low flux values together with very low 'A'.

A 27-day plot of the A-index shows recurrent peaks due to persistent sources of particle emission from the Sun and coronal holes. High peaks mean disturbed conditions, magnetic storms and possibly auroral effects, and if recurrent these can be predicted from the 'A' plots. However, an SID caused by a flare cannot be so predicted since it is a 'new' event and the rise in the A-index follows a day or two later when the effects of the flare disturb the magnetic field. All that can be said is that such events are more likely to appear near a peak in the solar flux plot. 27-day calendars are also of importance in VHF auroral communication (see p73).

The above should serve to show that with not too much effort an amateur can form quite a reliable view of the likely trend of radio conditions a month or more ahead, and with the help of the WWV reports be aware of what is happening more or less currently.

For those interested in the trend of the solar cycle a three-month running mean plot of the provisional sunspot number (published in *RadCom* about two months later) gives a good idea how things are going.

Since the geomagnetic field is affected by emissions from the Sun it also shows cyclic behaviour although this is not so clearly defined. Geomagnetic activity, ie the incidence and intensity of disturbances, ionospheric storms and auroral activity, certainly increases as the sunspot peak approaches but may continue to rise for a further two to five years before reaching its own peak.

The geomagnetic peak is currently taking place and activity will remain high for some years. Many books are available on this topic [24–26].

Computer programs such as MINIMUF, MINIPROP and IONSOUND are available; these give propagation predictions based on solar data.

Perhaps the easiest method of all is just to read the propagation predictions available each month in *RadCom,* news bulletins and the Internet!

The beacon network

Propagation predictions are all very well but if you have just switched on your radio and want to work DX, it's obviously better to know what is happening right now, rather than what has been forecast. This is where beacons can help.

On each HF and VHF band there are several beacons situated in strategic locations all over the world which can provide an excellent clue to band conditions, and these are proving valuable in both amateur and professional propagation research. A list of these beacons is given in Appendix 8.

Of particular interest are the beacons operated by the Northern California DX Foundation (NCDXF) in cooperation with the IARU. These beacons operate on 14.100, 18.110, 21.150, 24.930, and 28.200MHz. At the time of writing, 17 beacons are active, with the 18th and final due to become operational soon. Each beacon transmits in turn at various power levels from 100 watts down to 100mW, with each transmission repeated every three minutes. A full transmission schedule can be found on the NCDXF Web site (www.ncdxf.org) from which it is also possible to download software for automated monitoring of the beacons.

The beacons on 28MHz all have their own frequencies in the 28.190–28.225MHz sub-band, and run continuously.

Never transmit on, or very close to, a beacon frequency or in a beacon sub-band, even if you can't hear anything – your signals will cause severe interference to others straining to catch the faint signals.

DX ON THE BANDS BELOW 30MHz

Many factors, both technical and social, combine to give each amateur band a distinct 'character' and the bands below 30MHz are no exception. Each has characteristics which appeal to its own enthusiastic users. Those with limited facilities often find it worthwhile to specialise in DX working on one particular band.

This section takes a look around the various bands with this in mind, and gives some indication of the type of DX activity going on so that the newcomer knows what to look out for. Some hints on equipment and antennas are also given in the cases of 136kHz, 1.8MHz and 3.5MHz as these bands pose particular technical problems for the DXer.

IARU Region 1 has recommended band plans for the 3.5–28MHz bands which set aside segments of each band for use by various modes, and these are shown in Appendix 7. Although voluntary in the UK they are mandatory in some Region 1 countries and should be observed at all times. It is recommended that the portions dealing with the bands in use be memorised.

It should also be noted that, following a CCIR (International Radio Consultative Committee) recommendation, it is an accepted convention to transmit *lower* sideband *below* 10MHz and *upper* sideband *above* 10MHz.

136kHz (2200m) band

The 135.7 to 137.8kHz band, allocated to amateurs since early 1998, is still at an exciting stage where records are being broken regularly as new stations in new countries gain access to

it. An EIRP of 1W is allowed in most countries and theory would suggest that a range of around 1000km is the maximum that could be achieved. Fortunately, experience has shown that 2000km contacts are possible at good strength, so there is hope that further distances, perhaps even trans-Atlantic, can be worked.

One could consider any station in another country to be 'DX' on 136kHz but, to put a figure on it, most LF operators would feel very satisfied with an 800km+ CW QSO. Propagation is more stable than on MF with very long distances, such as G to OH, being worked in broad daylight. Darkness does bring considerable signal enhancement but with it often comes a similar increase in noise. Some of the best DX QSOs have been made in the early mornings when the atmospheric noise has died down and the local noise is still low.

At the time of writing there are 18 countries active on 136kHz, all within range of each other. Several more European countries are awaiting the issuing of licences and a further number of countries, mainly in Eastern Europe, have the allocation but no known activity.

As with any band, one has to work hard on all aspects of the station to be capable of working DX. The wavelength of 2km or so leads prospective operators to imagine that a 500m long wire would be the absolute minimum antenna! Although one or two lucky operators do have access to such things, most stations cope very well with their 160m or 80m antennas suitably modified. The secret is in that EIRP limit. The station with the huge antenna may only need 100W to get his 1W EIRP but the station with the small antenna can achieve the same result with a kilowatt or so. As with 160m, a good earth system is essential and earth stakes or radials should be used in conjunction with any existing underground metalwork such as water pipes.

Important as radiating a good signal may be, no DX will be worked unless it can be heard. Reception is undoubtedly the most challenging aspect of LF operating.

The main enemy of the LF DXer is noise – local noise from switching power supplies etc, QRM from nearby Loran or broadcast transmitters and atmospheric noise. Atmospheric noise, due to lightning, is bad during the summer months and can make the band unusable. Nothing much can be done to reduce the effect of the noise on the signal as it is propagated in the same way. This is the main reason that most DX is worked between October and March.

Other noises can usually be reduced by the use of directional receive antennas such as active loops. It must be borne in mind however, that the TX antenna will re-radiate the noise which it picks up. It is therefore often necessary to throw the TX antenna off-resonance whilst receiving on a loop.

The 136kHz band is only 2.1kHz wide and most of the CW activity is concentrated within 1kHz. If you are not to be forced to close down every time a local station comes on the band, very narrow CW filters must be used. Many modern receivers have the ability to use two CW filters at different IFs to give variable bandwidth. They may also have DSP. This is a useful addition but is not a substitute for tight IF filtering. If your budget doesn't run to the latest amateur transceivers, some commercial 'selective level meters' are available as surplus and, having 100Hz or narrower filters, can make the basis of a good LF receiver.

Many receivers which cover the 136kHz band have poor sensitivity and virtually no front-end selectivity at LF. A sharp band-pass filter followed by a well-designed 10dB preamp will bring them up to scratch. If all your efforts with the receiving

If you're going to work 173kHz be prepared for some big coils! Here is the lower section of HB9ASB's helical antenna for this band, showing the winding packets. The plastic waste-paper basket loading coil can be clearly seen in the foreground. For more details see reference [8]

system have been successful then the German transmitter at 138.8kHz should be an S9+ signal, whereas the daytime band noise should be about S3 in a 250Hz filter. A useful guide to conditions is the Greek RTTY transmission on 135.8kHz which should be heard easily at night.

There is no official band plan for 136kHz but a 'gentleman's agreement' exists between LF operators. Normal CW is the primary mode in use and occupies the lower part of the band, up to 137.4kHz. The quite slow speeds used (about 7–18wpm) will pass through the narrow receive filters necessary to pick out weak signals. Fast CW becomes unintelligible when heard through such filters and it is best to call quite slowly in order to have the best chance of being heard afield. There is no set calling frequency and many stations are crystal controlled so it is necessary to tune the band after calling CQ. With a narrow filter, this can take a long time. If you are replying to a DX station's CQ call, off his frequency, give a long enough call to allow him to find you.

The second most popular mode on LF is 'slow' CW. This is sent by computer and has a dot length of around three seconds. There are two versions, one is absolutely as normal CW, just very slow; the other uses a frequency shift of a few hertz to distinguish dots from dashes, which are both of the same length! Both these modes require the use of a spectrum analyser program to view the received signals on a PC screen. The most popular of these is Spectrogram by Richard Horne which is available free on the Internet. These slow CW methods have a signal-to-noise advantage of several decibels over normal CW and have allowed many of the less-powerful stations to make long-distance contacts. Slow CW is used at the top end of the band, above 137.6kHz.

Data modes such as PSK31 and Clover have also been successfully employed although the amount of activity is small. Data activity is centred around 137.5kHz.

So far there are two awards available to LF operators. The first is the RSGB 136kHz award for working five DXCC countries. The second is the Peter Bobek award and this is rather more challenging. It is to be awarded to the first pair of stations to make a QSO across the Atlantic from Europe to either the USA or Canada. At present neither the USA or Canada have an LF allocation but this could change soon. Slow CW techniques will probably be the only way of making such a QSO.

1.8MHz (160m) band

It is difficult to define 1.8MHz DX in terms of distance, but 3000 to 5000km would generally be considered to qualify. 160m propagation is not dissimilar to that experienced on the medium wave broadcast band, with little sky-wave propagation during daylight hours, but with long-distance working possible throughout the hours of darkness. In years gone by, achieving DXCC (100 countries worked) on 160m would have been considered an almost unattainable goal. However, recent years have seen a substantial increase in activity on the band which, combined with improvements in antennas and receiver technology (and in some instances, such as in the UK, increased power limits) have led to scores of 200 countries or more becoming relatively commonplace. A well-equipped UK amateur prepared to lose sleep in the cause of 160m DXing could nowadays expect to work 100 countries or more in a season, and large contest stations have been able to work DXCC in a single weekend. Even with 10W, many stations have worked all continents.

DX propagation on 160m requires that the whole path lies in darkness, though there can be significant signal enhancement at dawn and dusk due to ionospheric tilting. For example, the path between Europe and Australia often opens for about 10–15 minutes before and after Australian sunrise. Such openings will not occur every day, and openings can be very short (just two to three minutes) with signals peaking at good strength and then falling away very rapidly indeed. To some extent, propagation will depend on ionospheric activity. In particular, signal paths through the auroral zone will be subject to strong attenuation when auroral activity is high. In the northern hemisphere, signals from the north also arrive at much lower wave angles than those from the south. To achieve these wave angles with a horizontal antenna would require antenna heights of 250ft or more, so vertical antennas are generally favoured by 160m DXers.

On shorter paths (eg Europe to North America), the path can be open for several hours, although signals may still peak around sunset in North America and around sunrise in Europe. Having said this, peak time for paths between Europe and Central America is often around local midnight at the mid-point of the path, and therefore around 0200 or so European time. North-south paths, such as that between Europe and South Africa, frequently peak around midnight European time. In broad terms, DX working during the winter period tends to be more productive for stations in the northern hemisphere due to the longer hours of darkness and lower static levels. However, the converse is that this is the worst period for activity from the southern hemisphere and European DXers are increasingly recognising the need to be alert during the summer for activity from southern Africa, South America and even Australia.

Because of the uncertainty surrounding 1.8MHz propagation, and the way in which signal strengths can vary significantly over a short period, successful 1.8MHz DXing demands regular monitoring of the band, often at unsociable hours. For this reason it is not uncommon for 160m DXers to work together, each taking turns in monitoring, and agreeing to alert one another if a rare station becomes workable.

As mentioned above, 1.8MHz DXers typically use some sort of vertical antenna, which may be an inverted-L, a loaded vertical, or perhaps a shunt-fed tower. All such systems require an excellent earth system if reasonable efficiency is to be achieved. Details of suitable antenna systems can be found in references [4–9]. As ground-wave communication is more effective with vertically polarised antennas, the use of a vertical antenna will also enable good daylight distances to be achieved on 160m in addition to the DX capability.

The main limitation with vertical antennas is that they are

more prone to noise pick-up than horizontal antennas. Serious 160m DXers therefore use separate receiving antennas. This may be a small tuned loop or, if space is available, a Beverage antenna. Neither is suitable for transmitting purposes, but both can give significant signal to noise enhancement on received signals.

Receiver performance is important on 160m, not so much in terms of absolute sensitivity but more in respect of being able to receive weak DX signals in the presence of very strong local signals on adjacent frequencies. Good intermodulation performance is therefore crucial. Although less common than it used to be, many DX stations and DXpeditions will operate 'split frequency', and in some cases split-frequency operation is essential due to differing frequency allocations (for example, although they now have an allocation lower in the band, Japanese stations typically operate between 1907.5 and 1912.5kHz, but will usually listen around 1830kHz for European stations). The ability to operate 'split' is therefore essential. Most DX operation takes place on CW, and good-quality narrow CW filters make reception of weak signals that much easier. SSB DX working is, however, becoming more common as station improvements compensate for the inherent disadvantages of the mode (wider bandwidth and lower average power, resulting in significantly poorer signal-to-noise ratio than with CW, all other factors being equal).

160m has long been regarded as the 'gentleman's band', with users helping each other out and extending courtesy to one another. To an extent this is still true, though increasing levels of activity and competition have taken away some of the exclusiveness previously felt by 160m enthusiasts. Activity in the major contests can be especially frenetic and, although these are an opportunity for the newcomer to increase his tally of DX, they can be somewhat intimidating. Better, perhaps, to cut your teeth on working relatively common DX, for example European stations working the US, from where there is plenty of activity.

The IARU Region 1 Band Plan restricts SSB operation to frequencies above 1840kHz (which, given that LSB is used, means that the carrier frequency should be no lower than about 1842.5kHz), and the majority of CW activity takes place below 1840kHz. Because of the narrowness of the band, this split tends to be treated with flexibility in the major contests, but most Region 1 stations adhere to it at other times to the benefit of all users.

It should also be noted that, although 160m frequency allocations have been brought much more into harmony than was previously the case, there are still differences from country to country. The one band segment which is common to most users is 1810 to 1850kHz, which is why most DX activity takes place in this part of the band.

On CW, operating speed is very much dependent on conditions. If signals are strong each way, there is no reason why normal speeds of operation should not be used. Some stations advocate slower speeds when signals are weak, though often it is better to operate at normal speeds but with several repeats in the hope that at least one transmission will be heard through the noise or fading. Remember that band openings can be of short duration so, in fairness to other operators, contacts should be completed in the shortest possible time. Equally, it is unreasonable to work a rare station several times, on consecutive nights for example, when others are trying to make a first contact.

Most rare stations and DXpeditions nowadays will make at least some effort on 160m, although some are more prepared

than others to cope with the more demanding operating and lower QSO rates than on the other HF bands. In some cases it can be productive to work a rare station on one of the higher bands and make a schedule for a 160m contact, but not all DX operators will be prepared to do this. In any case, do not waste their time unless you feel there is a good chance of making a contact at the time and on the frequency agreed.

Unfortunately there are few reliable propagation indicators for European stations on 1.8MHz, but activity is high enough nowadays for the presence and strength of DX stations to be used instead. The signal strengths of local and semi-local stations provide no guide at all to DX conditions. Careful listening can provide clues, and the occasional directional CQ at what ought to be an optimum propagation peak (for example a grey-line propagation path) can sometimes yield a call from a rare DX station.

Although significant parts of 160m are now primary amateur allocations, UK amateurs are still restricted to lower power levels above 1850kHz and must avoid interference to other non-amateur band users. In the segment 1810–1830kHz there continue to be non-UK commercial users and, again, amateurs should exercise restraint and avoid interference to these other services.

1.8MHz has always been regarded as a specialist DX challenge. The late W1BB, the first station to gain DXCC on the band, took many years to achieve his goal. Even though DX working on the band is now much more commonplace, 1.8MHz remains the most challenging of the HF bands, requiring persistence and perseverance to achieve competitive results.

3.5MHz (80m) band

This is a most useful band for local and near-European working during daylight hours and for longer-distance contacts after dark. In winter, particularly during the years when solar activity is low in the sunspot cycle, it is an excellent DX band, capable of providing worldwide communication on CW and SSB. In summer, transequatorial contacts with Oceania, the Pacific and Africa are possible and there are openings to other parts of the world. Timing is all-important and static can be a limiting factor during the summer months. Particular attention should be paid to the use of 'grey-line' or 'shadow-line' paths (see p44).

In Europe and some other parts of the world the band is allocated to amateurs on a shared basis, with military and commercial stations and specialised point-to-point services taking many of the available frequencies. This sharing results in high levels of interference from on-frequency blanketing, and also from receiver-generated spurious signals and cross-modulation caused by high-powered non-amateur stations working off-frequency. While many countries allow their amateurs to operate between 3.5 and 4.0MHz, Europe, including the UK, uses 3.5–3.8MHz.

For CW DX working, the bottom 30kHz are used, with the lower 10kHz being most favoured in accordance with the IARU Region 1 Band Plan. As the USA restricts the use of the 3500 to 3525kHz segment to Extra Class licensees, it is necessary to monitor above this segment for contacts with American amateurs who do not hold the higher licence. Novice and Technician Plus licensees are restricted to the band segment 3.675 to 3.725MHz. As with other bands, DXpeditions and 'rare' stations often use split-frequency working, listening for calls in a specified band of frequencies away from their transmit frequency.

The majority of DX working on phone is in the upper 20kHz

A mobile antenna can make a good choice for those planning 3.5MHz operation from a restricted site. This Texas Bugcatcher has been fitted with an earth plate, available from the manufacturers. For more details, see reference [8]

of the European band (3780–3800kHz) although the band plan actually allows the upper 25kHz for this. When the band is open, DX contacts should always be given priority in this segment and ideally no local QSOs should take place at all to allow faint DX signals to be heard. Local contacts in this DX 'window' should also be avoided whenever the band is capable of supporting DX traffic to/from anywhere in Europe. In practice this means around the clock in December and January and from two hours before sunset to two hours after sunrise for the rest of the year.

For contacts with countries which are not permitted to use this segment, it is the custom to operate in the upper 10kHz of the band that is allocated to the DX country. Just as for CW working, the USA phone-band allocations are based on the class of licence held, and many of their amateurs are not allowed to work below 3775 or 3850kHz. For contacts with these stations it is necessary to work cross-band, viz non-USA stations below 3775kHz and USA stations above this frequency. At peak times when this band is open there is always substantial interference from a large number of stations working in a relatively small segment and it is often good practice for European stations to operate below 3750kHz and nominate the frequencies where they are listening for calls.

The introduction of the five-band DXCC awards and Worked All Zones has encouraged many more DX-minded operators to use the band and this has resulted in very substantial pile-ups of stations whenever a new or rare country appears. Some of these operators work split-frequency (particularly on CW), but on telephony the trend has been to use the list system (see 38). Many 3.5MHz operators dislike the list system, but it has become an established part of the scene and there is little that can be done, other than to persuade the DX station to adopt split-frequency working.

With simple antennas at moderate heights it is possible to take advantage of midwinter conditions and work into North America and other parts of the world during the hours of darkness. For more consistent DX working, some form of low-angle antenna is required and this is often the base-fed vertical radiator operated against ground. While a λ/4 vertical is capable of excellent performance, it can be quite inefficient unless the ground system is extensive, and being omni-directional and vertically polarised means it picks up more man-made noise and atmospheric static than a horizontal antenna.

A further disadvantage is of course that it provides no directional discrimination against interference from amateur and non-amateur signals. Many amateurs who have plenty of ground area available use a vertical antenna for transmission in conjunction with highly directional low-angle receiving antennas such as the Beverage type. This is an ideal combination but not at all practical when space is at a premium.

In addition to the conventional vertical, another type of semi-vertical antenna is now finding favour with many users. This is the sloping dipole (usually called a *sloper*), and this is basically a vertical dipole mounted from a single support and sloped at an angle of approximately 45°. To reduce the height of the support mast, the lower half of the dipole is often bent so as to run parallel with the ground in the direction of the support mast. Slopers exhibit some directivity towards the slope and radiate at the low vertical angles suitable for long-distance working. It is possible to use more than one sloper mounted from the same support mast, and to phase the feed-lines to enhance the directivity characteristics and to provide electrical rotation of the polar pattern. Vertical antennas including slopers radiate little or no high-angle signals; they do not therefore perform well under daylight conditions when high-angle radiation is required.

Horizontally mounted dipoles and simple wire antennas radiate at too high a vertical angle unless the support height is at least λ/2 above ground. As few amateurs have the facilities to erect antennas in excess of 40m high, recourse is made to other configurations that work satisfactorily at lower support heights. The pulled-out single quad and delta-loop, particularly when corner-fed for vertical polarisation, have found favour with many operators and support heights of 13–15m have been proved to be effective for DX working. Although smaller non-resonant loops have been suggested by G6XN and others in *Radio Communication*, it is probably easier to use the 1λ loop and extend the horizontal sides to make up for the shorter vertical sections of the loop. If the support height and mast structure is not a problem and there is plenty of ground area available, then almost any type of horizontally polarised low-angle antenna will give good results. Some amateurs take the provision of specialised antennas to the extreme and erect complex phased arrays and rotary beams. In the USA it is even possible to purchase complete four-element rotary Yagi antennas for the band.

Earlier in this summary, mention was made of receiver-generated spurious signals and cross-modulation problems. Unfortunately, many of the older solid-state transceivers and receivers suffer from these defects as a result of non-linearity and insufficient dynamic range in the RF and mixer stages. Often the user may be unaware of the problem, as strong off-frequency and out-of-band commercial signals appear in-band, perhaps on a completely different frequency, due to the non-linear mixing process in the receiver. As most receivers have more gain than is required at 3.5MHz, passive preselectors comprising several loosely coupled ganged tuned circuits will usually provide some improvement. Front-end attenuators are also helpful in reducing cross-modulation [13–15], but they are not usually as effective as the passive preselector in clearing up the spurious signals. Some valve receivers also suffer from the same problems and benefit from the addition of the passive extra tuned circuits and or a front-end attenuator.

The number of countries that could be contacted on the band was once relatively small in comparison with those available on the higher frequency bands. The Five-Band DXCC Award and Worked All Zones dramatically changed the situation and it is now much easier to achieve quite high country scores. A substantial number of operators have been able to contact in excess of 100 countries and there are some who have more than 200 countries confirmed for 3.5MHz CW and telephony working. A low-angle antenna system, a good receiver and an amount of single-minded persistence are needed, but for the newcomer to amateur radio and the operator who wishes to try a new band, DXing on the 3.5MHz band can be most rewarding.

7MHz (40m) band

The widely differing propagation conditions that exist on 7MHz during a 24-hour period, and the differences between winter and summer conditions, make this a very useful band for both local and DX working. In many ways there is a marked similarity to 3.5MHz, with signal attenuation through ionospheric absorption being the main limiting factor for DX working. Other common factors are the 'grey line', dusk-dawn openings (see p44) and the band's shared nature, resulting in high levels of interference from non-amateur stations, although this situation has improved due to IARU/ITU action to remove 'intruders'.

During daylight hours, ionospheric absorption is high and the skip zone is very short or is non-existent, thus limiting communication to local contacts and distances of a few hundred kilometres. Around sunset, there are long-haul grey-line openings and, because attenuation is less than on 3.5MHz, these are longer in duration and signals are usually stronger than on the lower frequency. After dark, the skip zone lengthens and good medium-distance contacts are possible with North and South America, Asia and Africa. Because of the varied conditions that are found on the band, many operators regard 7MHz as the key to making a winning score in the RSGB Commonwealth Contest and other multi-band CW events.

The full band of 7.0–7.3MHz is not available to amateurs in Region 1 who are generally restricted to the 7.0–7.1MHz segment. The band is shared with broadcasting and at night, particularly during the winter months, the level of direct interference is high. This, together with cross-modulation problems and the likelihood of receiver-generated spurious signals from off-frequency broadcast stations, limits the frequencies available for amateur use. Under these crowded conditions, receiver performance is critical and the parameters outlined for 3.5MHz equally apply to this band. The use of front-end attenuators and or passive preselectors will help to alleviate the problem. The use of rotary directional antennas is becoming commonplace as, apart from providing extra gain, the semi-unidirectional pattern helps to reduce interference. While low-angle radiation is desirable, dipoles, loops and other horizontal antennas are very popular and are capable of providing good DX performance. Ground planes, base-fed verticals and 'slopers' are also widely used.

The IARU Region 1 Band Plan recommends that 7.0–7.04MHz is used for CW and that 7.04–7.1MHz is shared between phone and CW with a small segment for data. As a result of the high interference levels, most CW DX operators use the bottom 10kHz during the peak broadcasting hours up to midnight GMT. Later in the night and in the early morning hours, the interference is less and more of the band can be used for CW DX operation. For phone working the situation is worse and it is often a case of finding a gap between the broadcast stations. The segment between 7.07 and 7.09MHz is popular, and often intercontinental telephony contacts are possible on these frequencies during the peak broadcasting periods.

US phone operation commences at 7.15MHz, so that working US stations on phone requires split-frequency operation. Finding a clear listening frequency in what, in Europe, is a broadcast band, can be difficult at times, though matters improve around dawn as the UK starts to lose propagation to Europe and the US can be heard more easily.

US phone operation commences on 7.15MHz for Advanced and Extra Class licensees and 7.225MHz for General Class. US Novices and Technician Plus licensees may use 7.1–7.15MHz for CW.

10MHz (30m) band

Allocated for amateur use at the 1979 World Administrative Conference, almost all countries now permit their amateurs to use 10.100–10.150MHz on a shared basis, though some administrations have yet to release the band for amateur use. Most operation is on CW and IARU Region 1 has recommended that phone operation and contests be excluded. Contacts on the band do count for a number of awards such as IOTA, DXCC, Commonwealth CC, WAB and IARU Region 1.

Propagation reflects the characteristics of the adjacent 7 and 14MHz bands. In daytime the skip shortens as MUF rises but a skip zone generally persists all day except under exceptional sunspot maximum conditions. Absorption is lower than 7MHz. Under night-time conditions propagation can become worldwide as absorption falls, being best towards the dark sector and being enhanced at the onset of sunset or sunrise. Grey-line effects are very noticeable, offering worldwide propagation between, for example, the UK and the Far East or South Africa. During periods of sunspot minima long-range propagation can persist all day, such an effect being noticeable in the last sunspot minimum – this allowed propagation between the Far East and the UK in the mid-afternoon (UK time). Under such conditions the MUF may not rise much above 10MHz at any time of day.

Long-path propagation, especially to VK/ZL, occurs most mornings under undisturbed conditions. This path starts at around sunrise and can extend for up to two hours afterwards. The closure of this long path is often signalled by a short enhancement of Central American signals, followed by rapid closure of both paths. In order to exploit 10MHz propagation some form of grey-line calculator is useful (see p44).

A major problem on 10MHz is commercial interference. This confines most amateur activity to a series of narrow 'slots' centred on 10.102, 10.104 and 10.110MHz, for example. These 'slots' tend to move under the pressure of commercial stations. Interference tends to limit the development of amateur activity since stations operate on the band for a limited period and little regular usage occurs, despite the favourable propagation charcteristics of the band. Fortunately, most DXpeditions now make an effort to operate on 30m, and it is remarkable to note the long hours the band is open to many parts of the world.

The characteristics of the 10MHz band mean that it is suited to antenna and propagation experimentation, Compact directional antennas for reception and a better understanding of the many propagation modes will allow good DX results to be achieved by the average CW amateur.

14MHz (20m) band

This is the band which carries the main load of intercontinental communication throughout the whole of the sunspot cycle. There are very few days even at the time of sunspot minima and in the middle of winter when propagation is not available for at least some time into each continent. A result of this is that for much of the time the band is very congested, and this situation is made worse by the ITU regulation which permits stations of the fixed service in the CIS and other countries to use the 14,250–14,350kHz segment on a shared basis with amateurs.

The IARU Region 1 Band Plan recommends that 14,000–14,099kHz be reserved for CW use – with the area 14,070–14,099kHz being used for data communications. In the USA only stations with Extra Class licences are allowed to use 14,000–14,025kHz.

Particular care should be taken to avoid causing interference to the beacon chain on 14,100kHz. Packet operation should take place in the data sub-band, preferably in the upper half.

DXpedition stations often tend to use CW frequencies which are multiples of 5kHz above the lower band limit but rarely go above 14,050kHz, and 14,025kHz is perhaps the most widely used. They listen for replies a few kilohertz either side of their transmitting frequency and usually announce their tuning procedure.

The part of the band below 14,150kHz is not available to phone stations in the USA and is therefore often used by non-USA stations working each other or by non-USA stations working 'split-frequency' and listening in the American band for callers (this is often helpful when interference is heavy). At the lower end of the segment there are very often nets of French speaking stations – including those from all over the world as well as France itself. DXpeditions operating from areas other than those controlled by the USA usually transmit in this part of the band, and listen for callers from the USA above 14,200kHz and from elsewhere below that frequency in areas as announced from time to time. The most used DXpedition frequency is 14,195kHz, with 14,145kHz as second choice. The USA phone segment starts at 14,150kHz and the first 25kHz is reserved for those who have Extra Class licences. The Advanced Class phone section commences at 14,175kHz and the General Class at 14,225kHz. SSTV signals from all over the world will be found around 14,230kHz (see Chapter 9). The area above 14,300kHz is very often used by USA stations running 'phone patches' with servicemen overseas in those countries with which FCC regulations permit third-party communication. This area is also favoured by a number of special interest group nets.

18MHz (17m) band

Allocated for amateur use at the 1979 World Administrative Radio Conference but only released for service some years later, this band is now very popular with some excellent DX around; most DXpeditions provide some operation on it. However, although commercial multiband beams are available for the band, many stations still use dipoles, especially if they have already put up, say, a triband Yagi for 14/21/28MHz, and this does limit the number of strong signals around.

As already mentioned earlier, propagation is similar to the 14 and 21MHz bands, and the band is a useful 'half-way' house – some stations have been known to move up from the crowded 14MHz band on days when the MUF makes this possible.

The IARU Region 1 Band Plan recommends that the segment 18.068–18.100 is reserved for CW use only, with the area 18.101–18.108 for data modes and CW. Phone operation is allowed from 18.111–18.168MHz.

21MHz (15m) band

This is a favourite band with many HF operators, with more space than the crowded 14MHz band, but significantly more reliable than 28MHz except perhaps near sunspot maxima.

The IARU Region 1 Band Plan recommends that the segment 21,000–21,149kHz be reserved for CW use with the area

21,080–21,120kHz for data operation. DX on CW, although fairly evenly distributed in the first 75kHz or so, tends to be heaviest around 21,025kHz, this being a popular frequency for DXpeditions. It is the calling frequency of the FOC (First Class Operators' Club) and is the upper limit of the CW frequency allocation exclusive to the Extra Class licensees of the USA. Contest stations usually operate in the first 50kHz of this segment although in the bigger contests they may spread up to the proximity of 21,100kHz. The rest of the Region 1 CW subband falls within the USA Novice and Technician Plus allocation (21,100–21,200kHz) and is mainly used for working the USA. When conditions are good this part of the band is very crowded and is therefore one of the simplest indicators of propagation into North America.

Moving up the band, SSB signals commence around 21,151kHz with German and French language stations showing a preference for DX work around the lower end. When conditions are good SSB signals may be heard from the low-power Australian Novice stations in the segment 21,151–21,200kHz. As it is outside the USA phone allocation, '151 to '200kHz is often used by DX stations particularly wishing to work Europe, including Canadians, Central and South Americans. When DX propagation is poor but short skip is possible this part of the band is used for inter-European working.

The USA phone segment commences at 21,200kHz and when conditions are good very strong signals are heard from USA Extra Class licensees who have the exclusive use of the first 25kHz in their country. The segment '250 to '350kHz is usually very crowded at weekends and tends to attract high-powered stations. The station with more modest means is therefore advised to operate in the '350 to '450kHz segment when working into the USA. The Advanced Class phone section starts at 21,225kHz and the General Class at 21,300kHz.

SSTV signals are occasionally heard around 21,340kHz and some 'phone patching' takes place in the area above 21,400kHz, the top 50kHz often being used by Americans living abroad to work stations back home.

Over the years both USA and non-USA DXpedition stations have tended to use 21,295–21,300kHz for their operations, but recently DXpeditions operating from areas other than those controlled by the USA seem to favour frequencies around 21,195kHz and 21,245kHz for this purpose, announcing their listening frequencies for various areas from time to time.

24MHz (12m) band

Allocated for amateur use at the 1979 World Administrative Radio Conference, this band is admittedly not very popular during years of sunspot minimum. Like the 18MHz band, it also suffers from a lack of beam antennas in use. However, it can provide communication on days when the 28MHz band is quiet, and should not be overlooked.

The IARU Region 1 Band Plan reserves 24.890–24.919MHz for CW operation, with data operation from 24.920–24.928MHz. Phone operation takes place from 24.931–24.990MHz.

28MHz (10m) band

The 28MHz band is on the borderline between HF and VHF. Indeed, some authorities allow the band to be used by operators who might elsewhere be regarded as licensed for VHF only. Of course the delineation between HF and VHF, specified by definition at 30MHz, is man-made and Nature draws no such firm division. It is this location in the spectrum which gives the band, sometimes described as "the band of surprises", the variability which is one of its attractions.

As at VHF, high power is by no means necessary to operate long ranges. It is generally accepted that if a path exists at all then 10W, say, in the CW mode and little (if any) more of SSB will achieve satisfactory communication. Indeed, contact with almost all the USA call areas has been made with 200mW. The smaller size of antenna elements for this band allows the construction of compact, efficient installations. It is interesting to note that the simple ground-plane performs well, especially for the sporadic-E (Es) or short-skip conditions which can produce good coverage of Europe from the UK during the summer (May–September). It will be seen that these features, coupled with the wide frequency range available, make it a very good band for relatively simple and inexpensive stations.

The band is much more dependent on solar activity than the 14 and 21MHz bands, but it is to be regretted that the majority of operators desert it during the years around sunspot minimum. Those who continue to use it through these periods are often surprised at the paths which appear from time to time. To work 100 countries during the lean years can be a stimulating challenge. While, as in the lower HF allocations, propagation from the UK tends to swing from the east in the morning to the west in the afternoon, paths to many parts of the world will often exist simultaneously and WAC in five minutes working is by no means a rare feat.

Although, again speaking generally, propagation requires daylight over most if not all the path, DX contacts may still be made during the hours of darkness. It may well be that the few contacts recorded under these conditions would be increased in numbers if operators did not take such a dogmatic view of 28MHz propagation possibilities, but checked the band at times when experts say there will be no signals and also transmitted a 'CQ'. The same suggestion holds good also for the years around sunspot minimum. As an aside, evening propagation to the southern states of the USA can sometimes happen when the operators at both ends point beams to the South Atlantic. Frustration occurs when the distant station does not realise this is necessary and turns to the direct bearing, and then considers the contact lost. Apart from E and F propagation, the band performs in a similar manner to that of 50 and 144MHz and signals can be ducted over quite long distances under 'lift' conditions. Extended ground-wave contacts over distances of several hundred miles are commonplace via ducting.

The IARU Region 1 Band Plan for the 28MHz band follows a similar pattern to those for the other HF bands. That is to say, the lower frequencies from 28.0–28.05 and 28.15–28.20MHz are reserved for CW, with a band of 100kHz centred on 28.10MHz allocated for CW and data shared use. The remainder of the band, 28.20–29.70MHz, is available for both CW and phone, though, as might be expected, CW operation centres on the portion below 28.20MHz. A small 10kHz band centred on 28.68MHz is recommended for SSTV working. The band plan recommends that beacons use the 28.190–28.225MHz segment. This segment should never be used for normal communication. The FCC permits Novice and Technician licensees to use 28.1–28.3MHz for CW and data, and 28.3–28.5MHz for CW/SSB, and restricts other USA phone stations to above 28.30MHz. During years of high solar flux this last division causes non-American phone operators to spread either below 28.30MHz or well up the band, eg 28.90MHz and upwards, to avoid USA interference. Towards sunspot minimum, random phone operation tends to be conducted between 28.50–28.80MHz.

In recent years the popularity of the IOTA Award (see later) has led to an increase in activity from remote islands. This is the NL7TB/P DXpedition site at Ushagat Island, 200 miles south of Anchorage, Alaska

Stations using low-power AM and FM may be heard above 28.80MHz. The RSGB recommends that FM activity should take place between 29.60–29.69MHz, with 29.6MHz as the calling frequency. In the USA 29.6MHz is also the FM simplex calling frequency, and there are 29MHz FM repeaters with outputs at 29.620, 29.640, 29.660 and 29.680MHz (inputs are 100kHz lower in each case).

A specialist aspect of the 28MHz band is concerned with amateur space service communication and the sub-band 29.30–29.50MHz has been recommended for amateur satellite downlink use (see Chapter 7).

The availability of synthesised FM CB equipment that can be easily converted for 29MHz operation has encouraged many mobile operators to use the band both for DX and local working.

LOW-POWER DXING

In a real sense most amateur radio operation is low powered when compared against the usual commercial power allowances. The maximum power allowed varies from country to country and in some cases according to the class of the licence, amateurs in many parts of the world looking with envy on the kilowatt 'big guns' of the radio amateur world. However, there has always been within the hobby a group of amateurs who specialise in low-power (QRP) working. This aspect of the hobby has been increasing in popularity in recent years, as readers of the amateur radio journals will have noticed.

Victor Brand, G3JNB, has this QRP station constructed from popular kits

There are good reasons for this movement towards using low power. Running low power gives more challenge and increases the reliance on actual operating skills to achieve reasonable results. W9SCH has called QRP operation "the sportsman's amateur radio" and certainly many experienced DX operators have turned to low-power working for the challenge.

Some operators offer an ethical explanation for their low-power work. In these days of crowded amateur bands, it is more reasonable to use only the minimum power requirements to complete a contact. It is difficult to be a bully with only a few watts.

Finally, many amateurs see low-power DXing as an inexpensive approach to the hobby. Although the modern tendency is perhaps to consider spending a couple of thousands of pounds on commercial equipment, it is possible to achieve good results on the amateur bands by spending only a few pounds on low-power equipment. It must be admitted that there is a certain feeling of glee when working a station with a kilowatt and a quad antenna with an inexpensive 2W transmitter, but this is not to suppose that QRP operators rely upon the power and sophistication of the other man's equipment to maintain a solid contact. Many QRP operators are keen to work other low-powered stations and this type of contact forms a large part of such operators' aspirations.

What is QRP?

Opinion as to what constitutes low-power operation varies. In the USA powers of under 100W are considered to be QRP. The G-QRP Club uses 5W DC input as the power limit under which a station may be said to be QRP, while for contests IARU Region 1 defines QRP as 10W input and QRPp as 1W. In the USA the term 'QRPp' is often used for powers of 5W and less.

An input power of 5W or less would appear to place severe restrictions upon what might be achieved on the amateur bands, but the power level is not the disadvantage it may appear at first glance. Obviously the station is weaker, but in theoretical S-point terms the reduction is not so great. Assuming 6dB per S-point, a power ratio/dB chart will show that reducing power four times will only give a 6dB, or one S-point drop in signal level. Therefore a reduction from 150W DC input to 37.5W

gives in theory only a single S-point loss. A station using 5W input can expect to be 15dB down, or about two and a half S-points, on a station using 150W input. Naturally these are only theoretical figures and are variable by such factors as efficient matching and high-gain antennas. In terms of increased signal strength a simple increase in power is therefore an inefficient method of obtaining results.

What can be achieved with such low power?

Recent results by low-power operators have shown that they consider their restricted power as no real barrier to real DX working. QRP awards are earnestly sought after and some individual results are worthy of mention. In the annual G-QRP Club Winter Sports between 26 December and 1 January 1992, Chris Page, G4BUE, operated as GB0QRP using 5W output from a Ten Tec Argonaut II. In that short period he had 476 two-way QRP QSOs, where both stations ran 5W or less. 62 DXCC countries were worked including: 8P, BV, HI, J7 (7MHz), KL7 (10MHz), KP4, SU (10MHz), VP5, VS6, VU and ZA. AA2U, NG1G, W3TS and N4AR were all worked two-way QRP on eight bands. No description of QRP DXing is complete without a mention of George Burt, GM3OXX, who has worked over 200 countries using less than one watt from all-home-built equipment with a simple wire antenna.

Equipment for low-power operation

It is probably true to say that the majority of QRP operators use home-built equipment. Circuits for such equipment appear in the various amateur radio journals, and especially in the G-QRP Club journal *Sprat*. Such equipment can be built to suit the amateur's expertise and pocket [30–36].

Low-power operation is mainly on CW, so simple equipment is capable of useful results. The serious operator, however, really requires a VFO-controlled transmitter and a good-quality receiver, if two-way QRP contacts are contemplated. He has to use extra cunning and skill to secure contacts, therefore simple-to-use equipment with full break-in facilities and incremental receiver tuning is ideal. Transceivers reduce the time required to call a station, and today's crowded bands demand a bandwidth filter, if only of a simple audio type. The low-powered station must not be hampered by difficult-to-use, inefficient equipment.

The recent increase in QRP operation has produced and even been fostered by several commercial transceivers. The Heath Company produced the HW7, HW8 and HW9 transceivers which are still available on the second-hand market. The Ten Tec Company in the USA produced a range of direct-conversion transceivers in the 'seventies with the 'PM' prefix coding. Ten Tec also produced the Argonaut range of SSB/CW transceivers and currently manufacture the Scout, a modular transceiver with a QRP option. MFJ Enterprises manufacture a small range of CW QRP transceivers and the Travel Radio, a 20m SSB transceiver. A more recent manufacturer of QRP equipment is Index Laboratories with their QRP PLUS, an all-band, all-mode, synthesised transceiver. In the UK many QRP operators used equipment built from kits produced by companies such as C M Howes, Lake, Kanga Products, Walford Electronics and Hands Electronics.

Antennas for low-power operation

In QRP circles the keenest discussion probably occurs on the subject of antennas and their matching. This subject should be close to the heart of all radio amateurs but, when using reduced power, valuable extra gain may be achieved by the antenna, and correct matching is essential to reduce losses of precious power. Anyone contemplating QRP operation is advised to read the standard works on antennas mentioned in Chapter 2.

An interesting fact about QRP operation is that most of the best-known and successful operators use only simple wire antennas. GM3OXX uses a dipole to work Australasia with 2W and G8PG and G3DNF have a history of good results only using various types of simple wire antenna. Good results have been obtained with random lengths of wire, but efficient matching between transmitter and antenna is essential. An antenna tuning unit is regarded as an important item for the low-power operator and many designs have been published in QRP journals, although circuits designed for high-power use are also used. The G-QRP Club have produced the *G-QRP Antenna Handbook*, available from the RSGB, which contains antennas and antenna matching circuits used by many QRP operators.

Another important item is an SWR meter or a meter for indicating forward and reverse antenna current. There are several models on the commercial market, although this is an easy-to-build item of equipment. Solid-state transmitters are invariably difficult to tune to optimum output by metering the current input to the power amplifier. It is more usual in QRP work to tune for maximum output to the antenna and match the transmitter and antenna by reading the lowest attainable reflected power.

Low-power operating modes

Most QRP operating is CW, with its obvious advantages of simplicity and effective range-to-power ratio. SSB operation has, however, come into more general use with the production of commercial SSB QRP equipment. Equipment is readily available for SSB QRP operation and several operators have achieved DXCC with 5W or less.

In recent years many more constructors have built their own SSB equipment. Hands Electronics and Walford Electronics have both produced kits for building SSB transceivers. In the USA there has been considerable interest in the single-signal direct-conversion boards designed by Rick Campbell, KK7B. These use phasing techniques to produce good performance SSB transceivers for any band.

Operating with low power

All radio amateurs ought to study the propagation patterns outlined earlier in this chapter to make full use of prevailing conditions, but for the QRP operator these conditions may make the difference between establishing a contact or not. In fact low-power transmissions are used in some forms of propagation study.

A useful rule of thumb is to use the highest-frequency HF band that is open for which one has equipment and a good antenna. The adverse effects of absorption from the D layer is an inverse square of the frequency in use during the hours of daylight. After sunset the MUF (maximum usable frequency) falls, making 3.5MHz and 1.8MHz the most useful night-time

and winter-evening QRP bands, although some useful work can be done if one dares to take low power on the crowded 7MHz band.

DX and long-range work bring into play the problems of radiation angles and skip distances from the very active F layers. Changes in the F layers as reflecting planes are very dynamic and only experience coupled with a general knowledge of propagation patterns can really determine the optimum paths for QRP work at various times on the HF bands. Although several paths may appear to be open from listening to incoming signals, many long-distance paths seem to show optimum propagation conditions in only one direction. The operator has to determine which outgoing path is open and this may well not be the one with the strongest incoming signals.

An example common to many operators is observed in the east-west path. On 14MHz, the 'your sunset, his sunrise' period appears to give the best results. However, the movements of the F2 layer appear to follow the Sun's path very closely. It is common to find that when the 14MHz band first begins to open to the west, the first weak signals appear to be the easiest ones to work. It is for example easier to work the first low-strength USA stations early in the evening with 2W on 14MHz, than to work S9-plus USA stations when the band appears to be really open (see also p47). It is also easier to work transequatorial than transpolar. Some QRP operators keep a propagation log – not a bad idea.

It might be said that operating technique and skill has to increase in inverse proportion to the available RF power. The key words for QRP operation are 'patience' and 'cunning' – the operator has to win his contacts. Like any good DX operator he must spend a lot of his time listening to the signals on the band, getting to know who is there and who they are working. A survey of DX operators conducted some time ago showed that they spent about 95% of their time listening and only 5% transmitting. Naturally this applies to operators who are looking for specific contacts, but the ratio of listening to transmitting should be high for reasonable results with low power. Another general tip is not to 'think QRP' and expect to do badly on the bands, but to assume that one is 'in with a chance' for most contacts.

Low-power operation often takes place around the international QRP frequencies of 3560, 7030, 14,060, 21,060 and 28,060kHz plus 1850kHz in the UK.

Calling CQ can be a waste of time with low signal strength; it is far better to call a specific station either in response to his CQ or at the end of an existing contact. When QRP is being worked both ways in skeds or on the international QRP frequencies, then calling "CQ QRP" is reasonable. Otherwise look for strong stations as they are probably the ones with the beams and sophisticated equipment best suited for the reception of low signal strength.

'Tail-ending' is a common practice for most QRP operators. Obviously it is wise to look for a station in contact with another station in a geographically close country. This indicates that a path does exist to your area and if a beam is being used it will be pointing in the right direction. Tail-ending involves speed, for the required station may decide to change frequencies and must be 'caught' while still listening on the frequency. Allow for the closing exchanges between the two stations in contact. The contact may close with an extra exchange of "73" or dit-dit, and if the station you are aiming for is working a local station, you may not be able to hear the local station due to the skip zone. This requires some judgement as to when to make the tail-end call. Full break-in facilities are a great asset in tail-ending. Breaking in to an existing contact is not advised. Not only is this bad manners, but the chances of a weak signal breaking through are slight. When calling the required station, either match his operating speed or send slightly slower to allow for clear reading of the signal. Sending above the speed of the other station is just 'showing off' and may frighten off the other station or render a weak signal unreadable.

It is not a good practice to use 'QRP' as a suffix when making a call. In the UK it is illegal to sign '/QRP'. If a specific station is required and is calling for DX, the technique of a rapid call followed by "QRP RPT PSE" sometimes brings results.

Conduct the contact in the usual way, but maintain only short overs, unless it is clear that one's signal strength is high with the other station. If possible avoid giving the other station information about one's power until his report has been received. It is surprising how much better reports are if the other station does not believe he is working a low-power station. Expect strength reports of S4 to 7 as normal – very few S9s will be given to QRP stations, except for the conventional and inaccurate 599 given in rapid contest working.

One of the real delights of QRP working is the kindly response it evokes from many operators. Remarks like "FB UR QRP SIGS" or "GUD SIGS FER QRP" are very common. Some operators are even known to 'turn down the wick' at their end and attempt to return the QRP signal! Do not be ashamed to be using low power as it usually gains respect from other operators.

Working DX in literal terms of 'long distance' is a matter of propagation study and use of conditions, dealt with earlier. DX in terms of rare stations is another matter in that the competition is greater. Rare stations attract a lot of attention and one may have to contend with the DX pile-up of calling stations. A station with low signal strength has obvious disadvantages in such situations, but may still be able to compete.

In a DX pile-up, the low-power station is probably at the end of the line, and must use all the usual techniques of frequency shifting and rapid calls. Many DX stations call by areas and this may give the QRP station a better chance. Do not call out of turn, which is not only bad practice but pointless with low power.

The DX pile-up can provide an extra bonus for the QRP operator. While half the amateur world is hunting after one station, move up or down the band and see which stations may be present on a clearer frequency. Some stations which are not DX for a normal station, but are DX for QRP, may be seeking contacts away from the pile-up.

'DX' may for the QRP operator take on the more literal meaning of distant stations or apply to prefixes yet to be worked with low power. This naturally begins with the extra patience required to seek specific stations or call areas. Look for the 'big' DX station, for he has the resources to work a low-power station. Listen to see if the required station is working other stations in one's own geographical area, and what sort of reports these stations are being given. Take into consideration that your signal will be two or three S-points down. In a band with a high noise level the problems are even more acute.

It is important to call the station on frequency, because a DX station with a 600Hz CW filter may not be able to hear a station calling off frequency. Try to zero beat with the station. This is a simple matter with independent transmitter and receiver systems, but with transceivers receiver incremental tuning (RIT) or VFO offset is important. It is then vital to know, by experience or experimentation, the degree of offset offered by such a control, which may vary from band to band. The

careful operator will know where his signal is located in respect to the other signals he can hear.

Snappy operating procedure may be vital. Full break-in facilities can enable the operator to call quickly and gain the initial attention of the other station, and perhaps even monitor other calling stations. Imitate the speed and methods of the required station. If he is making quick calls such as "QRZ DE XXXX", reply with a simple "XXXX DE G3--- BK". Although a long slow call is suitable for low-power operation, in quick operating situations another station may have gained the contact before one's own call has been sent a couple of times.

Always note the time and frequency of a required DX station and try the same time and frequency again. This often seems to work. With low-power operation, one frequently has to weigh available time against probability in securing a contact with a particular station. Having failed to secure a contact after several attempts, realistically assess the chances of one against the possibility of finding other suitable contacts on the band. The QRP enthusiast must expect to use more time per contact than other operators, so available time for operating must be used to the best advantage.

QSL card returns may on the other hand be somewhat easier for the low-power station. Most DX stations are happy to work a low-power station and will be pleased to confirm the contact. Boldly mark the card "QRP" and "TWO WATTS INPUT" or whatever.

QRP organisations

There are many international clubs serving the low-power DX operator [37–40]. The G-QRP Club now has over 5000 members in 54 countries and publishes a quarterly journal, *Sprat*. In order to promote low-power operation these clubs also organise awards and contests – some further details are given in the relevant section of this chapter and in Chapter 5. *RadCom* has a bi-monthly QRP column.

HF AWARDS

As QSL cards from DX stations come flooding in (hopefully!) the operator naturally begins to consider progress towards one of the many attractive operating awards and certificates available from national societies and other organisations all over the world.

Depending on its 'difficulty', the acquisition of one of these certificates can be a real milestone in an amateur's operating activities, and something to display proudly on the shack wall for many a year.

The first step is to obtain a copy of the complete rules from the organiser of the award. If this is a national society then the addresses given in Appendix 2 may be used for this purpose. When requesting rules of overseas awards always include a large self-addressed envelope and three IRCs. All correspondence should include return postage. Allow a reasonable time for a reply, especially to a query about rules – the awards manager has to do this work in his spare time!

The rules should then be read carefully. Some awards require special application forms but even if this is not necessary, a fairly standard sort of layout is always helpful to the awards manager dealing with the claim. He is invariably doing the job on a voluntary basis and a clear statement of exactly what is being claimed cuts down the amount of work involved. Always print your name and address in block capitals and if possible type the application. An easy and clear method of application is shown in Fig 10.

```
John Smith, G9ZZZ
4 High Street
Notown
Blankshire
England
                          Application for the ZZZ award class 1

Endorsed all 14MHz, all 2X ssb

Date           GMT      Station    His signals    My signals    Band    Mode
13 August 99   1234     G7AS       59             59            14      2X ssb
15 August 99   0750     G7CX       55             46            14      2X ssb
     .          .        .          .              .            .        .
     .          .        .          .              .            .        .
23 August 99   1453     G7NN       59             56            14      2X ssb

Certified that I have complied with my licence regulations

3 September 99                                     John Smith, G9ZZZ

                                                   John Smith
```

Fig 10. Suggested layout for awards application

Full details of the RSGB HF awards and brief notes on some of the other popular DX awards are given below but it should be realised that awards are available from most national societies and a complete listing is well beyond the scope of this book. However, these are often given on the national society's web site (see Appendix 2).

There are a variety of certificated and trophy awards offered for the low-power operator by the G-QRP Club (see p56). The most popular of these are the QRP Countries Award, which begins with a basic 25 countries confirmed and the G2NJ Trophy, which is awarded on a three-year cycle for various QRP achievements. The awards and contest manager for the G-QRP Club is G8PG. The American QRPARCI also have an awards scheme, the most popular being the 1,000 Miles Per Watt Award. The awards manager is WA8CNN.

For details of VHF, UHF and microwave awards available to UK amateurs, see p82.

RSGB awards

DX Listeners' Century Award (DXLCA)

This award may be claimed by any shortwave listener eligible under the General Rules who can produce evidence of having heard amateur radio stations located in at least 100 DXCC entities (see p63). Stickers are available for every 25 additional countries confirmed. Submit a list in radio prefix order with the callsign and country name.

Endorsements are available for hearing 100 countries on 5, 6, 7, 8 and 9 bands (they need not be the same countries on each band).

Commonwealth Century Club (CCC)

This award may be claimed by any licensed radio amateur eligible under the General Rules who can produce evidence of having contacted, since 15 November 1945, amateur radio stations in at least 100 Commonwealth call areas on the list current at the time of application.

The certificate holder may claim, on payment of a contributory charge, a handsome plaque with a plate detailing name, callsign, date and number of the award. Additionally, an amateur providing evidence of having contacted all the Commonwealth call areas on the list current at the time of application may claim the Supreme Plaque in recognition of the magnitude of the achievement, again on payment of a contributory charge.

Notes:

(a) Credit for South Georgia and the South Sandwich Is will only be given for contacts with stations using a VP8 callsign. Credit for Antarctica and the South Orkney and South Shetland Is will only be given for contacts with stations using a callsign issued by a Commonwealth government.

(b) Where, very occasionally, a contact is made with a station using a callsign legitimately outside the geographical area to which the prefix normally applies, it will count for the actual area from which the operation took place. The evidence submitted will need to be clear.

5 Band Commonwealth Century Club (5BCCC)

This award, available in five classes, may be claimed by any licensed radio amateur under the General Rules who can produce evidence of having effected two-way communication, since 15 November 1945, with the requisite number of amateur radio stations located in the call areas listed, using all five bands, 3.5, 7, 14, 21, and 28MHz. Each station should be located in a different call area per band. The five classes are for contacts as follows:

5BCCC Supreme	500 stations
5BCCC Class 1	450 stations
5BCCC Class 2	400 stations, with a minimum of 50 on each band
5BCCC Class 3	300 stations, with a minimum of 40 on each band
5BCCC Class 4	200 stations, with a minimum of 30 on each band

Certificates will be issued to winners of all classes. Additionally, as in the case of the CCC, winners of the Class 1 award will be eligible to claim a handsome plaque suitably inscribed on payment of a contributory charge, while winners of the Supreme Award will be able to claim an engraved plaque on payment of a contributory charge.

WARC Bands Endorsement

A holder of the basic award who can provide evidence of contact with the required number of call areas on the 10, 18 and 24MHz bands may claim a WARC Band 'sticker' endorsement. This is available in five classes as follows:

GENERAL RULES FOR RSGB HF AWARDS

The following general rules and conditions apply to HF certificates and awards issued by the Radio Society of Great Britain and should be read in conjunction with the conditions which govern the particular award programme:

Applicant eligibility

1. Claimants from the UK, Channel Is, and Isle of Man must be members of the RSGB and, as proof of membership, should provide a recent address label from *RadCom*. Applicants from elsewhere need not be members of the RSGB.
2. Claimants may be either licensed radio amateurs or short wave listeners. All certificates, but not special plaques, are available on a 'heard' basis to listeners.

Claim eligibility

3. Each claim must be submitted in a form acceptable to the HF Awards Manager. Where application forms are provided for particular award programmes, these should be used although a computer generated form including the same headings will generally be accepted. Each claim must include the following signed declaration: *"I declare that all the contacts were made by me personally from the same DXCC country and in accordance with the terms of my radio transmitting licence, and that none of the QSLs have been amended in any way since receipt. I accept that a breach of these rules may result in disqualification from the awards programme. I further accept that the decision of the HF Committee shall be final in all cases of dispute."*
4. All claims from within the UK, Channel Is and Isle of Man must be accompanied by QSL cards. Claims from elsewhere must also be accompanied by QSL cards but only in the case of those categories of award attracting a plaque. In all other cases a statement from the applicant's national society that the necessary cards have been checked will be accepted except that the HF Awards Manager reserves the right to ask to see some or all of the cards. For IOTA claims special rules apply (see the *IOTA Directory*).
5. Each claim must be accompanied by a fee of £3.00 or $US 6.00 or 9 IRCs per certificate or class of certificate. Applicants submitting cards for checking must include sufficient payment to cover their return. Cards will only be returned by air, recorded delivery (UK only), or registered mail (overseas) if adequate postage is enclosed. (For registered mail add US$ 4.50 or 7 IRCs).

Contact eligibility

6. All contacts must be made by the holder of the callsign.
7. Contacts may be made from any location in the same DXCC country.
8. Except where otherwise indicated, credit will be given for confirmed contacts made on or after 15 November 1945 on any of the 9 amateur bands below 30MHz.
9. Contacts with land mobile stations will be accepted, provided the location at the time of contact is clearly stated on the QSL card.
10. Credit will be given for two way contacts on the same mode and band, ie not cross-mode or cross-band. Certificate endorsements for single mode transmission and/or single band may be made on the submission of cards clearly confirming the mode or frequency of transmission, but the request must be made at the time of application. Special rules apply for IOTA.

Disqualification

11. The submission for credit of any altered or forged confirmations or, equally, bad behaviour on or off the air which is judged by the HF Committee to bring a particular programme into disrepute may result in disqualification of the applicant from all RSGB's award programmes. The decision of the HF Committee on this and other matters of dispute will be final.

Applications for awards (other than IOTA)

All claims, except IOTA, should be sent to:

RSGB HF Awards Manager,
F C Handscombe, G4BWP,
'Sandholm',
Bridge End Road,
Red Lodge,
Bury St. Edmunds,
Suffolk IP28 8LQ.

Prepare your application in accordance with the requirements of the award being claimed. Send QSL cards when required, and do not forget to enclose details of your name, callsign and full address as well as the certificate fee, adequate postage for the return of QSL cards, and for UK applicants, proof of RSGB membership. Payment may be made by cheque drawn on a UK bank or Eurocheque written in pounds sterling and payable to 'Radio Society of Great Britain'.

5BCCC (WARC)	Supreme	300 call areas
5BCCC (WARC)	Class 1	275 call arcas
5BCCC (WARC)	Class 2	250 call areas, with a minimum of 50 on each band
5BCCC (WARC)	Class 3	200 call areas, with a minimum of 40 on each band
5BCCC (WARC)	Class 4	150 call areas, with a minimum of 30 on each band

Note: on 10MHz credit will only be given for contacts on CW and data modes.

Top Band Endorsement

A holder of the basic award who can produce evidence of contact with the required number of call areas on the 1.8MHz band may claim a Top Band 'sticker' endorsement. This is available in five classes for 30, 40, 50, 60 and 70 call areas.

Credit for deleted Commonwealth call areas

Credit will be given for contacts with stations in Commonwealth countries at the time of contact, and additionally, with stations using a Commonwealth callsign in Antarctic and the S Orkney and S Shetland Is. Since 1945 some countries have left the Commonwealth, while others have joined. The list of call areas specifies the relevant dates. For 5BCCC (including endorsements) contacts with 'deleted' call areas made at a time when the countries concerned were in the Commonwealth may count in place of 'missing' credits on any band up to the maximum possible for that band. Deleted call areas do not count for CCC.

List of deleted Commonwealth call areas

Contacts with the following may count for a 'missing' credit if made before the date specified. A contact after that date may still count for one of the Commonwealth call areas (see brackets).

AC3	Sikkim	1 May 1975 (India)
P2, VK9	Papua Territory	16 September 1975 (PNG)
P2, VK9	Territory of New Guinea	16 September 1975 (PNG)

VO	Newfoundland, Labrador	1 April 1949 (Newfoundland, Canada)
VQ1	Zanzibar	25 April 1964 (Tanzania)
VQ6	British Somaliland	1 July 1960
VQ9	Aldabra Is	29 June 1976 (Seychelles)
VQ9	Desroches Island	29 June 1976 (Seychelles)
VQ9	Farquhar Group	29 June 1976 (Seychelles)
VR2, VS6	Hong Kong	30 June 1997
VS2, 9M2	Malaya	16 September 1963 (W Malaysia)
VS4	Sarawak	16 September 1963 (E Malaysia)
VS9	Aden	1 December 1967
VS9	Kamaran Island	1 December 1967
VS9	Kuria Muria Is	1 December 1967
ZC5	British North Borneo	16 September 1963 (E Malaysia)
ZD4	Gold Coast, Togoland	6 March 1957 (Ghana)

Applications for CCC and 5BCCC

Applications for 5BCCC should use the special application form available from the HF Awards Manager. The form allows space for callsigns to be recorded for contacts on each band. A check list of current Commonwealth call areas is also available and can be used for applications for the CCC Award. Please send an A5 size SAE for the check list and application forms.

Worked ITU Zones (WITUZ)

This award may be claimed by any licensed radio amateur eligible under the General Rules who can produce evidence of having contacted, since 15 November 1945, land-based amateur radio stations in at least 70 of the 75 broadcasting zones as defined by the International Telecommunications Union (ITU).

The certificate holder may claim, on payment of a contributory charge, a handsome plaque with a plate detailing name, callsign, date and number of the award. Additionally, an amateur providing evidence of having contacted all 75 ITU zones may claim the Supreme Plaque in recognition of the magnitude of the achievement, again on payment of a contributory charge.

5 Band Worked ITU Zones (5BWITUZ)

This card, available in five classes, may be claimed by any licensed radio amateur eligible under the General Rules who can produce evidence of having contacted, since 15 November 1945, the required number of land-based amateur radio stations located in the 75 ITU broadcasting zones, using all five bands, 3.5, 7, 14, 21, and 28 MHz. Each station should be located in a different ITU zone per band. The five classes are for contacts as follows:

5BWITUZ Supreme	350 zones
5BWITUZ Class 1	325 zones
5BWITUZ Class 2	300 zones, with a minimum of 50 on each band
5BWITUZ Class 3	250 zones, with a minimum of 40 on each band
5BWITUZ Class 4	200 zones, with a minimum of 30 on each band

Certificates will be issued to winners of all classes. Also, as in the case of the WITUZ, winners of the Class 1 award may claim a handsome plaque suitably inscribed, while winners of the Supreme Award will be eligible for the Supreme Plaque, both on payment of a contributory charge.

WARC Band Endorsement

A holder of the basic award who can produce evidence of contact with the required number of ITU zones on the 10, 18 and 24MHz bands may claim a WARC Band 'sticker' endorsement. This is available in five classes as follows:

5BWITUZ (WARC)	Supreme	210 zones
5BWITUZ (WARC)	Class 1	195 zones
5BWITUZ (WARC)	Class 2	180 zones, with a minimum of 50 on each band
5BWITUZ (WARC)	Class 3	150 zones, with a minimum of 40 on each band
5BWITUZ (WARC)	Class 4	120 zones, with a minimum of 30 on each band

Note: on 10MHz credit will only be given for contacts on CW and data modes.

Top Band endorsement

A holder of the basic award who can produce evidence of contact with the required number of call areas on the 1.8MHz band may claim a Top Band 'sticker' endorsement. This is available in five classes for 20, 30, 40, 50 and 60 zones.

Notes:

(a) The number of ITU broadcasting (ie land) zones recognised by the ITU is 75 zones (zones 1 to 75) and this therefore is the maximum score which can be claimed per band. However, the island of Minami Torishima (JD1) and Salas-y-Gomez lie outside the 75 zones in sea zones 90 and 85 respectively. For 5BWITUZ (including endorsements) contacts with this island may count in place of one 'missing' credit on any band up to the maximum 75 for that band. Contacts with Minami Torishima and Salas-y-Gomez do not count for WITUZ.

(b) In the case of the WITUZ and 5BWITUZ confirmations need not bear the appropriate ITU zone number but in order to count for credit they should give the location of the station in sufficient detail to place it clearly within one particular zone. Doubtful cases indicating possible overlap across two zones will not be given credit.

(c) The HF Awards Manager will use as his reference a list which is based on the *Radio Amateurs' Prefix Map of the World* published by Radio Amateur Call Book Inc, PO Box 2013, Lakewood, New Jersey 08701, USA. In the case of countries which encompass two or more ITU zones, eg USA, Russia and Brazil, zonal boundaries will generally follow the longitude/latitude grid lines as shown in the map. In the few instances of discrepancy between the map and the accompanying prefix/country list the decision of the HF Awards Manager will be final.

Applications for WITUZ and 5BWITUZ

Applications for WITUZ and 5BWITUZ should use the special application form available from the HF Awards Manager. A check list of ITU Zones and a map are included with the application form. Please send an A5 size SAE for the check list and application form.

IARU Region 1 Award

This award, available in three classes, may be claimed by any licensed radio amateur eligible under the General Rules who

Worked all Continents (WAC)

This award, issued by IARU headquarters, may be obtained by any licensed radio amateur in the UK, Channel Is or Isle of Man who is a member of the RSGB and can produce evidence of having effected two-way communication with amateur radio stations located in each of the six continents – North America, South America, Europe, Africa, Asia and Oceania.

Applicants should send QSL cards to the RSGB HF Awards Manager who will certify the claim to the IARU headquarters society (ARRL) for issuance of the award. They should also enclose a self-addressed stamped envelope for return of the cards, and proof of RSGB membership.

All contacts must be made from the same country or separate territory within the same continent. Various endorsements including 'all 1.8MHz' are available. In addition both a 5 and 6 Band WAC may be claimed.

can produce evidence of having contacted amateur radio stations located in the required number of countries whose national societies are members of the Region 1 Division of the International Amateur Radio Union (IARU).

The three classes are for contacts as follows:

Class 1 All member countries on the current list
Class 2 60 member countries
Class 3 40 member countries

Members of IARU Region 1 (as of August 1999) are:

Albania	Iceland	Oman
Algeria	Iraq	Poland
Andorra	Ireland	Portugal
Austria	Israel	Qatar
Bahrain	Italy	Romania
Belarus	Ivory Coast	Russian Federation
Belgium	Jordan	tion
Bosnia	Kenya	San Marino
Botswana	Kuwait	Senegal
Bulgaria	Latvia	Sierra Leone
Burkina Faso	Lebanon	Slovakia
Croatia	Lesotho	Slovenia
Cyprus	Liberia	South Africa
Czech Republic	Liechtenstein	Spain
Denmark	Lithuania	Swaziland
Djibouti	Luxembourg	Sweden
Egypt	Macedonia	Switzerland
Ethiopia	Mali	Syria
Estonia	Malta	Tajikstan
Faroe Is	Mauritius	Tanzania
Finland	Moldova	Tunisia
France	Monaco	Turkey
Gabon	Mongolia	Turkmenistan
Gambia	Morocco	Uganda
Germany	Mozambique	Ukraine
Ghana	Namibia	United Kingdom
Gibraltar	Netherlands	Yugoslavia
Greece	Nigeria	Zambia
Hungary	Norway	Zimbabwe

Islands on the Air (IOTA)

The IOTA programme was created by Geoff Watts, a leading British short wave listener, in the mid-'sixties. When it was taken over by the RSGB in 1985 it had already become, for some, a favourite award. Its popularity grows each year and it is highly regarded among amateurs worldwide.

IOTA is an award programme designed to encourage contact with island ststions worldwide. The general principle is that IOTA 'counters' are groups of islands rather than individual islands – though there are many exceptions where a group is just a single island. There are 18 separate awards currently available, graded in difficulty. They may be claimed by any licensed radio amateur eligible under the General Rules, who can produce evidence of having made two-way communication, since 15 November 1945, with the required number of amateur radio stations located on the islands both worldwide and regional. Many of the islands are DXCC 'entities' in their own right; others are not, but by meeting particular eligibility criteria also count for credit. Part of the fun of IOTA is that it is an evolving programme with new islands being activated for the first time (currently 935 of the 1172 groups listed have reference numbers).

The basic award is for working stations located on 100 islands/groups. There are higher achievement awards for working 200, 300, 400, 500, 600 and 700 islands/groups. In addition there are seven continental awards (including Antartica) and three regional awards – Arctic Islands, British Isles and West Indies – for contacting a specified number of islands or groups listed in each area. The IOTA Worldwide diploma is available for working a set number of islands in each of the seven continents. A Plaque of Excellence is available for confirmed contacts with at least 750 islands/groups. Shields are available for every 25 further islands/groups.

The rules require that, in order for credit to be given, QSL cards need to be submitted to nominated IOTA checkpoints for checking.

A feature of the IOTA programme is the annual Honour Roll

RADIO SOCIETY OF GREAT BRITAIN
ISLANDS ON THE AIR

IOTA AFRICA

This is to certify that

has satisfied the Council of the Radio Society of Great Britain that
he/she has made a two-way radio contact with amateur radio
stations on island groups qualifying for this award.

Endorsements: No:

IOTA Committee: Date:

which encourages continual updating of scores. This appears in *RadCom* and in the RSGB *IOTA Directory and Yearbook*.

If 'island chasing' appeals to you (and it can become compulsive!), then order the *IOTA Directory and Yearbook* from RSGB which gives full information on the awards, the essential worldwide list of islands, the honour roll and a 'most wanted island list'.

The address is:

RSGB IOTA Programme,
PO Box 9,
Potters Bar,
Herts,
EN6 3RH

136kHz Award

This award is to recognise achievements in both transmission and reception on the 136kHz band, and to stimulate experimentation and station improvment.

The award is available in three categories, with endorsements for additional countries heard/worked.

The basic award is for confirmed contacts on 136kHz with five countries ('entities') from the ARRL DXCC/WAE entity list.

The SWL award is for confirmation of SWL reports from five countries. The SWL award may also be claimed by amateurs working cross-band to stations transmitting in the 136kHz band.

The third category is for cross-band contacts where the station claiming the award has worked five countries by transmitting on the 136kHz band and receiving stations on other amateur bands.

Cross-mode contacts will be allowed for this award. The categories of the award may not be mixed but awards from some or all of the categories may be claimed and endorsed concurrently.

Once the basic award has been claimed, it may be endorsed in steps of each additional five countries worked or heard.

Other awards

DX Century Club Award

This ARRL award is the best-known and most popular DX award in the world. It may be claimed by any ARRL member in the USA and possessions and Puerto Rico, and all amateurs in the rest of the world, who can produce evidence of two-way communication with at least 100 geographical locations known as 'entities' as defined on an official list – these used to be called 'countries' but are now called 'entities' because they are often not countries in their own right. The DXCC entities are given in Appendix 4. Changes take place from time to time and these are noted in *QST*, *RadCom* and other magazines and DX news-sheets.

There are 12 separate DXCC Awards available, plus the DXCC Honor Roll:

(a) **Mixed** (general type): Contacts may be made using any mode since 15 November 1945.

(b) **Phone**: Contacts must be made using radiotelephone since 15 November 1945. Confirmations for cross-mode contacts for this award must be dated 30 September 1981 or earlier.

(c) **CW**: Contacts must be made using CW since 1 January 1975. Confirmations for cross-mode contacts for this award must be dated 30 September 1981 or earlier.

(d) **RTTY**: Contacts must be made using radioteletype since 15 November 1945 (Baudot, ASCII, Amtor and packet radio count as RTTY). Confirmations for cross-mode contacts for this award must be dated 30 September 1981 or earlier.

(e-j) **160 metre, 80 metre, 40 metre, 10 metre, 6 metre, 2 metre**: Contacts must be made on the respective band since 15 November 1945.

(k) **Satellite**: Contacts must be made using satellites since 1 March 1965. Confirmations must indicate satellite QSO.

(l) **Five-Band DXCC (5BDXCC)**: This is available for working and confirming 100 current DXCC entities (deleted entities don't count for this award) on each of the five bands: 80, 40, 20, 15 and 10m. Contacts are valid from 15 November 1945. The 5BDXCC is endorsable for 160, 17, 12, 6, and 2m bands.

(m) **Honor Roll**. This top award, available in mixed, phone and CW variants, requires that you have worked and confirmed a number of DXCC entities within 10 of the total number on the current list, eg if there are 326 current entities, you must have 317 confirmed. There is an even more exclusive variant, the #1 Honor Roll – you must have *all* current entities confirmed!

Full details of the rules and an up-to-date entity list are given in reference [42], obtainable from ARRL direct or from RSGB HQ. Complete details of the award are also given on the ARRL web site (www.arrl.org). Note that QSL cards for this award have to go direct to ARRL and cannot be certified by the RSGB HF awards manager.

Worked (Heard) All Britain

This award scheme, which is open to licensed amateurs and short wave listeners, is based on the geographical and administrative division of the UK. QSL cards are not required, only log entries, and special record books are available to assist in the claiming of awards. Full details and checklists of all the areas and counties for all the WAB awards are contained in the *WAB Book* [43].

Worked All Zones

This award is issued by *CQ Magazine* in the USA for confirmation of contacts with stations in each of the 40 CQ DX Zones (these are given in Appendix 3). Full details of the award and a zone map are available from *CQ Magazine* [44]. A five-band award is also available.

DX ON THE VHF AND UHF BANDS

A common belief seems to be that the HF bands are for DX working and the VHF bands are for local working – basically line-of-sight plus the odd 'lift' to perhaps the Continent. On a superficial and every-day level this view may be true, but further consideration shows how misleading it can be. Auroral contacts on 144MHz between the UK and Eastern Europe are quite common, Southern Africa has been worked from the Mediterranean on 144 and 432MHz, while the number of stations that have worked all continents on these bands by bouncing signals off the Moon, approximately 400,000km away, is gradually increasing. That is real DX by anyone's standards!

There is also an extra and fascinating dimension to DX work above 30MHz, this being the variety of propagation modes possible. HF band DX operation relies almost completely on ionospheric F-layer reflection, but on VHF/UHF modes used include tropospheric refraction, sporadic E-layer reflection, transequatorial E-layer reflection, auroral reflection, meteor scatter, field-aligned irregularity and moonbounce. The last four modes require special operating techniques to make full use of them, and have tended to attract enthusiasts accordingly.

In addition amateur satellites provide reliable intercontinental communication in the 145, 435 and 2400MHz amateur satellite bands (see Chapter 7).

Perhaps, however, the most enticing aspect of VHF/UHF DX is that it is still being explored. Countries continue to be worked for the first time from the UK and there is always a feeling that one might just be in the right place at the right time. This is especially true of the 50MHz band since it became available to all British amateurs on 1 February 1986 – so far (up to the end of 1999) over 175 countries have been worked from the British Isles stretching as far as Australia, there are some 64 countries active in Europe and most countries within Region 1 have now gained access to the band, this being due in large part to the hard work of the UK Six Metre Group.

BASIC EQUIPMENT CONSIDERATIONS

The power now used by serious DXers on VHF and UHF is comparable to that used on HF so, with the widespread use of beam antennas with 10–15dBd gain, really high ERP is possible. Consequently simple antennas such as dipoles or ground planes are rarely used, although sometimes conditions are so good that DX can be worked even with these.

Horizontal antenna polarisation has been used for most serious DX work since experiments carried out in the 'forties showed in general horizontal polarisation was superior

(although just occasionally vertical polarisation was better). The use of vertically polarised antennas for DX work is confined to FM. Some stations have crossed Yagi antennas so that a choice of either polarisation is possible.

The siting of the antennas is just as important as the type of antenna used, and a ground-plane antenna sited on a chimney top clear of all obstructions may give better results than a beam antenna located indoors. The station site is much more important on VHF (and especially on UHF) than on the HF bands. Obviously a hill-top location is best but good results can be obtained in low-lying areas provided all directions are clear of obstructions. In general stations in reasonable locations should be able to work similarly located stations on SSB or CW at ranges up to 300km under any conditions unless there are significant obstructions in between, eg a range of hills. Results do depend on the band used, obstructions having considerably more effect at higher frequencies. A nearby hill which merely causes some attenuation on 144MHz may give rise to considerable difficulty on 432MHz and be sometimes almost an impenetrable barrier on 1.3GHz. However, too much should not be made of this as sometimes propagation is enhanced by 'knife-edge' diffraction round or over an object.

The feeder connecting the equipment to the antenna should have a low loss at the frequency in use, and this is very important at VHF and especially at UHF. For example, standard TV coaxial feeder has a loss of about 3dB per 10m at 433MHz while even UR67 (which is considered a high-quality cable at HF) will lose about 1.3dB per 10m at this frequency. Low-loss coaxial cables for UHF work are available and should be used if significant feeder lengths are necessary; these ultra-low-loss feeders are expensive along with their associated connectors but are well worth the investment if you want to work the ultimate DX.

Because the external noise level is much lower on VHF and UHF than on HF, the noise factor of the receiver front-end is important if good results are to be obtained and many enthusiasts fit low-noise preamplifiers to their transceivers. However, this should not be done indiscriminately because it may make the receiver more susceptible to cross-modulation from unwanted strong signals. Furthermore, the receiver noise factor may already be satisfactory, especially if the station is located in an electrically noisy urban area.

It must be emphasised that a beam antenna is the ideal form of noise-less RF amplifier, and therefore the best antenna possible should be erected before any money is spent on improving the receiver front-end (or buying an external amplifier for the transmitter).

OPERATING PRACTICE

In general this is not very different to that found on the HF bands. Pile-ups can occur in good conditions and because these can be short-lived it is good procedure to limit the contact to the exchange of signal reports and location details. Such details as equipment, name etc can be left to the QSL card. In this way more stations get a chance to work the DX station.

When conditions are good, call CQ sparingly, if at all, unless a very good signal is available. If a DX station does appear it will probably be called by a string of UK stations. A UK station with an average signal calling "CQ DX" will simply cause interference and not achieve the desired result. The old adage "If you can't hear 'em you can't work 'em" is just as true on VHF and UHF, and all too often one can hear stations calling "CQ DX" on the same frequency as distant stations which they cannot hear. Listen carefully in several directions if you have a rotary beam before calling CQ DX!

One feature of VHF/UHF operating which is rarely found on the HF bands is the use in Europe of *calling frequencies* specific to each mode. The idea is that these frequencies provide a 'meeting place' for operators using the same mode. In the UK, the 50 and 144MHz band plans identify particular calling frequencies (eg 50.150 and 144.300MHz for SSB, 50.090 and 144.050MHz for CW). Once contact is established, stations change to another frequency (the *working frequency*) so that others can use the calling frequency.

This system works well in poor-to-moderate conditions, when activity appears to be low. Operators leave their receivers tuned to the calling frequency, obviating the need to constantly tune the band as used to be the case in the distant past when transmitters were crystal controlled. But even though the band may seem empty entire contacts should not be conducted on the calling frequency as there may be other distant stations calling.

When conditions are good or a contest is in progress there is little point in using the calling frequencies, and in these cases it is probably better to call CQ elsewhere. For example, in a major sporadic-E opening on 144MHz the congestion around 144.300MHz is often so severe that it is difficult to complete an SSB contact. This is further aggravated by the poor operating procedures adopted by some amateurs, such as the use of unfamiliar phonetics causing longer than necessary contacts.

LOCATION INFORMATION

On the HF bands location details are given by reference to well-known cities but on VHF/UHF a locator system is used. This is necessary for contest purposes where scoring is based upon distance and similarly for awards and records. The approved IARU locator system was designed by Dr John Morris, GM4ANB [45, 46]. It was initially known as the *Maidenhead locator system* since it was tabled at a meeting of IARU Region 1 VHF managers in that town in April 1980. It came into official use on 1 January 1985.

The system is based upon three sizes of what are commonly (if incorrectly) called *squares*. The largest, called *fields*, are each 20° wide from west to east, and 10° high from south to north, thus dividing the Earth into an 18 by 18 grid. These fields are given two-letter indices, 'AA' to 'RR', with the first letter specifying the longitude and the second the latitude, the origin being at the South Pole at 180°W. Thus the field AA runs from 180° to 160°W, 90° to 80°S. North of this is AB, east is BA and so on. The field covering most of the UK (50° to 60°N, 20° to 0°W) is then IO. These fields are divided into 100 squares, each 2° wide and 1° high, labelled from '00' in the southwest corner to '99' in the northeast.

The final division is of these squares into a 24 by 24 grid of sub-squares, each one being 5′ wide by 2.5′ high. The sub-squares are labelled using two letters, again the first specifying the longitude and the second the latitude, starting from 'AA' in the southwest corner, and running to 'XX' in the northeast. A full locator thus consists of two letters, two numbers and two letters, with a typical reference being IO91IP.

Fig 11 shows how the locator is built up from fields, squares and sub-squares. It may be noted that all of the longitude-defining characters – the first, third and fifth – run from west to east, while the latitude defining characters – the second, fourth and sixth – go from south to north. In addition, at all levels the east-west size, in degrees, is always twice the north-south size.

In distance terms, each sub-square is about 4.6km from south

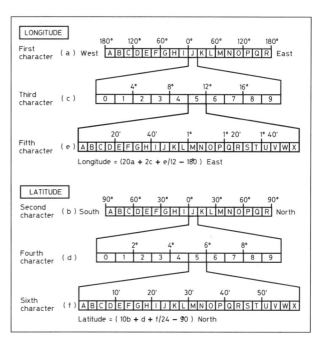

Fig 11. The locator system

to north. The east-west size varies with latitude, but in the middle of Britain (55°N) it is about 5.3km. If you take the accuracy of the system as being the farthest you can get from the middle of a sub-square without actually leaving it, this gives a maximum error of about 3.5km (at 55° latitude), which is quite adequate for most normal operation.

You can find out your locator by consulting one of the Internet web sites which will calculate your locator from your latitude and longitude. If you don't yet have Internet access, the following step-by-step procedure can be used to calculate your locator, using nothing more than pencil, paper and a simple calculator.

First of all you must find your latitude and longitude. These can be derived from the popular 1:50,000 series of Ordnance Survey maps but you should note that the blue grid lines do not run exactly north-south and east-west. Along the top and bottom margins of each map the longitude is given and along either edge the latitude, both at 1′ intervals. In the main body of the map the corners of 5′ by 5′ squares are identified by faint blue crosses. By pin-pointing your location within the appropriate square and pencilling in the sides you will be able to determine your latitude and longitude with great accuracy by proportion. The latitude should be rounded down to the next half of a minute south and the longitude to the next whole minute west.

Having established your latitude and longitude the procedure is as follows:

1. First deal with the longitude. Convert the longitude to decimal degrees. This is done by dividing the minutes part by 60 and adding the result to the degrees part.
2. If you are east of Greenwich, add 180.
3. If you are west of Greenwich subtract the value from 180.
4. Divide the result by two.
5. Now divide the value obtained by 10, and note the figures to left of the decimal point. These give the first letter of the locator, on the basis of 0 = A, 1 = B, 2 = C and so on, as shown in Table 1.
6. Multiply by 10, and note the single digit immediately before the decimal point. This is the third character of the locator.

Table 1. Number-to-letter conversion table for hand calculation of a locator

0	A	4	E	8	I	12	M	16	Q	20	U
1	B	5	F	9	J	13	N	17	R	21	V
2	C	6	G	10	K	14	O	18	S	22	W
3	D	7	H	11	L	15	P	19	T	23	X

7. Take just the fractional part of the value (the part to the right of the decimal point), and multiply by 24. The figures to the left of the decimal point give the fifth character of the locator, once again using Table 1.
8. Now follow a similar process for latitude. Convert the latitude to decimal degrees.
9. If you are north of the equator, add 90.
10. If you are south of the equator subtract the value from 90.
11. Divide by 10, and take the second letter of the locator from the digits to the left of the decimal point, using Table 1.
12. Multiply by 10, and note the digit to the left of the decimal point. This is the fourth character of the locator.
13. Take the fractional part, and multiply by 24. The figures to the left of the decimal point give the last letter of the locator, again using Table 1.

As an example, take Edinburgh Castle. From the map, its latitude is somewhere between 55°56.5′N and 55°57′N, and its longitude between 3°11′W and 3°12′W. Rounding to the next half minute south and whole minute west means that latitude 55°56.5′N and longitude 3°12′W are taken. Verify, using the above method, that its locator is IO85JW.

The intermediate results you should obtain are:

1. 3.2
2. No action
3. 1768
4. 88.4
5. 8.84, ie 'I'
6. 88.4, ie '8'
7. 9.6, ie 'J'
8. 55.94166667
9. 145.9416667
10. No action
11. 14.59416667, ie 'O'
12. 145.9416667, ie '5'
13. 22.59999984, ie 'W'

On the air the system should be called simply 'locator'. For example, you might say "My locator is IO83QP", or ask of another station, "What is your locator?". On CW the recommended abbreviation is 'LOC', so that the CW equivalent of "What is your locator?" is simply "LOC?".

The various divisions of the locator, fields, squares and sub-squares, have just those names. A VHF DX chaser might claim to have worked 300 'squares', or, for the real enthusiast, '20 fields'.

While the six-character locator system is accurate enough for VHF/UHF use more accuracy may be necessary for identifying locations and calculating distances and antenna bearings when operating in the microwave bands above 24GHz. At the IARU Region 1 conference in Belgium in September 1993 an RSGB proposal for an extended locator system was approved. The 5′ wide by 2.5′ high sub-squares are divided into 100 'micro-squares' each 30″ wide by 15″ high labelled from '00' in the southwest corner to '99' in the northeast. Thus a full micro-locator would be IO91IP37 and at 55°N latitude, the maximum error is reduced 10-fold to about 350m.

THE VHF/UHF BANDS

The term 'VHF' applies to frequencies between 30 and 300MHz, and 'UHF' normally means 300MHz to 3GHz, but in amateur practice frequencies above 1GHz are normally termed 'microwaves'.

The ITU VHF/UHF allocations for Region 1 are 144–146MHz and 430–440MHz. Most Region 1 countries permit 144–146MHz operation but there are several with restricted 430–440MHz Amateur Service allocations (see Appendix 7). Notwithstanding ITU allocations, national administrations may permit amateur service operation in additional bands. Starting with the UK in the early 'eighties, amateurs in the majority of Region 1 countries have been granted permits to operate in the 50–52MHz band. In some countries, such as the UK, normal licensing conditions apply but in others there are restrictions on the width of the band, transmitter power, modes, antennas and times of operating.

Amateur allocations in Regions 2 and 3 are broadly similar to those in Region 1, with the addition of an ITU allocation between 50 and 54MHz. In Region 2 the bands 220–225MHz and 902–928MHz may also be allocated for amateur use. It should be noted that many of the amateur bands above 30MHz are shared with other services, as shown in Appendix 7.

The four VHF/UHF bands of particular interest to UK amateurs are those at 50, 70, 144 and 430MHz, commonly referred to as 'Six', 'Four', 'Two' and 'Seventy' respectively. The 70MHz band is a special allocation granted by the UK administration in the 'fifties. The initial allocation was only 200kHz wide, but was later increased, then subsequently decreased to the present 70.0–70.5MHz. The two higher bands are those allocated to the Amateur Service by international agreement.

Because of the wide variety of activities taking place on VHF and UHF, the band plans are necessarily much more complex than those for the HF bands. The main aim of the band plans is to separate incompatible transmission modes, and so let everyone get on with the business of communicating by his or her chosen method. The outlines of the 144 and 430MHz band plans are laid down by international agreement in the IARU after much debate and discussion. The 50 and 70MHz band plans were devised by the RSGB, and follow a similar outline to those for the higher bands.

Although the band plans are purely voluntary arrangements this may soon change. Calling CQ on CW on 145.7MHz is quite legal within the terms of the amateur licence, but is very unlikely to result in a contact, and certainly will not make any friends. Similarly, the beacon sub-band may seem completely empty and an ideal spot for a contact, but even a short transmission there could well cause serious annoyance to somebody many miles away who is intently listening in the noise for a distant beacon. There is room on the bands for all of the many activities, provided a little care, consideration and respect for the interests of other operators are shown. The band plans act as a guide in this. For example, by having specific calling frequencies for data communications, packet operators can concentrate their listening and calling on known frequencies, and avoid interference both to and from other modes. All VHF/UHF users are recommended to keep a copy of the band plans to hand near the rig.

50MHz (6m) band

28MHz has already been described as having many of the characteristics of a VHF band. By the same token, 50MHz can take on many of the features normally associated with an HF band, especially at times around the sunspot maxima when F2 propagation is often possible and worldwide contacts may be made. 50MHz is the only band in the spectrum that exhibits all known propagation modes, ie 'tropo', ionoscatter, sporadic E, aurora, F2, TEP and EME.

One phenomenon noted during the peak of sunspot cycle 22 was anomalous beam headings, particularly on the path between the UK and Japan. Most openings were scatter paths and were in the direction of the Indian Ocean from the UK but this ionised path showed up several times and produced excellent results. It was also discussed during cycle 22 as a possible TEP tunnel effect, where the signal enters the TEP ionised zone and is reflected within the zone and exits at some other point. Just like sporadic E the phenomena will be difficult to prove. 28.885MHz has become the recognised worldwide 'talkback' frequency on which the dedicated enthusiasts meet to discuss such phenomena, conditions and openings.

Now we enter the Internet – since cycle 22 great technological advances have taken place and the Internet can, and does, provide an up-to-date worldwide platform for discussions, skeds, reports and much more. The worldwide packet cluster is now available via the Web which means more DX can be worked during the short openings on the band due to the alerts being posted on various DX pages.

The dedicated 50MHz pages on the Internet [65, 66] also provide us with up-to-date news items, beacon news which is revised weekly, and European and Asian TV frequency listings that can alert the DX operator to where the opening is going to occur. There are also recorded audio clips, expedition news, instant aurora warning networks that even send you an e-mail to let you know what is happening and coming soon will be the introduction of e-mailed QSL cards – this is a real technological breakthrough which will help the true DXer get the most out of the VHF/UHF bands. The Internet is not a substitute for DX despite its linking to VHF sources, but a tool to work alongside with to gain the ultimate DX.

As mentioned earlier, over 175 countries have been worked from Britain, and even several UK Class B licensees have achieved the famous ARRL DXCC certificate for working and confirming 100 entities on the band, but alas only 225 of these prestigious awards have been issued throughout the world.

The IARU Region 1 Band Plan for 50MHz is given in Appendix 7, but it must be appreciated that amateurs in some countries cannot adhere to it due to local restrictions – France and Andorra which have allocations above 50.200MHz fall into this category. Although 50.110MHz is the agreed intercontinental calling frequency, very many operators selfishly use it for local contacts, often obliterating weak DX stations from outside Europe. Unfortunately newcomers to the band may pick up this bad habit from the more-experienced operators who should know better. 'Intercontinental' means 'outside of your own continent' and not continental contacts. Although the band plan is a gentleman's agreement it is likely that very soon it will be written into BR68 as a condition of the UK/CEPT Amateur Radio Licence.

The UK Six Metre Group, along with other worldwide 50MHz organisations, has adopted an operating code of practice – this is very useful and can be found at: http://www.uksmg.org.

70MHz (4m) band

It is unfortunate that because the 70MHz band is available in only a very few countries, its full potential cannot be recognised. However, there are several operators in other countries

A multi-mode transceiver for 70MHz – the Yaesu FT847

who have suitable receiving equipment and are interested in cross-band working. Most commonly frequencies around 28.885MHz are used for talkback, but other bands have also been employed. There have been a few 70 to 144MHz cross-band contacts made from the UK to the Continent by meteor scatter, where the more modest power and antenna gains in use on the lower band are compensated by the greater propagation efficiency. Skeds for MS contacts are usually made on the 14MHz VHF net, as described later in this chapter. The ultimate cross-band achievement is to complete a 50 to 70MHz contact with a station on the other side of the Atlantic, and very few such contacts have been made.

When arranging cross-band skeds, care should be taken over the choice of the 70MHz frequency as this band is used for broadcasting and other purposes in various parts of the world. The advice of the distant station should be sought on this point.

The future of the 70MHz band now looks brighter. In its review of the 29.7–960MHz part of the spectrum the European Radiocommunications Committee [47] concluded: "It would therefore seem possible to agree a limited amateur transmitting facility of at least 100kHz centred on 70.2MHz. It is also hoped that the existing beacon network between 70.0 and 70.15MHz can be maintained and extended". The time scale for adopting such proposals is likely to be long as the ERC has suggested its recommendations be implemented by 2008. However, as was the case with 50MHz, some countries may be prepared to grant special permits to their amateur licensees in advance of any general ITU agreement at a future World Radio Conference. Poland and South Africa have indicated that they are willing to allocate 70MHz on a secondary basis from the year 2000 and, with the recently introduced SSB equipment from major manufacturers such as the FT847, 70MHz will now grow like 50MHz did 15 years ago. Slovenia (S5) is currently active and has been worked on AM/FM/CW and SSB from the UK.

144MHz (2m) band

This was arguably the most popular of all amateur bands in the 'eighties, largely because it was the lowest band allocated to 'VHF-only' licensees in many countries; most countries have now allowed lower bands like 50MHz to be used by VHF-only licensees, and so activity has declined. However, under good conditions it is possible to make contacts from the UK deep into Europe, to North Africa and down to the Canary Islands in the North Atlantic. When an opening of one sort or another occurs the high level of activity ensures there will usually be someone on the other end of the path keen to make a contact. As well as the normal CW, SSB and FM, many specialist communication techniques and propagation modes are in regular use, and this is reflected in the complexity of the usage part of the band plan.

The bottom 35kHz is devoted to moonbounce and it is

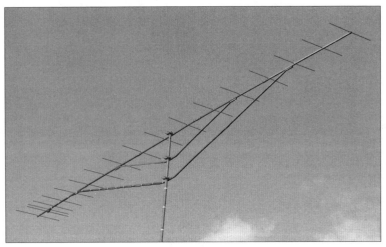

A long Yagi antenna for DX work on the 2m band

important that this region be kept clear unless the Moon is well below any European horizon. It is in regular use by stations using high-gain antennas and sensitive receivers to copy weak signals which may be inaudible on most normal equipment. The CW-only segment extends up to 144.150MHz, with a calling frequency at 144.050MHz. Most CW activity takes place between 144.035 and 144.075MHz, and during openings, especially of the auroral type, there are usually many good DX contacts to be made in this part of the band. In particularly intense auroras the entire CW exclusive section can become quite congested. There is a CW activity period from 2000 local time every Monday evening.

The most popular DX mode is SSB for which the calling frequency is 144.300MHz. To avoid overcrowding a substantial change of frequency is recommended once contact has been established on 144.300MHz. When activity is high, such as during openings and contests, the concept of a calling frequency tends to be dropped and stations may be heard calling CQ anywhere from 144.15 up to 144.35MHz or higher. In these circumstances contacts are made in a similar manner to those on the HF bands. Cross-mode CW/SSB working is rare but can prove useful in getting through interference. The SSB and CW exclusive segment continues up to 144.4MHz, the higher frequencies being more popular for local SSB working and nets, such as those involving Worked All Britain enthusiasts. The bottom 500kHz of the band includes several meteor scatter and field-aligned irregularity calling frequencies and working segments, which should be avoided by stations not participating in such activities.

The beacon band falls between 144.400–144.490MHz, and most European countries have one or more beacons in this region, as shown in Appendix 8. These beacons are used by keen DXers to help evaluate propagation conditions, and by careful monitoring may often give a valuable forewarning of an impending opening. This section should be regarded as strictly 'out of bounds' for transmitting.

The non-channelised all-mode segment runs from 144.5 to 144.800MHz and it is here that the specialist communication modes, including RTTY, fax and SSTV, are to be found. Each of these modes has a specific calling frequency which should be avoided by other operators.

The sub-segment 144.800 to 144.990MHz is designated for data modes, mostly the DX cluster network and packet radio in practice.

The top megahertz of the band contains allocations for FM simplex, repeaters and satellite communications, which are dealt with in Chapters 6 and 7.

430MHz (70cm) band

This used to be very much an enthusiasts' band, most of the equipment being home-built, but now there is a wide selection of excellent commercial equipment available, from all-mode transceivers through high-power amplifiers to antennas. Although path loss, which is frequency dependent, is greater than on the lower bands, this is compensated for by the higher antenna gains in common use. Most terrestrial contacts are made by tropospheric propagation, although the better-equipped stations complete auroral and meteor scatter contacts. Sporadic-E mode does not exist at 430MHz, the highest reported sporadic-E frequency being in the 220MHz region in the USA.

The anatomy of the DX section, 432.00–432.99MHz, is very similar to that of 144MHz, with CW, SSB, all-mode and beacon sub-bands, as well as special allocations for moonbounce and data modes. Many of the comments in the preceding section apply equally to 430MHz. Operating practice is also similar to that on 144MHz.

A common practice during lifts is for operators who are interested in 144MHz, 430MHz and perhaps 1.3GHz to call on the lower band and announce that they are "QRV on 70cm" or 23cm or both. If the other station is interested then a frequency on the higher band is agreed. This procedure has the advantage of allowing beam headings to be accurately determined, which is particularly useful when the narrow beamwidths in common use on the higher bands are considered. More and more new technology is appearing on the market place these days with DSP (digital signal processing) being just one new addition – DSP is extremely useful for weak-signal working on any VHF/UHF band.

1.3GHz (23cm) band

The 1.3GHz band is the lowest-frequency 'microwave' band but is included here as a borderline case, as it exhibits the characteristics of both the VHF/UHF bands and the higher bands. Propagation losses are higher than on 430MHz, and unless relatively efficient equipment is used results can be rather disappointing. Nevertheless, many dedicated 1.3GHz operators have made contacts of well over 1500km and some have worked over 100 locator squares.

Commercial equipment is readily available for this band but is of rather low power (a few watts). For serious DX work some form of PA is essential and raising the power output to 20–30W will effect a useful improvement in transmitter range in average conditions. Such power levels are easily obtained from the 2C39 family of planar triodes [48]. For the more ambitious the Siemens YL1050 can produce 600W RF output from 40W drive in linear mode [49]. Some Russian valves have now also appeared on the surplus market and, although they are quoted as good up to 1GHz, they will work well but at a reduced efficiency on 1.3GHz. These valves are triodes and need very little in the way of a power supply – the largest of them is called a 'GS35B' and has an enormous anode cooling fin. This valve will deliver 1500W at its 1GHz rating.

On the receive side, commercial transverters usually have some form of RF preamplifier built in, but most converters do not, and certainly require an external preamplifier for good results. Many stations employ masthead preamplifiers and these can often be of great benefit when long runs of feeder cable are unavoidable.

As with the lower bands, a good antenna system is of great benefit. Corner reflectors, and small Yagis and dishes, are unlikely to give very good results in average conditions. The more serious operators use multiple, stacked Yagi arrays or dishes (greater than 2m diameter). Since coaxial cables are considerably more lossy at 1.3GHz than on lower frequencies, the best-quality cable you can afford should be used. For example, an 11m run of RG214 cable will have a loss of 0.78dB at 144MHz, 1.52dB at 432MHz but almost 3dB at 1.3GHz. This means that only one-half the power generated by the PA will reach the antenna and that your receiver noise figure will be degraded by 3dB, too. So-called 'low-loss' UHF TV cable should not even be considered at 1.3GHz, and Heliax feeder is a good choice for this band.

Operating on 1.3GHz is a mixture of making contacts directly on the band and setting them up from 432MHz. The newcomer is strongly advised to seek 1.3GHz contacts from 432MHz initially, as random calls on 1.3GHz outside contests, activity nights or lifts will most likely go unanswered. Also, arranged contacts have the advantage that antennas can be aligned accurately beforehand.

The IARU Region 1 band plan is quite detailed. As far as DX activity is concerned, most operation is around 1,296.200-MHz, referred to as the *narrow-band centre of activity*. Two points are important, however. The segment 1,296.000 to 1,296.025MHz should be left clear for moonbounce operating, and no transmissions should be made in the beacon segment (1,296.800 to 1,296.990MHz). Contrary to the practice on the lower bands, all activity, local or DX, SSB, CW or FM, is horizontally polarised.

More information on the 1.3GHz band is given in the microwave DX section later.

VHF/UHF PROPAGATION MODES

As mentioned previously, signals at these frequencies can be propagated a number of different ways and it is more convenient to discuss VHF/UHF propagation by mode rather than by band, especially as enthusiasts often tend to be attracted to particular modes rather than one particular band.

It should be noted that anomalous propagation conditions which affect one band will not necessarily affect the others. Furthermore, an 'opening' will usually have decreasing effect, the higher the frequency, but this is also not necessarily true and in many cases operators have found signals at 432MHz louder than the 144MHz ones during a 'lift'. Also it has been noted that on several occasions openings on the higher bands outweighed the lower bands due to thermal inversions and tropospheric ducting.

Knowledge of VHF/UHF propagation and when openings are likely to occur is essential for the DX operator.

Also most useful is an all-mode receiver which can tune the whole band for indications of good propagation; most modern equipment now has this built in as an MUF (maximum usable frequency) monitor. The following sections discuss briefly each mode from an operating point of view. A much more detailed account of the mechanisms of VHF/UHF propagation is given in reference [50] to which the interested reader is referred.

Tropospheric propagation

In the early days of VHF it was thought that propagation was only possible over line-of-sight paths and at distances beyond

this attenuation was rapid; however, experience in the early 'thirties showed that this was not always so and sometimes much longer distances could be covered. These effects were soon related to atmospheric conditions and it was realised that radio waves were being bent back to Earth by the troposphere, hence the name.

The relation between weather and tropospheric propagation is complex but in general it may be said that the main requirement is a temperature or a humidity inversion. Normally the temperature of the atmosphere decreases with increasing height above Earth, but in abnormal circumstances it may increase over part of the distance so that VHF waves which are normally lost in space are bent back to Earth, sometimes at a range of hundreds of kilometres.

Since first published in the early 'seventies *Dubus Magazine* has included 'Top Lists' which now include the bands from 50MHz up. Examination of the 144MHz list for mid-1994 shows the average best distance claimed (usually known as 'ODX') by the top 80 stations to be just over 1500km, with four operators claiming contacts in excess of 2000km.

Clearly the DX enthusiast will wish to forecast the good conditions. Two essential aids are:

1. *Weather map (synoptic chart)*. This may be found on television or in a newspaper. The former is preferred as it is more likely to be up-to-date. Be careful to distinguish between actual maps showing the situation a few hours before and forecast maps, which are not necessarily accurate, as the weather is notoriously difficult to predict.
2. *Barometer/barograph*. Some means of observing changes in atmospheric pressure is required. A barometer, or better still a barograph, is suitable, though a surplus aircraft altimeter will be satisfactory, providing this is of the correct type (ie not a radio altimeter).

What to look for

Good conditions are usually associated with stable weather patterns and this means areas of high pressure (*anticyclones*). Therefore, look for anticyclones on the map and a barometer reading high. It will soon be found that the appearance of a high-pressure system does not produce good conditions, but when it starts to decline things may start to happen, especially if the anticyclone is short-lived. However, good conditions can occur when the atmospheric pressure is steady or even increasing.

It must be added that anticyclones may form over the British Isles and decline without any significant effect on conditions. Conversely DX can appear when the weather map appears to be a series of depressions, though an opening under these conditions is usually short lived. A complete understanding would require a study of the atmosphere in great detail.

How to identify enhanced conditions

The simplest method is to monitor known distant signals and the most useful of these are the beacons. Appendix 8 gives a list of European beacons in the 50MHz to 1.3GHz part of the spectrum operational at the time of going to press. A beacon list is also published in the annual *RSGB Yearbook*. A quick check of signal strengths of the beacons on the bands of interest will usually give a clue to the state of conditions. This is especially useful on 144MHz where many beacons are operating. It should be added that the beacons may be observed at enhanced strength when no other signals can be heard, either because there is no activity or because stations which are operating are beaming in other directions.

Other distant stations, eg television transmissions from Europe, may give a clue to conditions. These used to be difficult to identify but a detailed study has taken place during the last 10 years and now the 48 and 49MHz TV video signals can be identified by their own personal offset frequency. Sometimes unusual patterning on the screen of the domestic UHF TV set or an announcement from the broadcasting organisation concerning 'Continental interference' may indicate that an opening is taking place.

Some points to observe:

1. All bands are not necessarily affected equally. It is possible for propagation to be enhanced at 144MHz but not at 432MHz and vice versa.
2. Propagation does not vary uniformly. It can be good to, say, the south but poor to the north simultaneously.
3. Skip effects are noticeable at times. It can be possible for a station 300km to the west to be working DX several hundred kilometres to the east which is quite inaudible. This is very frustrating and the only thing to do is to monitor carefully and hope for a change.

Sporadic-E (Es) propagation

This is the name given to intense ionisation of the E-layer which happens sporadically. The mechanism has been studied for many years and the general conclusion seems to be that there is no single cause of this phenomenon [51]. In the northern hemisphere, Es is most likely to occur between May and August with a minor peak in late December, and in the southern hemisphere Es is most likely to occur between November and January with a small peak in June/July. It is a reverse pattern and maybe solar driven with other additions.

Es gives rise to propagation over distances from a few hundred to over 2000km at frequencies up to about 200MHz, so that it affects the 50, 70 and 144MHz bands. However, the ionisation has to be very intense for 144MHz propagation to occur such that on occasions when Es may be observed continuously on 28MHz for several days, it may reach 144MHz for only a few minutes during this time.

Particularly in June and July, the 50MHz band is often full of Es signals on a daily basis, sometimes supporting multi-hop propagation from the British Isles to the Eastern Mediterranean, the Middle East and across the Atlantic to North America. For example, there was a superb multi-hop Es opening to North America in June 1994 when JY7SIX (Jordan) made a multi-hop contact with W4 at 9600km. This event occurred within about a year of sunspot minimum so F-layer propagation can be discounted.

Several other major openings have occurred on 'Six' in recent years during June and July between Sweden and Japan, Greece and Japan. Several rare DX stations located in the Caribbean were reported as being worked into the UK during the summer months. The 70MHz band is often open for lesser periods in these months but still provides a stepping stone and MUF monitor to 144MHz openings.

The most interesting Es band is 144MHz and since openings can be very short, from a few seconds to at most a few hours, it is desirable to know when Es is likely. This involves careful monitoring of signals in the approximate range 28–118MHz and a general-coverage receiver, preferably able to receive both AM and FM, is desirable. Programmable scanning receivers (eg Icom IC-R7100) are ideal for this purpose, but most current transceivers now incorporate this frequency span (FT847, FT100, IC746, IC706; even the new FT90R with

its air-band VOR frequencies is a very useful addition to any shack for 144MHz Es monitoring).

Various groups of signals can be monitored as follows but please be careful when monitoring some of these frequencies, especially about storing them into any memory locations as this is not legal. Should you be visited by the Radiocommunications Agency serious consequences may arise if unauthorised frequencies are found stored in your equipment.

28MHz band. European signals will always be audible when Es is around. When ionisation is intense, signals are extremely strong and the shorter the distance, the more likely it is that higher frequencies such as 50MHz are affected.

Band I TV video signals (48–68MHz) can be useful but are often difficult to identify, especially the vision on air due to multiple signals being received. This can be overcome by using a receiver with SSB reception – the offset frequency can clearly be monitored and recognised as all the European TV Band 1 frequencies are now listed on the Internet (http://user.itl.net/~equinox/tv.html). Small portable TV sets are useful monitors if you can get one but most modern British sets are UHF only so they are useless for 50MHz MUF monitoring. It is now possible to buy multi-standard portable TV receivers in the UK and even surplus spectrum analysers are becoming so realistically priced that their use can be put to other situations.

68–73MHz. This band is still used by a few countries in Eastern Europe for FM broadcasting but all have agreed to discontinue transmissions in the future. When Es is present, many of these stations may be heard (eg Radio Gdansk in Poland on 70.31MHz). If they are very strong, careful attention should be paid to 144MHz, especially if they are audible as early as 0630 GMT. If they are weak or fading, 144MHz is unlikely to open to these areas but it may do to somewhere else.

Band II (88–108MHz) is used by most European countries for VHF FM broadcasting. The sudden appearance of DX stations from Italy or Spain indicates that 144MHz may well open in those directions. In some cities in the British Isles several dozen UK FM stations are always audible and the appearance of a DX station may pass unnoticed. There can be some confusion since there may be local stations serving ethnic communities broadcasting in foreign languages (eg London Greek Radio on 103.3MHz).

Operating notes

As 144MHz Es openings are often very short-lived, efficient operating is essential and an exchange of report and QTH locator is the most that should be attempted. It is possible for a station to appear, calling CQ, be worked and disappear totally in less than a minute. Avoid using the SSB calling channels – there is no hope of establishing an Es contact and then changing frequency as is normal practice with other propagation. All too often a station fades out during a contact and there is usually nothing to be gained by waiting in the hope it will reappear – it is best to 'cut your losses' and look for someone else.

The ionised reflecting layer, which may be relatively small in area, can move very rapidly and also tilt. For example, Maltese stations may be very strong one minute then fade out to be replaced by Sicilians which are in turn replaced by Sardinians – all in the space of a few minutes.

CQ calls should be brief, perhaps 10–15 seconds, and used with discretion – if an operator is in an area of high activity, any DX station is likely to have many UK stations calling him and be unlikely to reply to a call. Conversely, if an operator is isolated, he may have an opening all to himself and a pile-up of DX may call him. In this case one will want to work as many as possible, so just exchange reports with each station but give the locator on every third or fourth contact – the waiting pile-up will have heard it already.

However, always remember that sporadic-E can be very localised and it is possible for a DX station which is S9 to one operator to be inaudible 10km away – and that a CQ call may well pay dividends, especially just as the band is opening.

Auroral propagation

The natural phenomenon known in Europe as the *northern lights* has been studied and used as a propagation medium by amateurs for more than 50 years. Radio auroral events cause great excitement

The strange light of an aurora

among VHF operators, who are able to work DX stations at distances far exceeding their normal tropospheric range. During strong events stations located in southern England are often able to make contacts with Finland and Estonia by beaming their signals towards the auroral reflecting zones.

All auroral openings occur after a solar flare has released energy from the Sun. The GB2RS news bulletins in the UK and Internet news pages often give details of sunspot activity and flares that are usually associated with these eruption holes on the disc of the Sun. A solar flare releases tremendous amounts of energy across the entire electromagnetic spectrum from X-rays to radio waves.

The Sun emits ionised gas continuously and this is termed the *solar wind*. During flares bursts of energetic charged particles stream outwards from the Sun and spiral towards the Earth via the solar wind. These particles are divided by the Earth's magnetic field and then follow the field lines to regions known as the *auroral zones*. These zones are oval shaped and extend outwards from the poles to a radius of 23° on the night side of the Earth and to 15° on the daylight side. Visual auroral sightings indicate where the charged particles impinge on the Earth's upper atmosphere, ionising the E layer at a height of 110km.

The number of auroral openings in any year is dependent on the solar activity. Some areas of the Sun can remain active for several weeks, causing repeats of events 26 to 28 days after the initial aurora. This is due to the period of rotation of the Sun and events like these are known as *solar repeats*.

Warning signs of impending auroral events

Large increases in the solar noise levels on 95, 136 and 225MHz can be measured during flares. These sudden ionospheric disturbances are often followed by short-wave fade-outs on the 14–28MHz bands, causing disruption of amateur and commercial HF communication (see p46). This is caused by cosmic particles and X-rays ionising the D-layer which absorbs rather than reflects radio signals. These particles complete the 150 million kilometre journey from the flare region on the Sun in less than 15 minutes. It should be noted that short-wave fade-outs occur which are not accompanied by auroral events, but fade-outs followed by large magnetic disturbances almost always signal an impending aurora.

A good and reliable indication of a forthcoming auroral event is known as *pre-auroral enhancement* and is familiar to HF band operators. This effect is particularly obvious during periods of poor propagation such as in the summer months at sunspot minimum. A typical example occurred on 6 April 1995

when, during a sustained period of mediocre conditions on 14MHz, the band became full of very strong signals from Australia and New Zealand with the propagation akin to that during sunspot cycle peaks. The following day HF band conditions had collapsed and there was a very intense aurora enjoyed by VHF operators on the 50, 144 and even 430MHz bands.

Changes in the Earth's magnetic activity are measured in three directions – horizontal, vertical and declination. All three change prior to and during auroral events but amateur observers have noticed that the angle of declination measurements give the first geomagnetic auroral warning signs. Declination is the angle measured between geographic and magnetic north. Fig 12 shows the magnetic disturbance recorded on 1 May 1978 on a chart recorder connected to a magnetometer set up in the garden of the home of Dr Gadsen of Aberdeen University. The chart shows a fairly quiet initial period with small normal daily changes in magnetic activity, and then at midday it shows the start of a *sudden commencement geomagnetic disturbance*. These magnetic effects are measured in professional observatories and a daily equivalent planetary amplitude figure (Ap) is issued which gives the average level at several different locations. These changes in magnetic activity which take place on a worldwide basis are caused by the slower-moving particles which take one or two days to complete the journey from the Sun to the Earth's magnetosphere. Large-scale radio auroral events took place on 1 May 1978 in three distinct phases – in the afternoon, late evening and again after midnight – the chart in Fig 12 shows the changes in magnetic activity preceding and accompanying the aurora.

It is quite easy to make a simple magnetometer to monitor deviations in the Earth's magnetic field using readily available components. The sensor can be accommodated in a 35mm film spool case containing a small bar magnet sitting in some oil and suspended by a cotton thread. Full details of the construction of a sensor and a simple magnetometer will be found in reference [52].

Shortly after the commencement of a magnetic storm the comparatively slow-moving particles ionise the E layers and align along the Earth's field lines. VHF radio signals beamed towards the auroral regions are reflected and refracted by the moving area of auroral ionisation. This moving reflector causes frequency shift and spreading, making all auroral signals sound distorted and difficult to copy. Morse signals are transformed into a rough hissing note and sideband voice transmissions vary from a growl to a whisper. The amount of frequency change on

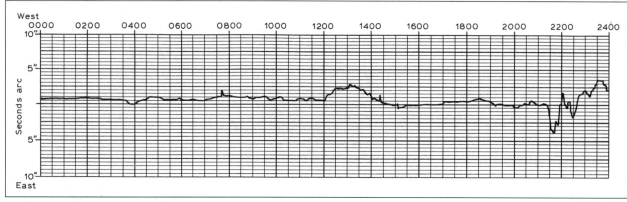

Fig 12. Magnetometer recording taken on 1 May 1978

signals varies proportionately to the frequency band used – this effect is known as *Doppler shift* and can be as much as 1.5–2kHz HF or LF of the actual transmit frequency on 144MHz.

Auroral operation

The beacon stations closest to the auroral zones are usually the first auroral signals to be heard in the south. The Swedish beacon SK4MPI (144.412MHz) and the Lerwick beacon GB3LER (144.445MHz) in Shetland are good auroral indicators as is the new Faroes beacon OY6SMC (50.035MHz). Auroral events are usually first noticed by amateurs in Norway, Sweden, Finland and northern Russia, who will be on the band making contacts before the auroral reflections extend to the south. Due to their northerly locations, they see far more visual auroras and participate in more radio events than stations in southern England. Amateurs have noticed that the larger the change in geomagnetic activity, the further south the area of auroral ionisation extends, and stations as far south as Italy make auroral contacts in major events.

If you hear someone calling "CQ aurora" and you have not operated in an auroral opening before, resist the temptation to call indiscriminately, and listen only. There are going to be hundreds of future auroras and this is your chance to learn the entirely new operating techniques required for auroral contacts. First select horizontal polarisation and beam between north and east at 45°. Tune the beacon band between 144.412MHz and 144.445MHz; it will take some time to get used to the rough-sounding keying of the beacons which will be slightly off their usual frequencies due to Doppler shift. If you hear GB3LER and SK4MPI check whether DL0PR on 144.486MHz is readable. If these three beacons are heard the aurora is extending to at least as far as Germany, and is therefore a large-scale event which will probably last for a few hours and may repeat later in the evening. Turn the beam between north and east on each beacon heard and it will be noticed that different beam headings give peak signals for different countries. Generally the farthest DX is worked with the beam well to the east.

Next tune the SSB section between 144.160MHz and 144.400MHz and try listening to an experienced local station who is working auroral DX. Due to the distortion he will be speaking slowly, using correct phonetics and possibly end-of-transmission tones. Remember that you will hear the local station direct but the DX stations will be replying via the aurora and will be slightly off his frequency. A typical SSB auroral contact starts like this:

"CQ aurora, CQ Aurora, GM8FFX, Golf Mike Eight Fox Fox Xray calling CQ aurora . . ." (repeated slowly several times) "

and GM8FFX listening." pip (end of transmission tone). "GM8FFX, GM8FFX, LA2PT calling. Lima Alpha Two Papa Tango, LA2PT calling GM8FFX . . ."

End-of-transmission tones which give a low-frequency pip are very helpful in auroral openings when signals are weak – the tone readily identifies the end of each station's transmissions and could be a 'K' tone or 'pip' tone.

Independent receiver tuning (IRT) is a 'must' for auroral reception as the amount of Doppler shift often changes in the middle of a contact. It will soon be seen that SSB contacts are difficult due to the distortion, and contacts tend to be limited to exchanging reports, names, and locators. Many amateurs also exchange and log the beam headings used at both ends of the contact as a study of these figures can reveal the particular area of ionised E layer being used. During weak auroral events SSB operators in England, Northern Ireland, North Wales and Scotland can work each other and operators with better facilities can contact Norway and Sweden. During strong auroral events SSB stations all over the UK and north-west Europe can work each other.

A new operator listening in the CW section between 144.020 and 144.120MHz will hear a great many rough-sounding hissing CW signals during a strong event. Experienced operators are used to the strange AC sounding notes and contacts are completed quickly and efficiently. The letter 'A' is added when calling CQ and is also added after the readability and signal strength report, in place of the normal tone reports which are not sent during aurora openings as no signal sounds T9.

The best auroral DX is always worked on CW, just as on other propagation modes – CW is easier to copy in weak signal conditions and contacts are therefore completed much faster. Some auroral operators use full break-in technique on CW, allowing them to listen for stations breaking in during the sending of a CQ. During the large-scale aurora on 1 May 1978 GM4COK worked more than 150 stations in 20 countries on 144MHz CW.

Auroral openings can occur in three separate phases in a single 24-hour period. The first phase can start as early as 1300 GMT but usually takes place between 1500 and 1900 GMT. The second phase can occur between 2100 and 2300 GMT and a third phase can run from after midnight till 0600 GMT. Very few auroral contacts take place around 2000 GMT and there is often a fade-out between the evening phase and the after-midnight session.

Some auroras have no afternoon phase and start in the evening, often continuing again after midnight. Some auroras have no afternoon or evening phase, only starting after midnight.

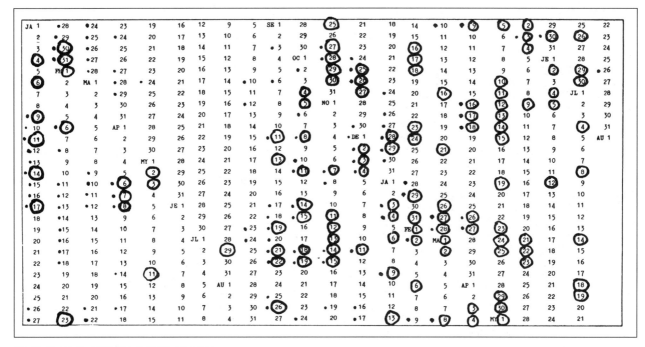

Fig 13. Example of 27-day auroral calendar

These are almost always weak events, sometimes heralding a larger occurrence on the next day.

Due to the Doppler shift and distortion on signals, auroral contacts can only be made on CW and SSB. High power is not essential but helps greatly in weak events. During strong events almost anyone can participate – mobile-to-mobile contacts have been made from southern England to Scotland using halo antennas and 10W SSB transceivers.

Signals reflected from the auroral curtain on 144MHz do not change polarity, and high-gain horizontal antennas give the best results. Operators who are blocked to the south and south-east enjoy auroral openings as they can work stations, normally unheard due to the obstructions, by beaming well to the north of the direct path.

It has already been noted that the amount of Doppler shift is proportional to the frequency band in use. For this reason auroral signals on the 50 and 70MHz bands are easier to read and have less distortion than on the 144MHz band. Inter-UK signals are generally stronger on 50 and 70MHz and the openings start a little earlier and finish a little later than on 144MHz.

Auroral contacts have been made on both CW and SSB on the 432MHz band but signals are about 40dB weaker than on 144MHz, and the Doppler shift can be as much as 4kHz. No auroral contacts have yet been made on the 1.3GHz band but as power levels increase and receivers improve this will doubtless occur soon. Professional studies reveal that radar reflections have been received at over 3GHz.

Predicting auroral events

Before the advent of the packet radio network most operators relied on the telephone to alert others about auroral openings. These links were quite successful. Nowadays serious DXers monitor the DX PacketCluster and the Internet warning systems which provide a real-time record of events as they happen.

Magnetometers are live on the Internet showing you what is and what will be happening, and full daily solar geomagnetic reports are also to be found on the Internet.

Many operators keep special 27-day auroral calendars on which they record both the visual and radio events which are reported in *RadCom* and on the GB2RS news bulletins. The auroral calendar in Fig 13 which started on 1 January 1977 and records auroral events until 1 August 1978 shows the correlation between the visual and radio events and demonstrates the fact that auroras often repeat 27 days later. A study of Fig 13 shows that in the period 1 August 1977 to 1 August 1978 UK observers recorded 70 visual aurora and 100 radio events. These numbers are considerably higher than during the same period for the previous two years, thus reflecting the increase in solar activity as Cycle 21 got under way.

The auroral warning calendar is very easy to use – simply circle any radio events and dot any visual auroras. An operator who was on for the 21 September 1977 event would have been prepared for the repeats which occurred on 18 October, 14 November and 11 December 1977. As the calendar shows, many events repeat in 26 to 28 days, often three or four times, proving that this method can be used successfully. It is also interesting to look at the calendar with hindsight – with the exception of 8 December 1977 the aurora of 18 September repeated consecutively 27 days later seven times. Did we miss an aurora on 8 December or was the event of 4 January and its subsequent repeats unconnected?

The fact that amateurs note, record and make radio contacts via the aurora is of great interest to professional scientists studying the auroral phenomenon. Predicting correctly the date of an aurora and analysing the ensuing results can be almost as interesting as working to Russia on 144MHz during an event. The fact no two auroral events are ever the same and that it is impossible to predict which countries will be worked are all factors which add to the attraction of working DX via the aurora.

Meteor-scatter propagation

Meteor scatter (MS) is a DX propagation mode which is open to exploitation by most serious VHF operators; it should not be regarded as the province of a few specialists. However, it does require a higher level of station organisation and operational competence than random tropo DX chasing. The information here should enable operators to fulfil the few special requirements to explore this very rewarding DX mode.

What are meteors? Meteors are particles of rocky and metallic matter ranging in mass from about 10^{-10}kg to larger than 10kg. About 10^{12} are swept up by the Earth each day. At an altitude of about 120km they meet sufficient atmospheric resistance to cause significant heating. At 80km all but the largest are totally ionised, and this ionisation can be used to scatter radio signals in the range 10MHz to 1GHz.

Amateur MS operation

Most amateur MS operation takes place on 144MHz although 28, 50 and 70MHz are also good bands. Operation on 432MHz is also marginally possible, but the path losses approach those experienced in EME operation. Commercial MS links tend to use the low VHF region, as the path availability is greater than at the higher frequencies.

The distances which may be covered by typical MS operation are similar to those possible via sporadic-E. Assuming most signals are reflected from a region at an altitude of 110km, the maximum range possible with an antenna exhibiting a main lobe at 0° elevation is about 2300km. Typical amateur antennas have a main lobe at 2–5° and thus the ranges to be expected are somewhat less.

Most MS contacts take place during periods of more-or-less predictable meteor activity – the so-called *showers*. Rather fewer contacts are made via sporadic meteors during the intervals between.

MS QSO procedure

The intermittent nature of MS propagation means that special operating procedures are necessary. Within IARU Region 1, they are the subject of international agreement (see panel on the next page), and should thus be employed. Failure to do so has resulted in not a few lost contacts.

On CW, high speeds are employed. During skeds, speeds from 200 to over 2000 lpm are in use. In random MS work 800 lpm is the recommended maximum speed. Most operators use memory keyers to send and tape recorders to receive but sophisticated computer software is also available to deal with these speeds.

Equipment

Meteor-scatter contacts are possible with low power, particularly on 50MHz but on 144MHz higher ERP is necessary for consistent success. 100W RF at the feed point of a 10–14dBd gain antenna should be aimed for.

A genuine system noise figure of less than 2.5dB is highly desirable. Many commercial transceivers, especially the earlier models, have inadequate front-end performance but a new generation of DSP transceivers has helped somewhat. Masthead preamplifiers are used by serious operators but care should be taken to avoid too much gain which could degrade the performance of the transceiver.

Frequency setting is of paramount importance. A tolerance of ±500Hz is demanded on CW and 200Hz on SSB. A stability of better than 100Hz/h is also expected. This implies that no matter what technique is used for frequency calibration, the frequency standard employed should be standardised rather frequently; most of the crystals used in old amateur frequency measurement systems are likely to vary over a range of about ±10ppm around room temperature. Today, modern transceivers employ much

more stringent frequency-derived systems which are very accurate and stable.

Most MS signals will be quite weak so the antenna system should be as large as possible within environmental, structural and financial constraints. Long Yagis are popular, particularly the commercial 13- and 17-element models, although smaller antennas are capable of good results. With a high-gain array, such as a box of four 17-element Yagis, received signals will be much stronger than those from a single 9-element antenna. However, many MS operators have found that a too 'sharp' array often misses signals coming into the broader capture angle of a smaller single Yagi. Experienced operators tend to favour two or more stacked Yagis since these give extra gain while retaining a broader E-plane lobe.

The timing requirements of MS technique are stringent but not too difficult to meet. A radio-controlled clock or watch is probably the most popular timepiece now found in many shacks. The prices of these clocks/watches incorporating radio receivers phase-locked to radio transmitters such as MSF and Droitwich are now very reasonable. There are radio receiver boards available for plugging into your PC which will ensure the computer's clock is always spot on. Mains-driven clocks are to be avoided as they have poor short-term stability. On CW, two other items of equipment are required; a means of sending repetitive messages at the speeds involved and a means of decoding the received CW. For many years programmable electronic memory keyers have been used to store and send the various messages used during a sked. Cassette tape recorders, modified for variable speed, are used by most operators to record received bursts for play-back at readable speed during transmit periods. During 1995, some keen MS operators began using a digital tape recorder made by DF7KF which easily decodes CW at speed in excess of 2000 lpm.

Designs for memory keyers suitable for MS work have appeared regularly in amateur radio journals and literature. LA8AK has contributed circuit ideas to the 'Technical Topics' column in *RadCom* and commercial keyers are available like the Super Keyer III that now has non-volatile memory included. More recently dedicated MS operators have employed their station computer to control their MS activities. Some operators use their PCs to receive CW in WAV files which obviates the use of mechanical tape recorders. Modern MS operation is a far cry from the early days of multi-speed reel-to-reel tape recorders and endless tape loops for sending!

As the advancement of PCs has been astonishing during the past few years so have the technical aspects of development. The sound card in the personal computer can now offer a full TX and RX for high-speed CW contacts and even switch the radio for you!

Arranging skeds

Once the procedures used for MS have been understood, and the necessary equipment gathered together, the intending MS operator has to find someone to work! There are really two choices: to make a first contact directly on the frequencies set aside for random MS operation or to arrange a sked with an active MS station. Of the two, the latter is the approach most likely to be fruitful.

In order to maximise the chance of success, skeds should be arranged at optimum times. The sporadic meteor rate peaks at 0600 but with shower meteors it is possible to be more specific. A knowledge of the propagation mechanism and the astronomical co-ordinates of the individual showers makes its possible to calculate the optimum time for any given path.

IARU REGION 1 METEOR SCATTER PROCEDURES

Scheduled and random contacts

There are two types of meteor scatter contact, scheduled or random.

(i) A scheduled contact is where two interested stations have arranged in advance upon the mode, frequency, timing and duration of the test. This may be done by correspondence, or via the European VHF net which is active around 14.345MHz.

(ii) Non-scheduled contacts are made by calling CQ or responding to a CQ call and then following the IARU procedures. These are called *random contacts*.

Timing

It is recommended that stations use 2.5 minute periods on CW and one-minute periods on SSB. This period gives quite satisfactory results. However, improving technical standards make it possible to use much shorter periods. With scheduled contacts you can arrange for any time period you wish but it is recommended that the periods are kept to one minute or less, especially during major showers. The use of 'break' procedures within scheduled contacts is very effective. On SSB this could be every 15 seconds if so desired.

(i) All MS operators living in the same area should, as far as possible, agree to transmit simultaneously in order to avoid mutual interference.

(ii) If possible northbound and westbound transmissions should be made in periods 1, 3, 5 etc, counting from the full hour. Southbound and eastbound transmissions should be made in periods 2, 4, 6 etc. UK stations have chosen by default to transmit during the second period.

(iii) Start times should be arranged to be on the hour, eg 0000, 0100, 0200 etc. This makers the best use of everyone's operating time. It can indicate how much time a station may have before the next scheduled contact.

Scheduled duration

Scheduled contacts are usually of one or two hours duration although during shower periods this can be reduced to 30 minutes or less. Every uninterrupted schedule period must be considered as a separate test. It is not permissible to break off and then recommence at some later time.

Choice of frequency

(a) Scheduled contacts

The frequency selected for scheduled contacts should avoid popular transmission channels, taking into consideration the mode and band plan. For example CW schedules could be arranged to run between 144.130–144.150MHz, and SSB from 144.150–144.190MHz or from 144.410–144.450MHz.

(b) Non-scheduled contacts using CW

The frequency used for CW calls should be 144.100MHz. Contacts resulting from such CQ calls should take place in the range 144.101–144.126MHz. The following procedure should be used by the caller to indicate during the CQ on which exact frequency he will listen for a reply and continue the QSO.

(i) Make sure the selected frequency it is clear of traffic and QRM.

(ii) In the call, immediately following the letters 'CQ', a letter is inserted to indicate the frequency that will be used for reception when the CQ call finishes. This letter indicates the frequency offset from the actual calling frequency used: CQA = 1kHz from calling frequency; CQB = 2kHz from calling frequency; CQC = 3kHz from calling frequency, all the way to CQZ = 26kHz from calling frequency. For instance, "CQE G4ASR CQE G4ASR" would indicate that G4ASR was listening on the calling frequency plus 5kHz. In all cases the letter used will indicate a frequency *higher* than the CQ frequency. Contacts will therefore take place in the segment 144.101–144.126MHz.

(iii) At the end of the transmitting period the receiver should be tuned to the frequency indicated by the letter used in the CQ call. If a signal is heard on this frequency and identified as an answer to the CQ call, the transmitter should be moved to the *same frequency*. The entire QSO procedure will then take place there.

(c) Non-scheduled contacts using SSB

At the 1993 IARU Region 1 Conference it was agreed that 144.195–144.205MHz and 144.395–144.405MHz should be used for SSB operation. In an attempt to spread out activity no specific calling frequency has been mentioned. However, during non-shower periods it will generally be expected that stations will call on either 144.200MHz or 144.400MHz. During major meteor showers, operation should be anywhere within the 10kHz segments, having first ensured that the frequency is not in use.

CW transmission speed

Speeds up to 2000 letters per minute (400wpm) or higher are now in common use. For non-scheduled work a speed of more than 800 letters per minute is not recommended. In scheduled tests the speed should always be agreed before the test. Note that in some countries the national PTT requires that callsigns to be sent at a slower speed at the end of each transmission. Check that the message being sent is correct and readable before and during transmission.

QSO procedure

(A) Calling procedure

Scheduled contacts start with one station calling the other, eg "UV1AS G4ASR UV1AS G4ASR . . .". For non-scheduled operation the call is in the form: "CQ G4ASR CQ G4ASR . . .". On CW the letters 'DE' are not used unless required by the national PTT.

(B) Reporting system and procedure

The report consists of two numbers as shown below:

1st number (burst duration)	2nd number (signal strength)
2: bursts up to 5 seconds	6: up to S3
3: bursts of 5–20 seconds	7: S4 to S5
4: bursts of 20–120 seconds	8: S6 to S7
5: bursts over 120 seconds	9: S8 and stronger

A report is sent only when the operator has positive evidence of having received the correspondents or his own callsign, or parts of them. It is given as follows: "UV1AS G4ASR 38 38 UV1AS G4ASR 38 38 . . .", and should be sent between each set of callsigns, three times for CW, twice for SSB. The report must *not* be changed during a QSO, even though a change of signal strength or duration might well justify it.

(C) Confirmation procedure

(i) As soon as either operator copies *both of the callsigns and the report* he can start sending a confirmation report. This means that *all* letters and numbers have been correctly received. Confirmation is given by sending an 'R' before the report: "UV1AS G4ASR R38 R38 UV1AS G4ASR R38 R38 . . .". Stations with an 'R' at the end of the callsign could possibly send it twice, eg "UV1AS G4ASR RR38 RR38 UV1AS G4ASR RR38 RR38".

(ii) When either operator receives a confirmation message, such as "R38", and all other required information is complete he must confirm with a string of 'R's, inserting his callsign after every eighth 'R': "RRRRRRRR G4ASR RRRRRRRR G4ASR . . .".

When the other operator has received 'R's the contact is complete and he may respond in the same manner, usually for three periods.

(D) Requirements for a complete QSO

The requirements for a valid contact is that *both* operators *must* have copied *both callsigns*, the *report* and also an 'R'

(ROGER) to confirm that the other operator has done the same.

Missing information (CW)

If a confirmation report is received at an early stage in the contact, the other operator has all the information he needs. The following strings may then be used to ask for missing information:

BBB Both callsigns missing
MMM My callsign missing
YYY Your callsign missing
SSS Duration and signal strength report missing
OOO Information incomplete
UUU Faulty keying or unreadable

The other operator should respond by sending only the required information. This approach must be used with great caution to prevent confusion.

Meteor scatter using SSB

Contacts are conducted in the same way as on CW. Letters are generally spelt in the ICAO alphabet but may be spoken without phonetics during a schedule. The letter 'R' in confirmation reports is pronounced "Roger".

Procedures on 50, 70 and 432MHz

Most MS activity on the 50MHz band uses SSB. The MS calling frequency for SSB is 50.350MHz and that for CW is 50.300MHz. The standard periods of 1 minute for SSB and 2.5 minutes for CW are still used.

There is infrequent MS activity on the 70MHz band. The recommended MS calling frequency is 70.150MHz. However, most activity is with DXpedition stations who generally operate on pre-arranged frequencies in the SSB segment of the band.

Similarly, 432MHz activity is very low and all tests are scheduled.

An example of a computer listing for a meteor shower is shown in Fig 14. This gives at the top of the table the name of the shower, dates between which the shower usually occurs, approximate date(s) of the maximum, zenith hourly rate (the relative activity of the shower) and the astronomical co-ordinates (right ascension and declination). Explanation of these terms is given in reference [53].

In the body of the table, the information tabulated at each hour is the position of the radiant in the sky (in altitude and azimuth) and, for each of eight reciprocal bearings, the relative effectiveness (presented as a histogram normalised to a maximum of 10 units) and the optimum antenna offset in degrees in the direction indicated. This last parameter is the deviation required from the great-circle path for conformity to the correct reflection geometry, and is illustrated in Fig 15. It is the same for both stations. A complete set of such tables, covering all

```
QUADRANTIDS        JAN 1-6        MAX. JAN 4.5    ZHR 110    RADIANT AT RA,DEC  232.0      50.0

HRS    ALT     AZ          N-S                NNE-SSW              NE-SW                ENE-WSW
 0    20.5   31.8E   XXX         7W   X          21WNW  X          16SE   XXXX        6SSE
 1    26.0   40.5E   XXXXX       7W   XX         15WNW  X          38SE   XXXX       11SSE
 2    32.5   48.7E   XXXXXX      8W   XXXX       14WNW  X          45NW   XXX        18SSE
 3    39.8   56.4E   XXXXXXX    10W   XXXXX      14WNW  XX         32NW   XX         32SSE
 4    47.8   63.8E   XXXXXXXX   12W   XXXXXX     16WNW  XXX        28NW   X          50SSE
 5    56.3   71.1E   XXXXXXX    15W   XXXXXX     19WNW  XXXX       28NW   X          52NNW
 6    65.2   78.7E   XXXXXX     20W   XXXXX      23WNW  XXXX       30NW   X          46NNW
 7    74.4   87.5E   XXXX       29W   XXXXX      31WNW  XXXX       36NW   XX         46NNW
 8    83.6  104.2E   XX         44W   XX         44WNW  XX         45NW   X          49NNW
 9    86.1  119.1W   X          50E   X          52ESE             55SE              56NNW
10    77.1   90.9W   XXXX       33E   XXXX       34ESE  XXX        39SE   XX         47SSE
11    67.9   81.1W   XXXXXX     22E   XXXXXX     23ESE  XXXXX      26SE   XXXX       34SSE
12    58.9   73.4W   XXXXXXX    16E   XXXXXXX    16ESE  XXXXXX     18SE   XXXXX      23SSE
13    50.3   66.0W   XXXXXXXX   13E   XXXXXXXX   12ESE  XXXXXXX    13SE   XXXXXX     16SSE
14    42.1   58.7W   XXXXXXX    10E   XXXXXXXX    9ESE  XXXXXXXX    9SE   XXXXXXX    11SSE
15    34.6   51.0W   XXXXXX      9E   XXXXXXX     7ESE  XXXXXXXX    7SE   XXXXXXX     8SSE
16    27.8   43.0W   XXXXX       8E   XXXXXXX     6ESE  XXXXXXX     5SE   XXXXXXX     6SSE
17    22.0   34.5W   XXX         7E   XXXXX       5ESE  XXXXXX      4SE   XXXXXX      4SSE
18    17.4   25.5W   XX          7E   XXXX        4ESE  XXXXX       3SE   XXXXXX      3SSE
19    14.1   16.1W   X           9E   XXX         4ESE  XXXX        3SE   XXXXX       3SSE
20    12.3    6.4W              18E   XX          5ESE  XXX         3SE   XXXX        2SSE
21    12.1    3.4E              29W   X           7ESE  XXX         3SE   XXXX        2SSE
22    13.4   13.2E   X          10W   X          14ESE  XX          5SE   XXXX        3SSE
23    16.3   22.7E   XX          8W   XX         55WNW  XX          8SE   XXXX        4SSE
24    20.5   31.8E   XXX         7W   X          21WNW  X          16SE   XXXX        6SSE

HRS    ALT     AZ          E-W                ESE-WNW              SE-NW                SSE-NNW
 0    20.5   31.8E   XXXXXX      4S   XXXXX       4SSW  XXXXX       4SW   XXXXX        5WSW
 1    26.0   40.5E   XXXXXX      6S   XXXXXXX     5SSW  XXXXXXXX    5SW   XXXXX        5WSW
 2    32.5   48.7E   XXXXXX     10S   XXXXXXX     7SSW  XXXXXXXX    6SW   XXXXXXXX     7WSW
 3    39.8   56.4E   XXXXXX     14S   XXXXXXXX   10SSW  XXXXXXXXX   8SW   XXXXXXXXX    8WSW
 4    47.8   63.8E   XXXX       22S   XXXXXXX    14SSW  XXXXXXXX   11SW   XXXXXXXXX   11WSW
 5    56.3   71.1E   XXX        34S   XXXXXX     21SSW  XXXXXXXX   16SW   XXXXXXXX    14WSW
 6    65.2   78.7E   X          46S   XXXX       30SSW  XXXXXX     23SW   XXXXXXX     20WSW
 7    74.4   87.5E              55S   XX         43SSW  XXXX       34SW   XXXXX       30WSW
 8    83.6  104.2E   X          54N              55SSW  XX         50SW   XX          46WSW
 9    86.1  119.1W   X          53N   X          51NNE  X          49NE   X           49ENE
10    77.1   90.9W              56N   XX         46NNE  XXX        38NE   XXXX        34ENE
11    67.9   81.1W   X          49S   XX         46NNE  XXXX       32NE   XXXXX       25ENE
12    58.9   73.4W   XXX        37S   X          50NNE  XXXX       28NE   XXXXXX      20ENE
13    50.3   66.0W   XXXX       25S              54SSW  XXXX       28NE   XXXXXX      17ENE
14    42.1   58.7W   XXXXX      16S   XX         38SSW  XX         30NE   XXXXXX      15ENE
15    34.6   51.0W   XXXXX      11S   XXX        22SSW  X          39NE   XXXXX       14ENE
16    27.8   43.0W   XXXXXX      7S   XXX        12SSW             49SW   XXX         14ENE
17    22.0   34.5W   XXXXX       5S   XXXX        7SSW  X          20SW   X           18ENE
18    17.4   25.5W   XXXXX       4S   XXXX        5SSW  XX          9SW               38ENE
19    14.1   16.1W   XXXXX       3S   XXXX        3SSW  XX          5SW   X           21WSW
20    12.3    6.4W   XXXX        2S   XXXX        3SSW  XXX         4SW   X            8WSW
21    12.1    3.4E   XXXX        2S   XXXX        2SSW  XXX         3SW   XX           5WSW
22    13.4   13.2E   XXXX        2S   XXXX        2SSW  XXXX        3SW   XXX          4WSW
23    16.3   22.7E   XXXX        3S   XXXXX       3SSW  XXXX        3SW   XXX          4WSW
24    20.5   31.8E   XXXXXX      4S   XXXXXX      4SSW  XXXXX       4SW   XXXX         5WSW
```

Fig 14. Example of the computer printout for a particular shower

Table 2. Principal meteor showers

Shower name	Limits	Max	ZHR	N–S	NE–SW	E–W	SE–NW
Quadrantids	1–5 Jan	3–4 Jan	110	02–06(W) 11–16(E)	11–17(SE)	23–03(S) 15–17(S)	00–05(SW)
April Lyrids	19–25 Apr	22 Apr	15–25	22–02(W) 06–10(E)	23–03(NW) 08–11(SE)	03–06(N)	22–01(SW) 05–08(NE)
Eta Aquarids	1–12 May	3 May	50	03–04(W) 10–11(E)	04–09(NW)	05–11(N)	08–12(NE)
Arietids	30 May–18 Jun	7 Jun	60	04–08(W) 11–15(E)	05–09(NW) 14–16(SE)	08–12(N)	04–06(SW) 10–14(NE)
Zeta Perseids	1–16 Jun	9 Jun	40	05–10(W) 13–17(E)	06–11(NW) 15–17(SE)	09–14(N)	07–07(SW) 11–15(NE)
Perseids	20 Jul–18 Aug	12 Aug	95	23–04(W) 09–13(E)	08–17(SE)	11–01(S)	18–04(SW)
Orionids	16–27 Oct	22 Oct	25	00–03(W) 07–09(E)	00–04(NW)	03–06(N)	05–08(NE)
Taurids S	10 Oct–5 Dec	3 Nov	25	02–05(E) 20–22(W)	20–01(NW)	22–03(N)	00–05(NE)
Geminids	7–15 Dec	13–14 Dec	110	04–09(E) 20–01(W)	22–02(NW) 05–09(SE)	01–04(N) 03–07(S)	03–07(NE) 19–23(SW)
Ursids	17–24 Dec	22 Dec	15	—	07–01(SE)	00–24(S)	16–09(SW)

usable showers for locations in the southern part of the UK, is obtainable from RSGB [55].

An abridged version of these tables is given in Table 2. After the shower name the dates of the usual limits are given, followed by the date of the maximum and the zenith hourly rate. Optimum times are given for four pairs of reciprocal bearings. Although space does not permit the precise listing of the antenna offsets given in Fig 14, a suitable approximation is to beam 10° off the great-circle path in the direction indicated in parentheses after the optimum times (see Fig 15).

One of the most comprehensive suites of public-domain MS software is that written by OH5IY [56]. It features the usual predictions of the peak times of the major and many minor meteor showers, and times when reflections will be optimum for the required path. The program can be used to key the transmitter at speeds of 100–9999 lpm, receive and playback the CW if the PC has a sound card and Microsoft Windows operating system. It features a comprehensive sked editor, automatic logging and will even wake you up with an alarm call when it is time for a sked in the middle of the night! Like all such software, the various files are subject to continuing development and updating. Within days of a major shower, the observed radio data for it is posted on the Internet.

There are several ways of making MS skeds. The most convenient method is to arrange a test via e-mail or the Internet reflectors or finally if you do not have this access than the European VHF Net on the 14.MHz band. This is an informal meeting place for VHF DX enthusiasts and is centred on 14.345MHz. It gets very busy in the run-up to and during major showers. On Saturdays and Sundays 14.345MHz should be left clear for those arranging EME skeds (see following section).

When arranging an MS sked, a certain minimum amount of data needs to be exchanged:

(a)   The date
(b)   Times of start and finish (2h duration is usual)
(c)   Length of transmit and receive periods
(d)   The frequency
(e)   Which station transmits first
(f)   CW speed

Further reading

The discussion of MS propagation given here is necessarily rather superficial. Much has been written about MS propagation in both professional and amateur journals. Some sources are given at the end of this chapter [53, 54].

DX WARNING NETS

The most common lament heard on 144MHz in the aftermath of a sporadic-E or auroral opening runs along the lines: "Well, the band was absolutely dead and then I heard a very strong Italian station calling CQ. I called him, and we gave each other 5 and 9, sent our 73s, and signed off. I tuned around to see what else was about, and the whole band was full of 'G' stations calling CQ and working all of the DX! Where do they all come from?"

The second most common lament is heard from operators who have come on the air just after the end of the opening: "I missed it all as usual. It's very annoying, because I have been in the shack all day, tidying up but I didn't have the rig switched on."

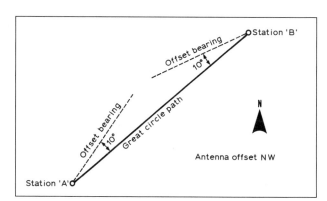

Fig 15. Illustration of antenna offsets

Many operators now rely on the Internet's VHF reflectors and newsgroups, while others watch the DX cluster on the packet network for news of tropospheric openings, sporadic-E and auroral events. This is a very efficient and informative use of packet radio as it gives a running commentary on what is happening without costing you a fortune connected up to the Internet for hours on end!

MOONBOUNCE

Probably the most challenging type of VHF/UHF DX communication is that using the Moon as a passive reflector, popularly known as *EME* (Earth-Moon-Earth) or *moonbounce*. The first known use of the Moon as a relay was by the United States Navy which set up a circuit between Washington, DC and Hawaii using 400MW ERP. The first amateur EME contact was on 1296MHz in July 1960 between W6HB and W1BU. The majority of EME work takes place on the 144, 432 and 1296MHz bands although there is some activity now growing on 50MHz and so far four British stations have special research permits for 30dBW (1kW) to allow EME and other propagation tests. The microwave bands up to 10GHz are attracting more and more experimenters as PCB microwave construction increases.

The logistics

The mean distance between the Earth and the Moon is 385,000km. The Moon's diameter is 3476km so a little simple trigonometry shows it appears a mere 0.52° wide as viewed from Earth. Bearing in mind the three-dimensional polar diagrams of typical amateur antenna arrays, it is obvious that only a very small amount of ERP will illuminate the disk. Moreover the Moon is a sphere, not a flat mirror, so only radio waves that hit the middle region will be reflected back to Earth. In other words, the path loss is enormous. To put some figures on it, the minimum round-trip path loss at 144MHz is 251.5dB rising to 270.5dB at 1296MHz. Even so, it is possible to work a few of the biggest EME stations on 144MHz using a few hundred watts and a single long Yagi under favourable conditions.

For stations without antenna elevation capability, EME communication is only possible around moonrise and moonset. At such times, depending upon the height of the antenna above ground, advantage can be taken of so-called 'ground gain' whereby the apparent gain of a horizontally polarised antenna can be up to about 5dB more than the free-space figure. This is equivalent to increasing the size of an array from a single Yagi to four – and all for free!

Arranging EME tests

The best time to arrange EME tests is when the Moon is nearest to Earth, known as *perigee*. At *apogee* (furthest from Earth) there is an extra 2dB path loss to overcome. However, this is not the only consideration since the background sky temperature and/or Sun noise at new Moon periods may mitigate against perigee operation. Unfortunately there can be periods lasting several years when conditions are far from ideal, particularly on 144 and 432MHz.

Publications such as *Dubus Magazine* and the *432 and Above EME Newsletter*, edited by K2UYH [57], identify favourable sked weekends and include details of active EME stations. Skeds can be arranged by letter, telephone or e-mail, but most EME operators check into the EME net on 14.345MHz on Saturdays and Sundays. This starts around 1600 UTC for those interested in 432 and 1296MHz and from 1700 UTC the

144MHz devotees take over. A contact can only be attempted when both stations can 'see' the Moon over their radio horizons. These periods are known as *windows* and are best derived from computer programs such as those written by WA1JXN [58], VK3UM [59] and F1EHN [60].

EME QSO procedure

There are internationally agreed operating procedures for EME work and, as with MS operation, timing is all-important. On 144MHz 2-minute periods are used and on 432MHz and above, 2½ minutes. The convention is that the station whose call appears first on a sked list transmits first then in following 'odd' periods. In the absence of a formal list the convention is that the furthest east station transmits first on 144MHz but on 432MHz it is the one furthest west.

There is a special TMOR reporting system for EME work. On 144MHz these letters denote the following:

T	Signals just detectable
M	Portions of calls copied
O	Both calls fully copied
R	Both calls and 'O' report copied.

On 432MHz and above, these reports have slightly different meanings

T	Parts of calls copyable
M	Both calls fully copied weakly
O	Both calls copied comfortably
R	Both calls and 'M' or 'O' report copied.

In favourable conditions RST reports are exchanged.

At the start of a sked both callsigns are sent for the complete period of 2 or 2½ minutes, depending on the band. Only when both callsigns have been copied completely may you send a report which will be 'O' on 144MHz, and 'M' or 'O' on other bands. At this stage the first 1½ minutes on 144MHz, or the first 2 minutes of the period on other bands, will be devoted to sending callsigns and the final 30 seconds the reports. Reports are acknowledged by sending "RO" on 144MHz and "RO" or "RM" on 432MHz and above and when this is received by the other station a series of 'R's is sent.

As in MS work, random EME contacts are made especially during contests. If you hear a station calling CQ, reply by sending his and your own calls. If the calling station does not copy your complete call he will reply with "QRZ", "???" or perhaps "YYY", the latter indicating he has not copied your call. In this case, send your call for the whole of the next period until you receive an 'O' or 'M' report. If you receive "GGG" it means the other station needs your grid, eg IO91, for scoring purposes in a contest.

Equipment

As mentioned earlier, on 144MHz it is possible to work a few of the stations using very large antenna systems with a few hundred watts and a single Yagi. However, if serious EME operation is contemplated you must be prepared to make a considerable investment in time and money. On the receiving side it is essential to achieve the lowest possible system noise figure and this usually means fitting a high performance low-noise preamp at the masthead. To avoid damage to the preamp it is vital to incorporate a properly-designed, fail-safe transmit/receive switching system – this is known as *sequential switching*, so that the transmitter power is removed before going into receive, and vice versa.

The antenna array should have as much gain as possible

combined with a clean polar diagram. On 144 and 432MHz Yagi antennas are the most popular because of their ready availability. Thanks to the research by DL6WU, DJ9BV, K1FO and others, the design of optimum performance Yagis is now available to anyone with a PC and there are several Yagi design and performance analysis programs in the public domain. Collinear arrays have been used with success by some VHF operators and have the advantage of not being as 'deep' as an equivalent array of long Yagis. On the microwave bands dishes are favoured and there are plenty of graphs and tables in the handbooks to enable you to design your own.

G3PHO/P at Alport Height, near Matlock, Derbyshire operating on 10GHz narrow band

Supporting a large EME antenna array should be regarded as an engineering project, the safety of the complete structure under all weather conditions being of paramount importance. A weak point can be the steering system, commonly referred to as the *az-el rotator*. Obviously the rotator(s) must be able to deal with the weight and inertia of the load imposed so it is essential to ascertain from the supplier that it/they will cope. The system will have to be carefully calibrated so the array can be aimed accurately at the Moon. Some operators adapt the screwjack systems from satellite TV dishes to elevate their EME antennas. The Japanese are very active with large antenna systems on EME and have developed a mighty elevation chain-driven rotator, but at a price!.

On the transmitting side the aim should be to deliver the maximum licensed power to the antenna array, usually the transfer relay at the masthead. The lowest loss feeder that can be afforded, for the power involved, should be used; there is not much point in generating lots of RF at the output of the PA then wasting 25% of it in the feeder! Such feeder and the appropriate connectors are not cheap if purchased new but sometimes surplus lengths become available from commercial sources. Some surprising bargains can also be found at radio rallies by way of hefty relays and connectors.

There are some proven designs of power amplifiers and associated power supplies and control circuits in the literature such as those by W1SL and K2RIW for 144MHz and 432MHz respectively [61, 62]. Commercial amplifiers are available capable of delivering many hundreds of watts, but if you have the time, skill and a proper understanding of RF engineering, it is very satisfying to build your own amplifier. Surplus valves are fine as long as they are new – don't go out and buy well-used ones and expect them to perform cleanly in your newly designed amplifier. There are many designs on the Internet, including several tried and tested designs by G4ICD which include the Russian surplus tubes GS35B (1.5kW @ 1GHz), GS31B (1kW @ 1GHz). The prices of these tubes are under £100 each (1999) – compare this with Eimac's 8877 (3CX1500) at over £600! All the latter tubes are triodes and offer simple construction of the power supply and RF circuitry.

Summary

Interest in EME communication is growing steadily. Many regular operators are those who have been, and still are, MS enthusiasts. Both modes have similar equipment and operating requirements. However EME operation can be very frustrating. Having arranged a sked with a much-sought-after station, conditions may be ruined by an unpredicted aurora, high winds or static rain. Or you can copy your sked partner perfectly but he hears nothing from you, and vice versa, due to Faraday rotation. Sometimes you hear nothing from your partner then learn later it was because he had a local power failure.

DX ON THE MICROWAVE BANDS

There are 11 internationally allocated amateur microwave bands extending from the 1.3GHz (23cm) band to the 241GHz (1mm) band. Most, but not all, of these bands are available in the UK. All the bands below 24GHz are Amateur Secondary, shared bands, whilst many of the bands above this (including part of the 24GHz band) are Amateur Primary or Amateur Primary (Exclusive).

Since most of the microwave bands are many tens (and sometimes hundreds or thousands) of megahertz wide, is difficult to tune the whole of the band in order to find stations to work. This has led to the use of 'preferred' frequencies or 'preferred' sub-bands and, just as there are voluntary bandplans in all the bands below 1GHz, so there are similar bandplans in all the amateur microwave allocations. The narrow-band sub-bands (or 'segments') were, at one time, based upon either multiples of 144MHz or multiples of 1152MHz, since the microwave frequencies were derived either by multiplying from 144MHz or 1152MHz. It is becoming increasingly difficult to use some of these harmonic relationships, as they no longer lie within the amateur allocations in many countries.

Full details of the current microwave band usage (such as calling and working frequencies for terrestrial and space/EME use) are given in Appendix 7 and are generally very similar to those in the VHF and UHF band plans given there.

Nowadays, modern microwave equipment uses mixing and linear amplifying techniques (as well as multipliers in local oscillator chains) in both receivers and transmitters – *transverters*. Multimode transceivers at 28MHz, 144MHz or 432MHz (or occasionally at 1296MHz) are used as very sensitive and stable intermediate frequency receivers and transmitter signal sources. It should be noted that the direct use of 28MHz as an intermediate frequency is not recommended because the small frequency separation between the local oscillator injection frequency and wanted and unwanted mixing products makes it extremely difficult to produce a 'clean' output signal by filtering the mixer products to suppress both the LO and unwanted (image) product. In this case, it is usual to

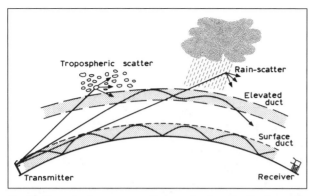

Fig 16. Trans-horizon propagation mechanisms in the lower atmosphere

introduce a 28/144 or 28/432MHz low power transverter between the 28MHz transceiver and the microwave transverter.

MICROWAVE PROPAGATION

Ionospheric propagation never occurs above 1GHz. Indeed, the upper limit appears to be somewhere around the amateur 70cm band (432MHz). There are a few isolated reports claiming to have observed ionospheric propagation at 1.3GHz but these are unconfirmed and very doubtful.

Free-space transmission losses, measured in decibels, increase with the square of the frequency (in megahertz) as well as with the square of the transmission path distance (in kilometres). In addition, above about 23GHz, atmospheric absorption by water, water vapour and oxygen peak in certain frequency bands, adding greatly to the free-space losses. As a consequence, some microwave bands are useable for contacts over very limited distances.

All the tropospheric propagation modes observed at VHF and UHF occur at microwave frequencies. It is becoming apparent, as more regular fixed-station activity takes place, that 'lifts' or 'openings' may occur more frequently than at VHF or UHF because the vertical extent of the tropospheric anomaly can be smaller to have the same effect. That is, the ability of a 'duct' of a particular thickness to propagate radio waves with low-loss (attenuation) is related to frequency – amongst other things! DX of 1000km or more has been worked quite frequently.

In addition, over-the-horizon forward-scatter propagation occurs all of the time because of normal atmospheric irregularities (turbulence) and, often spectacularly, by cloud, rain, hail, snow and aircraft scatter, most of which are on too small a scale to noticeably affect VHF/UHF propagation. Another propagation mode, allied to tropospheric ducts, is the formation of a low-level *super-refraction* layer over large stretches of water, for instance across the North Sea.

Fig 16 illustrates some of the trans-horizon atmospheric propagation mechanisms, briefly discussed in the paragraphs above, and in the earlier section on VHF/UHF propagation modes. The principal difference between VHF/UHF propagation and microwave propagation is one of scale: the effectiveness of an atmospheric 'duct', or the forward scatter properties of the atmosphere itself and of hydrometeors (rain, hail, snow), are all functions of frequency. A shallow duct (such as a super-refraction duct over calm water in settled weather) which will not support VHF or UHF propagation may well act as a near-perfect 'waveguide' for microwave signals. Similarly, large rain-drops, hail or snow, such as those typically associated with

thunder cells, can provide a very effective, if short-term, scattering medium particularly in the 10 or 24GHz bands. The is much valuable information on microwave propagation in reference [69], although much remains to be discovered!

It can be seen, therefore, that there are several apparently conflicting factors affecting microwave propagation. On the one hand, settled weather conditions can lead to the formation of ducts and, on the other, turbulent, unsettled weather can enhance scatter modes. Again, at higher microwave frequencies, atmospheric water vapour absorption can increase attenuation and yet super-refraction layers over water (where the humidity is obviously high) can enhance propagation! It is these apparent contradictions which perhaps add to the fascination and challenge of microwave operation.

MICROWAVE EQUIPMENT

Unless efficient equipment is employed – and operated well – results on the microwave bands can be disappointing in terms of distance worked. Until a few years ago, 100km was considered DX on the 23cm band, for instance. There is still very little commercial amateur gear designed to operate on the microwave bands and what there is appears to be relatively inefficient in terms of low power output and sub-optimal receiver sensitivity. There are also few commercially available high-gain microwave antennas, with the exception of parabolic dishes designed for satellite TV.

Man-made noise and noise emanating from natural sources such as thunderstorms, electrical discharges and sparking decrease in intensity as the frequency increases. This makes it possible to use microwave receivers with very low noise figures to detect really weak signals: this is simply not possible at HF and VHF because the noise 'heard' by the antenna is greater than the noise generated within the receiver itself. A few years ago microwave receivers were very noisy, completely masking the fact that noise levels drop with increasing frequency. Modern gallium arsenide field effect transistors (GaAsFETs) produce so little noise that many decibels of thermal noise emanating from the Sun, Moon or Earth can be detected by amateur receivers, even as high as the 10GHz band, where sub-1dB receiver noise figures are attained. This is a level of performance which could not have been dreamed about 10 years ago, let alone in home-built equipment!

Because of these factors, a well-equipped amateur microwave station will generally consist of home-built, optimised receive and transmit converters operating in conjunction with a good, all-mode VHF or UHF transceiver used as the intermediate frequency, usually at 144MHz but occasionally at 432MHz.

Transmit power amplifiers and low-noise receive pre-amplifiers are usually mounted at or very close to the antenna in order to minimise feeder losses and the degradation of receiver noise figure resulting from these feeder losses. The antenna system itself will invariably be designed and constructed to give as much gain as can be engineered and managed: fortunately high-gain microwave antennas are neither unmanageably large nor obtrusive.

Since the advent of satellite TV broadcasting in Ku-band (11 to 13GHz), GaAsFET devices and other components, such as microwave monolithic integrated circuits (MMICs) and surface mount devices (SMDs), as well as complete low-noise amplifiers, block converters and 'set-top' tuners, have become readily and cheaply available. Since Ku band is close to the amateur 10GHz band, the dish antennas, feeds and low-noise block converters can be easily modified to operate in the latter band. Similarly, the set-top tuners cover the range 950 to 1700MHz, which encompasses the amateur 1.3GHz band, and can also be quite easily adapted to amateur use. Such equipment is, however, better suited to amateur wide-band operation, such as ATV, rather than narrow-band operation.

Having said that amateur construction and operation can be greatly aided by the availability of components and equipment, it is still quite difficult to construct or modify equipment for either band, since most modern equipment is constructed on printed circuit board using components which are, in conventional terms, very small. They are thus fairly easily damaged by excessive heating (whilst soldering) or static voltages (whilst handling), although many amateurs new to microwave construction and operating have successfully built, aligned and are operating quite elaborate equipment with a minimum of resources.

THE MICROWAVE BANDS
The 1.3GHz (23cm) band
Solid-state transmitters up to 40–50W, or valved amplifiers to full allowed amateur output, are in common use. Antenna gains, from Yagi or quad-loop Yagi arrays up to about 26dBi are in use for terrestrial QSOs whilst dish gains up to 35dB or more are in common use for EME contacts. Receiver noise levels are well below 1dB.

Terrestrial DX up to 1000–1200km by tropospheric propagation is possible and not uncommon. With the exception of over-water super-refraction paths (West Coast USA to Hawaii, or across the Great Australian Bight), longer paths are easier via the half-million mile Earth-Moon-Earth path! Future uses include amateur satellite Mode L operation where the Earth-to-space uplink is in the 23cm band and the space-to-Earth downlink is in the 70cm (433MHz) band.

The band is popular for ATV and there are growing numbers of ATV and speech repeaters in this band. There are also quite a lot of propagation beacons (between 1296.800 and 1297.000MHz) which help to indicate when tropospheric propagation is good.

Some further information on the 1.3GHz band is given in the VHF/UHF section of this chapter.

The 2.3GHz (13cm), 3.4GHz (9cm) and 5.7GHz (6cm) bands
These three 'intermediate bands' are considered together since comparatively little use has been made of any of them. Transmitter power output levels drop with increasing frequency, either because solid-state power devices become progressively more expensive for a given power output, or because conventional disc-seal valves become less efficient and are definitely 'running out of steam' above 3.4GHz. Receiver noise levels are below 1dB.

At 2.3GHz, distances up to around 860 to 900km for terrestrial contacts are possible, beyond which the EME path is easier! The corresponding terrestrial distance for the 3.4GHz band is a little less, between about 820 and 840km: at 5.7GHz it is about 800 to 820km.

Loop-Yagi antennas are practical up to about the 3.4GHz band. Beyond this, dimensions and constructional tolerance become too critical to make them worthwhile. Consequently, above 3.4GHz, parabolic dish antennas predominate.

Present uses are limited. 2.4GHz and 10GHz beacons have been 'flown' on amateur satellites Oscar 9 and 11 (UOSATs):

it is intended that some of these bands will be used for amateur satellite communications, probably starting with the Phase 3D satellites, the first of which is to be launched in early 1996. The next two bands up, 10 and 24GHz, are also likely to be included in these packages which are intended to be in highly elliptical orbits, in order to mimic 'near-geostationary' communications capability (see Chapter 7).

The 10GHz band
This has, for many years, remained a very popular microwave band. This is mainly because very simple, low-powered, wide-band FM (speech) transceivers are very easy and inexpensive to construct. These have been based on surplus intruder-alarm doppler units used as both the receiver local oscillator/mixer and the transmitter at power levels in the range 1mW to 30 or 40mW. The techniques are simple and serve as an admirable introduction to microwave operation over short distances from fixed locations, or longer line-of-sight paths when operated portable.

A small dish antenna, typically 600mm diameter, may have a gain of 35dB or more and a beamwidth of one or two degrees. This makes accurate antenna pointing important: at the same time, the antenna gain on both receive and transmit is such that a 10mW station 'sounds' like a 10W station.

Since the advent of satellite TV broadcasting (at 11 to 13GHz), much surplus equipment has appeared on the amateur market. This has proved to be very easily modified for ATV use at 10GHz and so the band is much used for this mode. Normally distances of up to 150km, line-of-sight, with either WBFM speech or ATV, are possible.

At the same time as satellite TV broadcasting started to expand, amateur-designed narrow-band (linear) transverter designs, suitable for all-mode transmission and reception, appeared. In the UK, the local oscillator designs of G4DDK, the transmit and receive converters and power amplifiers of G3WDG and the transverter control circuits of G3SEK or G4JNT have enabled the home construction of very effective equipment for this band. Transmit powers up to 1W, solid-state, or 5 to 30W using travelling wave tube amplifiers (TWTAs) are within the grasp of many home-based stations. Receiver noise figures can, again, be below or around 1dB. Coupled with enough antenna gain (say a 4 or 5m dish), EME contacts are possible.

Under good tropospheric propagation conditions, even a modestly powered home station with, say, 1W output power, a 2dB receiver noise figure and a small (600mm) dish antenna can work distances up to around 1000km. Fig 17 shows some of the paths worked from the UK, on the 3cm band, mainly due to favourable tropospheric conditions in late October 1994. There have been similar openings since, and the world record for terrestrial tropospheric DX is almost 2000km. Worldwide intercontinental DX contacts have been made between North America, several European countries (including the UK), Australia and New Zealand.

The 24GHz band and higher
Not enough work has yet been done to prove the real worth of these bands, although fairly long-distance terrestrial communications ought to be possible on the 24 and 47GHz bands and worldwide ranges, via amateur satellites, will be possible. Although the technology exists, most amateurs are several years away from being able to afford the devices or learn the skills necessary to exploit the 'millimetre' bands. Work has started, with low powers (currently well below 1W), high

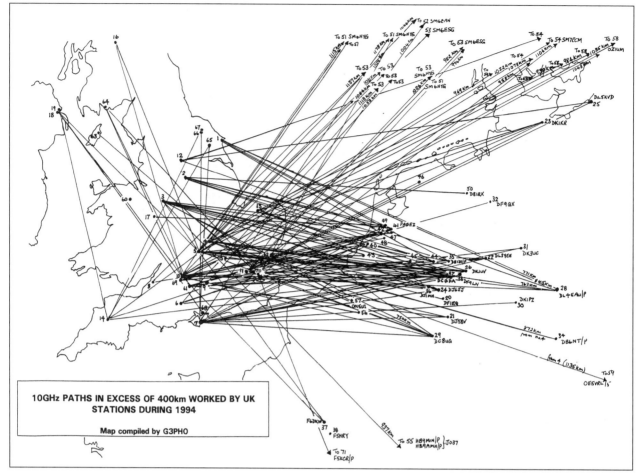

Fig 17. Paths in excess of 400km worked from the UK during 1994

antenna gain, narrow-band (CW) techniques and is yielding promising results!

VHF/UHF/MICROWAVE AWARDS

Naturally the awards and certificates for HF operating (see p58) can also be gained for work on the higher frequencies, though this is usually much more difficult. For example, a Worked All Continents award is easy for the average 14 or 21 MHz operator to acquire, but a 432MHz WAC is another thing altogether, implying the use of moonbounce techniques, and possibly years of technical refinement.

There are, however, some awards available specifically for the VHF, UHF and microwaves. In the UK the RSGB issues various awards which are also available to non-members. The rules, current at the time of going press, for these are shown below but please remember they are subject to change and before applying you should check wth the RSGB VHF/UHF Awards Manager.

RSGB 50MHz Countries Award

The initial qualification for this certificate is proof of completed two-way QSOs on 50MHz with 10 countries. Stickers will be provided for increments of every 10 countries worked. Only contacts with countries permitting 50MHz operation can be considered.

Rules

1. All contacts must have been on or after 1 June 1987.

2. QSL cards submitted must be arranged in alphabetical order of the countries claimed, and a checklist enclosed.

3. Stations are eligible for awards in the following categories:
   (a) Fixed stations.
   (b) Temporary location or portable operation (/P).
   Categories cannot be mixed.

RSGB 50MHz DX Certificate

This certificate takes into account the considerable potential for cross-band working when transmitting in the 50MHz band. There is therefore no stipulation on the band used for the incoming signal. The initial qualification is confirmation from 25 different countries of a successful QSO with transmission from the applicant's country taking place within the 50MHz band. Stickers will be provided for increments of 25 countries confirmed.

Rules

1. All contacts must have been on or after 1 June 1987.

2. QSL cards submitted must be arranged in alphabetical order of the countries claimed, and a checklist enclosed.

3. Stations are eligible for awards in the following categories:
   (a) Fixed stations.
   (b) Temporary location or portable operation (/P).
   Categories cannot be mixed.

RSGB 50MHz Squares Award

The 50MHz Squares Award is intended to mark successful VHF achievement. The initial qualification needed for this

GENERAL RULES FOR RSGB VHF/UHF AWARDS

1. Awards are available to licensed amateurs and listeners (on a heard basis). All claims must be fully supported by QSL cards. For the various squares awards, these cards must also bear the IARU (Maidenhead) locator details. A card without an IARU locator originally printed on it is acceptable provided that it bears some adequate form of positional information (for example old QTH locator or latitude and longitude), in which case the IARU locator square designation should be clearly added to the card by the award claimant.

2. For all awards with a fixed station category, the applicant must state that all the contacts were made from the same location. In the case of an amateur moving home location he/she may apply for the award using QSL cards gained from more than one location, but the award will be endorsed as 'gained from more than one location'.

3. Endorsements such as all CW, all SSB, all auroral contacts, or all contacts made during the first year of being licensed may be made on application. The appropriate information must be contained on the QSL cards and a declaration signed when applying for endorsements.

4. The charges for awards in 1999 were: RSGB members £3, US$6 or 12 IRCs. UK residents who are not RSGB members £6, US$12 or 24 IRCs. Overseas applicants who are members of their national society £6, US$12 or 24 IRCs. Other overseas applicants £9, US$18 or 36 ircs. Where applicable proof of membership of their national (IARU approved) society is required, eg recent *RadCom* label or photocopy of membership card. There is no charge for stickers to update levels of achievement, however, if a new certificate is requested, the charges above will apply.

5. All claims must be submitted to the RSGB VHF/UHF Awards Manager, Tony Jarvis, G6TTL, 31 The Downings, Herne Bay, Kent, CT6 7EJ. Application forms may be obtained from this address by sending an SAE (A4 or A5 size preferred)..

6. For the safe return of the QSL cards, adequate postage and a self addressed envelope must be sent with the application.

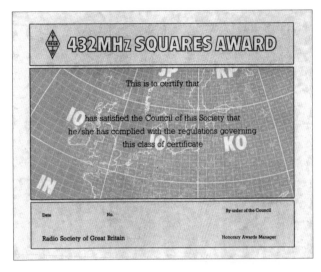

70MHz	20/4	144MHz	40/10	432MHz	30/6
70MHz	25/6	144MHz	60/15	432MHz	40/10
70MHz	30/8	144MHz	80/18	432MHz	50/13
70MHz	35/8	144MHz	100/20	432MHz	60/15
70MHz	40/8	144MHz	125/20	432MHz	70/15
70MHz	45/8	144MHz	150/20	432MHz	80/15
70MHz	50/8	144MHz	175/20	432MHz	90/15
		144MHz	200/30	432MHz	100/15
		144MHz	225/30	432MHz	110/15
		144MHz	250/35	432MHz	120/18
		144MHz	275/35	432MHz	130/18
		144MHz	300/40	432MHz	140/20
		144MHz	325/40	432MHz	150/20
		144MHz	350/45	432MHz	160/20
		144MHz	375/45	432MHz	170/23
		144MHz	400/50	432MHz	180/25
		144MHz	425/50		
		144MHz	450/50		

Rules
1. All contacts must have been after 31 December 1978.
2. Eligible countries are those shown in the countries list printed in the latest edition of the *RSGB Yearbook*.
3. Stations are eligible for awards in the following categories:
   (a) Fixed stations.
   (b) Portable stations, any location.
   (c) Mobile stations, any location.
   Categories cannot be mixed.
4. QSL cards submitted must be arranged in alphabetical order of the QTH squares claimed, and a checklist enclosed.

VHF Countries and Postal Districts Awards
The following awards, intended to mark successful VHF/UHF achievements, are available:

	Requirement	
Title of Award	Countries	Districts
50MHz Standard Transmitting	12	60
50MHz Senior Transmitting	25	90
70MHz Standard Transmitting	3	45
70MHz Senior Transmitting	6	80
144MHz Standard Transmitting	9	65
144MHz Senior Transmitting	15	100
432MHz Standard Transmitting	3	40
432MHz Senior Transmitting	9	70

certificate is proof that 25 different locator squares have been worked with complete two-way QSOs within the 50MHz band. Squares in any country will qualify provided that operation from that country is formally authorised. Additional stickers will be provided when proof is submitted for increments of 25 squares.

Rules
1. All contacts must have been on or after 1 June 1987.
2. QSL cards submitted must be arranged in alphabetical order of the QTH squares claimed, and a checklist enclosed.
3. Stations are eligible for awards in the following categories:
   (a) Fixed stations.
   (b) Temporary location or portable operation (/P).
   Categories cannot be mixed.

4-2-70 Squares Award
The 4-2-70 Squares Awards are intended to mark successful VHF/UHF achievement. Initially, a certificate and one sticker will be issued. Further stickers will be issued as additional locator squares are claimed. The title of each award gives the number of locator squares and countries needed to qualify for the award. For example, to obtain the 144MHz 40/10 award you must have QSL cards confirming contact with 40 locator squares including 10 countries on 144MHz. The following awards are available:

1.3GHz / 20	20
1.3GHz / 25 etc (up to 80)	80
2.3GHz / 5	5
2.3GHz / 10, 15, 20, 25 etc	as 1.3GHz
3.4GHz / 5	5
3.4GHz / 10, 15, 20, 25 etc	as 1.3GHz
5.7GHz / 5	5
5.7GHz / 10, 15, 20, 25 etc	as 1.3GHz
10GHz / 5	5
10GHz / 10, 15, 20, 25 etc	as 1.3GHz
24GHz / 5	5
24GHz / 10, 15, 20, 25 etc	as 1.3GHz

Countries + Counties:

Two-way contact with three countries and 20 UK counties on 1.3GHz, 2.3GHz, 3.4GHz, 5.7GHz, 10GHz and 24GHz. For the purposes of the Award a county is defined as that current at the time of introduction of the award.

Rules

1. All claims must be fully supported by QSL cards carrying the relevant IARU Locator information or Country-and-County information.
2. All contacts must be made after 31 December 1978.
3. Eligible countries are those listed overleaf.
4. Stations are eligible for awards in the following categories:
   (a) Fixed stations;
   (b) Portable and mobile stations. (The applicant must state that the operation was from one site, defined as being anywhere within a 5km radius of the point.)
   Categories cannot be mixed.
5. QSL cards submitted should be listed and arranged in IARU QTH locator alphabetical numeric order.

Microwave Distance Award

The following distance awards, intended to mark achievement on the microwave bands, are available.

1.3GHz   for the first contact made beyond a distance of 600km
2.3GHz   for the first contact made beyond a distance of 500km
3.4GHz   for the first contact made beyond a distance of 400km
5.6GHz   for the first contact made beyond a distance of 300km
10GHz   for the first contact made beyond a distance of 150km (Basic Class)
10GHz   for the first contact made beyond a distance of 300km (Intermediate Class)
10GHz   for the first contact made beyond a distance of 600km (Advanced Class)

On the following bands a certificate and sticker will be issued for the qualifying distance. Subsequently claims will be rewarded wiht appropriate stickers for the incremental distance as shown below.

24GHz   for the first contact made beyond a distance of 100km, then increments of 50km.
47GHz   for the first contact made beyond a distance of 25km, then increments of 25km.
76GHz   for the first contact made beyond a distance of 40km, then increments of 20km.

Rules

1. Stations are eligible for awards in the following categories:
   (a) Fixed stations;
   (b) Portable stations (/P any location);
   (c) Mobile stations (/M any location).

| 1296MHz Standard Transmitting | 3 | 30 |
| 1296MHz Senior Transmitting | 6 | 60 |

Listener Awards to be on an 'as heard' basis.

Supreme Award (for fixed stations only) for holding: 3 Senior awards or 2 Senior + one 1296MHz Awards.

Rules

1. Starting date for this award is 1 January 2000.
2. All contacts made after 1 January 1990 are eligible.
3. Eligible districts are listed in the contest section of the *RSGB Yearbook*.
4. Eligible countries are listed in the prefix section of the *RSGB Yearbook*.
5. For QSL cards not showing a 'district code', the claimant may add the appropriate code providing suitable other location information is on the card.
6. Stations are eligible for awards in the following categories:
   (a) Fixed stations;
   (b) Portable stations (/P any location);
   (c) Mobile stations (/M any location).
   Categories cannot be mixed.
7. Each different confirmed contact with a station in a Scottish district may count up to a maximum of three per district. Belfast (BT) may count up to six contacts.
8. The 'county and country' based awards will continue to be available until 31 December 2002, the rules for which were published in the 1999 edition of the *RSGB Yearbook*. These will continue to be valid towards the Supreme Award.
9. The VHF Committee reserves the right to modify these critieria as necessary.

Microwave Award

The following awards, intended to mark achievement on the microwave bands, are available. Successful applicants will initially receive a certificate and one sticker; further stickers will be issued as later claims are received.

Locators:

Award	Two-way contact with locator squares
1.3GHz / 5	5
1.3GHz / 10	10
1.3GHz / 15	15

FURTHER INFORMATION

SWL clubs

[1] International Short Wave League; details from Honorary Secretary: Mr J M Raynes, G16436/G0BWG, 267 Pelham Road, Immingham, Lincs DN40 1JU. E-mail: john. g0bwg@freeserve.co.uk. Web page: www.geocities.com/ CapeCanaveral/Hall/6248/.

[2] International Listeners' Association; details from Trevor Morgan, GW4OXB, 1 Jersey St, Hafod, Swansea SA1 2HF. E-mail: gw4oxb@net.ntl.com. Web site: websites. ntl.com/~gw4oxb/index.htm.

HF antennas

[3] 'HF DX – the inside story. Part 1: Antennas, rotators and towers', Ian Buffhan, G3TMA, and Bob Whelan, G3PJT, *Radio Communication* June 1992.

[4] *Radio Communication Handbook*, 7th edn, ed Dick Biddulph, M0CGN, RSGB, 1999, Chapter 13.

[5] *Practical Wire Antennas*, John Heys, G3BDQ, RSGB, 1989.

[6] *HF Antennas for All Locations*, 2nd edn, Les Moxon, G6XN, RSGB, 1993.

[7] *HF Antenna Collection*, ed Erwin David, G4LQI, RSGB, 1991.

[8] *Backyard Antennas*, Peter Dodd, G3LDO, RSGB, 2000.

[9] *ARRL Antenna Book*, ARRL.

[10] *The ARRL Antenna Anthology*, Vols 1–3, ARRL.

[11] *All About Beam Antennas*, Bill Orr, W6SAI, and Stu Cowan, W2LX, RPI.

[12] *All about Cubical Quad Antennas*, Bill Orr, W6SAI, and Stu Cowan, W2LX, RPI.

HF receivers etc

[13] 'HF DX – the inside story. Part 2: Receivers and transmitters', Peter Hart, G3SJX, and Ian White, G3SEK, *Radio Communciation* July 1992.

[14] 'Dynamic range, intermodulation and phase noise', Peter Chadwick, G3RZP, *Radio Communication* March 1984.

[15] 'Modern receiver front-end design', Ian White, G3SEK, *Radio Communication* April-July 1985.

HF DX information

[16] 'HF DX – the inside story. Part 3: Information and software', Don Field, G3XTT, *Radio Communication* August 1992.

[17] *The DX Magazine*, produced by VP2ML, PO Box 150, Fulton, CA 95439, USA.

[18] *Les Nouvelles DX*, produced (in French) by F6AJA, 515 Rue de Petit Hem, F-59870 Bouvignies, France.

HF propagation

[19] 'HF DX – the inside story. Part 4: Operating and propagation', Martin Atherton, G3ZAY, *Radio Communciation* August 1992.

[20] *Radio Communication Handbook*, 7th edn, ed Dick Biddulph, M0CGN, RSGB, 1999, Chapter 12.

[21] 'Some new insights into the mechanism of the sunspot cycle', F M Smith, G8KG, *Radio Communication* July 1976.

[22] 'The DXer's crystal ball', *QST* June/August 1975.

[23] 'A breakthrough in simplifying ionospheric propagation forecasts', *CQ* March 1975.

[24] *The Shortwave Propagation Handbook*, George Jacob, W3ASK, and Theodore J Cohen, N4XX, CQ Magazine.

[25] *Radio Wave Propagation*, F C Judd, G2BCX, Heinemann.

[26] *Radio Propagation Handbook*, Peter N Saveskie, Tab Books.

General HF DXing

[27] *Secrets of Ham Radio DXing*, Dave Ingram, K4TWJ, Tab Books.

[28] *The Complete DXer*, Bob Locher, W9KNI, Idiom Press, Box 583, Deerfield, IL 60015, USA.

[29] *Low Band DXing*, John Devoldere, ON4UN, ARRL.

QRP

[30] *QRP Power*, ARRL, 1996.

[31] *G-QRP Club Circuit Handbook*, 1983.

[32] *G-QRP Club Antenna Handbook*, 1992.

[33] *Solid State Design for the Radio Amateur*, Wes Hayward, W7ZOI, and Doug DeMaw, W1FB, ARRL, 1977.

[34] *QRP Notebook*, Doug DeMaw, W1FB, ARRL, 1991.

[35] *W1FB's Design Notebook*, Doug DeMaw, W1FB, ARRL, 1990.

[36] *QRP Classics*, ARRL, 1990.

[37] The G-QRP Club. [Up to 5 watts] c/o John Leak, G0BXO, Flat 7, 56 Heath Crescent, Halifax, HX1 2PW.

[38] QRP Amateur Radio Club International, c/o Michael Bryce, WB8VGE, 2225 Mayflower, N W Massilon, OH 44647, USA.

[39] CW Operators QRP Club, c/o Kevin Zietz, VK5AKZ, 41 Tobruk Ave, St Marys, SA 5042, Australia.

[40] OK QRP Klubu, c/o Petr Doudera, OK1CZ, Ul Baterie 1, 16200, Praha 6.

Awards

[41] RSGB HF Certificates and Awards leaflet.

[42] ARRL DXCC List (CD216) and form (CD164).

[43] Worked All Britain Award, Membership Secretary, Brian Morris, G4KSQ, 22 Burdell Avenue, Sandhills Estate, Headington, Oxford, OX3 8ED. Web site: www.users. zetnet.co.uk/g1ntw/wab.htm.

[44] Worked All Zones Award, CQ Magazine, 76 North Broadway, Hicksville, NY 11801, USA.

VHF/UHF

[45] 'Locator system for VHF and UHF', J Morris, GM4ANB, *Radio Communication* November 1980.

[46] 'The new locator system', J Morris, GM4ANB, *Radio Communication* October 1984.

[47] Detailed Spectrum Investigation Phase 2, 29.7–960MHz, Section 10.4.3 Spectrum Issues, European Radiocommunications Committee. See also *Radio Communication* June 1995, pp30–31.

[48] *VHF/UHF Handbook*, ed Dick Biddulph, G8DPS, RSGB, 1997, Chapter 4.

[49] '1kW power amplifier for 23cm', Karl Schötz, DL9EBL, *Dubus Magazine* 3/1993.

[50] *VHF/UHF DX Book*, ed Ian White, G3SEK, DIR Publishing Ltd, 1992, Chapter 2.

[51] 'Sporadic E studies', Jim Bacon, G3YLA, *Radio Communication* May–August 1989.

[52] 'Plotting of magnetic deviation and aurora', D J Smillie, GM4DJS, *Radio Communication* February and March 1992.

[53] 'Meteor scatter: theory and practice', T Damboldt, DJ5DT, *VHF Communications* 4/1974.

[54] 'The astronomy of meteor scatter', J R Matthews, G3WZT, *Radio Communication* May 1981.

[55] *Meteor Shower Data*, RSGB (a computer list).

[56] MS-Soft suite of meteor scatter programs written by Ilkka Yrjölä, OH5IY, Jukolantie 16, FIN-45740 Kuusankoski, Finland. Available on the Internet at ftp.funet.fi in the pub/ham/VHF-work/ directory. The current filename is mssof42f.zip.

[57] *432 and Above EME News* edited by Allen Katz, K2UYH, Engineering Dept, Trenton State College, Trenton, NJ 08650-4700, USA. E-mail address a.katz@ieee.org. Web: http://www.nitehawk.com/rasmit/em70cm.html.

[58] MOON.BAS High Performance Moon Tracking Program written by Lance Collister, WA1JXN. CP/M and PC versions available.

[59] EME.EXE Moon tracking suite of programs written by Doug McArthur, VK3UM, 'Tikaluna', 26 Old Murrindindi Road, Glenburn, Victoria, Australia 3717. Versions written for PCs with and without math co-processors.

[60] TRACKING.EXE and SETUP.EXE written by Jean-Jacques Maintoux, F1EHN, 24 rue de Villacoublay, F-78140 Velizy, France. Now ported to Windows 3.x. Ready-made tracking interface board available.

[61] 'The W1SL 144MHz power amplifier', John Nelson, GW4FRX, in *VHF/UHF DX Book*, ed Ian White, G3SEK, DIR Publishing Ltd, 1992, Chapter 8.

[62] 'The K2RIW power amplifier', John Nelson, GW4FRX, in *VHF/UHF DX Book*, ed Ian White, G3SEK, DIR Publishing Ltd, 1992, Chapter 10.

[63] G3SEK's 'In Practice' – Dr Ian White has some excellent items here on VHF/UHF that must not be missed: http://www.ifwtech.demon.co.uk/g3sek.

[64] VHF/UHF news, beacons, solar data: A very comprehensive collection of all data associated with the VHF/UHF spectrum: http://user.itl.net/~equinox/.

[65] 50MHz news bulletin along with amplifier designs, beacons, daily DX: http://user.itl.net/~equinox/50dx.html

[66] The UKSMG 50MHz news pages: http://www.uksmg.org/notice.htm.

[67] The 70MHz UK site of GM4AFF/GM4ZUK – this site provides you with up-to-the-minute data on the band: http://www.70mhz.org/.

There are of course many more VHF/UHF sites on the Internet and the above are just a few that will also provide links to more informative sites with detailed information.

Microwaves

[68] The UK microwave pages are handled by Peter, G3PHO. These show construction, news updates, and much more: http://www.g3pho.free-online.co.uk/microwaves/.

[69] *The Physics of Microwave Propagation*, Donald C Livingston, 1970, Prentice-Hall Inc, Englewood Cliffs, NJ, USA.

5 Contests

CONTESTING is one of the most popular activities in amateur radio, and has been for a very long time. For many amateurs, it is a real incentive to improve operating skills and station efficiency, while others enjoy being part of a team and competing with antenna arrays and equipment which they could only dream about in a home environment.

In this chapter HF and VHF/UHF contest operating are discussed separately but there is of course much in common and the would-be contest operator will benefit by reading both sections. There is also a section on amateur radio direction finding, an increasingly popular form of contesting which combines physical effort and radio skills, and which can be enjoyed by young contestants without transmitting licences.

HF CONTESTS

Most national societies around the world organise HF contests for their members, catering for all levels of operating ability and sophistication of equipment, and several societies also organise international contests.

The prime object is to encourage entrants to develop their operating skills, study and obtain a better understanding of propagation and experiment with equipment and antennas with a view to improving the overall performance of their stations. A contest is, by definition, a competitive activity but there is a wide range of different scoring schemes so that it is not always the entrant with the greatest number of QSOs who wins. Many events use a system of 'multipliers' so that, for example, the final score is made up of the QSO points multiplied by the number of countries worked on each band. Some RSGB contests use bonus points for new countries rather than multipliers.

Most modes, frequencies and power levels are catered for, with the exception of 10, 18 and 24MHz which it has been agreed through the IARU shall remain free of contests. Almost without exception the prizes offered for these events are non-monetary, usually consisting of certificate or a trophy that the winner holds for one year.

During international HF contests, stations will be active all over the world, which provides a unique opportunity to test new antennas and equipment. If you are working for an award that requires contacting stations on islands, in countries or rare locations then these events do offer an excellent opportunity to establish contact with such places. There is no doubt that the main international events create a lot of worldwide interest and band activity. Under these circumstances, confirmation of a contact can take longer as the contestant will have to prepare his log before tackling QSL cards and may have had several thousand contacts.

Contests fall broadly into two categories – single operator and multi-operator. The former, as its name implies, is for stations operated by one person only for the duration of the

Chris Swallow, G3VHB, operating as G3VHB/P in the RSGB National Field Day Contest, June 1994

contest. Most organisers define this quite closely, insisting that all of the station operating including logging and check listing is done by one person only. Forward planning is essential and is dealt with in detail later. Some people can operate for 48 hours without sleep, while others find difficult to concentrate after operating for 18 hours. Most experienced contest operators can operate for a 24 hour period with only brief breaks to attend to nature's demands!

Multi-operator contests usually include a separate section for group entries. Callsigns belonging to the different contest groups are regularly heard in all the major events. While multi-operator contests may lighten the load on the individual, they bring fresh problems of their own – particularly organisation. The majority of contests specify CW or SSB, with some major events having separate sections on different weekends catering for each mode. There are differing opinions on the relative merits of CW and SSB in contests. Many people find that higher scoring rates can be achieved on SSB, while to others CW is more personal, less tiring and more of a test of individual skills. Certainly SSB contests tend to require more extensive equipment and antenna systems than CW events.

RSGB HF contests

The policy of the Society's HF Contests Committee is to organise events that cater for all sections of the membership – annual calendars of both RSGB and international contests are shown on the next page.

Although quite a number of contests do require contacts outside the operator's own country, there are several of the RSGB events that feature inter-UK working. For RSGB events the contest exchange usually consists of a signal report followed by an incrementing serial number – though there are exceptions.

Among the most popular events are *The Affiliated Societies*

HF CONTEST CALENDAR

Month	RSGB contests	International contests
January:	Cumulatives on 1.8, 3.5 and 7MHz Affiliated Societies Contest	CQ 1.8MHz CW DX Contest REF CW Contest
February:	First 1.8MHz Contest 7MHz CW Contest	ARRL DX CW Contest VERON CW and SSB Contests REF SSB Contest
March:	Commonwealth Contest	ARRL DX SSB Contest CQ WPX SSB Contest
April:	Ropoco 1	European Sprint SSB
May:	—	ARI International CW and SSB European Sprint CW CQ WPX CW Contest
June:	National Field Day Summer 1.8MHz	All Asia DX SSB Contest IARU HF Championship
July:	Low Power Contest IOTA Contest	—
August:	Ropoco 2	All Asia DX CW Contest WAE DX CW Contest
September:	SSB Field Day	WAE DX SSB Contest Scandinavian Activity Contests
October:	21 and 28MHz SSB Contest 21 and 28MHz CW Contest	CQ WW DX SSB Contest VK/ZL Contests European Sprint SSB European Sprint CW
November:	Second 1.8MHz Contest Club Calls Contest	CQ WW DX CW Contest
December:	—	ARRL 28MHz CW and SSB Contests Canadian CW and SSB Contests

Contests (AFS) which take place annually in January. Members of RSGB affiliated societies operate from their own homes but their logs are grouped together into teams and submitted as a combined club score. In the CW event five members constitute a team, and for SSB there are three in a team. Each AFS contest lasts for four hours.

An ideal series of contests for potential contest operators wishing to gain experience and improve their operating skills are the *Cumulatives*. Out of a series of three sessions, two count for points. Logs from all sections are always welcomed by the adjudicator for checking purposes. These events start early in January on weekday evenings on 1.8MHz and at the weekends on 7 and 3.5MHz – all on CW. As there is a training element to these events, when checking is completed, the adjudicator may write to entrants who have lost a number of points offering constructive advice. Contestants with considerable experience of contest operating are encouraged to offer advice on-air during the event.

Some contests allow separate entries for the specialised interest of QRP and the RSGB recognises this group by running the *RSGB Low Power Contest* on 3.5 and 7MHz CW in mid-July. Another QRP contest is organised by the German DLAGCW CW activity group which takes place twice a year in January and July. This group also runs a short 'Happy New Year Contest' on 1 January which features a QRP section.

Worthy of special mention are the *RSGB ROPOCO Contests*. In these events accuracy of information in the exchange is of primary importance rather than operating speed. The contest exchange consists of RST plus the postcode. It is essential to ensure that codes are copied correctly at the time that they are sent, for the code received in one contact is passed on in the next contact, and so on. The postcodes thus rotate, hence the name: ROtating POst COdes.

In contrast another spring event is the *RSGB Commonwealth Contest* (formerly BERU) where the accent is more on quality than quantity. This contest is limited to 24 hours and is a fascinating contest with its requirement to 'dig out' various Commonwealth call areas on different bands.

During the summer months two events – *RSGB National Field Day (NFD)* and *SSB Field Day* – take place. The history of NFD goes back to the 'thirties and it has evolved from a low-power event which encouraged the use of home-brew transmitters and receivers to one designed to recognise the use of modern commercial equipment but at the same time retaining the requirement to operate under emergency conditions, ie without mains power supplies and away from permanent buildings. NFD is split into two major sections, giving scope for groups of all sizes to compete. There is also a QRP section in which operation is limited to 12 hours. In the open section the only limitation on antennas is that they must not be higher than 65ft whereas in the restricted section only one antenna is permitted with no more than two supports up to 36ft high. The general strategy remains the same for all sections. Carefully study the rules to allow maximum latitude of operation. Learn as much as possible from the results of previous years – this is important to a club trying to improve its overall position in the results table. For example, it is apparent that the outright winners of NFD rarely win an individual band award. This leads to the conclusion that some equality of operation between different bands must be established, although in some instances the leading station on 14MHz has also been the overall winner due to the large potential score that can be realised from contacts with North America.

September's SSB Field Day runs on similar lines to NFD but the scoring system includes country multipliers. In the open section there are no restrictions on antennas and it is left to the ingenuity of individual groups to set up the most extensive effective arrays possible, though complexity does not always pay off!

Following the example of NFD a restricted section has been introduced, again catering for the smaller group. The dates and times of both field days are co-ordinated within IARU Region 1 and the key to success is to work a high proportion of European portables.

The *IOTA Contest* at the end of July has grown in a few years into a major international event which attracts over a thousand entries. There are sections for island stations, phone, CW and mixed modes, and a special category for DXpeditions which form a major feature of this contest. Those working for the various IOTA awards find the contest an excellent way of picking up some of the rarer island groups.

The *Club Calls Contest* (CCC) is a club-based contest in November in 1.8MHz, designed to encourage use of this band and also to allow clubs to 'air' their distinctive club callsigns. Class B members also have an opportunity of entering a club HF event under supervision.

International contests

Turning now to the international scene, the main event of the spring is the *ARRL DX Contest*. The object here is to contact only stations operating from Canada and the USA, which at first sight may not seem terribly interesting. However, this is not the case – American states and Canadian provinces on each band provide multipliers which adds a great deal of interest. Furthermore, it is apparent that North American stations comprise probably the most efficient contesters in the world and this makes contacting them in quantity a real pleasure. Quantity it really is too. In this event, 3000 QSOs on one mode is quite normal and there is every prospect of higher totals to come with improved band conditions, antennas, receivers and station technology in general. Yet another major spring event is the *WPX Contest* sponsored by *CQ* magazine in which prefixes serve as multipliers. There are separate CW and SSB events which are very good for multi-operator participation.

Undoubtedly the biggest event of the year is the *CQ Magazine World Wide Contest*, with the phone event being held on the last full weekend in October and the CW section on the last full weekend in November. This is a contest where you contact anything and everything. There are separate sections: single-operator with or without packet, multi-band, multi-operator or single band, QRP – in other words there's something for everybody.

Logging and computers

It is important not to get the idea that you must have a computer in order to take part in contests. Several top-flight operators use paper in the major contests, and for casual contest operation or for someone trying their hand at contesting, it is better to log on paper if that is how logging for non-contest operation is carried out at your station. However, it is clear that more and more amateurs are turning to the use of computers to maintain the station log, using such programs as Turbolog or Shacklog. Most station logging programs have a contest mode which is fine for the beginner or casual contester, but to the get the most out of the computer a dedicated contest logging program is required.

Undoubtedly the leading contest program in use in the UK is Super Duper (SD) written by Paul O'Kane, EI5DI. It is simple to use yet provides a wealth of features and supports most international contests and all RSGB contests. For use in the major international contests such as CQ Worldwide a better choice might be one of the American programs: CT by K1EA, TR by N6TR or NA by K8CC. All these programs are DOS-based. Probably the leading Windows-based contest logger is WriteLog. This software and a PC sound card is all that is needed to try RTTY contesting.

All the contest logging programs can keep the log, enter the time of QSO using the computer's internal clock, update the outgoing serial number automatically and check for duplicate contacts in

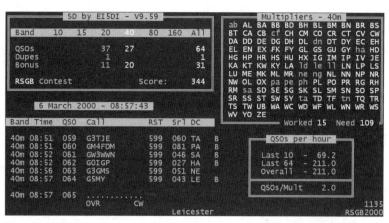

The EI5DI Super-Duper contest logging program *(screenshot courtesy EI5DI)*

milliseconds. They include a CW sender which (with the addition of a simple interface) will key the transmitter and send the outgoing exchange, as well as CQ calls and other programmable messages. The more sophisticated programs offer statistical analysis of your performance as it happens, often on a band-by-band basis. If your software is programmed with the scoring rules for the contest all you have to do is enter the callsign and exchange. The computer determines whether it is a bonus or multiplier contact, works out the points to be claimed and often updates a running total display on screen. For multi-operator (or single-operator 'assisted') categories, the computer can be connected to a packet TNC to enable DX 'spots' from a local PacketCluster to be fed to the screen automatically.

Using a computer to log contest QSOs in real-time can provide all the above benefits but there are a number of potential problems. For example, computers and displays can radiate RF interference and may be sensitive to the field from your transmitter – you may well need to do some screening and filtering, and cables should be kept as short as possible. Power cuts and hardware faults are not uncommon – make sure that the program you want to use writes contact data to disk after every QSO, be prepared to disable any disk caching utilities and copy the data frequently to floppy disk. You need to be fairly proficient in using a keyboard: 'hunt and peck' is usually not fast enough. Also, everyone makes keyboard errors and in the heat of a contest these may be worse than normal, bringing the possibility of mis-typing or even accidentally deleting information. Take time to learn how to use the program you finally choose and become completely familiar with it, before the contest starts. Try to start off in one of the smaller events.

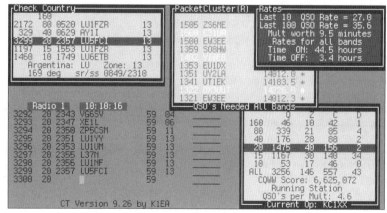

This is the CT contest logging program *(screenshot courtesy K1EA Software)*

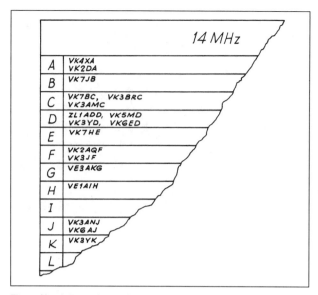

Fig 1. Check list suitable for contests with low QSO totals

Some contesters prefer to log on paper during the event but use a computer for dupe-checking and log submission afterwards. Most of the programs mentioned have a 'post-contest' mode. This avoids many of the problems of real-time logging, but can introduce another – transcription errors. Ensure you enter the callsigns correctly, as mistakes here cost points. Duplicate contacts may not be picked up because a zero was typed the first time, and a letter 'O' the second. Not all the programs check callsigns for validity as they are entered. If the program prints a dupe sheet, please remember to send it in with the log. Finally, always double-check the printed log against the original. Time spent doing this is well worthwhile.

Anyone who has a PC installed in the shack should be prepared to have a go at contest logging on computer. While it may seem daunting and there is a always a 'panic' when trying to work out how to erase or correct a wrong log entry for the first time, almost everyone finds they can adjust and the mountain of contest paperwork becomes a thing of the past. Perhaps the main advantage is that, when the contest is over, the entry can be printed, copied onto disk or attached to an e-mail message and sent off immediately, thereby doing away with the need to spend many hours writing up the entry by hand. In the past, many high-scoring contest entries have failed to appear in the results listings because after all the hard work of the contest itself, the entrant ran out of time doing the paperwork.

Of course it is still perfectly acceptable to log contest QSOs using paper and pen. Unless you are unhurried during the contest and your writing is neat, it will usually be necessary to re-write the entry onto contest log sheets (available from RSGB HQ) after the event. It will probably also be necessary to keep a duplicate check list ('dupe sheet') during the contest and, if applicable, a list of multipliers worked.

In an event such as the Commonwealth Contest check sheets are essential as it is normal practice to call stations as opposed to calling CQ. As the contact total in this event is quite small the form of check sheet shown in Fig 1 is quite suitable. In this example stations are entered by the last letter of their callsign, which some find an easier way of spotting a callsign on a check sheet. On the other hand, many use the first letter of the suffix, so that the checker can scan the correct part of the check list as soon as the first suffix letter is copied. This type of dupe sheet is satisfactory for any contest with low contact totals.

In contests where the contact total is high (such as the CQ, ARRL and WAE Contests), if the check is to be of any use for reference during the event itself, the listing can be sub-divided. If this is not done, there may be so many calls listed under each heading that it will be impossible to find a particular call quickly and maintain a reasonable contact rate. It should be mentioned that this type of check list is unsatisfactory for high-contact-rate contests, as the requirement of entering each call down on the list can be self-defeating. In this case it may be preferable to contact everyone and sort out the duplicates after the contest. Before things get really hectic programmable keyers can help in giving a little time to keep on top of the paperwork but it is in this situation where the computer really comes into its own.

In some contests where almost all of the contacts are made in response to CQ calls such as the ARRL Contest a check list can probably be dispensed with, it being quicker to work the occasional duplicate than to check every call. Of course duplicates in the log must be clearly marked after the contest. These problems do not exist in multi-operator contests when a log keeper will keep all the paperwork up to date. However, if only one second op is available then they can be more productively employed looking for multipliers, especially if a second receiver is permitted in the rules.

Almost all contest organisers accept entries on disk or via the Internet. This cuts down on postage costs but be sure to include all the information required by the organiser. Take care to submit Internet or disk entries in the format specified by the organiser. Many do not accept .BIN files or other coded formats. Entries should normally be plain ASCII. Mark the submitted log file with your callsign; for example, G0AAA.LOG would be easily identifiable, whereas if many log files all called CQWW.LOG are received, the organiser will have to do additional work to discover who submitted each log.

Organisation and strategy

Whatever the mode, whatever the contest, the key to success lies in detailed preparation, with the emphasis on detail. This really does mean thoroughly going through everything which is associated with the contest. Some first-time participants do not pursue contest operating further because during their first event they find their scores are low compared with the other UK stations. What they do not realise is that the leading stations may have well spent a whole year preparing for that single event, testing equipment and antennas, analysing results and planning strategy.

What is meant by 'strategy'? Basically this is the combination of operating technique, band changing, multiplier hunting, timing of meals, shack organisation, sleeping arrangements and a host of other factors which, when taken together, mean that you will at least make one more point than your nearest rival and thus win the contest. Take a look at the results of a few contests. It is surprising how often there is a very narrow margin between first and second place. Although 'Murphy's Law' can play its part, the difference is usually due to attention paid to relatively small details in station performance.

Rules

Contest rules for individual events are published by the organisers, and are summarised in magazines such as *RadCom*, *QST*, *CQ* and *NCJ* (*National Contest Journal*). Summary rules are available on the packet cluster and on the Internet. The contesting web site www.contesting.com is an excellent source of information about contesting from a US perspective, and it links to a comprehensive contest calendar. On the main RSGB site www.rsgb.org there is a link to the HF Contests Committee page and this contains rules, results and claimed scores for a selection of the Society's contests.

Contests organised by the RSGB are shown in *RadCom* in the regular feature 'Contest', while a summary of some other events appears in 'HF News'. There is also the RSGB Contesting Guide which appears in the October issue. It is recommended that a full set of rules be obtained by any competitor considering a serious entry.

It may seem obvious to say "read the rules", but many participants do not *read* them, they glance through them or read them after the event! Read carefully so that a clear understanding is obtained of exactly what the organisers require as well as what is not required! For example, some contests automatically disqualify entrants with more than 2% duplicates. Make sure the points and multiplier system is understood. Some contest organisers request copies of an entrant's dupe sheet if the number of contacts exceed a specified number. It is always helpful to the adjudicator to include it – it does make checking considerably easier. Note the closing date for submitting entries and to whom the log has to be sent. Again organisers vary, some request that the log be posted by a certain date, others that the entry must be received by a certain date. In the latter case it is important to check that the mailing method will ensure that the entry arrives in time. The post office can give you an *approximate* time that a letter will take. Having taken the trouble to participate and prepare your log, it may be worthwhile including a stamped self-addressed post card (airmail if abroad, with IRCs rather than stamps) acknowledging that your entry has arrived safely and in time.

For many contests it is now possible to access an Internet site in order to check that your entry has been received and classified in the correct category.

Propagation

For HF contests, knowing when bands are likely to be open to various parts of the world at different times is a basic necessity and the reader is advised to become familiar with the information presented in Chapter 4. A set of detailed predictions for the month of the contest is ideal. A variety of computer programs are available for propagation predictions which allow you to enter radiation angle, estimated solar flux number and a choice of either long-path or short-path predictions. Use may also be made of the grey-line DX path. Originally this method was developed for 3.5 and 1.8MHz DX but the principle applies equally well to the other HF bands. There are both computerised and manual methods of calculating the path parameters. These calculations can be given visually using a DX Edge available from the RSGB (see Chapter 4). Remember that 28, 21, 14MHz, and to a certain degree 7 and 3.5MHz, all follow a similar pattern, but the lower the frequency, the shorter the opening. Note that European contacts can sometimes be made more easily by beaming due south, particularly on 21 and 28 MHz. The frequencies of the HF beacons should be known for they can be an excellent guide to band conditions. Beacon frequencies are given in Appendix 8.

Antennas and equipment

Antennas are a crucial part of the contest station, the one item that offers the greatest scope for ingenuity and experimental work. The antenna used must be suitable for the particular contest in mind; for example, it's no use trying to work a lot of Europeans on 7MHz with a ground plane. A dipole fairly close

Using a trailer-mounted telescopic tower and beam will greatly simplify antenna erection. This is G3VHB (left) and G3LNS raising the TH7 beam for G3VHB/P in NFD 1994

to the ground would provide a greater contact rate. Conversely the low dipole would be of little use in trying to work into Australia where a good low-angle radiation pattern is required. Similarly, forward gain might be beneficial under some circumstances, but a good front-to-back ratio could be just as important to reduce European QRM for DX contacts. The subject is vast and there are several excellent books available from the RSGB dealing with all aspects of the subject.

During the contest there may be several stations using high-gain antennas, erected just for the event, that can overload your receiver. Even with a top-grade receiver a step attenuator in the receive antenna line can make all the difference between 100% copy and none at all, particularly on 7 and 3.5MHz. Experiment with equipment layout to find the arrangement that suits you best. It is apparent from viewing several top contest stations that none are exactly the same. The arrangement of receiver, transmitter, keyer and computer so that operating is comfortable and everything comes to hand easily is very personal.

Multi-operator, multi-band contests have one additional facet. *All* of the equipment must be capable of operating simultaneously without degrading receiver performance on each band. Commercial filters are available which will help to cut down inter-station interference, or they may be home-brewed.

Many operators enter their first contest from their home station only to find shortcomings in the station configuration. It is important that the station equipment is capable of virtually continuous operation for the duration of the event. Other shortcomings may become very apparent during a long period of operating. For example, the keying monitor may be too loud or too soft. The antenna rotator may go out of sync. The VFO tuning knob may be just a little too high from the operating table for comfort during a long spell of operating. The chair which for occasional sessions in the shack seemed ideal can become very uncomfortable after 36 hours. It is surprising how few chairs are really comfortable for long periods.

A Field Day station in a riverside setting

It must be the aim of every serious contest operator to use the best equipment that is available. A thorough examination of all the equipment should be made in the months prior to the event – do not leave everything to the last minute! If the station for the contest station is to be made up of equipment borrowed for the occasion from friends or club members, check that the system works well in advance and that everything is compatible.

Regular maintenance is also very important. Most serious contesters maintain and experiment with antennas during the summer months and service all their equipment during the winter. If there is the slightest doubt about an item of equipment - check it. It may take a whole day to bring down and check a beam only to find that everything is working fine but at least your mind has been put at rest and the confidence that everything is working well is worth many extra points.

Some maintenance procedures that should be carried out regularly are to examine all coaxial fittings and ensure that the weather-proofing is still intact, and to check all feeders for excessive loss. Checks for any EMC problems on all bands should of course be made regularly.

Club entries

There may be a club member who is prepared to make an existing station available for a club entry in one or more contests, or the club may be fortunate to have a purpose-built club shack. More typically, a club's first attempt at contesting will be in a field day or other portable operation, and operating in the field brings additional challenges.

Contest operating can be an enjoyable club activity – potential exists for large antenna arrays to be used that would be impossible from most amateurs' home locations. Even a modest operation will depend for its success on the planning and organisation that is applied prior to the event.

One of the first decisions to be made is the attitude of club members – do they want to enter for the fun of the event and gain contest experience or are they entering as a dedicated band of enthusiasts who are prepared to go to some lengths to win? Entering to win can be very hard work, requiring a lot of time and dedication. Entering for fun is as easy as you make it, but more ambitious members of the group may not be satisfied. Achieving a balanced approach is a special skill required of a club contest manager.

The most difficult problem that groups always have to face is the question of operators. A club may for example have two really proficient CW contest operators, but many a lengthy committee meeting has been held to try and decide how long these two should operate when balanced against the need to satisfy the other members who have come along to help put the station on the air and provide equipment. One person should have overall responsibility for the group effort. It is up to him or her to ask other members to take on particular aspects of the organisation as needed. Responsibilities can be divided in several ways, eg site and accommodation, transport, power, eating arrangements, antennas, antenna erection and equipment for use on each band. Decide on how big a contest you are going to tackle. A multi-band 48 hour contest needs a lot of operators and equipment for a successful entry. A more modest effort to gain experience is advised for a group just embarking on contest operating as a club activity. Next find a good site and obtain permission to use it. There is no doubt that location and type of soil does play quite a significant role in the efficiency of an HF installation.

Produce a budget of the expected cost of participation. Hire of generators and the fuel to run them can be quite expensive. Some groups are able to meet these additional expenses from their subscription income but often an extra levy will be needed on those taking part – make sure that this is acceptable before proceeding. All the 'gear' that is to be used should be tested as an entity before the contest. This often shows up surprising problems which can be dealt with in advance. Accommodation for equipment and people should be considered, especially if you are operating from an exposed site. Tents may not be adequate under severe weather conditions – would it be possible to hire or borrow a caravan? If so how will you get it on site and retrieve it later?

Safety

Portable operating generally means that antennas should be erected as high as practicable, usually supported by means of a tubular mast. Select the antenna you intend to use carefully – small tri-band beams can give quite good performance but for superior gain the quad is hard to beat.

Now for the mast itself, the erection of which may well be the most dangerous activity contemplated in the contest – so do have sufficient help and plan carefully exactly what you are going to do and make sure everyone helping understands what is expected of them. If you have not erected a mast before or have a limited amount of help, do not be too ambitious. Choose thick-wall scaffolding poles, *not* the thin-wall TV antenna masts which tend to buckle very easily. The tubular section should be joined with external sleeve clamps, *not* the interior expanding type of connector. When using a sleeve with a bolted flange (like the Jaybeam type) arrange the flanges to oppose the bending during lifting, or better still weld two more ribs at 90° to the bolt flange to provide additional rigidity. The person in charge of the group erecting the antennas should check personally that all of the bolts have been tightened. It is worthwhile using shakeproof washers. A gin-pole is practically essential, and if properly constructed will simultaneously ease erection of the mast and increase the overall safety of the operation (Fig 2).

As a rule of thumb, the gin-pole should be a third of the length of the mast. If a base plate is being used it should be securely fixed to prevent slipping when lifting the mast. There should be a ring of four guys for each section of the mast. Calculate the length of guys beforehand and have them already prepared and clearly marked – allow a reasonable amount of spare for handling and tying off. The guy stakes should be placed on a

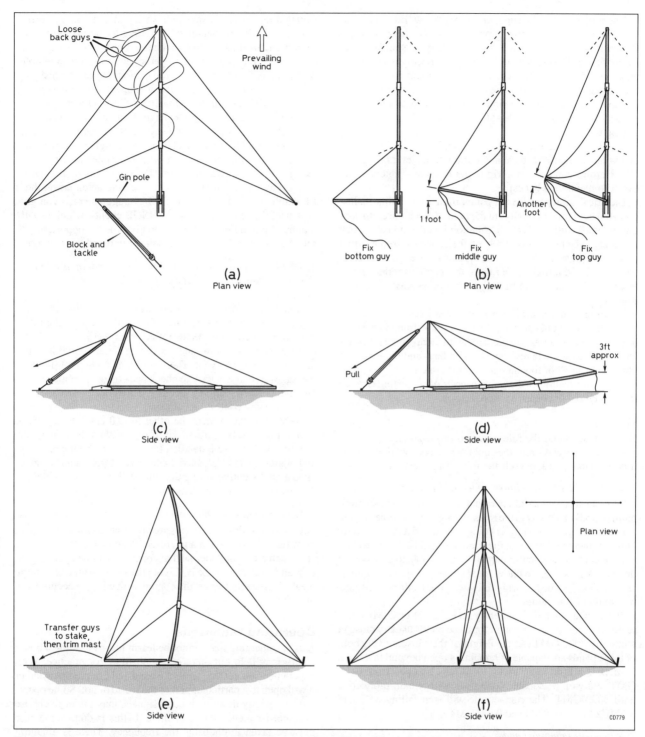

Fig 2. Erecting a mast using the gin-pole method. The mast typically consists of three scaffold poles giving 60ft, with another pole for the gin-pole. Guys are 8mm polypropylene. (a) Overhead view showing the mast and gin-pole laid out on the ground prior to erection. (b) Setting the pre-bend; the amount varies the top load. (c) Side view showing the gin-pole being raised using the block and tackle. (d) As the guys take the strain, the mast should be slightly bent as shown. The antenna can now be fitted to the top of the mast. (e) With the mast in position, the gin-pole guys are transferred to stakes and adjusted to remove the bend (The other two sets of guys are not shown for clarity.) (f) The mast in final position. Note the procedure should be reversed to lower the mast, ideally lowering *away* from the wind

circle with a radius not less half than the height of the mast. Use substantial lengths of angle iron driven into the ground at an angle away from the mast. The depth required will depend on the soil composition but usually if they are driven in about 60/90cm they should be adequate for most installations. Strips of reflective material to mark the low end of guys and stakes may prevent one of the helpers driving or walking into them in the dark. Be especially careful in placing guys and stakes if you are on public property and near to public footpaths.

If there is a wind blowing always raise and lower the antenna into the wind. The force that can be applied to a beam at 15m by a strong breeze may take the guys right out of everyone's hands. Be prepared to accept reduced height rather than risk the loss of the entire installation to a powerful gust of wind.

Use gloves to handle rope, and everyone working close to the mast should wear some form of protective head gear. As shown in the plan view in Fig 2 the side guys should be securely tied to the stakes though they will probably need adjustment once the mast is up. The back guys should be tied to the back stakes rather than risk the mast going 'over the top'. The gin pole must *always* have side guys and these should be tied to the side guy stakes. It is best to allow some slack when side-guying the gin pole.

If the mast does get out of control and starts to fall, call to everyone on site to stand clear and if possible, let it fall gradually. Heroics could mean a trip to the hospital – antennas are easier to mend than bones!

Electrical safety is also very important when using mains-powered equipment literally 'in the field'. Check the earth continuity of all earth leads on distribution boards and cables that are likely to be used. Do not rely on the generator frame sitting on the wet grass as the safety earth return – use a proper earth spike and bond it to the generator – this can also reduce interference. All gear should be fully enclosed against misplaced fingers.

Never erect masts near overhead power lines.

Finally, remember that weather can play strange tricks on high exposed locations. Go prepared with plenty of warm clothing as well as some in reserve, sleeping bags and the means of obtaining some hot food under extreme conditions.

More information on the safety aspects of contesting is given in the VHF/UHF contest section later on.

Strategy

Having understood the rules, studied propagation for the period in question and sorted the equipment aspect out thoroughly, it is advisable to ask oneself the following questions.

1. When will I start and finish; when will I take rest periods?
If the contest is over a short period with no restrictions on minimum periods of activity in the rules, the answer is simple. Where a 36 or 48 hour DX contest is being considered a careful study must be made to determine the time that will be most productive for contacts and multipliers. It is impossible to give advice on exactly how long to allow for sleep as people vary in their requirements, some requiring only two hours, others needing a minimum of four hours.

If you plan a single-band entry on either 21 or 28MHz this really does not pose a problem for these bands usually close in winter by 2000 GMT. At the height of the sunspot cycle both of these bands are capable of remaining open beyond that time. 14MHz is a little more tricky for it may be open from 0400 GMT and with reasonable conditions will remain productive until 0000 GMT. The contact rate will start falling on 7 and 3.5MHz between 0200 and 0400 GMT.

2. When will I change bands?
The majority of 'big guns' and DXpeditions change bands either on the hour or the half-hour. After studying the propagation forecast a plan can be drawn up with the object of obtaining as many multipliers or countries on each band and being on a band at the right time to yield a maximum number of contacts. Having prepared a band plan, stick to it as far as multipliers are concerned, even if it means leaving a pile-up on 14MHz to look for multipliers on 21MHz. Conversely be prepared to take advantage of short openings on 28MHz.

3. Who has won the contest over the past two years? What equipment and antennas were used?

Detailed results are always published and additional information can usually be obtained by writing to the operators concerned enclosing a stamped addressed envelope. After a big event, the top contesters often get together to exchange notes on equipment, antennas and propagation. Do not be disheartened if for example G4 . . . has a four-element wide-spaced Yagi on 7MHz at 50m – obviously he will be able to work DX that others cannot hear – but even the best operators make mistakes. While he is pounding away on 7MHz an opening may occur on 21MHz which yields an even higher scoring rate! There is much that can be done with experimental wire antennas and phased arrays on 3.5 and 7MHz to bring up the station rating. Pay particular attention to receiving antennas on these LF bands. If you have the space, Beverage antennas can give very useful front-to-back ratios. RSGB events, which usually require a running serial number, do give you the opportunity to see how last year's leaders are doing compared to your score.

4. What was the winning score? How many contacts and multipliers? What were conditions like? How did other UK stations fare?
Again, the majority of this information will be published and a few hours studying the results will usually produce the information required. The '3830' reflector on the Internet is a forum for reporting scores and discussing contest conditions immediately after the event. There is also an archive of previous messages so that it is easy to check back on the previous year's comments when planning an entry. The reflector can be found from the www.contesting.com site.

As well as the leaders the other top 10 entrants are worth examining closely, particularly if one of them is known personally. In this case a detailed knowledge of both equipment and planning can be gleaned. Scoring rate targets can be calculated as an incentive during the contest. Tables 1 and 2 and Fig 3 show typical scoring plans and rates.

5. What should I do about meals?
Food and non-alcoholic drinks are important. A lot of contesters have fizzy drinks or tea and coffee. Sandwiches for breakfast and lunch are a good idea so that you can continue operating on CW and listening on SSB. Light snacks are better than large meals. It is useful to start the contest with a large vacuum flask of coffee.

Operating technique

This is something that cannot be learnt just by reading a book; theory may help but there is no substitute for practice. With experience you get the 'feel' of a band and can predict whether it will open to a particular part of the world or not. Some operators, especially those with good signals, may prefer to choose one spot frequency and stay there. If this is done some care must be taken in choosing the frequency. There is a natural tendency to move up the band to avoid what appears to be an area full of interference. Tighten up the selectivity and you will be able to find a clear spot. When looking at the lower portions of the bands do not forget that the lowest sections are restricted to those holding Extra Class licences in the USA.

A conscious decision must be made about when to call and when not to call CQ. A general rule of thumb is that more points will be made calling CQ and getting replies, than by calling stations. This obviously presupposes that you have a reasonably commanding signal or that you are operating from a rare spot and that everyone will be looking for you. There are however exceptions to these rules. The aim of some contests is

Table 1. 1998 RSGB 7MHz CW Contest

	G0IVZ		G2QT (op G4TSH)	
	QSOs	Mult	QSOs	Mult
1500–1600	74	26	68	24
1600–1700	51	7	48	7
1700–1800	36	3	30	2
1800–1900	38	5	28	4
1900–2000	50	5	28	4
2000–2100	28	5	24	5
2100–2200	30	7	31	10
2200–2300	34	7	34	7
2300–0000	20	3	18	2
0000–0100	28	—	11	2
0100–0200	31	3	30	1
0200–0300	38	2	30	—
0300–0400	27	—	24	1
0400–0500	12	—	22	2
0500–0600	14	—	17	2
0600–0700	19	—	17	3
0700–0800	26	1	13	1
0800–0900	27	3	26	2

Table 2. 1999 CQ Worldwide CW Contest 21MHz single band (G4BUO)

Time	First day			Second day		
	QSOs	Zones	Countries	QSOs	Zones	Countries
0547–0600	2	1	2	3	—	1
0600–0700	40	11	26	28	1	3
0700–0800	73	4	14	66	—	4
0800–0900	91	3	13	66	—	1
0900–1000	70	2	8	72	—	—
1000–1100	77	—	8	43	1	3
1100–1200	53	6	14	39	1	1
1200–1300	114	—	5	49	—	—
1300–1400	107	—	1	40	—	1
1400–1500	115	2	4	44	1	1
1500–1600	109	2	3	84	—	1
1600–1700	106	1	2	82	—	1
1700–1800	95	1	3	86	—	1
1800–1900	109	—	2	77	—	—
1900–2000	113	—	2	56	—	1
2000–2100	41	1	7	12	—	2
2100–2200	15	2	3	5	—	1
2200–2300	4	—	—	1	—	—

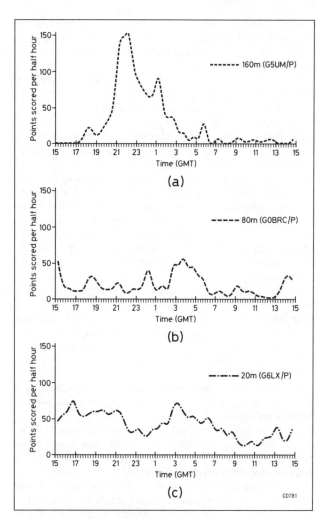

Fig 3. Scoring rates for three stations in NFD

to work stations in one particular region and only these should call CQ. Furthermore under some band conditions, or towards the end of a contest, there may be little response to CQ calls. These are the times when it is necessary to systematically search a band for multipliers or for additional points. Remember it is extremely bad practice to find an apparently clear frequency and start calling CQ. Listen on the frequency for at least 10 seconds and then ask "Is this frequency in use?" on SSB or send "QRL?" on CW. Only if no one replies can a short CQ be tried. If the frequency is in use then move away and repeat the procedure. While no-one has an absolute right to occupy a channel it is generally understood that the station who was using the frequency first has the right to remain there. For this reason it is unwise once a frequency has been found and contacts are being made to give it up, unless changing bands or looking for multipliers.

As already mentioned, in most contests there is a bonus or multiplier system, often based on the number of countries worked. It is always useful to have in mind before the contest starts just how much value in time (or contacts lost) each multiplier has. For example in the ARRL DX Contest, one multiplier is equivalent in terms of points to about 15 contacts, or at a peak scoring period to about seven to eight minutes operating time. In the CQ DX Contest where multipliers are countries and zones, each multiplier is worth about eight contacts

or roughly five minutes operating time. These figures give some idea of the 'worth' of a multiplier and at peak contact times it does not pay to take time off to look for multipliers. With an event lasting longer than 24 hours then it is sound advice to go for contacts the first day and concentrate a little more on quality during the second day. Scoring rates or the number of contacts you should achieve each hour will vary from contest to contest and from year to year. However, it is helpful to establish a target to try and achieve per hour. Probably the maximum scoring rate using a 'G' call is in the region of 120–150 per hour on CW and maybe twice that on SSB. These rates apply to the peak periods when the band is open to the USA. Again, most logging programs display the current rate.

The scoring rate is not necessarily directly proportional to the speed of sending – QRS for slow stations will save time in the end. The degree of abbreviation and callsign repetition enter into the equation. The speed of sending should be such that contacts are dealt with efficiently but not at the expense of having to meet continual requests for repeats from other stations. The more a callsign is repeated, the more stations hopefully will join in the 'pile-up' waiting for a QSO. The bigger the 'pile-up', the harder it becomes to pick out calls – so there is a question of balance with the aim to just keeping the 'pot' simmering and not boiling vigorously! Time spent listening to well-run DXpeditions can be very instructive in learning techniques for handling pile-ups whilst maintaining some semblance of order. Large pile-ups from Japan can be a problem both on SSB (with the difficulty of phonetics) and on CW with weak signals. Do not be tempted to withhold your callsign in an effort to control the pile-up.

Often operators can benefit by considering if all parts of their transmissions are really necessary. It may be polite to say "73 and good luck in the contest" but it will certainly lower your scoring rate without adding any essential information to your or his log. After all, the other side of a contact knows the number that he sent and does not want it repeating back to him unless there is a doubt. Accuracy is of great importance and should not be sacrificed in the interest of speed – again it is a question of striking the right balance. Note that in CW contests it is quicker to send at 20wpm giving all the information once than to repeat everything at 35wpm. Many operators are now using memory keyers or DVKs to give CQ calls and reports. At first sight these may appear to be gimmicks but they do save time and give an extra few seconds to deal with the log and check lists.

When tuning the band and calling people (known as 'search and pounce') use your computer or dupe sheet to verify you have not already worked the station on that band before calling them. On SSB do not be tempted to call using the last two letters of the callsign. This is bad practice and some top contest stations will completely refuse to work callers who only send 'last two'.

Listen to the top contest stations working and note their style and the number of contacts that they achieve in an hour. If possible arrange for a local operator to record you operating in a contest so that you can analyse your own technique. Listening to a recording highlights the mistakes that you may be making. You can only work stations one at a time but it is essential that you work them faster than the opposition!

Finally, know and observe the IARU Band Plan – do not call non-contest stations for points and remember at the end of the day it is only a hobby – apply the rules of sportsmanship to your operating.

Analysis
Within 24 hours of completing a contest, make notes for next year while everything is still fresh in your mind. It is indeed a rare person who can recall months later *all* the little details which need to be improved to make the operation more comfortable and efficient. Indeed, you may prefer to keep a sheet of paper handy during the contest so that any thoughts can be jotted down immediately.

Graphical analysis of the scoring rate can also highlight peak times and enable you to examine the times when the scoring rate plummeted – this is where the analysis of one contest merges into the preparation for the next. The rate at which contacts are made is of prime importance; if displayed graphically it is easier to get an overall picture of your operation. See Fig 3. If graphs are drawn for several years then you have a better base on which to formulate a contest plan showing when to change bands and look for multipliers. It is interesting to plot a contact rate graph and a scoring rate graph. This will illustrate the importance of concentrating on high-point-value QSOs, and may suggest when the best time to sleep is during a 48-hour contest.

Sending in an entry
Official forms should be used if specified, as it does make the adjudicators work so much easier. Don't forget the cover sheet and declaration. Like the rules, these two points may seem obvious but adjudicators – no matter how impartial they try to be – tend to be biased against the log submitted on narrow lined absorbent paper written with a green felt tip pen!

It is imperative that duplicate contacts are clearly marked and that points are not claimed for them. The check list or dupe sheet should be included with your entry to assist the adjudicator. Also if you get someone else to type your log – do check for transcription errors. Transcription errors are also common in computerised logs – at the very least make a cursory examination for nonsensical callsigns.

Comments and helpful suggestions for improving the contest are always welcome, while the odd anecdote makes a welcome break for the adjudicator. Last, but not least, ensure that you have put sufficient postage on your entry and that it is sent to the correct address. It would be a pity to see all that time and effort wasted.

VHF/UHF CONTESTS
Contesting on VHF is essentially great fun, just as it is on HF. It also poses a great many challenges for us, both as individuals and in working together as a team. Experience can only be gained through practice, improvement and refinement. It involves learning from failure and being sufficiently self-motivated to keep on going when the going gets rough and the odds are apparently stacked against you, particularly when it's wct, windy, you're tired, no one will do it the way you want and you know that you're right, the equipment's failed yet again and somebody forgot to pack the milk for the coffee and the bottle opener too! If it can go wrong, it usually will and does, but when it does go right, it's one of the best feelings in the world. You know that you have worked well and rightly deserve your success. You don't have to actually win a contest to feel like this – simply seeing your scores and placings improve, overcoming practical problems or exceeding personal goals can provide an enormous amount of satisfaction.

Contesting is very much about being in competition with yourself, rather than just being about beating fellow contesters. Don't expect to beat the major groups on your first attempt. They have worked hard to get to where they are and invariably possess the resources and experience which you are hoping to gain. However, always remember that they can and, at sometime in the future, will be beaten. Select the bands, contests and sections which are of interest to you and carefully maximise your available resources. Set yourself realistic targets and gain the satisfaction of seeing a steady improvement.

The first thing you need to do is to prepare. Read the rules, which for RSGB contests are published towards the end of each year in *RadCom*. These rules tell you what is and what is not acceptable in the operation and paperwork requirements associated with the wide variety of contests throughout the year. Then, read the individual contest rules. Alright? Now read them again. Have you noticed, for example, that the third and fourth sessions in the Backpackers require Rule 14e and not Rule 14c as applied in sessions one and two? It is small, subtle changes like this which facilitate the smooth running of contests, particularly when there may be more than one event taking place simultaneously.

As you read through this chapter you will, we hope, receive a great deal of information. In fact, there may be too much information to take in on a single reading and so we recommend that you return time and again and consider the issues raised. These may also form the basis for group or club discussions to see how they effect your decisions and reasoning in

RSGB VHF/UHF CONTEST CALENDAR

January	144MHz CW, 70MHz Cumulatives
February	70MHz Cumulatives, 432MHz Fixed/AFS
March	March 144/432MHz, 70MHz Fixed
April	144MHz Fixed Station Cumulatives, 1.3/2.3GHz Fixed
May	432MHz–238GHz, 432MHz Trophy, 10GHz Trophy, 144MHz, First 144MHz Backpackers
June	IARU 50MHz, 50MHz Trophy, 1st 50MHz Backpackers, 70MHz CW, 2nd 144MHz Backpackers, 432MHz FM
July	VHF NFD, 3rd 144MHz Backpackers, 2nd 50MHz Backpackers, 144MHz Low Power, 432MHz Low Power
August	432MHz Fixed, 70MHz Trophy
September	144MHz CW Cumulatives, 144MHz Trophy, 4th 144MHz Backpackers
October	144MHz Cumulatives,432MHz–238GHz IARU, 1.3/2.3GHz Trophies, 1.3/2.3GHz Cumulatives, 432MHz Cumulatives, 1.3/2.3GHz Fixed
November	1.3/2.3GHz Cumulatives, 432MHz Cumulatives, 144MHz CW Marconi, 6h 144MHz CW
December	144MHz Fixed/AFS, 70/144/432MHz Christmas Cumulatives

A typical VHF contest scene!

how you actually go about entering contests. Indeed, this is the main purpose of this section. There are many excellent books which deal with the theory, design, construction and operation of equipment and readers are advised to refer to these for further reading.

Sites

At VHF there is something of a myth that height above sea level is everything, but this is far from being the complete story. What is more important than plain elevation is whether the site has a clear unobstructed take-off in the important directions. This can mean that even a small rise of a few tens of metres can be a very good location if the surrounding land for a good number of miles is essentially flat. If you happen to live near to the coast a site on or very close to the water can be excellent, particularly on 70cm and higher, since you will be able to take advantage of the fragile marine ducts which sometimes form over water and break up very quickly when they hit land. This said, nothing is ever hard and fast, and these sorts of conditions are not present for so very much of the time, so under normal circumstances you could well be better off on a site which is further inland but higher. If, like many people, you are planning to use a site which is some way inland, then height certainly seems to become more important and you ideally want to be a good few hundred feet above the surrounding terrain. However, even away from the coastal plains of the country there are many flatish parts of the country where sitting on top of a relatively small dimple will pay big dividends.

Since there is no 5000ft mountain in the UK which has a clear shot for 360°, and which is only 10km from the coast in all directions – and even if there was it would already be well and truly booked by a major contest group – it is very important to try and decide what directions are important to you, and these may vary from contest to contest. For a major Europe-wide contest, the bulk of activity is likely to be in Central and Southern England and of course in the rest of Europe – for this

sort of event you will want to pick a site with a good take-off in those directions. However, a UK-only contest with postcode multipliers will lead you to wanting a different site, with a good take-off to most of the UK, and you may be prepared to sacrifice some performance into Europe. If you are planning to travel any distance, you will probably make your initial choice of site from looking at the OS maps; however, you cannot rely on this alone – you really need to go and take a look at your proposed site before the contest to check out how to get access etc. At this point a good hint is to take a small station with you – perhaps even just a mobile – and make sure that the beacons or even repeaters are as loud or louder at your proposed site than in the rest of the surrounding area. If you do not have any amateur station with you, you can get a fair idea from just using ordinary broadcast VHF stations on the car radio.

So, we've discussed how to find your ideal site – now how do you go about getting it? Unfortunately, the obvious technique of just driving to the site on the day of the contest and setting up is the wrong one. First you need to check that the site does not have anyone who regularly uses it for contests – contesting etiquette and common sense says that you do not just turn up and use someone else's site without first making contact with them to see if they are planning to use it. Just turning up very early at a site and claiming it on the basis that "I was here first" is a guaranteed technique for not winning friends and influencing people! Equally, don't pick a site that is only a few kilometres down the road from one which is likely to be used by another group – you are only going to cause both of you a lot of heartache and shortened tempers with all the mutual QRM which will be present during the contest, no matter (within sensible practical limits) how clean your transmitters are or how good your receivers are. The sensible exclusion zone around the 'big gun' portable stations is probably a few tens of kilometres in directions where they are likely to spend a lot of time beaming – even at 20km, these stations can be well in excess of 100dB above noise with you, and hardly any VHF receiver will take kindly to this level of abuse.

While on this subject, if you are going somewhere new, then you should also be conscious of the locals who live near your station. Particularly if you are travelling somewhere remote, many of them may not be accustomed to big signals on the band and may not have engineered their stations to cope with this much RF. Equally, the attitude "Well, it's not my fault – get yourself a decent receiver" will do nothing to further the

CODE OF PRACTICE FOR VHF/UHF/SHF CONTESTS

1. Obtain permission from the landowner or agent before using the site anfd check that this permission includes right of access. Portable stations should observe the Country Code.
2. Take all possible steps to ensure that the site is not going to be used by some other group or club. Check with the club and last year's results table to see if any group used the site last year. If it is going to be used by another group, come to an amicable arrangement before the event. Groups are advised to select possible alternative sites.
3. All transmitters generate unwanted signals; it is the level of these signals that matters. In operation from a good site, levels of spurious radiation which may be acceptable from a home station may well be found to be excessive to nearby stations (25 miles away or more|).
4. Similarly, all receivers are prone to have spurious responses or to generate spurious signals in the presence of one or more strong signals, even if the incoming signals are of good quality. Such spurious responses may mislead an operator into believing that the incoming signal is at fault, when in fact the fault lies in his own receiver.
5. If at all possible, critically test both receiver and transmitter for these undesirable characteristics, preferably by air test with a near neighbour before the contest. In the case of transmitters, aim to keep all in-amateur-band spurious radiation, including noise modulation, to a level of –100dB relative to the wanted signal. Similarly, every effort should be made to ensure that the receiver has adequate dynamic range.
6. Above all, be friendly and polite at all times. Be helpful and inform stations apparently radiating unwanted signals at troublesome levels, having first checked your own receiver. Try the effect of turning the antenna or inserting attenuators in the feed line; if the level of spurious signal changes relative to the wanted signal, then non-linear effects are occurring in the receiver. Some recent synthesised equipment has excessive local oscillator phase noise, which will manifest itself as an apparent splatter on strong signals, even if there is no overloading of the receiver front-end. Preamplifiers should always be switched out to avoid overload problems when checking transmissions. If you receive a complaint, perform tests to check for receiver overload and try reducing drive levels and switching out linear amplifiers to determine a cure. Monitor your own signal off-air if possible. Remember that many linear amplifiers may not be linear at high power levels under field conditions with poorly regulated power supplies. The effects of overdriving will be more severe if speech processing is used, so pay particular attention to drive level adjustemnt. If asked to close down by a government official or the site owner, do so at once and without objectionable behaviour.

general cause of contesting, or indeed your or anybody else's ability to use the site in the future. Some pre-contest planning, perhaps by contacting the locals, maybe via the local club, in order to explain to them what you are planning to do and perhaps to invite them to come and see the station could go a long way to turning a potentially unpleasant situation around. They will then feel more involved in the situation and also if you understand the potential problems, you may be able to avoid beaming your eight Yagis over the top of a particular local station's house more than absolutely necessary.

In all likelihood, the site you have chosen to use will be on private land – obviously you need to obtain permission from the land owner to use it. Even if the site is public land, be sensible – don't obstruct gateways and paths, and don't damage any of the surroundings – basically, observe the Country Code (see panel on the left).

Operating

Operating techniques vary enormously between individuals and between different contests and there is no single right and wrong way to do things. For example, 2m contests at busy times tend to be relatively frantic affairs, but 23cm events are usually quite relaxed. First read the rules and make sure that you know what is required of you during the contest. Check out what the exchange is and ensure that you have all the relevant information to hand. It is also worth knowing what the WAB details are for your site because there are almost bound to be a number of people who will ask you for these.

Throughout the event, you have a basic choice between calling CQ and tuning around for calls, and the relative amount of time you spend on each activity will depend upon what band you are using, how big your signal is, whether it is a multiplier contest, what conditions are like and what time it is in the contest, to name but a few factors. Certainly for the main contests on 2m and 70cm it is likely that, unless you have a very small station indeed, you will want to call CQ for some of the time. When you call CQ, choose your frequency carefully – obviously make sure that you aren't sitting too close to one of your local stations from whom you could suffer heavy QRM, but equally, make sure that you aren't going to have heavy QRM at the DX end of the QSOs. For example, if you are expecting to make a lot of contacts into Germany, don't sit 1kHz LF of one of the big German portable stations. If you find a particular frequency doesn't run well, then try another – you may well have some QRM at the far end which is too distant for you to hear.

While calling CQ make sure that you sound enthusiastic and that you are enjoying yourself even if you are frozen to the bone and finding it slow going – remember that you need to attract that casual station just tuning the band and make him/her just want to call you. Don't leave too long a gap between CQs – two to three seconds is usually about right – this is long enough for anyone who is a bit slow off the mark to go over to transmit, but not so long that someone who just catches the end of your call and who wants to know who you are before calling, gets bored and tunes onto the next station. When someone answers your call, say "Good morning" or whatever and pass the exchange clearly and concisely. If signals are reasonable, it usually isn't necessary to send the information more than once. It is worth passing the information in the same order as the columns appear on the log sheets (callsign, report, serial number, locator, county etc), and do spell out your locator phonetically. If you miss any bits, or are unsure that you have copied it correctly, do ask for a repeat or confirmation – it is better to take a little longer to get it right that to lose the points for the QSO. Finally, when all is done, say thanks and move onto the next QSO. There is a fine balance to be drawn between making the QSO very quickly and needing lots of repeats, and making the QSO too slowly so that if anyone is waiting to call you they will get bored and tune off. If more than one station replies to a CQ call it is worth asking the other caller to stand by – more often than not, they will do so. If you are uncomfortable with any of these ideas, before you call CQ, take a listen around the band, particularly to the big stations, and see how they handle their QSOs. As you are calling CQ, keep the antenna moving around – don't forget that VHF/UHF antennas often have quite narrow beamwidths and the more distant QSOs rely upon both your and the far ends' antennas being in the right directions.

At some point you may feel the need to stop calling CQ and take a tune round the band. This is likely to be when your QSO rate from calling CQ drops too low. A typical technique is to start from one end of the band, with the antenna pointing in one direction and work your way to the other; then nudge the antenna round a bit and starts at the bottom of the band once again. As you tune through the band you'll come across various stations that you may or may not have worked already. Pay particular attention to the weak signals – these may well be the DX and of course are worth more points. Turn the antenna enough towards the station you are calling to make the signal reasonable, but don't worry about getting it all the way round if you don't need to. Do check your dupe sheet or computer before calling them to make sure that you haven't had a previous QSO. A dupe sheet is just a list of callsigns of stations you have already worked on the band in a form which you can look up very quickly. If you are logging by computer you won't need to use one of these – the software will have a substitute function built-in – but if you are logging by hand, you will need to maintain a manual dupe sheet as well as your log. A popular way of making a dupe sheet is to have a piece of paper (preferably A3) ruled into 26 columns headed A-Z, and maybe even with a couple of spare overflow columns. As you work people, not only do you enter them in your log, but their callsigns also get entered onto the dupe sheet – in the column determined either by the first or last letter of their main callsign. It doesn't matter which option you choose – just so long as all the operators know which you are using. Don't forget the final suffix has no bearing on whether the contact is a duplicate – eg a contact with DB8KJ and DB8KJ/P are duplicates.

When you call a station, do use phonetics for your call, even if they are strong – of course you know what your callsign is, but they may not be quite sure. If you don't get a response immediately try again a couple of times – making sure that you don't have the RIT in, split VFO enabled, repeater shift enabled etc, and do try calling on CW if you are just too weak for SSB. If nothing succeeds at making a particular station hear you, it can be worth either storing their frequency in the other VFO or a memory (if you have these facilities), or simply writing the frequency down on a piece of paper and trying again from time to time in case they conditions improve, they turn their antenna more towards you, or maybe they have less QRM. It is often worthwhile alternating calling CQ and tuning over periods of some tens of minutes. After all – while you are tuning the band some new casual operators will have come on and you may want to net them with a few CQ calls. Equally, it can take some time for both your antennas and a particular DX station's antennas to become properly aligned.

One topic which arises from time to time in discussions between contesters is that of what to do when someone who you have worked before calls you. If you are using a computer to log, you will have immediate notification that he/she is a 'dupe' and will have to decide what to do about it. Some people take the view that it is always quickest just to work them again, others are less convinced and, certainly for SSB contests, will say something like "G0ZZZ, sorry we've worked, G4PIQ Contest". On CW the decision is less easy – it takes a bit longer to send an appropriate message, such as "G0ZZZ QSO B4 TU G4PIQ TEST". If all goes smoothly and the recipient of this message understands, this is definitely quicker than working the station again. However, this all comes horribly to grief in those 10% of cases where either the other operator doesn't understand what you sent or, even worse, enters into a debate with you as to whether you really have worked before or not,

The Backpackers mini-contests enable you to have fun contesting with the minimum of equipment over a four-hour period. Here Backpackers trophy winner Julian Ross is operating G0LBO/P from the summit of Coniston Old Man

and what was the previous QSO number and what was the time, and so on. . . ! When this happens you immediately regret not just working them, since the rhythm of the pile-up has been broken and the QSO rate has been smashed by this long QSO. Sometimes when a 'dupe' calls, a second sense says "This one is going to be trouble" and you work them anyhow but, to say the least, this is a somewhat unscientific criterion!

Paperwork

Any contest entry consists of the logs themselves and some supporting summary information. Of these two parts, writing up the logs themselves is by far the most time-consuming part. There are two main options for doing the logs – either use a computer or do it by hand. We'll cover the computing option in more detail later but, if you are forced to write up the logs by hand, please use the standard RSGB log sheets – LSVHF for VHF contests (25 contacts per page), and use separate sheets for each band. Also, please *do* take the time to write the logs up after the event. Even if you use these forms to scrawl on during the contest, unless you are a really remarkable operator, they will be full of crossings-out, corrections, illegible scribblings, and so on. It doesn't do anything for the image of your log (not to mention the adjudicator's sanity and patience) if he/she receives a log like this. If your log is neat and tidy, you are much more likely to be given the benefit of any doubt when it's 3am and the adjudicator is struggling to finish the checking before the next committee meeting!

The other part of the entry is the supporting information. The most obvious piece of this is the summary sheet – RSGB Form 427 (one needed per band, and any version from 1986 onwards is acceptable). However, for some contests, additional information is required. If the contest is a multiplier event, you also need to submit a list of the multipliers worked on each band, along with the callsign and serial numbers of the QSOs for which multipliers were claimed. In case one of the contacts you have claimed as a multiplier is disallowed, where you have them, you should include details of more than one QSO with each multiplier. The RSGB VHF Contest Committee have a form MUL1 which provides a useful template if you wish to use it. For multiband VHF contests, there is no longer a requirement to submit a 4422 Summary Sheet. For cumulative contests, please include a note of your scores from the individual days either on the back of the cover sheet or on a separate piece of paper. There is no problem with using computer-generated forms with a similar format as the official ones, provided they carry all the required

G3XDY in action in the 1997 RSGB VHF NFD operating G4MRS/P on 23cm

information clearly. Sample forms are given in references [1] and [2].

If, once the contest is over, you switch off brain, automatically churn out the paperwork, and just put it in the post without checking it, you are doing your entry a grave disservice. Dave Lawley, G4BUO, former Chairman of the HF Contest Committee, has said that in his experience about half the points lost by people are given away after the contest has finished, often through careless writing-up of the log. The classic mistakes are made in mis-reading other people's handwriting – 'P's for 'D's, 'L's for 'C's etc, and anyone who has every tried to decipher one of G4PIQ's hand-written logs will know exactly why he prefers a keyboard! These sorts of mistake can only really be avoided by careful checking of the logs afterwards, and preferably by the generators of the spider-scrawl handwriting themselves. Other errors are just plan silly though – watch out for F stations in JO16 (the oil and gas fields in the North Sea) instead of JN16 (Central France), or UK stations in IO01 (about 400 miles off the west coast of Ireland) rather than JO01 (eg Essex). Slightly more subtle, and even more common examples are PA stations in parts of JO31 which are definitely in Germany, or ON stations in parts of JO10 which are in France. After adjudicators have been on the band a few years and have checked a few contests, they develop a sixth sense of spotting these sorts of errors (even the subtle ones), so it is worth your while taking the time to check your log carefully.

Finally, the ultimate sin must be mentioned – that of having unmarked duplicates in your log. We all end up making more than one contact with a few stations on each band in a contest, but it is essential to the health of your score that you mark them in the log as duplicates and claim zero points for them. If you don't, you will lose 10 times your claimed score, and indeed if there are an excessive number (five or more) in the log you can be disqualified. Also, don't forget that DB8KJ and DB8KJ/P count as duplicate contacts. You may think it unlikely that anyone, and particularly the top contest groups, would allow these to creep into their logs. However, astonishingly, in VHF NFD during two recent years, the claimed scores in one or other of the sections in VHF NFD have been very close, and the winner has been decided by one station having unmarked duplicates in their logs. On a third occasion, the only reason that it did not happen was that the first and second placed stations *both* had unmarked dupes!

Computers

Computers are perhaps the area in which the greatest development has occurred in contesting in the last few years. It is by no means essential to use one, but the vast majority now do so and reap the benefits. Perhaps the first question to ask is – what does computer logging give me which paper logging doesn't? The cynic may reply – nothing other than more equipment, something else to learn about, and more boxes to blow up! Most take rather the opposite viewpoint to this and find that the little extra effort required to organise computer logging for a contest pays big dividends both during and after the event – the vast majority of logs arrive in electronic form these days. The biggest bonus from using a computer must be that you don't have to spend hours after the event transcribing your real-time logs into something which is acceptable to the contest adjudicators. Many 'old hand' contesters hate doing any form of paperwork and find it almost impossible that to believe that they used to write up 500+ QSO contests by hand. Now at the end of the contest, all that is required is to make a fairly careful examination of the contacts in the file on the computer – make sure that everything looks sensible, and then copy the log onto a disk to post, or send it by e-mail.

The computer is also a big help during the event. Search-and-pounce operation is much easier than with a paper-based dupe sheet. All you have to do is to type a part of the callsign of the station you have heard into the computer, hit the appropriate key, and the machine will show you whether you have had a previous contact or not. For a single operator who cannot write in two places at once, keeping a paper-based dupe sheet while making contacts at a reasonable rate is always going to be a difficult exercise. This problem is, of course, eliminated with the computer where the dupe sheet is automatically generated as you log the contacts you make. Keeping track of multipliers is normally a similar problem, and again the computer can keep track of this for you automatically, with screens to indicate what multipliers are still required. The ability to send appropriate automatically generated CW messages from the keyboard is a great aid to keeping down the stress level in CW contests too.

You can get computer logging packages for many different types of computer, but here we are going to concentrate on those available for the PC. One of the accusations often levelled at computer logging is that you need to spend a lot of money on the computer. This is not necessarily the case, with the packages commonly being used at VHF being able to run on a simple ancient XT. Anything based on a 486 processor or earlier can be picked up second-hand very cheaply. What is probably more important than the raw power of the computer is whether it is quiet from an RF point of view. Machines vary greatly in this respect, both between models and with frequency for an individual computer. Unfortunately, the only really sure way to find out if a particular machine is going to cause a problem is to try it in the situation in which you intend to use it. There are quite a number of techniques for reducing the noise, much of which tends to escape on the connecting leads, and these solutions have often been covered in *RadCom* and elsewhere over many years. Do, however, make sure that the screening in your own receive system is up to the job – computer QRM at a much higher level than you deserve can easily occur if, for example, the braid of a piece of coaxial cable was open at one end. Immunity to transmitted RF also needs checking – some monitors have a crisis with large quantities of RF, and keyboards have been known to take on ghostly self-typing properties too. In one case, on 80m, you just had to lay your hands above the keyboard to make it start typing of its own accord!

Software is the other important consideration for computer logging. Assuming that you are not going to write your own, there are a large variety of possible packages to use, some freeware, some shareware and some commercial. Having taken a look at many of the alternatives, there are several packages which stand out as being the most developed and suitable for VHF contesting in the UK. They all have their good and bad points – you have to make a decision on what you want. Two of the commercial packages and are LOG by G3WGV, and SDV by EI5DI. LOG is a package which covers both HF and VHF contests, while SDV is a VHF variant of SD (Super-Duper – the HF version). You can check the September 1993 issue of *Radio Communication* for more details on Super Duper in a review, but its great virtue is the ease of use and one can well believe the various stories told of complete beginners to computer logging being very comfortable with the software within minutes. There is full window editing, so it is just a matter of moving the cursor keys up and down to change the details of a previous QSO, and SD is somewhat unique in that it won't let you do anything inconsistent, by, for example logging G4PIQ in Essex on 2m, but in Suffolk on 70cm. The automatic display of bearing and distance is particularly useful with the narrow antenna systems in use at VHF/UHF. Some VHF contests now use a multiplier scheme which consists of locator squares, counties and countries, and the latest versions of SDV have the option to cope with all of the variants. SDV supports all the usual features which help search-and-pounce operation such as 'check partial' (eg enter 'IQ' and it will list all the calls worked containing 'IQ') and a nice touch here is the ability to search either by suffix or prefix. Also, the normal multipliers and countries worked lists are well implemented, making it easy to keep and handle on your multiplier situation. The standard CW keying functions are all there too, and it all appears to require very little in the way of computer power since it is claimed that it will even run fine on an XT.

LOG by G3WGV has a more difficult user interface, but one which is based upon the 'standard' of the HF CT package by K1EA, making it easier for those accustomed to CT and its clones, but not necessarily simpler for beginners. One big difference between CT and LOG is the lack of a full window editor in LOG with the process for editing a previous QSO being rather messy. Most of the functionality is very similar to Super Duper although implemented differently but LOG supports some nice extra features for multi-operator contests, enabling you to look at the statistics of the operating, and it has a good front-end for the filing system. LOG is also able to support most of the RSGB events and will run on a simple computer platform. The old longstanding version of LOG does have a couple of omissions though – it cannot use locators as multipliers and so, during such events, you will have to keep track of squares worked manually, and edit the LOG file afterwards to show these multipliers, also it cannot score contests at 1pt/ km, so the log has to be re-scored after checking (for example, using a BASIC program to write kilometres into the comments field). A revised version was released in 1998 which is understood to correct these points.

Two programs which are completely free are TACLOG and G0GJV. The latter has been developed in co-operation with the VHFCC and has proven very popular. The display can appear a little confusing at first, although most beginners have little difficulty learning to use the wide range of features. A copy can be obtained by writing to the VHFCC.

Safety

This section attempts to shed some light on aspects of safety which both prospective and experienced contesters should consider. It is weighted towards portable operation but remember that many of the points, especially those addressing the erection of antennas, apply equally as well when operating from home or a club shack. Most new contest operators operate from home or their local club shack and it is important to think about safety in these cases: bad habits are difficult to break when operating in the field, so have fun and be safe.

What does 'safety' mean? Some people might think it means erecting and collapsing masts and antennas without anyone getting hurt. Others may think it applies to the possible use of high voltages in amplifiers on site. Similarly some consider that insurance policies form a safety net if things go wrong. The simple answer is 'safety' means all this and more.

First things first: this section is written from hands-on experience and hopefully covers most of the salient points, but please bear in mind safety and common sense cannot be separated. If you do not feel happy about something on a contest site, either tell someone or do something about it. Above all, if you are unsure about the safety of any aspect of your contesting activities don't do it: antennas, masts, tents, expensive radio equipment and even cars are all replaceable – lives are not. Please be careful in your contesting activities: your first mistake could be your last – don't fool around when someone's safety is involved.

Picture the scene: there you are, sitting on your 3000ft hill in Scotland with a superb take-off in most directions. You are congratulating yourself and your colleagues with your choice of site and how everything seems to have got off to an excellent start. The problem is you may have already made your first and most important mistake. Always remember to tell someone where you are going and for how long. It really is not too surprising how amateur radio kit doesn't work very well under a foot or more of snow. We are trying not to be too patronising here – you may think that with all that radio kit you could summon help in an instant. That may be true but think of it this way – a member of your team who has been injured could be in quite a bad way by the time the local mountain rescue team get him to the ambulance at the bottom of the hill. In other words, summoning help by radio is a cure but it is not a prevention. Note that if using private land it will always pay to seek the permission of the landowner – your health could suffer otherwise (but that's another story!).

So, you have let somebody know where you are and when you should be back, now on to the weather. High pressure is forecast so there shouldn't be any problem with the weather. Wrong! Temperature generally falls by 1° C per 300ft height gained and that is without wind chill. No need to bother you with the sums, but you can see that the summit of the hill will be a lot colder than the track running in the valley below. Add the wind chill and you would be amazed at just how easy it is to get caught out. So remember to take cold and wet weather gear with you, even if you are just out for a Sunday afternoon. Naturally, you cannot always rely on the British weather, the rain may not be belting down and sometimes it does get quite sunny. Therefore in addition to your cold weather gear, don't forget the sun cream. Most people have a good idea of how

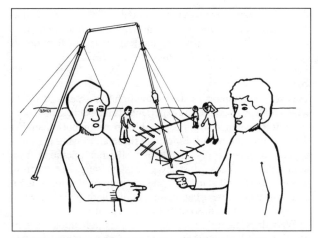

"But I thought *you* were giving the orders!"

their own skin copes with sunshine, but don't forget you could be out it in it for many hours. Something else that you should consider: it is possible to get sunburnt in all seasons and, especially in winter, wind burn leading to accelerated dehydration can be a problem. So always lean on the side of caution – experienced contesters would rather look silly wearing a floppy sun hat than get burnt on their bald spots!

Hopefully most readers will be well aware of the potential problems the weather can throw at you, but please remember that the weekend isn't over until you are tucked up safe and sound in bed at home. Please try to remember your journey home. For some this might be a five-minute stroll down a leafy lane in pleasant sunshine but for others it could be a nine-hour drive from one end of the country to the other. So take it easy – don't spoil the weekend by crashing the car on the way back because you are too tired.

Back to our mountain top. You can now start to put the station together, erecting masts, putting antennas together, erecting tents, connecting every bit of radio kit on site to your earth. "Earth?", you say. Yes, even in the middle of nowhere you should take an earth spike and connect the chassis of your precious radio to it. This will not only help to protect you from nasty RF burns, but also help to maintain a Faraday cage to prevent static build-up (or worse, static discharge).

Now to the antennas and masts. Before you do anything, look and think. Obviously no one in their right mind would put up a mast near overhead power lines, would they? What about the person who was seen flying a kite only 20 or so yards from a three-phase overhead power line! No, of course we are all too sensible to do anything like that. However, don't forget to think about the worst case: what would happen if the mast did come crashing down? Is it near a road? Is it near a building? What about the car park? Is there a footpath nearby – the general public may not appreciate a stake with several guys lashed to it knocked into the path. Is there likely to be any livestock roaming about? Sheep, cattle and other animals can be quite destructive; it's not their fault, they just see a new tree with funny branches that just has tested by rubbing against it or by nibbling at the guys. You might think that your presence on site would deter any of this activity – not so: one group was operating from a farmer's field when someone spotted the rear end on a cow sticking out of an unmanned tent. The 4m gear inside was rather messy by the time the cow had finished; another group had a cow push its head clean through the sidewall of a tent. Comments about problems with livestock and guy ropes crop up regularly on the coversheets.

Other points to think about: many good sites are on coastal cliffs, with obvious dangers. One point might, however, be worth noting: when operating from a coastal cliff you should consider informing the coast guard. Low-flying helicopter training sorties can't be ruled out. Also take time to think about the lights in use at night from such sights – it is easy to misinterpret coastal lights from off shore.

Inland there are different problems. Some National Parks have military training ranges and the local soldiers are not likely to take to kindly to a strange antenna, generator and tents overlooking their firing range. Check the local firing range timetable before setting out.

Now to the mast and antennas themselves. Do you have the right tools for the job? Nothing is worse than slipping with a screwdriver that is too big and putting it through your hand. What about heavy duty footwear? Trainers and the like do not provide much protection from a sledge hammer; heavy walking boots are better but steel toe-capped boots are best. Hard hats are cumbersome but could just save you from a nasty head injury. They are available with chin straps for those who might have them blown off in a gale or to stop them falling off when bending over to work on something at ground level.

Erecting the masts could and should be the subject of many more words than this section can cater for. Just remember that guy ropes can stretch, stakes can loosen, knots can untie themselves. Always watch the mast, guys, stakes and other people. Never fool around when lifting a mast into the sky – this is potentially the most difficult and dangerous time of the whole weekend.

Should there be any children, cats, dogs, or other unwanted distractions about, assign somebody to look after them during this time. If they want to help, tactfully point out the dangers and don't allow inexperienced people to get involved.

Always discuss the details before erecting the mast. Nominate someone (who should know what they are doing) to oversee the operation. This person should stand back so he or she can get a good view of the shape of the mast. Try to agree a set of simple instructions that are short but meaningful, eg "Pull in on your bottom guy, Richard . . . OK" instead of "Erm . . . I think that bottom one needs to be a bit tighter, hmmm maybe if the wind shifts . . . just a minute, I'll check . . . oh yes, but not too much now".

When lifting masts using gin-pole methods, remember the forces exerted on the top guy to the gin-pole. This guy can easily stretch, causing the mast to sag under its own weight, leading potentially to catastrophic consequences. If erecting a mast taller than 45ft, with antennas mounted close to the top, you should consider using steel guy wire to the gin-pole. Steel doesn't stretch, but does have safety implications all of its own. If you use steel securing wires, ensure that they are fastened off properly with shackles and turn-buckles.

Never use steel rope to pull the gin-pole or any other lifting arrangement. Always use good-quality rope, and always wear gloves. Even in the height of summer, a pair of thick industrial gloves or heavy-duty leather gloves are essential for handling ropes. Don't worry about feeling daft, or anybody making fun – always wear gloves when handling ropes. Naturally the gloves should not remove all sensitivity from your hands because you will still need to tie knots, but fingerless gloves are, in general, not recommended.

Once your mast with antenna(s) has been lifted into place and secured you can breathe again. No apologies for repetition – operations involving masts and antennas needs careful thought and planning. The front cover of *Radio Communication* a while back showed a picture of a substantial mast with antennas in the process of falling over. Talking sometime later to the team involved, they said that the only explanation they had was that as the mast went up a guy rope had snagged on a bench. This caused it to stretch, the mast went out of true and the rest is history. Luckily this group are amongst the most experienced in the UK and were able to cope with the problem, though not without damaging elements of the mast and several antennas. Even more luckily, as the photograph confirms, nobody was under or near the mast and everybody was watching it. The point of highlighting this is not to embarrass the group in question, but to emphasise that erection of masts can go wrong, even with the best groups. Do take care when erecting your mast and antenna. Think very carefully about where to stand even if you use something as (apparently) simple as a trailer tower. Winching wires can, and have, snapped while luffing (moving the tower between horizontal and vertical). Tower sections are extremely heavy, and potentially deadly when in free fall.

So now your mast is up, the antennas are fine, the radio equipment is ready, so off you go . . . well not quite yet. The generator, if not already fired up, needs some thought. Generators should be relatively straightforward, but not always. Always use a funnel to fill the generator, even if it is diesel, but especially if it is petrol. Of course if you are using a petrol generator ensure that strict fire precautions are observed. Always use high-quality mains cable to connect to the equipment. This will not only ensure less losses, but is less likely to suffer from someone accidentally driving over it. Don't forget that generators are heavy beasts – if you are having one delivered to site try to get it in the correct place first time. Nobody wants to put their back out by moving a generator a few yards.

That's enough about the safety aspects of generators – what happens if you are using batteries (for low-power or Backpacker-type contests)? Most batteries these days are intrinsically safe, but it always pays to check. Try to ensure that there is no possibility of an accidental short-circuit as some batteries can leak toxic gases or even explosively vent electrolyte if a short-circuit is applied; protection need be nothing more than a fuse right at the terminals.

So now your masts are safely up in the sky, equipment working nicely, generator humming away well and you're 100 QSOs in front of the next station on the band. That's just how it should be. You can relax and enjoy a nice beer or two. Fine, but please remember the effects of alcohol. This is not the place to preach about the latter, but one major UK VHF contest group insists on a 'dry' site. This is because they nearly dropped a mast on someone a few years ago after one too many pints in the pub *(Actually, we were stone cold sober, but it made us think – G8GSQ).* Of course we all like the odd pint or two, but be sensible. Think of the dangers, and remember the law.

Has all this sounded rather negative? It is difficult to treat a subject such as safety in anything but a serious manner. Above all safety on site is the application of common sense. Being aware of the potential dangers and doing something about them can help to ensure an enjoyable and successful weekend.

Further information on the safety aspects of contesting is given in the HF contest section.

Generators

Finding a source of power at any portable site is a major consideration, particularly at VHF where good sites may be quite some distance from a mains electricity supply. If you are just running a few tens of watts you may well be able to get away

with battery power, but more substantial stations require an inordinate amount of effort to ensure a ready supply of battery power – particularly if you are at the top of a mountain! The obvious alternative is a generator, and these can be extremely reliable and behave just as you expect. However, equally, they can be an immense source of frustration and the cause of hours of lost operating. Chris Parry, G8JFJ, has sent these very useful thoughts on this subject, focusing some of the problems which can (but don't always) occur!

The basic choice is between diesel or petrol (although gas power is also feasible) but nevertheless all the options are expensive to buy or hire! Most sets produce 220VAC or 240VAC, although battery charging 12VDC or 24VDC sets produces an attractive alternative solution for modest power stations. Petrol or gas engines can both stop in driving rain, and have been known to lose regulation upwards, which is potentially dangerous. (Chris can vouch for this, having used one petrol generator which suffered from major overshoot on its throttle control loop – as you went back to receive and the main load came off, it would shoot up to 350V – this destroyed a rig and bits of an amplifier very quickly!) Putting a box in line between the generator and equipment which will cut the load if voltage outside the limits 200-270V is detected is an excellent insurance policy.

You should bring the generator up to its governed speed before connecting the load, and also isolate the load before stopping the generator. One computer PSU failure was definitely traced to waveform stretching as the machine slowed down. Again, an automatic switchbox would have avoided this situation. 10% voltage regulation is an everyday problem for portable contest groups, and both the receiver and the transmitter reservoir capacitors need to be rated to cope with the potentially high voltage on receive. An alternative is to switch in a fan heater on receive (a popular winter-time solution). The flywheels of Japanese generators are often inadequate for smoothing the pulsating torsional load presented by a big SSB transmitter, and this causes poor voltage regulation at syllabic rate.

Both petrol and gas engines use spark-ignition which can cause potential EMC problems, and the storage and transport of upwards of 20 gallons of petrol needs to be considered. 'Hot' filling of petrol generators can be very hazardous, especially in windy weather. At least one station lost many hours of operating recently with a trip down the mountain to hospital after an operator spilt petrol all over a hot generator.

Diesel sets can be difficult or impossible to start after long periods of storage unless they are 100% healthy, and summer-grade fuel can freeze in March. Diesel is very prone to develop bacteriological contamination during storage, particularly in the summer, and especially when stored in clear or white vessels/pipework. Diesel sets are bigger and heavier than similarly-rated petrol sets, but are far more economical and, as an extra bonus, fuel can be tax-exempt 'red' diesel, available from boatyards etc. Anything bigger than 5kVA is best trailer-mounted in order to reduce the manhandling hazards in rain or snow.

Generators of all types need to be sited away from the main antenna lobe, downwind, and where cables do not cross footpaths.

One big issue is how big a generator do you need? In general, a much bigger one than you might at first expect! This because most transmitters have capacitor-input power supplies which draw current only on waveform peaks, causing high copper losses in the alternator. As a 'sizing' guide, aim for a generator with a VA rating fourfold greater than the required transmitter output PEP. Even with a generator significantly bigger than this, with most valve amplifiers you should not expect to see the same power output on the generator as you do at home, mainly because the HT will be down. As an example, on a 240V mains supply a typical amplifier using a pair of 4CX250s has the EHT supply sitting at 2.6kV off load, and the amplifier will deliver about 600W – on a 9kVA 230V generator it is down to 2.3kV, and will only deliver 450W, and on a 2.5kVA petrol generator you are down to 300W out! The only real solution to this problem is either to have much bigger amplifiers than you need, or to use those with choke-input supplies which take current over the whole mains cycle. A good example of an amplifier with a choke input supply is the old Tempo 2002 which is almost as effective on a generator as it is on the mains – it isn't known what the newer 2002A is like, however. Some forms of switched-mode supplies can also be a good option.

Cable size may be less significant than expected, because of the high source impedance of the generator as compared with domestic mains. Ex-building-site alternators are often 220VAC brushless types and produce a square-wave EMF. Better types with proper sinewave output exist, but these have brushes, and may therefore be less reliable and present possible EMC problems. Beware of heavy oil consumption – smoky generators could run out of oil and seize if run continuously for 24 hours. Also beware of vibration – avoid the temptation to get 240VAC out of a 220VAC mechanically governed set by raising the governed speed by 10%. The engine mountings etc may have a high Q and be ineffective (or worse) at anything other than the intended mechanical frequency.

Many alternators have the neutral connected to the frame, but others have the L-N output fully insulated. Chalk hilltops in summer present earth conductivity problems, even assuming it is possible to drive in an earth stake, and in all cases it is necessary to think carefully about safety in terms of fusing and RCCB/ELCB performance, especially if using multiple generators as part of a single installation (see previous section). If you don't fully understand this side of the system, get good advice from someone who really does!

One final point is that of physical safety – when using a crank handle, fingers and thumb must go on the same side to avoid risk of injury in event of kickback, and all belts, chains and shafts must have guards.

Equipment

Contesting provides an ideal opportunity to develop a wide range of personal skills by encouraging innovation and experimentation, and this applies equally whether you are running a modest, low-power station, stretching the limits of what a single operator can achieve, or if you are part of a team effort. Very few successful contest stations are put together without some parts being custom manufactured especially for the purpose, be they mechanical or electronic.

Although this is not a technical book, some basic points will be discussed here as, unlike better-quality HF gear, VHF equipment 'off-the-shelf' is often unsuitable for contest operation and may need some modification.

Receiver basics

There are too many factors to consider when looking at receivers in detail, but two of them are most significant in the context of a simple contest station; how well you can hear very weak signals and how well the receiver copes with very strong

G8ZRE shows all the equipment needed for a successful Backpackers operation

signals. The following notes are a very simplified view of these two factors, as they apply to a 2m SSB receiver.

Weak-signal reception
Weak-signal reception is ultimately limited by the background noise, (sometimes called the 'noise floor') of the receiver system. This noise comes from two sources; the inherent noise in the receiver and both man-made and atmospheric noise picked up by the antenna.

The noise from the antenna varies with frequency (eg 6m is noisier than 2m) and quite dramatically with location; sitting overlooking miles of built-up area is likely to give you plenty of man-made noise to hear. On 2m a reasonable antenna can pick up enough noise to over-ride the inherent noise in the receiver. In this case, weak signal reception is limited by the signal strength arriving from the ether and even the very best preamp will make only a slight improvement. As a simple test, switch to SSB and connect a good dummy load to the antenna socket. Connect an AC voltmeter to the speaker socket and set the volume control so that the meter indicates about 50% full scale (the reading might not be completely steady). Replace the dummy load with the antenna, tune to a quiet spot and read the meter. If the reading has increased by more than 20%, then weak signal reception is limited by the antenna noise and a preamp will be of little or no use. This test won't give reliable results if you have a really-low-noise GaAs FET preamp or input stage and your antenna impedance is much different to the dummy load. In practice, if you have such a set-up, external noise will be the limiting factor anyway and the potential inaccuracy is irrelevant.

A smaller antenna (dipole, halo, HB9CV or small Yagi) will collect less noise and the inherent noise in the receiver then becomes very significant. Many modern receivers are well designed and very little improvement is possible, but equally many can benefit from a *modest gain* preamp. This has the effect of improving the signal-to-noise ratio (making the signal more readable), even though the speaker and S-meter may indicate that the noise level has increased.

Try some organised tests with a local station who can reduce power to give you a barely readable signal (this avoids propagation variations). If there is no clear improvement from adding a preamp, leave it out!

Strong-signal problems
The first draft of this section quickly assumed the proportions of a book, discussing IMD, intercept point, reciprocal mixing,

cross modulation and blocking. This indicates the complexity of the subject, and this section is, therefore, extremely simplified. Those people who wish to pursue the matter further are advised to consult the amateur press which contains numerous articles giving greater detail.

Unfortunately, most reasonably priced (and some expensive) multi-mode VHF transceivers have been designed with little consideration for the need to work in the presence of off-channel strong signals. When such signals are present, such as on a crowded contest band, various symptoms can occur. This includes strong signals appearing to splatter widely (even over the whole band), the receiver not working when tuned near a strong signal, the background noise jumping up and down or just being much higher than normal. All of these can be caused by a faulty transmitter, but they are almost always due to the receiver being overloaded.

How to spot overload? Firstly, look at the S-meter when tuned to a strong signal. If it is pinned on maximum, then the receiver is likely to be overloaded. Change the antenna direction, antenna or use attenuators to reduce the signal strength to S5–S6. If the problems reduce a lot, or disappear, then the problem is receiver overload. If the problems remain, then call the station and politely explain the problems and the tests carried out at reduced signal strength, and ask that the transmitter be checked. If the problems persist, write to the contest adjudicator.

What to do about it? Overload problems escalate dramatically with signal strength and most solutions will involve reducing the level of the interfering signal. Some sets can be improved by modification or replacement front-end units (see below). The inherent problems of a poor receiver cannot be placed on anyone except the manufacturer, so other solutions have to be pragmatic:

(a) Live with it.
(b) Make sure that any preamp is really needed and be prepared to live without it.
(c) Make a more directional antenna so that the wanted signal can be peaked and/or the unwanted signal nulled out.
(d) Find another site.
(e) Modify your existing radio.
(f) Buy a new (more modern) radio, having checked with the manufacturer that the receiver has been designed to avoid or significantly reduce such problems.

Commercial transceivers
A great many sets, especially of the 'older' generation, can possess a relatively high level of synthesiser noise in comparison to more modern radios. Under normal conditions operating from home, there may have been no previous problems noticed but as soon as you venture to a good location with anything like a reasonable antenna under contest-style conditions, you will notice that very strong signals may appear to either spread across the band or may be audible on more than one frequency. This fact is compounded if a high-gain preamp is placed in series with the rig. This is not an attempt to play down what are, and have been popular radios; what we are trying to do is to give you some idea of the type of phenomena you are likely to experience when a really strong signal appears on the band. If you do experience these problems, firstly make sure your noise blanker is switched off, disconnect your preamp, then your antenna and tune across the 'offending' station to confirm the effect before rushing in and making a complaint.

Unfortunately, it takes a lot of time and effort to build a receiver which will cope with strong signals as it is usually the mixer or IF stages which easily become overloaded. On the other hand, even the 'big boys' get it wrong occasionally and if you genuinely believe that a station is too wide, politely make a complaint and make appropriate notes in your log book. The other station must do the same and checks are made by the adjudicator. If the problem persists, ask other stations, particularly those who are not so local, to listen to confirm the effect. If they do, they should also make and log a similar complaint. Complaints are discussed by the VHFCC at their meetings and can lead to a variety of actions. An errant station might be penalised, or encouraged to ensure that the problem does not recur.

If you are in the fortunate position of needing to buy a radio (and being able to afford it!), it may be worth searching out equipment reviews in magazines and consulting books such as the *Buyers' Guide to Amateur Radio* (sadly now out of print) written by Angus Mckenzie, G3OSS. This particular book gives an account of a wide range of equipment, much of which is 'out of date' now but which is still capable of doing useful work. A wide variety of 'old' radios appear advertised in the 'Members' Ads' column of *RadCom* or turn up from time to time as trade-ins on retailers' shelves – but they don't stay there for long! In particular, FT221, FT225 and IC202 are still held in some regard, especially if updated by fitting the appropriate MuTek front-end board. Undertaking such a modification will make a significant improvement to both the sensitivity and strong signal handling performance, making the radio more competitive. It has to be said that most of the proliferation of multi-band radios (especially those covering HF as well as VHF) brought out during the 'nineties are designed with many compromises and suffer terribly when subjected to a crowded contest band.

Another way round the problem is to transvert, usually from 28MHz, thereby making use of the excellent technical characteristics and features of most of the more modern commercial HF transceivers. When you read about the superb technical specifications on offer, it's a great shame that these are not generally replicated in standard VHF equipment. At the end of the day it is all about cost, market forces and the numbers of units sold. It has always been assumed that the average VHF operator (if one exists) is likely to be less demanding about his equipment if its main use is as a talk-box to work across the other side of town, particularly as the generally low levels of activity do not place the same rigorous demands on performance as they do on HF. It is only when people, such as the VHF DX/contesting community, place increased performance demands on the equipment does this limitation become a problem. In real terms, we are generally a low percentage of the amateur market and must therefore be regarded as being economically non-viable.

There are several excellent commercial transverters currently available, particularly the SSB Products range with designs extending into the microwave region. The LT series are among some of the best around and variants of these are available in kit form. These are not cheap but you do get what you pay for in terms of performance capability. The MuTek range also have a good reputation with the Mark II series of transverters in particular being specifically designed to be 'bomb-proof'. Of course, you can build your own and there are several very good designs available in the amateur press, particularly those published in *VHF Communications* and the *VHF/UHF DX Book* [3]. Kits may also be available from the usual specialist suppliers. Always take care to ensure that the design will cope with strong signals; with modern semiconductors and diode ring mixers, it is relatively straightforward to put together a receiver chain where the strong signal performance is determined primarily by the HF radio that is used.

Modifications

A little has already been mentioned about replacement front-end boards in VHF transceivers which, because of their increased performance, in most circumstances, can do away with the need for a preamp. Assuming you are using an unmodified radio of vintage years and you believe it to be a little 'deaf', the addition of a preamp can work wonders. There is, however, the considerable drawback that the dynamic range of your radio, its ability to perform under strong signal conditions, will almost certainly suffer quite badly. Those of you who are more technically minded may wish to modify the first stage of their radio by doing away with lossy diode switches, slug-tuned cores and rather dated front-end devices possessing high noise figures. As a first step, consult the circuit diagram and try replacing the first RF device (if a dual-gate FET) with either a BF988 or BF981. This should lower the level of internal receiver noise, making weak signals more audible. Then, bypass the antenna switching diodes and note a further improvement (don't transmit!). If you are suitably impressed, like many of those who have successfully completed similar modifications, purchase a miniature change-over relay and wire it in place of the existing components. Of course, don't even think of delving into your rig if you are not confident about what you are doing. Needless to say, any warranty previously given will no longer be valid. Another way round the problem is to use a mast-mounted preamp, but this can lead to increased expense, problems in switching and additional weight. They are also prone to blow at the most inconvenient of times, particularly when you do something stupid in the middle of nowhere! However, if you can interface and sequence everything correctly they can be of enormous benefit, if not essential, particularly on the higher frequencies.

Linearity and output power

It should be remembered that a radio advertised as producing 10W output will in fact often produce considerably more. All too familiar are the horror stories of the 10W radios (producing anything up to 16W) driving '10W input' solid-state amplifiers into distortion and non-linearity. What you need to do is to measure the input versus output power of your amplifier. In doing so you will notice how little power it will require to drive it to something like 70% of its stated output power – it will probably require a further doubling of the previously measured drive power to produce the stated output. Probably even this power is lower than the stated amplifier input. In the meantime, driving your amplifier to around 70% of its rated output will almost certainly give you one of the cleanest signals on the band. This is even more important given that on battery power you will not be able to realise the 13.8V that your radio was aligned on. Given that the gain and linearity of a transistor amplifier is affected by variations in voltage, you should always run your equipment below its rated output to maintain a level of clean output.

As far as valve linears are concerned, it is beyond the scope of this book to discuss the various issues and particular advantages these have to offer – suffice to say that they are the norm in high-power contests. At full legal output, they offer the most cost effective and efficient way of generating high power levels and are far superior, when driven correctly, than any solid-state alternative that is generally available to the amateur. However, great attention must be paid to voltage generation, regulation and associated issues which have been discussed earlier. Readers are invited to read the excellent sections written by John Nelson, GW4FRX, on the design, building and operating of valve linears as published in reference [3].

Feeders

At the end of the day you can only work what you can hear. Losses encountered before the receiver will limit its ability to hear weak signals. Therefore, use a good-quality feeder between your antenna and radio. Don't be fooled – many of the cheaper cables marked as 'RG58U' are rather thinly braided – fine if you're working on HF but particularly lossy even with short runs on VHF. UR67 is certainly better, even more so when terminated correctly with good quality N connectors. There are several alternative low-loss cables currently being advertised which have significantly lower loss than UR67 while still maintaining much of the latter cable's flexibility. It is certainly worth considering these, particularly when feeding single or group antennas from a common feed point above the rotator. Its also worth checking to make sure that there are no special connector requirements and that every effort is taken to ensure a good waterproof bond with the cable. One very popular brand was nick-named 'hosepipe' because it carried water almost as efficiently as RF!

Look out at rallies and radio car boot sales for some LDF250, or even better, LDF450 coaxial cable. These cables are of particularly low loss, meaning that more of your valuable transmit power reaches the antenna and that more of those incoming weak DX signals make it to your receiver. Unfortunately, they are not very flexible and even in medium-size lengths can be a little on the heavy side. The cost of these cables new is highly expensive and such an heavy financial commitment would be unrealistic for the majority of amateurs. It is because these cables are used so extensively commercially that odd runs do thankfully appear on the amateur market at prices comparable to new UR67. Don't forget that open-wire feeder, when correctly terminated, can be of particularly low loss and is very light weight (ask the 70cm EME enthusiasts). Again, this is where there is considerable scope for experimentation. Read the amateur press for further information.

Antennas

This is probably going to be one of the main areas were contests are either won or lost. Quite simply, the more metalwork that you can get airborne to capture/radiate those precious radio waves, the more effective and competitive your station is likely to become. However, herein lies the problem – you'll have to compromise between performance and what you can actually carry, erect, and support safely and keep in the air under windy conditions.

There are numerous designs available for the constructor as well as several commercial models. Gain, directivity, front-to-back ratio, matching, feeding, stacking, baying, weight, length and rotation techniques will all have to be considered when deciding on the optimum arrangements. Some contests impose restrictions on antenna size and height; here we are not expecting massive, multiple arrays which require several acres of land and an army of workers. The 'no tower' and support restrictions will have told you that, but it is perfectly within reason to be able to run two or more antennas or to erect long-wire directional arrays, which can prove particularly effective if you are operating from the extremes of the country. Alternatively, a single nine-element Yagi will perform well on 2m, particularly from a high spot with a good take-off. It is always highly interesting to note in the results what stations have been using and to observe changes in fashion, particularly when new designs are either published or come onto the market. Offbeat designs such as collinears, quagis, slopers, and rhombics can also work very well. Look out for details in the amateur radio press, particularly some of the American publications. There's even a lot of fun to be had by stacking a couple of HB9CV antennas a half wavelength apart. In fact, this aspect of stacking and baying antennas is of particular interest to contest stations.

A typical 10GHz Cumulative Contest portable scene. G4EQD/P (left) and G3PHO/P on Houndkirk Moor, near Sheffield

Quite simply, the stacking and baying of antennas is a highly effective way of increasing gain. However, this increased gain is at the expense of concentrating greater energy into the main lobe of the antenna, thereby reducing its beam width. However, if you reduce size of the forward lobe, either in the vertical or horizontal plane, you are restricting the area of its overall coverage and will therefore have to turn the antenna more frequently to beam-up on the weaker stations. This is particularly the case if you bay two or more antennas side by side. One way round the problem is to stack the antennas above each other, thereby reducing the vertical beamwidth. In fact, this makes a great deal of sense because you are concentrating more ERP back towards the horizon where it's going to be of greater use rather than firing a proportion of your valuable radiated energy into the sky or ground where it will be wasted. At the same time you still retain the original (wider) horizontal beam width. Stacks of up to eight antennas high, correctly phased, are common practice in major events, particularly on UHF.

Some groups prefer to stack and bay their antennas in a box-of-four formation. This arrangement is certainly popular when up to eight big Yagis are used, say on 2m, particularly with groups who have access to mobile towers as this formation is more easily accommodated above the head unit bearing. Another way around the problem is to use more than one mast support, effectively creating two or more independent antenna systems. However, even as this increases the flexibility of the system allowing the antennas to be beamed in different directions, it requires extra care in the general layout and phasing of the antennas and a great deal more hardware and man-power to erect and maintain the system. In addition to the main forward lobe, all antennas possess side and back lobes. With careful stacking it is possible to achieve a little more than the theoretical 3dB additional increase in gain as these lobes are either nulled or combined. However, the more antennas you have, the greater the feeder loss, which can easily be sufficiently high so as to outweigh the advantages of any increase in antenna gain likely to be achieved. In any case, all antennas perform better the higher they are able to radiate in free space. In this way you may find that a smaller, light-weight antenna appears to out-perform one with a higher quoted gain mounted at a lower height. Don't forget that given half a chance, strong winds will do nasty things to large antennas mounted high above the ground! The whole operation is something of a balancing act between what is desirable and what is achievable given the available resources and the nature the environment and local conditions. However, there is little doubt about the overall effectiveness of big arrays when correctly matched as they really do give added fire-power (reciprocated on receive) to a contest station.

With a little reading you'll soon become more familiar with the basics of antenna performance and design. However, a word of warning: be suspicious of many of the high gain figures quoted in both amateur and commercial literature. Often, gain is quoted as a single point or isotropic radiator, rather than gain over a dipole giving rise to some confusion. To convert, subtract 2.1dB from the isotropic figure. Also, running some of the more recent computer simulation software reveals that some of the quoted gains are impossible to achieve on the specified boom length. Several families of designs have set standards for reliable, achievable performance. These include the NBS and DL6WU designs, the latter working well for long boom lengths. It should be borne in mind that many of the more recent published designs are attempting to squeeze out the last morsel of gain and are often primarily aimed at those whose main interest is EME. Quite frankly, although desirable in terms of performance, these types of antennas are not at all essential for a good contest station. The difference in gain between an old, slightly corroded, non-optimised antenna and a new, shiny, computer designed one is likely to be in the regions of one decibel or thereabouts, a tiny fraction of an 'S' point and completely imperceptible during the general noise and confusion of a contest. However, there is still considerable scope for experimentation, particularly in the areas of manufacturing for portability and in the combining, stacking and baying of more than one antenna. Depending on where you are, and how you are running the contest, the modern computer optimised designs with ultra-clean polar patterns can be a positive disadvantage! For more information, consult references [3] and [4], and also the DUBUS publications.

AMATEUR RADIO DIRECTION FINDING CONTESTS

There is one sphere of amateur radio activity that offers a unique opportunity for combining radio skills with the diverse skills of navigation and (for those so inclined) cross-country running,

The start of an event with competitors trying to get their first bearing

made the more interesting by a competitive atmosphere. These are the elements which combine to make amateur radio direction finding (ARDF) contests, and those who respond to the challenge of ARDF find themselves taking part in a most enjoyable activity. This section seeks to explain how these contests are organised and the techniques which are used by those taking part in them.

More comprehensive information can be found in the *Amateur Radio Direction Finding Manual*, which is published by the RSGB – see reference [5]. The following pages summarise ARDF activities in three parts: 1.8MHz, 144MHz, and IARU Rules contests.

1.8MHz contests

Historically, ARDF in the UK has developed mainly on the 1.8MHz band and regular contests have taken place since the 'twenties. Currently, there are several clubs which organise programmes of events – on Sunday afternoons, weekday evenings, and even at night. These may involve finding just a single hidden transmitter, but sometimes there are up to four. Each year, the RSGB sponsors a series of qualifying events leading to the National Final in September, and much of the following reflects these events. The general principles apply equally, however, to other 1.8MHz contests which various clubs organise under local rules.

An attempt to summarise the main points of a set of rules is made below.

1. Starting and finishing times are quoted.
2. The number of hidden transmitters, their callsigns and frequencies should be given.
3. The transmission schedule should be specified, and if 'random' transmissions are used the minimum duration of each transmission and maximum gap between transmissions should be given. If Morse signals are to be used as an aid to identification, this should be stated.
4. Provision should be made for approximate bearings to be given in the event of transmissions being inaudible to some competitors at the start.
5. The power input to the PA of each transmitter should remain constant throughout the contest. The antennas will be directly connected to the transmitters, which will not be operated by remote control.

6. The hidden stations should be located at least 50 yards (46m) from any inhabited building and must be directly accessible to competitors without them entering, crossing or trespassing upon property in private occupation.
7. The hidden stations must be at least 50ft (15m) from any public highway.
8. Transmitter locations and the start should be covered by a specified map sheet, details of which are published before the event.
9. A competitor may be accompanied by a team, providing it does not consist of more than three others.
10. Only one portable receiver capable of being tuned to the 1.8MHz band may be carried by any competitor's party during the event but a second portable receiver may be available for use only as a reserve.
11. The rules must state how the winner is to be determined. Usually this is the first contestant to locate all stations, in any order, but there may be variations. In some contests, bearing accuracy and other factors are taken into account.

The competitor's equipment

Obviously, the first item of equipment required is a suitable receiver. The main features of a 1.8MHz ARDF receiver are listed here, but the reader who wishes to consider detailed circuit layouts is advised to consult the various articles in which such circuits have from time to time been published [5–7].

To be suitable for 1.8MHz contests a receiver should:

1. Cover the frequency range 1.8–2.0MHz.
2. Have a directional antenna (frame or ferrite rod).
3. Have a telescopic vertical antenna for 'sense' purposes (more about this later).
4. Have the circuitry effectively screened, so that the signal pick-up is on the antennas only, and not on the internal wiring.
5. Be capable of handling a wide range of signal strengths. This is because the directional antenna will only pick up a small signal from a distant transmitter – thus requiring a substantial amount of gain – but there must be adequate gain control or attenuation to prevent the receiver from overloading when it is operating in close proximity to the transmitter.
6. Be reasonably selective.
7. Incorporate a BFO.
8. Be portable and preferably of rugged construction.

The competitor will also need a compass, which is normally mounted on the receiver (care being taken to keep it well away from any ferrous components, and especially any loudspeakers or meters which may be carried). He will also require a map of the area covered by the contest (normally a single sheet of the Ordnance Survey 1:50,000 series) with a board to rest it on, a 360° protractor, ruler and pencil. Unless the range of the contest is very limited, as in pedestrian events, some means of transport will of course be essential.

Competitor technique

Some competitors say that direction finding is a science; others assert that it is an art. The truth probably lies somewhere between!

At the start of a contest, each competitor is given the frequencies and callsigns of each transmitter. He will also be advised of the transmission schedule, and whether Morse signals will precede telephony signals as an aid to identification. If this is the case, he will use his BFO to identify the stations but will probably wait for the continuous carrier of the telephony transmission before attempting to take an accurate bearing – still using the BFO.

It has already been mentioned that a DF receiver should incorporate two antennas. The main directional antenna (frame or ferrite rod) is capable of giving quite accurate bearings (to within a degree or two) but, as the set is rotated through 360°, the operator will note that there are *two* positions at which the received signal falls to a minimum (these are referred to as *null points*). If a frame antenna is used, the null points occur when the frame is at right-angles to the bearing from the transmitter, and if a ferrite rod is used they occur when the rod is in line with the bearing from the transmitter. Using the directional antenna alone, the operator cannot tell in which of two directions (180° apart) the hidden station lies. However, if he now switches in a signal from the vertical antenna and combines this with a signal from the main antenna, there will be only *one* signal maximum and one null as the set is rotated through 360°. The receiver is thus able to *sense* the direction of the transmitter positively, so the vertical antenna is referred to as a *sense antenna*.

The null point which occurs with the sense circuit in operation will be at 90° to the null points observed when using the directional antenna alone. With experience of a particular receiver, the operator can tell the direction of the transmitter with respect to the maximum and null positions.

Note that the competitor, having used the sense facility to determine the general direction of the transmitter, will not attempt to obtain a precise bearing in this mode, but will disconnect the sense circuit and obtain an accurate bearing using the appropriate null with the directional antenna alone. This is because the maximum and null obtained with the sense circuit in operation are very broad, unlike the nulls obtained with the directional antenna alone, which are quite sharp.

Finally, after carefully positioning the receiver to take the bearing (and preferably after two or three attempts, to reduce the chance of errors) the competitor can read the compass and note the bearing. It will be easier if the operator has an assistant (or navigator) to record the bearing, especially in multistation contests where the bearings of the other stations have to measured in a similar fashion. The navigator can then transfer the bearings to the map, using the protractor and ruler.

Assuming that there is only one hidden station to find, the competitor now has a line on the map indicating the direction of the transmitter. What he does not know is how far along the line the transmitter lies, or how accurate the bearing is. He might make an estimate of the distance based on the strength of the signal, but this is far from reliable since cunning organisers will often arrange for a very weak signal from a station within a few hundred metres from the start, or equip a distant station with a very effective antenna to ensure a strong signal. As regards bearing accuracy, errors of 2° or 3° may normally be expected, but occasionally quite large errors may arise. Much can be done to reduce the risk of serious error by choosing a good spot to take the bearing, as far as possible from any metallic objects which will re-radiate the signal. Wire fences, telephone wires and, above all, overhead power lines should be avoided like the plague!

How then does the competitor close in on the transmitter? The standard technique is to select a point some distance to one side of the plotted bearing, and take a second bearing on a subsequent transmission. In choosing the site for the second bearing, the seasoned competitor will probably make an estimate of where he thinks the transmitter lies and will choose a spot which will give him a good (ideally right-angled) cross on

the original bearing. If he has a reliable crystal ball, he may perhaps notice that his bearing passes through a patch of woodland that looks a promising site from the map, and may even risk going there without taking a cross bearing. It is usually advisable, however, to take two, three or even more bearings before becoming sure that the area of the hidden station has been pin-pointed.

Note that as the competitor takes bearings on subsequent transmissions, the relative signal strengths of the received signals at different locations can be used as a guide to the proximity of the transmitter – though caution must be exercised in case the transmitter antenna has marked directional characteristics. Also, as the competitor nears the transmitter the effect of bearing errors becomes less significant. For example, a bearing error of 5° results in a bearing which passes 2.6km from a transmitter at 30km, but an error of less than 0.2km when the transmitter is 2km away.

If the transmissions are on a random schedule, it is of great benefit to the competitor if he has some means of listening for transmissions while travelling in his car. The ideal arrangement comprises a loaded whip antenna and a monitoring loudspeaker amplifier which can be used in conjunction with the DF receiver. The RSGB rules also permit the use of a fixed monitoring receiver in the competitor's car.

There is an additional technique which can sometimes be very useful if the receiver operator has a driver. If, for example, the transmitter is hidden in a wood with a road passing alongside it, the competitor can use his receiver, on its directional antenna, to follow the bearing as he is driven past. Such bearings taken inside a car are of course not accurate – but as the vehicle passes the transmitter the operator will note that the bearing 'turns'. This tells him the best point at which to enter the wood, and furthermore by noticing how sharply the bearing turns he may deduce how far inside the wood the transmitter lies.

When the competitor has located the area of the transmitter with sufficient confidence to abandon his car and set off on foot, the way in which his receiver is used changes since there is no longer much point in taking compass bearings and plotting them on the map. The main priority is to get to the transmitter as quickly as possible, so if the competitor can run rather than walk, so much the better! It will probably be obvious in which direction the competitor should go when leaving the road, but in some instances he will need to wait for a transmission and take a 'sense' before setting forth. He will then run towards the transmitter as directly as he can, while taking maximum advantage from any paths which may exist. He may leave his sense circuit switched on, or alternatively may run forward in the direction indicated by the null of the directional antenna, in which case he will be conscious of the turning of the bearing if he runs along a path which passes fairly close to the transmitter, thereby indicating the point to leave the path and dive into the undergrowth.

In any case, the signal strength will increase perceptibly as the transmitter is approached, and most receivers will probably overload as they reach the transmitter antenna. The competitor now knows he has arrived – well, almost!

Almost, because the transmitter and its operator are probably well concealed, and because the organiser may have deliberately provided it with an antenna several wavelengths long, in which case the receiver will lead him to the points of maximum radiation which occur every half-wavelength along the antenna. He may therefore have to resort to a detailed search of the area, or to find the antenna wire and follow it to its source.

Sometimes, especially if an elaborate antenna system has been used, the receiver will lead its operator time and time again to a particular spot where the transmitter is obviously not sited. The trampled undergrowth may well indicate that other contestants have suffered the same problem. The best avenue open to the competitor is to control his frustration and try another approach from a new point some distance away on the next transmission. If that fails, methodical searching is the only alternative.

When the transmitter has finally been located, the competitor hands his entry paper (or card) to the operator, who will record his time of arrival. The competitor then leaves the site, taking care not to disclose its whereabouts to any other competitors in the vicinity.

So far only a single-station contest has been discussed. When two or more transmitters are involved the basic technique is of course unaltered. However, the contestant has to decide in which order to hunt the stations. There can be no set guidelines here – every contest situation is different. A competitor may decide to play for safety and select for his second bearings a site which is likely to give reasonable cross bearings on all the transmitters. Alternatively, he may 'play a hunch' on the location of one particular transmitter, and give scant attention to the others until later. The worst situation which can arise is to discover, on the second transmissions, that none of the stations can be heard. The competitor is then on his own since all the others have dispersed to different points, and he has to decide a new policy. When this situation arises, it is most likely that at least one station is near the start.

Transmitting equipment

Attention will now be turned to the other aspects of ARDF contests, as viewed from the organisational side. The main item of equipment is the transmitter, and basically any transmitter which is capable of operating in the 1.8MHz band and being powered from a battery may be used. The elaborate apparatus used in the shack is not, however, appropriate and it is normal practice to use a fairly simple transmitter.

The emphasis should be on portability and reliability. To achieve this, attention should be given to weather protection. Rechargeable batteries, housed within the transmitter case, are practicable and crystal control is desirable. It needs to be capable of transmitting speech modulation (A3E) and Morse (A1A); this can be achieved with a conventional microphone and key, though more sophisticated equipment using electronic keying and stored speech is also used.

Contest organisation

When setting out to run a contest, the organiser has to bear in mind the standard of difficulty which is appropriate. If the contest will attract an entry of novices, a fairly simple standard is needed – but the organiser of a three-station national final should make it tough. Several factors combine to determine the overall difficulty of a contest, and careful consideration must be given to each one.

The first things to fix are the start and the transmitter site or sites. These cannot be decided in isolation since the distances between the start and transmitters, and between the transmitters themselves, as well as the nature of the route between them, must all be considered. It is necessary to ensure that the transmitters will be able to radiate adequate signals to the start.

The start itself should be arranged off the public highway, with adequate parking space for the expected number of competitors' vehicles, and somewhere which will not inconvenience members of the public. A large open space, free of overhead wires and fences, will help competitors to get good bearings – but a start which lies beneath high-voltage power cables may be deliberately chosen if it is desired to make the contest more difficult.

The selection and arrangement of the actual transmitter sites is the most important part of contest organisation, and is the main factor by which a contest will be judged. It is most important, however, that the sites chosen are directly accessible to competitors without them entering, crossing or trespassing on property in private occupation. Access via public rights of way is, of course, permissible. In some circumstances it will nevertheless be courteous for the organiser to advise local residents of his plans so that they are not alarmed when a horde of strange characters wearing headphones descends on the area!

A good transmitter site will be well away from the road and well hidden. A long antenna or one with several spurs off it will make the site much more difficult for the competitors, but things will be made easy for them if a simple vertical wire going straight up into a tree is used, as this will lead them directly to the transmitter. There are many ruses which the transmitter operator can deploy to hide himself and his equipment – probably the most common is to crawl into a thicket. Other ideas which have been exploited are holes in the ground, boats moored to river banks, stations in trees etc. Sometimes a transmitter operator need not be hidden at all – he may be disguised as a fisherman (with the antenna connected to the end of his rod and the transmitter in his basket), or the transmitter may be operated by a 'courting couple' in the back of a car. The contestant needs adequate skill to be confident of any decision to challenge in the latter case!

The broad principles of setting up a hidden station have been outlined above, but there are of course many finer points which will help to ensure a good contest. For example, a station may be on a river bank but far from a bridge – very frustrating for the competitors who arrive on the wrong side, unless they are good swimmers and have come prepared with a plastic bag which will keep the receiver dry!

The transmitter operator should always ensure that he leaves no obvious tracks leading to his hide, and the antenna itself should be concealed as far as possible. Fine wire is difficult to follow through woodland but care must be taken with the lead-in to the transmitter so that arriving competitors will not inadvertently break it. Care should also be taken that the operator's car is not parked conspicuously by the approach to the hidden site – this is a certain give-away. Arrangements should be made for someone else to drive him and the equipment to the spot and then drive off again.

In an important contest, it is worth arranging for radio contact between the hidden stations and the start officials, so that signal strengths may be checked prior to the start and also so that all parties may be kept advised of the progress of the contest. Ideally, a frequency band other than that in use for the contest should be chosen for control purposes. It is a wise precaution for a spare transmitter and power supply to be on hand during the setting up of a contest so that these can be rushed to the site in the event of an equipment breakdown.

144MHz contests

VHF direction finding, or *foxhunting* as it is more popularly called, is fun – the ultimate game of hide and seek! The participants (known as the *hunters* or the *hounds*) have to locate a hidden transmitter (known as the *fox*) by using a receiver and a directional antenna. The person who finds the transmitter or transmitters first wins the contest.

Anyone can take part in foxhunting: you do not need an amateur radio licence, since participants are only operating a receiver. VHF foxhunting also has advantages over other forms of amateur radio direction finding: both classes of full licensees can operate the hidden transmitter, and unattended transmitter operation is allowed without restrictions. More importantly, VHF foxhunting can be accomplished using ready-made receivers. The majority of amateurs own a 144MHz transceiver and most of these are suitable for use on foxhunts.

Types of events

Newcomers to foxhunting are recommended to attend events organised by the many established local DF groups throughout the country. There are also a number of annual VHF foxhunts designed to encourage national and inter-club competition, and some foxhunting weekends with organised camp sites. You could, however, set up your own foxhunting group. As a minimum for a competitive hunt, you would need two teams prepared to do the hunting and someone to hide and operate the transmitter.

Usually, VHF foxhunts are organised with one or two transmitters hidden in an area covering part, or all, of an Ordnance Survey Landranger map. This can cover an area of 40 × 40km, but is usually much smaller. Transmissions are made at regular intervals and the hunters take bearings on the signals using directional antennas. This is done by rotating the antenna while observing the receiver's S-meter and using a compass to measure the direction the antenna is pointing. This bearing is plotted on a map and the hunter then proceeds to another location somewhere along this plotted line to take another bearing.

The first few bearings will indicate the general area of the transmitter. Further bearings will lead either to the discovery of the transmitter which may be in a vehicle, or will indicate a particular area into which the hunter must enter on foot. Sometimes the transmitter can be cleverly hidden in dense woodland, or disguised as an innocent-looking object. The team that finds the transmitter (or both transmitters in the case of a 'double fox' event) in the shortest time, wins the contest.

Although most VHF foxhunts require the use of a car, some events are designed to be 'walking only' contests, in which case the designated search area is reduced appropriately. There are some events where searching in cars is practised over a relatively small area with very regular transmissions and low power. These are ideal for beginners.

Equipment required

Any 144MHz FM transceiver can be used for the hidden transmitter. All the operator has to do is either operate the transmitter manually or to arrange for the PTT connection to be controlled by a microprocessor or timer unit. Any legal level of power can be used (but note the licence conditions regarding unattended operation) and the type of antenna is optional.

As for the hunter's receiver, most people will use commercially available transceivers such as FM handhelds or small mobile/base station transceivers rigged for portable use. These should have an S-meter, although those without S-meters can still be used with some modification, or with the use of additional devices such as noise generators. If the rig has CW/SSB capability, these modes will help to give more accurate bearings on very weak signals. The ubiquitous Yaesu FT-290 has all these features and is a popular and successful foxhunting rig.

"Now where's that fox?"

Dual-band 144/430MHz handheld transceivers are also popular. These sets can be used to take bearings on the third harmonic of the fox frequency as the hunter gets close to the transmitting antenna. As harmonics are relatively weak, they are easier to take bearings on.

While these transceivers are certainly practical for foxhunting, problems can occur when you try to locate a powerful transmitter that may only be a short distance from you. This is because commercial transceivers are designed for long-distance communication where signals are expected to be very weak. As an RF gain control is absent on most modern 144MHz transceivers, the extremely strong signals near a transmitting antenna can overload the receiver and prevent you from seeing any movement of the S-meter. One remedy is to build a simple attenuator that can be switched into the antenna coaxial line.

Some hunters have also built devices known as *sniffers* which are often just 144MHz crystal sets with a DC amplifier and meter. They are basically very insensitive receivers that cope well within strong RF fields and can be switched into the directional antenna line when the hunter gets close to the fox.

Another serious problem is the breakthrough of strong RF signals directly into the receiver or through the antenna coaxial cable and connections. This can often be solved by placing the receiver in a metal container or wrapping it in aluminium foil. Anti-static component bags also work well. There is a solution for most problems, so almost any receiver can be used successfully – it's just a matter of experimenting and finding the best solution for your rig.

As for the directional antenna, a small beam such as the two-element HB9CV is highly recommended, since it has good gain for weak signals and is small enough to be taken through dense undergrowth and stored in a small car full of foxhunters!

You will also need a car (or team up with someone who has transport), a map, compass, ruler and pencil etc for plotting bearings.

With this basic equipment, you can start practising by taking bearings on local repeaters or amateur stations. You should also try taking bearings on a transmitter which is only a few metres away from you, so you can identify any problems your equipment might have when operating in a strong-signal environment.

Advanced hunting

After participating in a few competitive hunts, you will appreciate fully the technical nature of VHF foxhunting and the problems that can occur. In theory, your bearings should point directly at the transmitter, but factors such as the accuracy of the compass, antenna, and even the hunter himself can lead to errors! Also, VHF signals tend to reflect easily off hills and buildings, so you may end up taking a bearing on one of these 'bounces'.

One of the least understood problems of VHF foxhunting is that of *local phasing*. This occurs when a bearing is taken in a built-up area, and it can be caused by the signal bouncing off nearby objects like houses, cars and hedgerows.

The effect of receiving these numerous bounced signals varies. If you are lucky, the main signal will still be stronger than the bounces, but usually they will show up as several peaks on the S-meter, or they will 'combine' and give an S-meter reading that barely moves as the antenna is rotated. Very often, the bounce signals will 'phase' with each other so that a maximum S-meter reading is observed when the antenna is pointing at a reflected signal. This could lead the hunter off in the wrong direction.

One way to avoid this problem is to walk a few metres when taking a bearing, as this will help to average out the phasing effects and a predominant bearing should be observed, indicating the direction of the transmitter. However, the only reliable way around the problem is to move away from the area completely and find somewhere else, preferably high and in the open, and take another bearing.

You should always bear in mind that most bearings will not be completely accurate; you only use each bearing to indicate the general area to proceed to. To find the transmitter, you simply keep taking additional bearings, preferably at right-angles to previous readings, and continually narrow down the possible area for the fox to be in. Make a mental note of the maximum S-meter reading for each bearing. This will indicate that you are getting closer to the transmitter.

Do not be daunted by these problems, as they actually add to the interest of a foxhunt. You will be able to overcome these technical hurdles by experience and possibly by developing some additional equipment such as 'sniffers' and attenuators. Some ingenious work has been done by foxhunters and there is still scope for more development of DF hardware and techniques. Apart from being an absorbing and competitive challenge, foxhunting is also an ideal outlet for design and construction skills. This is an area where the average radio amateur has the chance of contributing some original and useful answers to technical problems. Computer buffs may be interested to know that any application of PCs to amateur radio direction finding problems will probably be pioneering work.

The future

At the moment, the rules used for a local VHF foxhunt depend on the club that is organising the event. However, it would be an advantage if all clubs could aim towards using standard rules. For example, if a standard frequency is used, then participants would be able to develop simple, lightweight receivers with a minimum of components. This could help more people get involved in the sport, particularly young people. To this end the

RSGB ARDF Committee has adopted a set of simple rules to be used as a recommended standard. These were published in the January 1995 issue of *Radio Communication*.

VHF foxhunting is steadily growing in popularity. Although still predominantly a local club activity, there are now some nationally advertised hunts held every year and even a foxhunting weekend which is held every July in the Forest of Dean. This event has a number of advanced foxhunts, a beginners 'walking only' session and even experimental hunts for IARU foxhunting (see next section).

IARU Rules contests

As stated earlier, ARDF in the UK began on the 1.8MHz band and later evolved to include the 144MHz band. In other parts of the world ARDF has developed along different lines, and a form of contest which is formalised in a set of IARU rules has become established in most other European countries as well as the Far East. These contests, also known as *foxhunting*, take place on the 3.5MHz and 144MHz bands.

National societies organise their own programmes of events leading to championships (sometimes open to members of other IARU societies including the RSGB), there are IARU Regional Championships, and a World Championships series is held every two years.

There are several major differences between these events and those UK activities described above. For those familiar with the sport of orienteering, there are a number of similarities. The contest takes place entirely on foot, normally in forested terrain, and is confined to the area of a special large-scale map (usually on a scale of 1:15,000) which is issued to competitors at the start. There are five transmitters, or foxes, located within the area. These use the same frequency, but each transmits for one minute in five in sequence. Thus, Fox 1 transmits for the first minute, then Fox 2 for the second minute etc until it is the turn of Fox 1 again. The transmitters are unattended, and identify themselves simply by repeating their callsigns in Morse code.

Non-directional antennas are used for the transmitters, and their locations are marked by a red and white marker flag, of the type used in orienteering competitions. Normally, there is a marking device attached, which is used to 'clip' a card carried by the competitor to prove that he has visited that fox. Alternatively, electronic devices are sometimes used to log progress on 'smart cards' carried by competitors.

Competitors participate in five classes: Seniors, Juniors, Women, Old Timers (over 40) and Veterans (over 55). Groups comprising not more than one from each class start at five-minute intervals, at the same point in the transmission cycle. Whereas seniors are required to hunt all five foxes, the other classes search for only four, but the fox to be omitted is different for each class. Certain rules specify the spacing of the foxes, and they should be arranged such that a line taking the shortest possible route from start to finish via all five foxes will be in the range 4 to 7km.

The location of the finish is shown on the map, and is also marked by a beacon which transmits continuously, sending its callsign on a different frequency to that of the foxes. Each competitor is timed at the finish, and the winner in each class is the one who has visited all the appropriate foxes in the shortest time.

There are other factors to consider, too. A maximum time limit is specified – typically this is two hours. A competitor who is approaching the end of the time limit may have to decide to head for the finish before finding all the foxes, or risk being 'out of time'. There are no restrictions on the order in which the foxes may be visited, but clearly there is an optimum order, and the determination of this order is a most important requirement for the contestant.

All the equipment carried is influenced by the need to use it during an energetic run through the forest, probably a wet forest(!), and participants dress accordingly. As well as an appropriate receiver for the band in use, most competitors carry a compass and pen, a means of keeping the map dry, and sometimes a map board and protractor arrangement.

Receivers can be relatively simple affairs, since the limited competition area means that signal strengths are fairly strong. However, a good dynamic range is necessary to cope with the dramatic increase in signal strength on final approach to the fox. For 3.5MHz, ferrite rods used in conjunction with short sense antennas are popular. There is more variety for 144MHz: many prefer the compact HB9CV antennas but Yagis are also used. Some receivers are actually built into the central spar of a Yagi. Whatever arrangement is preferred, it is never ideally suited for running through a forest. One method of easing the problem is to construct Yagi elements from some flexible material, such as steel rule tape. However, this has the problem of affecting compass readings if care is not taken.

The reader may have gathered from the above description that the ethos of IARU contests is very much about fairness, and there is no intention of deliberately hiding the foxes or using trick transmitting antennas, which often feature strongly in the 'fun' ethos of ARDF in the UK. Despite this, the basic techniques of taking bearings are just the same as described in the 1.8MHz and 144MHz parts of this section. Bearings and signal strength assessments on 3.5MHz are generally a fairly reliable guide to direction and distance, but on 144MHz it is a very different story

since reflections due to the terrain and vegetation can lead the competitor astray, especially in wet weather. Coping with these difficulties is just another part of the skill required.

The major Championship events are excellent international social occasions as well as providing competition of the keenest standard. The 3.5MHz and 144MHz contests take place on different days, and there are usually social functions associated with prize-giving ceremonies. It is clear that several countries have well-managed and successful national teams, whose members benefit from selection procedures followed up by regular training sessions to develop and perfect their techniques.

FURTHER INFORMATION

[1] *RSGB Yearbook*, ed Mike Dennison, G3XDV, RSGB, published annually.
[2] *VHF Contesting Handbook*, ed David Johnson, G4DHF, RSGB, 1995.
[3] *VHF/UHF DX Book*, ed Ian White, G3SEK, DIR Publishing, 1992.
[4] *VHF/UHF Handbook*, ed Dick Biddulph, G8DPS, RSGB, 1997.
[5] *Amateur Radio Direction Finding*, ed George Whenham, G3TFA, RSGB, 1994.
[6] 'HF direction finding', Chris Plummer, G8APB, *Radio Communication* June/July 1991.
[7] 'DF receiver for 160m' and 'DF transmitter for 160m', Peter Asquith, G4ENA, *Radio Communciation* May 1993.
[8] 'VHF direction finding', C J Seymour, G4NNA, *Radio Communication* October 1983.
[9] 'Foxhunting in the 2m band', Phillip Smith, GW1XBG, *Radio Communication* July 1995.

6 Mobile and portable operation

OPERATION away from the home station has always been an enjoyable aspect of the hobby – for some it offers an escape from EMC problems and indoor antennas, for others the excitement of using a really good site or a DX callsign. Today's radio amateur can use the same equipment at home, in the car, or out in the middle of a meadow, such has been the progress of miniaturisation and operating convenience. Each year this process continues, with more and more sophistication in smaller and smaller equipments. Truly, this is the age of 'go anywhere' amateur radio!

MOBILE OPERATION

About half of all amateurs in the UK operate stations fitted into their cars, and all bands from 1.8MHz to 1.3GHz are in use, with AM, SSB, FM and even occasionally CW! Mobile operating has particular attractions for those amateurs making regular car journeys to and from work, for gradually the companionship of a group of fellow travellers is formed which can ease the tedium of the journey and bring fresh interest to the hobby. Most local mobile-mobile work is carried out on VHF/UHF FM.

Alternatively some HF enthusiasts have fitted 100W SSB transceivers and loaded whip antennas to their cars, and can make intercontinental contacts while driving to work! The social side of amateur mobile radio is well catered for in the UK with many mobile rallies held during the summer months.

Equipment and antennas

Mobile transceivers have to contend with relatively inefficient antennas, high electrical noise levels and rapidly fluctuating signals. A high-sensitivity receiver is required, with efficient AGC and a noise blanker (SSB) or with excellent limiting characteristics and a correctly adjusted squelch (FM). The acoustic noise level will be high and a reasonable audio output of at least about 2W will be found necessary.

Adequate protection of the RF power amplifier against high VSWR is essential. Many mobile antennas exhibit high VSWR when wet, and if a loaded whip is used this may well have a narrow bandwidth, giving a high VSWR when out-of-tune. Many modern HF and VHF/UHF transceivers feature automatic ATUs which can be very useful in allowing the matching of a mobile whip which is slightly off-resonance. However, such radios are larger than those without ATUs and space considerations may necessitate the use of one of the separate remote-controlled ATUs which can be mounted elsewhere in the vehicle.

Mobile safety is paramount, and this demands a minimum of easy-to-operate controls with clear and unambiguous dials and meters, preferably illuminated at night.

The 12V power requirements should not exceed the spare capacity of the vehicle generation system. In practice this may, for example, confine the use of a 100W HF transceiver to daylight hours when the vehicle lights are not required. The average vehicle voltages fluctuate considerably and are 'spiky'; therefore internal transceiver voltage regulation and overvoltage protection are very desirable. Vehicle interference may be a problem and require suppression [1, 2].

A mobile transceiver should be small enough to fit in or under the dashboard – such transceivers are available in HF as well as VHF/UHF models from all the major manufacturers. Some HF transceivers are available with a remote head for the dashboard, allowing the rest of the rig to be tucked away out of sight. Security is increasingly a problem for equipment mounted in vehicles and such head units can be easily stowed in a glove box out of sight or even taken away from the vehicle. No matter what the attractions of larger equipments are, they will demand special mountings which will spoil the car interior and may lack operating convenience.

Rugged mechanical construction is essential, because a great deal of vibration will be encountered. Note that extremes of temperature and humidity are common in vehicles.

When planning a mobile installation, consideration should be given to the possibility of using the transceiver in other modes of operation, particularly as a portable station, for in this case requirements are in many ways similar. For example, if a hand-portable SSB or FM transceiver is available, all that needs to be done is to obtain an add-on RF amplifier and antenna, and leave these permanently in the vehicle (see later).

Recent editions of the UK Highway Code [3] have warned against the dangers of operating equipment while mobile. The present edition states (rule 127) "You MUST exercise proper control of your vehicle at all times. Never use a hand-held mobile phone or microphone when driving. Using hands-free equipment is also likely to distract your attention from the road. It is far safer not to use any telephone while you are driving – find a safe place to stop first." The Department of Transport, in a letter to the Society, adds: "The Highway Code is an advisory code of practice in that a failure to observe any of its provisions is not in itself an offence. Such failure, however, may be used in any court proceedings which may arise. Current legislation already places the responsibility on drivers to have proper control of their vehicles at all times. A motorist who fails to do so as a result of distraction or lack of concentration is liable to prosecution". It follows that radio amateurs who operate mobile stations while actually driving should not use hand-held microphones and take great care not to be distracted by the radio.

The basic necessity is to leave both hands free for driving and to allow the driver's head unrestricted movement, while keeping the mouth-microphone distance roughly constant. A boom microphone with headset has two disadvantages – it is conspicuous and the ear muffs will seriously impair the driver's hearing. One solution is to use a miniature condenser-type

An HF mobile antenna doesn't have to be too obtrusive. This is the Pro-Am antenna

microphone with a tie clip or alternatively mounted on a lightweight boom from spectacle frames. A suitable control circuit can be found in reference [4].

Whichever alternative is chosen, operators should bear in mind the recommendations of the Highway Code and the need to maintain good audio quality. This can only be achieved by making sure that the microphone is reasonably close to the operator's mouth and that the audio gain is not excessive. It may be that if the vehicle is excessively noisy the audio gain must be kept low and the operator must speak more loudly to compensate. It must be borne in mind when making adjustments in the relative quietness of a stationary vehicle that we all tend to speak more loudly when the vehicle is moving. There is a need to ask for reports over the air and to take action to improve the quality if unfavourable ones are received. It may not be possible to operate in a noisy car or lorry travelling at 70mph with all the windows wide open as the readability of the signal can become R3!

The best positions for HF mobile whip antennas are the rear wings or the rear bumper bar. Commercially made bumper mounts are available and can considerably lessen the risk of unsatisfactory mounting. Many modern vehicles have plastic bumpers and therefore care needs to be taken to ensure that an effective earth connection to the bodywork of the car can be obtained so that the performance is not compromised. If the car is fitted with a towing hitch, this can also serve as a very effective mounting point for a HF antenna. The ideal position for VHF or UHF mobile antennas is the centre of the roof, but this is not always possible or desirable for other reasons. Various types of mounting are available for those who do not wish to drill the car metalwork. These include gutter mounts, window-clip mounts, magnetic mounts and bootlip mounts. An alternative method of mounting HF antennas is the triple magnetic mount which is quite an effective way of mounting an HF whip, although difficulties may be experienced when travelling at high speed especially if there are gusty cross winds.

Whatever antenna is chosen, the fixing must be mechanically strong enough not to be damaged when the vehicle is travelling at high speed, or succumb to a blow caused by a low-hanging branch or other obstruction. Care should also be taken that the antenna does not project horizontally from the vehicle, even at speed, in such a way that it becomes a danger to other vehicles or pedestrians.

On the HF bands, the antenna system is the most critical component in any mobile installation, even more so than with fixed stations. The whip antenna with loading coil normally used has a low feedpoint impedance and a narrow bandwidth, and this involves careful tuning and matching for optimum results [5].

Most VHF FM operators use a $5\lambda/8$ or $7\lambda/8$ whip mounted on the rear wings, although in some cases this may have only marginal advantages over a $\lambda/4$ whip mounted on the roof. The most popular UHF antenna is the collinear, usually roof mounted. However, dual or even triple-band antennas are growing in popularity as they can cover two or three bands with only one mounting. All popular VHF and UHF mobile antennas are a good match to 50Ω coaxial cable, and no special matching circuit is necessary.

Expensive radio equipment left permanently attached to the vehicle should always be insured against theft, and the driver's ordinary insurance may not cover this eventuality. A burglar alarm should always be fitted and, if possible, the transceiver disguised or removed completely when not in use. Information on insuring equipment can be obtained from specialists such as Amateur Radio Insurance Services [6].

Operation

Mobile operation on the HF bands (and in the DX portion of the VHF and UHF bands) is little different to normal SSB fixed-station operation in that there are no specific mobile calling frequencies or sub-bands allocated. The HF SSB mobile operator is usually more interested in working DX stations in fixed locations rather than other mobile stations, and mobile-mobile contacts are fairly rare, except perhaps on 1.8MHz SSB and 29MHz FM.

As the frequency is raised, the effects of the terrain through which the vehicle is passing become more pronounced. On the HF bands this is usually confined to a relatively slow variation in signal strength but on VHF, and especially UHF, individual buildings and trees will markedly affect the radio path, giving rise to a rapid and characteristic signal 'flutter' when the vehicle is on the move. Sometimes a poor VHF signal may be obtained when the car is parked in an apparently good position on the side of a hill. This may be caused by the vehicle being in a 'null', and the solution may be to move the car forward a few metres and try again.

Most mobile-mobile working takes place on the FM sections of the 144 and 432MHz bands. 50 and 70MHz are also becoming increasing popular, especially as there are several 50MHz repeaters and it is also possible to work into the Continent during the sporadic E season on the DX portion of the band. The first UK 10m FM repeater has recently been licensed and this band offers the possibility of working DX relatively easily during the appropriate parts of the sunspot cycle and sporadic E season.

Operating in these sections is channelised, which is a considerable safety and convenience advantage for the mobile operator. Instead of fine tuning a VFO dial, he or she can click a rotary switch round to change frequency, and with practice this can be done without taking the eyes off the road. Many HF and VHF/UHF transceivers feature memories and these are extremely useful for mobile operators in that they allow fast, accurate frequency changes while on the move.

The use of FM is convenient in this case because one of its characteristics is that strong signals will completely obliterate weak ones in the receiver (the so-called *capture effect*). Therefore operation on the same channel (termed *co-channel working*) is often possible with two pairs of stations only 20km apart, without the annoying heterodynes typical of AM operation.

The channels are spaced 12.5kHz apart in IARU Region 1, and to avoid the necessity of giving long strings of digits when specifying frequencies on the air, they have been given identification codes. This supersedes the earlier S20, R0, SU20, RU0 system. Under the new system the coding starts at 145.000MHz which becomes RV48 (formerly R0), 145.500MHz becomes V40 (formerly S20), 433.000MHz becomes RU240 (formerly RU0) and 433.500MHz becomes V280 (formerly SU20).

To distinguish between simplex (non-repeater) and repeater channels the channel numbers are prefixed with 'V' for VHF (145MHz) or 'U' for UHF (433MHz). In addition 'R' is also used to denote a repeater channel. Lists of repeater channels are given in Appendix 9.

The normal way to set up a simplex contact while operating VHF/UHF mobile is to make (or answer) a CQ call on the mobile calling channel and to change frequency as soon as contact is made to one of the 'working' channels. For example, the mobile calling channel on the 144MHz band is V40 (145.500MHz) and the working channels are V16 (145.200MHz) to V46 (145.575MHz) although some of these channels are designated for particular priority uses such as emergency communication and RTTY – see Appendix 7 for the band plans, up-to-date versions of which appear in *RadCom* and the *RSGB Yearbook* [7]. This procedure does have its difficulties: one or more of the working channels may be occupied, but sometimes this is not apparent to both stations.

Using the standard 10–50W transceiver and a 5λ/8 whip it will usually be found that the range for mobile-mobile simplex work is very unpredictable in low-lying urban areas, and that mobile 'flutter' is a problem, particularly when both stations are on the move. Of course, one solution is to fit an add-on RF amplifier to boost the transmitter power, and perhaps also a preamplifier to improve the receiver sensitivity. Although these measures can give a useful increase in range, the extra expense, spectrum pollution and power consumption involved have led to a different and much more sophisticated technique being preferred – the use of *repeaters*.

REPEATER OPERATION

A repeater is a device which will receive a signal on one frequency and simultaneously transmit it on another frequency. Thus a low-power transmitter in a vehicle can transmit on the repeater's *input channel* and the signal will be faithfully reproduced on the repeater's *output channel*. Careful design has meant that repeaters can receive and transmit in the same band. This means that the same antenna on the car can be used for both reception and transmission.

In effect the receiving and transmitting coverage of the mobile station becomes that of the repeater and, since the latter is favourably sited on top of a hill or high mast, the range is usually greatly improved over that of unassisted or *simplex* operation (see Fig 1). Typically the effective range is increased from the order of 10–25km from the mobile station to something like 50km in any direction from the repeater, depending upon terrain and band used. Another advantage is that contact with other stations becomes more predictable. The coverages of stations A and B in Fig 1 continually change shape as the two stations pass through different terrain. Thus it is never easy for the mobile operator to estimate his or her simplex range. In contrast, the repeater service area is a known and much less variable factor. One further advantage is that mobile 'flutter' is usually diminished due to the superior antenna location at the repeater site.

How a repeater works

A repeater is an unmanned relay station and therefore requires an automatic system to control its operation. This logic system must ensure as far as possible that the repeater only relays signals intended for that repeater, and that those signals it does relay come up to an acceptable standard in respect of frequency, strength and deviation. For example, there is little point in the repeater relaying a signal which is so weak that it is unintelligible.

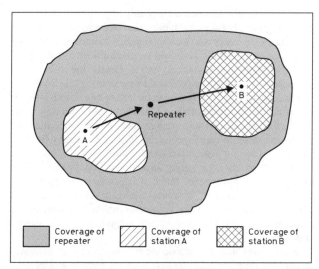

Coverage of repeater   Coverage of station A   Coverage of station B

Fig 1. The improved range of communication available between mobile stations using a repeater. The simplex coverage areas of stations A and B shown are constantly changing shape as the two vehicles pass through different terrain, and thus the two stations would have to be quite close before reliable simplex communication was possible

It would be very wasteful of power to have an FM repeater transmitter left on continuously if there were no signals being relayed, and so it is arranged that it is only switched on when a signal for relaying is present on the input channel. To ensure that signals which are not intended for relaying cannot accidentally switch on the repeater transmitter, UK repeaters on 2m, 70cm and 23cm conform to IARU Region 1 specifications which require a short audio tone *(toneburst)* to be sent by a user at the start of a transmission *(tone access)*. This tells the repeater to switch on its transmitter. Once this has been done other stations may *carrier re-access* the repeater indefinitely until it is no longer required. It will then automatically switch its transmitter off and another toneburst will be required if further use is to be made of it. 50MHz repeaters require a sub-audible tone to be transmitted continuously in order to access them. This is to avoid the problems of co-channel interference during periods of enhanced conditions with Continental repeaters which share the same frequency.

In order to make the best use of a repeater it is desirable to have knowledge of how the repeater control logic works in outline, and a typical sequence of events during a transmitting over is as follows. The repeater receiver is switched on continually and monitors the input channel, using a squelch system. When a signal appears on the input the control system determines if:

(i)   the signal has tripped the squelch (ie it is of adequate strength);
(ii) there was an audio tone (toneburst) of correct frequency and duration present; and
(iii) the received signal is correctly deviated.

If these initial conditions are satisfied (known as a *valid access)* the control system will switch the repeater transmitter on and allow the receiver audio output to modulate it. During the over, the control logic continuously monitors the incoming signal level, and if it falls at any time below the standard required for valid access, may switch off the transmitter or disconnect the receiver audio from the modulator. When the over is finished and the incoming signal disappears from the input channel, the repeater squelch will close and indicate to the control

logic that it must ready itself for the next over. After a short delay the repeater will signal it is ready by transmitting either a 'K' or 'T' in Morse code.

The short delay between the end of an over and the 'K' (or 'T') is quite important. During this interval the repeater will still relay any new signals on the input channel. Consequently this interval may be used by a third station to quickly announce its presence by giving its callsign. The use of 'Break' is unnecessary as the repeater users will already be listening for such a call. This practice is termed *tail-ending* and is a good way of inserting urgent or emergency messages between overs.

If the repeater has relayed a signal for more than a certain period of time (typically 2min on 144MHz and 5min on 432MHz) the control system may go into a *time-out* mode, close the talk-gate, and possibly transmit some form of *busy* signal until the incoming signal disappears off the input channel. This is done primarily to prevent overs from being too long, but is also a useful anti-jamming measure. It should be noted that some repeaters do not incorporate time-out devices. It is good operating practice to keep repeater overs to less than a minute whether or not time-out is fitted.

Equipment considerations

UK repeaters use the standard IARU Region 1 toneburst frequency of 1750Hz. The tolerance of this frequency is nominally ±25Hz and an accuracy of ±10Hz should guarantee satisfactory operation in all cases. Likewise the duration requirement varies but 500ms will satisfy all UK repeater requirements. The toneburst deviation should be set to ±1.5kHz. If you are using a modified PMR or a homebrew transmitter, do ensure that the transmitter, receiver and toneburst are operating correctly before you try use a repeater – *do not align equipment through a repeater*.

UK 144MHz repeaters have their inputs 600kHz lower than the outputs according to IARU Region 1 recommendations, while the 432MHz repeaters have their inputs 1.6MHz higher than the outputs (see later). Most modern commercial transceivers have the useful ability of being able to reverse this frequency shift at the touch of a button. For example, assuming the transceiver was set to channel RV62 (145.175MHz transmit, 145.775MHz receive) the frequencies would be reversed as soon as the REVERSE-REPEATER switch was depressed, becoming 145.175MHz receive, 145.775MHz transmit. It thus allows the operator to listen on the repeater input channel (the ability to listen on any frequency in use for transmission at the station is required by UK licence regulations).

50MHz repeaters use a 500kHz shift system with the transmit frequency 500kHz below the receive frequency. 1.3GHz repeaters use a 6MHz shift with outputs from 1297 to 1297.375MHz and inputs from 1291 to 1291.375MHz. There is just one 29MHz UK repeater licensed at present and this uses the standard worldwide system of output channels above 29.600MHz with 100kHz downshift. Access is by means of sub-audible tone (see below).

CTCSS repeater tones

CTCSS is is also available on many repeaters in the UK. It is an additional means of access on 144 and 433MHz repeaters but in the case of 29 and 50MHz repeaters it is the sole means of access.

The principle of CTCSS (continuous tone-code squelch system) is that a sub-audible tone is continuously transmitted in addition to the usual signal. Being below the normal speech frequencies, it does not affect the received signal.

CODE OF PRACTICE FOR REPEATER OPERATION

1. Avoid using a repeater from your base station; it is really for the benefit of the local mobiles. If you really must use it, use the lowest possible power and a directional antenna to avoid interfering with other repeaters on the same channel which you may not be able to hear. To be sure, use a CTCSS tone to access only the repeater you want.

2. Listen to the repeater before you transmit to make sure it is not in use. If you hear a local station you wish to call, listen on the input frequency to check whether the station is within simplex range before calling.

3. Unless you are specifically calling another station, simply announce that you are "listening through", eg "G1XYZ listening through GB3ZZ". One announcement is sufficient. If you are calling another station, give its callsign followed by your own callsign, eg "G2XYZ from G1XYZ".

4. Once contact is established:

   (a) at the beginning and end of each over you need give only your own callsign, eg "From G1XYZ";

   (b) change frequency to a simplex channel at the first opportunity (especially if you are operating a fixed station);

   (c) keep your overs short and to the point or they may time-out, and do not forget to wait for the 'K' or 'T' (if used);

   (d) do not monopolise the repeater as others may be waiting to use it;

   (e) if your signal is very noisy into the repeater, or if you are only opening the repeater squelch intermittently, finish the contact and try later when you are putting a better signal into the repeater.

5. If the repeater is busy, emergency calls may be made by tail-ending before the 'K', and announcing (a) that you have emergency traffic, and (b) which facilities you wish a station to provide. This will normally in most 'risk-to-life' situations be a telephone so that the other station can alert the emergency services. Do not reply to an emergency call if you cannot provide the services requested.

A repeater user who is on the border of more than one repeater's coverage area can now be selective. By transmitting the appropriate CTCSS tone, only one repeater will be activated rather than all the others on the same channel. The system operates in parallel with the usual 1750Hz access tone – either can be used to access a repeater.

In addition, a repeater only transmits its CTCSS tone when relaying speech, but not with its periodic idents. A suitably equipped amateur station can screen out the annoying idents, so making it more convenient to monitor the repeater.

The UK has been divided into 23 different CTCSS regions using 10 different tones, so that repeaters in the same area share the same CTCSS tone. The scheme is optional and will not be used to form 'closed repeaters'. Repeaters with the CTCSS facility available transmit the appropriate letter in Morse code after the callsign.

See Appendix 9 for a map of CTCSS regions and tone frequencies in the UK. The *Radio Communication Handbook* [5] gives more technical details of the CTCSS system and the circuit of a simple tone generator.

Using a repeater

The proper use of a repeater requires a high standard of operating ability and courtesy. Knowledge of the way in which repeaters work and confidence that one's own equipment is 'spot-on' does help, but also required is an ability to express oneself concisely; this being especially important on a repeater with a high level of activity. It must also be remembered that the purpose of repeaters is to facilitate mobile communication, and therefore mobile stations should be given priority at all times.

If it is required to test access into a repeater, the callsign and purpose of the transmission should be stated, eg "G1XYZ testing access to GB3ZZ". The repeater will respond with a 'K' or 'T' if access has been made. CQ calls are not normally made through repeaters; instead stations usually announce they are "listening through" the repeater, eg "G1XYZ listening through GB3ZZ". One such announcement is sufficient.

If it is apparent after setting up a contact that the stations are likely to be within simplex range of each other, the input channel should be checked. If signals are reasonable, the repeater should be vacated and the contact completed on one of the simplex channels. This is especially important if both stations are fixed.

Sometimes stations outside the repeater's service area will access the repeater successfully, but their signals will be very noisy and they may only open the repeater squelch intermittently. If this is the case, the contact should be terminated and another attempt made when a better signal into the repeater can be obtained.

If you wish to join an existing contact on the repeater then you should transmit your callsign in the gap immediately after one of the participating stations drops carrier but before the 'K' or 'T'. Don't worry if you are a little slow in doing this as the repeater will inhibit the 'K' or 'T' as soon as it realises another transmission is taking place. The use of "Break" on its own is unnecessary and illegal as an amateur should always identify his/her transmission using a callsign. To insert emergency or urgent messages, use the time before the 'K' or 'T' to announce your callsign and the problem, eg "G4AFJ/M emergency road traffic accident A46/A606 junction".

Because many UK repeaters will time-out after a few minutes, lengthy repetition of callsigns wastes the time available for each transmission. For example, "From G1XYZ" is quite sufficient at the beginning and end of each over.

A spell of listening will soon show that two common operating errors are timing-out and forgetting to wait for the 'K'. As a result, the repeater may eventually interrupt communication and the user, quite unaware of this, may spend up to a minute or so blocking the repeater to no avail. A simple time-out warning device may therefore prove useful.

There are two selfish attitudes which should be discouraged. The first is the practice of "taking another 'K'". This defeats the idea of sharing out the available air time to all users, ie it defeats the time-out of the repeater. The second bad practice is to use the gap before the 'K' to make a comment about the previous over. A good operator waits for his/her over before commenting because they realise that otherwise they are preventing the proper use of the gap to allow others to join in by inserting their callsign.

Certain repeaters, especially those outside the UK, may have different requirements, but if the rule 'listen before transmitting' is followed this should present no particular problem. Repeater groups usually have available literature with full details of their repeater and its facilities, which is well worth studying. Contact the keeper (see Appendix 9 for lists of keepers).

Several UK repeaters can link to one another and this is usually achieved by sending an appropriate DTMF tone to the repeater which activates the link. Further information on this is available from the RSGB repeater list [8] or the RMC web site which can be accessed via the RSGB web site [9].

The UK repeater network

UK FM repeaters are operational in the 50, 144, 432MHz and 1.3GHz bands, giving coverage of most of the country. All repeater callsigns are in the series GB3-plus-two-letters, and callsign identification is regularly given in Morse code at 12wpm.

All UK repeaters are designed, built and maintained by groups of enthusiasts under the overall management of the RSGB Repeater Management Committee which has full responsibility to the licensing authority for all aspects of repeater operation, including technical standards and frequency allocation etc. There are no 'closed' or private repeaters in the UK – all are available for general use. The annual cost of a repeater is some £500 and every regular user of a repeater should join the repeater group to support it and contribute to its funds.

Over 80 144MHz FM repeaters are licensed, using IARU Region 1 repeater channels RV48–RV63. The network was originally planned on a basis of non-overlapping service areas, but additional repeaters have been brought into service to cope with the increasing popularity of FM mobile operation, especially in the major cities where the amateur 'population density' is high.

432MHz repeaters far outnumber those on 144MHz in the UK. The IARU Region 1 frequencies are used, but the input and output channels have been transposed to avoid placing the output adjacent to the amateur fast-scan TV band. The channel number prefix 'RB' is used instead of 'RU' to denote the UK system. UK 432MHz repeaters were originally planned on a grid basis, each one nominally serving a 33km square. Consequently each has more of a local and community character than the VHF repeaters.

Several 1.3GHz repeaters are currently licensed. These use channels RM0 (input 1291.00MHz, output 1297.00MHz) to RM15 (input 1291.375MHz, output 1297.375MHz), ie 25kHz spacing and 6MHz shift. These repeaters mostly use horizontal polarisation and radiate a continuous carrier (to act as beacons) when not being used for talkthrough.

Details of all the above repeaters are given in Appendix 9. It should be emphasised that repeaters are experimental devices subject to modification and improvement, and therefore these details are subject to change. In addition new repeaters are being brought into service from time to time. Further details of these are published in *RadCom* and up-to-date lists including proposed repeaters are available as computer print-out from RSGB HQ [8] or on the RMC web site (accessed via [9]).

Problems with repeaters

From time to time in urban areas there is abuse or jamming (mostly on 144MHz band repeaters). The best advice is:

- Do not respond in anyway at all on the air to unlicensed transmissions or abusers.
- Do not approach suspected offenders as this can encourage further abuse and may prejudice investigations already underway.
- Help to gather as much information about the problem as possible.

For example, write down dates, times, and frequencies when the interference took place. Note any pattern of operation, suspected location of offenders, details of any bearings obtained with DF equipment etc. Tape recordings of the interference can be useful. Also details of other callsigns, names and addresses of those who have heard the interference as well.

A copy of this information should be sent to the repeater keeper (see the list in Appendix 9 for the callsign of the keeper). If problems persist, then copies of all correspondence and information should be sent to the Repeater Management Group Chairman c/o RSGB HQ. Further information on procedures to be followed can be obtained by contacting the Zonal Repeater Manager or the AROS Co-ordinator c/o RSGB.

MARITIME MOBILE OPERATION

For the fortunate operator who combines an interest in amateur radio with boating, it is possible to become an almost permanent one-man DXpedition, girdling the world and sharing the experience with friends back home and in the many countries visited. Even a yachtsman whose cruising aspirations take him no further than English coastal waters can sample the experience of being a rare call for, in spite of a considerable increase in the popularity of maritime mobile operating over the last few years, the /MM suffix is still sufficient to create additional interest at the 'other end'.

Before one can join this exclusive fleet, there are a number of obstacles which are peculiar to maritime operation to be surmounted, and here a look will be taken at one or two of the major ones. More information can be found in *A Guide to Small Boat Radio* [10].

Licensing

The UK licence permits operation in any vessel or vehicle, requiring the suffix '/M' or '/MM' to be added to the callsign. Written permission is required from the Master of a ship for the installation, use or modification of a station.

Before the proposed station can be operated, British Telecom International may have to inspect it. This inspection is done by the Ship Radio Inspection Officer and normally has to be arranged at or near a major commercial port. The licensee must have on board the same power- and frequency-measuring equipment as he is required to have at home.

Apart from inspecting the station and requiring the would-be operator to give an on-the-spot demonstration of the accuracy of, and his ability to use, the test equipment, BTI may also need to establish that the amateur station does not interfere with the ship's main transmitters and receivers. BTI may also require to see the ship owner's written agreement to the establishment of an amateur station on his ship. Not much of a problem when the owner is also the skipper and operator, but possibly more so if the ship happens to be a supertanker. As far as BTI are concerned, the same regulations apply to all kinds of ship.

Their inspectors may well also bring their own equipment to check compliance with licence conditions and levels of harmonic radiation, spurious emissions and unwanted sidebands etc.

The Class A or Class M5 (power limitation to 100W) licence allows operation on all bands but does not permit operation by anyone other than the licensee, even under supervision, unless the other person is in possession of an Amateur Radio Certificate. Radio silence must be observed if requested by the ship's Master.

Finally, when operating with a licence which does not permit third-party message handling, or with shore stations similarly restricted, do remember the terms of your own (and your friends') licences. When a long way from home it is all too easy to risk infringing your licence and embarrassing friends, not to mention the risk of losing for the boating fraternity a very valuable privilege.

Roger Wheeler, G3MGW, operating on his boat

Foreign ports and territorial waters

The UK licence does not authorise use while ships are alongside in foreign ports or in foreign territorial waters (unless the country authorises UK radio amateurs to operate under CEPT TR61/01 – see later). Maritime mobile operation has become much more popular in recent years, and on several occasions yachtsmen in particular have found themselves in trouble with local overseas radio regulatory authorities or police as a result of them using amateur band transmitters in foreign ports without prior permission.

If the country does not permit CEPT TR61/01 operation, it is wise to make contact with the local amateur radio association before operating. In smaller countries, quite often it will be found that they either operate the amateur licensing system on behalf of their telecommunications authority or certainly can arrange personal introductions which help avoid administrative delays. Experience has proved that advance application in writing (other than to the USA and major European countries) tends to be unsuccessful as many countries have no fixed policy on amateur maritime licensing, and the granting of permission to operate is dependent upon the reaction of the individual official on the spot. See the 'Operation in foreign countries' section at the end of this chapter for further information.

Equipment

Favoured equipment, both by yachtsmen and ships' radio officers, is the smaller portable transceiver which will operate on 12V DC. While not a problem in a large ship's wireless room, the climatic conditions encountered on board a small yacht, especially in the tropics, were certainly not envisaged by the manufacturers of the rig, who thought their equipment was destined to be tucked, warm and dry, beneath the dashboard of a car. Initially many yachtsmen make the mistake of locating the rig among their other radio gear around the chart table. Such a position is almost inevitably going to expose the equipment to a splash of sea water sooner or later. Better to put it inside a locker next to the chart table, and run a multi-way cable to a remote alternative operating position in the cockpit.

Spares

If a yachtsman is going to set off across the Atlantic working regular skeds with friends back home it is irresponsible to do so without both spares and some knowledge of the potential weaknesses of the particular 'black box' that will be aboard. While retailers and manufacturers will normally maintain that the gear they are selling just never goes wrong, they will usually become a lot more helpful if approached with a full explanation of just why the enquiry is being made!

If a /MM station goes suddenly off the air with equipment failure in mid-ocean, great concern is likely to be caused to friends who have been keeping skeds. Certainly, they know that the most likely cause is radio equipment failure, but when it actually happens they can hardly be blamed for fearing worse. In one case, the problem was transmitter PSU failure and, as the receiver section was still working, we could hear our friends becoming more worried as each missed sked went by, wondering whether to report our disappearance to the appropriate authorities. We were able to get back on the air by cannibalising other electronic equipment that was on board but ever since then we have shipped a small 'floating junk box' and also a spare transceiver.

Power supplies

Even a modern 150W PEP transceiver will represent quite a considerable drain on a yacht's batteries and at sea it is easy to become involved in quite long operating periods without thinking. Ideally a separate battery should be carried specifically for the rig, with suitable means to isolate it from the ship's main batteries. Here again, not a 'big ship' problem for in that case the ship's 110V AC supply will be available.

Antennas and earths

Due to its ability to operate with a perfect earth system a boat will put out a remarkably good signal with quite modest power. For instance, when working the UK from the West Indies, a yacht running 150W to a vertical dipole may well be given the same report as a shore station on an adjacent island which is running 500W to a two-element cubical quad! The problem is more that of the antenna being non-directional, and with the pile-up of callers who may be anxious for a contact there is no way to notch out any of the interference. In these circumstances the infallible solution is to plead a sudden navigational problem and switch off!

Shipboard antennas are normally vertically polarised and radiate an excellent low-angle signal. The favourite seagoing antenna is normally a vertical $\lambda/2$ dipole although if the boat is of steel construction then a $\lambda/4$ ground plane driven against the ship's hull may be better. This author has a 'floating antenna farm' with a three-way rotary coaxial switch giving access to:

(a) a $\lambda/2$ dipole;
(b) a 1λ ground plane (yes, full wave on 14MHz – the back stays to the mast are just 19.8m [65ft]);
(c) a $\lambda/4$ ground plane.

This makes it possible to change in mid-contact. It is quite often found that there will be long slow fading on one antenna and not on the other, and clearly there is quite a lot of interesting experimenting to be done.

Whatever the antenna it is essential to ensure a good match and an SWR of 1:1, as with limited power available it is essential that it all gets radiated. With a high SWR, transmitters with

Table 1. Maritime mobile nets

Net name	Frequency (kHz)	Time (UTC)	Controller
Jerry's Happy Hour (US)	21,402	0100	
Travellers	14,116	0300	Roy, VK6BO or Peter, VK6HH
Pacific	14,313	0400	Terry, ZL1MA
Canadian	14,115	0400	
Kenya	14,345	0500	Tony, 5Z4FZ
Anzus	21,205	0500	
South African	14,316	0630	Davina, ZS5GC
German	14,313	0700	Various
Pacific Inter Island	14,315	0800	Rick, NH2F
United Kingdom	14,303	0800	Bill, G4FRN
German	14,313	1030	Various
Indian Ocean	14,332	1115	Roy, VK6BO or Peter, VK6HH
South African	14,316	1130	Alistair, ZS5MU
Mississauga Canadian	14,121	1145	Various
Seanet	14,320	1200	
JA Okera	21,437	1220	
Transatlantic	21,400	1300	Trudi, 8P6QM
German	14,307	1630	Various
Swedish	14,303	1630	
United Kingdom	14,303	1800	Bruce, G4YZH, Sue, G0OEP, Rudi, G4FTO, or Gerald, G0THB
SW Pacific	14,315	2100	Tony, ZL1ATE
Indonesian	7090	2230	
United States	14,300		Various

transistorised output stages having high-SWR protection will not function at all.

Seagoing RF

On a merchant ship the antenna is usually well elevated and the operating position is adequately screened, but on a yacht the reverse is true and everything can easily become 'humming' with RF. Even if a linear amplifier can be powered, the stray resonant circuits provided by the stays, rigging and wiring on a yacht can be positively dangerous when excited by a strong RF field. It is therefore far better to keep the power level down and make a really serious effort to get maximum radiation and minimum reflected power. If part of a boat's rigging is used as a radiator, remember that rigging usually comes in symmetrical pairs. For example, if one of the mast stays is a convenient 15m long, its opposite number must be detuned-the favourite technique is to bridge across about 500mm with 1.2mm (18swg) copper wire with a 500pF mica capacitor in the middle.

Even with low power, other electronic equipment on board may seriously misbehave when subjected to RF fields – remember, the makers of equipment used in small yachts never had it in mind that their gear would have to tolerate a 14MHz SSB signal in close proximity. Electronic speed/distance logs

and quartz crystal clocks should be carefully checked out for RF tolerance before setting out on any major voyage.

Operating frequencies

Throughout the world, most /MM operation is carried out on the 14MHz band, irrespective of the stage in the sunspot cycle. Each region has its own net frequencies and sked times. Most English-speaking maritime mobiles in Region 1 and 2 check into a net on 14,303kHz at 0800 and 1800 (see Table 1). Other nets to note include the Indian/Madras stations operating between 0700 and 0800 on 7080kHz and at 1900 on 14,150kHz (all times Indian local), offering information for yachts approaching India.

If you are neither /MM nor keeping regular skeds with a /MM station, a wanted contact and subsequent card will be more readily forthcoming if it is borne in mind that the net operations should be allowed to proceed without interruption to their natural conclusion before an attempt to contact is made. Most /MM operators are always delighted to chat and reciprocate a card provided they not interrupted when talking to their friends in the Mediterranean – after all, exchange of information about harbours, fishing and RF in the rigging is to the /MM operator what life is all about!

SINGLE-OPERATOR PORTABLE OPERATION

'Going portable' single handed is one of the most interesting and satisfying challenges of amateur radio. To set up a completely self-contained station in the middle of nowhere is somehow to go back to fundamentals and rediscover the hobby anew. Each expedition is an adventure and, irrespective of whether the station is 'black-box' or 'homebrew', much will be learnt about radio communication, probably more than is realised at the time. The temptation to throw the rig and a makeshift antenna into the back of the car at the last minute should, however, be resisted. Some forethought and planning will pay great dividends – probably only owners of hand-portable stations can contemplate 'instant' activity! Note that this section only deals with the sort of equipment you might use on a 'fun' outing – if you are comtemplating serious operation for a contest you should of course read Chapter 5.

Transceivers

Table 2 shows the three main classes of portable transmitting equipment as used in practice. The first class consists generally of completely self-contained equipment in one enclosure and specially made for that particular type of operation. The second class covers a wide range of equipment, but the most suitable will often be mobile transceivers with add-on preamplifiers or power amplifiers if required. The third class comprises main station equipment generally delivering close to the maximum permitted power. The power supply to be used is an

Table 2. Choice of mobile and portable equipment in the UK

Operation	Power output range (continuous)	Power supply	Transportation	Alternative operation
Hand-portable (pedestrian)	Low (50mW–5W)	Dry batteries or nicad rechargeable batteries (6–12V DC)	Pocket or shoulder	Mobile (add-on PA) Base station (add-on PA)
Fixed portable (vehicle mobile)	Medium (5W–50W)	Vehicle battery (12V or 13.8V DC)	Vehicle	Base station (12V or 13.8V DC PSU)
Fixed portable (110–230V AC)	High (50W–100W)	Petrol generator	Vehicle	Base station

A 6m guy-less mast in position. For construction details see [14]

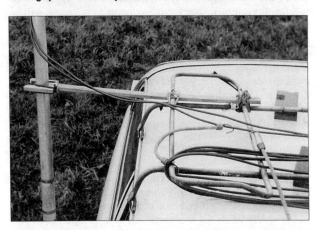

Pan-handle steady for mast. Make sure the roof-rack is well secured

important factor. The low-power equipments with under 5W input can be run off lantern batteries, but those contemplating this are reminded that mains electricity costs far less than battery electricity (up to about 1,000 times less in the case of a PP3!) and therefore a set of nicad batteries and charger will prove a good investment. The larger nicad batteries can be used up to about 20W peak transmitter input, and it is therefore quite feasible to run a 10W PEP SSB transceiver from this type of power source.

Higher powers will demand lead-acid batteries or a portable generator, neither of which are particularly convenient to carry about single-handed. If these power sources are chosen they should be maintained regularly. Generators should get the manufacturer's recommended maintenance and be exercised during lay-off periods. During the winter wet batteries should

be exercised and a freshening charge given each month. Discharge them into a load at a measured rate every so often, and then recharge at a known rate. All very elementary but all too often this can be overlooked.

Careful thought should be given to which type of portable operation is really intended and, if possible, equipment chosen that can also be used at the home station. For instance, if high-power portable operation (eg for contests) is the eventual goal, it may be worth considering purchasing a generator and using base station equipment from the outset, rather than wasting time and money building up a system of low-power transmitters and power amplifiers. On the other hand, if the main interest is in hand-portable operation with occasional mobile work, an attractive possibility is an add-on power amplifier, which may be left permanently connected in the vehicle ready for service.

Antennas

Antennas used for single-operator portable operation must be simple to transport and put up, robust, easy to tune, efficient and, perhaps surprisingly, occupy little space, for sometimes the roadside is the only accessible land.

Three popular antennas for HF portable work are the inverted-V dipole, the short loaded vertical and the ground plane. The dipole takes up the most space, but on flat ground is a good high-angle radiator and on steeply sloping ground can give excellent DX results [12]. The short loaded vertical is similar to the mobile antennas from which it was derived. Its main advantage is that it can be made into a multi-band system by various combinations of coils and antenna sections. The use of HF ground-plane antennas is somewhat controversial, with some operators claiming they give poor DX results, but nevertheless they continue to enjoy much popularity [11]. They are simple to make and erect, can be made self-supporting and require little space.

On VHF, omni-directional antennas are usually useless when located on top of a hill, for they will pick up all the locals as well as the DX. Far better results will be obtained with even the simplest beam antenna located a couple of metres high. The Yagi is easily the most popular beam antenna for portable VHF work, and usually the five-or six-element 144MHz versions are the largest that can be carried inside the average car, the boom length being the limiting factor. Other more portable antennas worth considering are the instant cubical quad [12] and the two-element HB9CV antenna [5]. If a roof rack is available, larger antennas can be carried, but high-gain systems are by nature heavy and unwieldy, and therefore not a good choice for the single-operator station. One tip when transporting the commercially available Yagi antennas having wing-nut element fasteners is just to loosen these rather than taking them off altogether. The elements can then be pivoted along the boom axis, which saves car space and leads to much quicker assembly once the destination is reached. Portable masts can be obtained ready-made or alternatively made up out of sections of aluminium tubing (see Chapter 5 and reference [13]).

Setting out

It is recommended that the portable station be completely assembled at home and soak-tested before any excursions are made. Remember that access to all the home station test equipment will not be available while out portable. Critical control settings should be noted down. Make sure the equipment will operate satisfactorily down to about 10V – sometimes relays are sticky at less than the ideal 12V.

Practise erecting the antenna in the garden several times until the routine is learnt, and test it using the transmitter if this is possible. Draw up a check list of items required for the portable station, so that these can be ticked off while loading the car. Make sure the equipment will not slide around and get damaged while in transit – an old rug or blanket laid over the car paintwork will protect it and the equipment from scratches. The load must be secure, even in an emergency stop or in sharp cornering, otherwise this could be dangerous. Do not neglect the 'inner man' when going out on a prolonged expedition. Comfort is an important item, and a car tends to get much colder than a tent at night. Even in the summer months the early mornings can be very cold in the UK, especially on high ground, and a sleeping bag and hot food is a necessity, not a luxury, in such circumstances.

Selecting a site
HF stations will probably work reasonably satisfactorily at most locations providing there are no significant obstructions within about 100m. Those using ground plane antennas should try the marshy ground near rivers which ensures good earth conductivity, while users of horizontal dipoles are recommended to try hillsides sloping in the desired direction[12]. VHF stations are best located on top of an open hill with a steep slope in the direction of the likely DX.

A few likely sites in the area of interest should be selected from maps before setting out, for most of these will be found to be unsuitable for various reasons (see Chapter 5) once they have been reconnoitred. With experience comes a feeling for good sites which will save much time. Once a suitable site is selected find out who owns the land and call on the owner to get permission to set up the station: quite apart from being common courtesy this may bring to light any local ordinances or restrictions. It may result in an even better spot being discovered as local knowledge is tapped. A side effect of such calls can be freedom from interference by authority in any of its uniformed guises since the residents will know what is happening. In the latter connection, it is recommended that the validation document be carried at all times to avoid possible misunderstandings.

OPERATION IN FOREIGN COUNTRIES
This is getting easier and easier, thanks to a relaxation of licensing regulations between many countries worldwide, and also the minimal Custom formalities in the European Union.

Whether you are planning a DXpedition (see later) or are just going on holiday and simply wish to try a little casual operating then CEPT operation (see below) is ideal if the country allows it because there are no formalities. If the country does not allow CEPT operation or if you are going to take up residence then you will need to apply for a foreign amateur licence which may be granted under reciprocal or unilateral licensing regulations.

Operation under CEPT T/R 61-01
In Europe a significant and welcome development was the introduction of a scheme whereby amateurs from countries which have implemented CEPT Recommendation T/R 61-01 can temporarily operate with the minimum of formality in any of those countries (see Chapter 1). CEPT licensing does not replace reciprocal/unilateral operation, as it is a short-term solution for portable or mobile operation, eg holiday operation from a hotel; however, it does also authorise operation from the station

of a radio amateur in the host country. If you wish to operate from a permanent address or for over three months, you still need to apply for a reciprocal/unilateral-type licence.

Operating privileges under CEPT are those of the country you are visiting so, for example, you cannot operate on 70MHz from the Netherlands because Dutch amateurs do not have that band. (Remember that even if the host country has the same band, it may not be the same frequency range!) However, the Dutch authorities permit full-power operation on 430–432MHz bu the UK authorities do not. UK Class A/B and B licensees are not allowed to operate on bands below 30MHz and Novice licensees are not allowed to operate under CEPT at all.

The callsign that you use is your own, preceded by the host country prefix, eg PA/G4FTJ. However, you may need to vary the host country prefix depending on your location (just as in the UK), and in some countries, such as Italy, it is dependent on whether you are operating a CEPT Class 1 station (UK Class A) or a CEPT Class 2 station (UK Class A/B or B). As you are mobile or portable, you will need to add the appropriate suffix '/M' or '/P'.

If you are interested in operation under CEPT in a particular country you should:

- read the CEPT sections of BR68 and Chapter 1 of this book;
- send for the RA information sheet RA247 [21];
- write or telephone the licensing administration of the country to ascertain the current licensing conditions for CEPT operation (and how to import amateur radio equipment). Addresses are given in RA247 and also in Appendix 2 of this book. Many countries have helpful leaflets in English on this topic.

More European countries are expected to implement CEPT T/R 61-01 but in the meantime some countries outside the CEPT area such as New Zealand and Peru have adopted the CEPT recommendations, giving an even wider range of countries to operate in with no formality. Recently the USA has also agreed to allow any visiting amateur to operate providing they have with them their licence document from their home licensing authority and other form of identification.

Reciprocal and unilateral licensing
Since the time in 1965 when the first UK reciprocal licensing agreement was approved, the facilities available to persons wishing to operate from foreign countries using this form of licence have become numerous. It is generally recognised that in Europe the Belgian national society, UBA, was the first to commence the pattern which today is commonplace. With the increase in business travel and the desire to visit foreign countries on holiday the demand for reciprocal arrangements has multiplied. A list of the countries with which the UK has a reciprocal agreement is given in the UK section of Appendix 2.

There are also countries whose legislation does not permit the conclusion of a formal agreement, or whose conditions for licensing are to a lower standard than that in the UK, but who are prepared unilaterally to grant temporary amateur licences; therefore UK amateurs visiting countries not listed and wishing to operate an amateur station there should write to the administration concerned well in advance of their visit to ascertain whether a licence can be obtained. Again, these possibilities are noted where known under the relevant country section in Appendix 2.

It is essential that the applicant should allow plenty of time for the issue and despatch of a reciprocal or unilateral temporary licence. There may be queries from the administration concerned which will take additional time to resolve. This does mean that a suddenly arranged business trip may not be the time to seek a reciprocal licence in the country concerned.

Typical documents that *may* be required when applying in person are:

- Your amateur radio licence (take a photocopy or two in case they wish to retain a copy).
- Your passport.
- A completed application form giving your temporary address.
- Details of amateur radio equipment to be used, eg brand, serial number, frequency coverage and power output.
- Proof that your stay is temporary, eg a visa.
- The current licence fee.

In some cases, you may even be required to supply a letter written by your own local police attesting to your good character. Follow particularly carefully any instructions provided by the issuing administration regarding payment.

If you are applying by post you may need to photocopy your licence and passport and in some cases get these notarised as true copies. In the UK the RA will issue, on request, a short letter confirming the existence of a current licence.

In some cases a reciprocal licence will be accompanied by a document showing the regulations for the amateur service in the country concerned. These should be carefully studied and strict compliance with the regulations is obviously required. Particularly do not assume, even in a Region 1 country, that the band allocations follow those of the UK.

Importation of amateur equipment

The customs treatment of amateur radio equipment varies not only between countries but also between different ports of entry or border crossings, and is an entirely separate matter to licensing. In practice there are minimum customs formalities between EU countries. Most non-EU countries will allow the temporary importation of a transceiver, power supply and antenna as part of your personal effects, but you should carry receipts and an inventory, complete with serial numbers. If in any doubt, contact the relevant embassy, licensing administration or national amateur radio society.

A carnet covering the equipment may be required and it is recommended that due to the continuing changes regarding carnets that up-to-date information be obtained from the AA or RAC. A number of countries will accept equipment in transit provided that it can be sealed and inspected when it is exported from the country.

See the 'Customs' section later for some further notes on this topic.

ORGANISING A DXPEDITION

A 'DXpedition' can be defined as a portable amateur radio operation from another DXCC country, IOTA island, or locator square, purely or mainly for the purpose of making as many contacts as possible. From this definition it follows that if you go on holiday to Spain and take your equipment, but only talk to your local friends on 20m, that is not a DXpedition.

This section is mainly about HF DXpeditioning. The sections on licensing, customs and general comments about equipment and antennas apply equally to VHF DXpeditions, the main difference being that – other than specialised 6m or moonbounce operations – most VHF DXpeditions are likely to be closer to home than their HF counterparts.

HF DXpeditions have grown in popularity – with both DXpedition operators and DXers – since the 'sixties, when pioneers such as Gus Browning, W4BPD, and Don Miller, W9WNV, activated a number of very rare and even 'all-time new' countries.

In those days, when 'portable' equipment consisted of separate, valved, transmitters and receivers that were physically large and heavy, even a one-man DXpedition was a major logistical challenge. In the late 'seventies the advent of single-box transceivers and transistorised equipment led to a greater number of DXpeditions by such people as Martti Laine, OH2BH; Erik Sjölund, SM0AGD, and the late Lloyd and Iris Colvin, W6KG and W6QL. In the 'eighties and throughout the 'nineties, DXpeditioning really took off and now, with tiny modern transceivers and a general relaxation in licensing in many countries, it has never been easier to put on a DXpedition.

There are two main types: the one- or two-man DXpedition, usually a 'holiday' operation, and the major multi-operator, multi-transmitter DXpedition. The only real difference is the amount of planning and organisation involved. We will discuss only small-scale DXpeditions in detail in this section. A new book, *DXpeditioning Behind the Scenes* [18] – written by the team which organised the UK's largest DXpedition thus far, the CDXC [19] 9M0C Spratly Islands operation – covers the planning, logistics and execution of a major DXpedition in great detail. Two operators taking a single transmitter means that one station can be kept on the air for much longer periods by sharing the operating time and sometimes modes between the two operators.

In the UK, we are fortunate in having three rarish DXCC entities on our doorstep which make excellent DXpedition 'targets'. These are Jersey, GJ; Guernsey, GU; and the Isle of Man, GD. Close at hand too are numerous destinations in holiday companies' brochures, many of which are attractive DXpedition destinations, such as Crete, SV9; and Malta, 9H. Further afield, but still package holiday destinations and easily activated, are such places as Barbados, 8P; the Gambia, C5; and the Maldives, 8Q.

The RSGB Islands on the Air (IOTA) programme [20] is becoming more popular every year, and this provides numerous island destinations, near and far, for the would-be DXpeditioner. In the British Isles alone, there are 28 IOTA island groups, ranging in difficulty of activation from Anglesey to Rockall, which has *never* been legally activated.

Licensing

The advent of the CEPT Licence (CEPT Recommendation T/R 61-01 – see also Chapter 1 and 'Operation in foreign countries' earlier in this chapter) has been a real boon for potential DXpeditioners. With the inclusion in 1999 of the USA and its many overseas territories it has become possible to operate from over 100 DXCC entities and numerous IOTA islands without the necessity of having to apply for overseas licences.

The CEPT Licence is intended for *temporary* portable operation only, and you need to take your UK licence and BR68 booklet with you. You should ensure that you are fully aware of the licensing restrictions in force in the country you wish to operate, eg the appropriate band limits, the maximum power permitted, whether 6m operation is allowed etc. See the 'Operation in foreign countries' section above for more details of CEPT operation. Don't forget to get hold of a copy of the RA247 leaflet from the Radiocommunications Agency [21].

Details of the countries which have agreed to CEPT Recommendation T/R 61-01 can be found in Appendix 2 of this book, but from a DXpedition point of view remember that there are many additional DXCC entities associated with these countries which can be attractive DXpedition locations. For example, Finland is a CEPT Licence country, but the licence is also valid in the Åland Islands, OH0, and Market Reef, OJ0; Portugal covers the Azores, CU, and Madeira, CT3; and the USA covers many overseas territories including the US Virgin Islands, the Northern Marianas, American Samoa etc.

The CEPT Licence covers the majority of locations for 'holiday' DXpeditions, the exceptions being long-haul destinations such as the Gambia, the Maldives, Malaysia, and most of the Caribbean islands (although the Netherlands Antilles, the French Caribbean islands of Martinique, Guadeloupe, St Martin and St Barthelemy, and the US Virgin Islands and Puerto Rico, *are* covered by the CEPT Licence).

Even if not included in the CEPT Licence, there are reciprocal or unilateral licensing agreements with the majority of countries which UK amateurs are likely to choose for a holiday DXpedition (see also 'Operation in foreign countries' earlier). However, there are exceptions: Thailand, Singapore, Indonesia (eg Bali) and Kenya are a few long-haul destinations which can be found in most holiday companies' brochures but where it is difficult to receive a licence for a holiday operation (in most cases licensing *is* possible for UK amateurs, but only for those in possession of a residence permit). Details of the countries where there are reciprocal or unilateral licensing agreements with the UK can also be found in Appendix 2.

In a few countries it is not enough simply to have obtained a licence. The Andaman and Laccadive Islands are good examples. Both island groups, which are separate DXCC entities, are politically part of India and both have tourist resorts on them. It is easy to travel there as a tourist and it's perfectly possible (though it can take a long time with considerable bureaucratic wrangling) for a UK amateur to obtain an Indian licence. However, a separate permit is required for operation from either group of islands, and this is not normally granted, either to foreign visitors or even Indian nationals. Naturally, it is for this reason that the Andamans and Laccadives are so rare, and are therefore the 'target' of many would-be DXpedition groups. One day, no doubt, a group will find a way of obtaining the necessary permit and licence and these islands will again be activated; until then they remain among the rarest DXCC entities in the world.

Customs

Customs regulations vary from country to country and it is beyond the scope of this book to go into detail. There are no customs formalities between European Union (EU) countries, so 'holiday DXpeditions' to any EU destination should not present any difficulties. These countries now have a 'Blue Channel' in addition to the more familiar Red and Green customs channels for travellers from within the EU.

Most 'tourist-friendly' countries will permit the temporary import of a limited amount of amateur radio equipment as part of the traveller's 'personal belongings'. There is a limit, however, and while a single, obviously used, transceiver, power supply and multi-band vertical is likely to be considered personal belongings by all but the sternest of customs officials, several boxed new transceivers, linear amplifiers, beams, masts and 100m drums of coaxial cable turning up without the necessary importation documents will certainly not be so considered. If you are planning a multi-operator DXpedition with that sort of equipment count, do your homework beforehand and find out what paperwork is required.

Often a *carnet* is the simplest: *all* the equipment must be listed in an inventory and it *must* be re-exported to the country of origin. It's important to note that such paperwork does not come cheap. In a major multi-operator DXpedition the cost of customs clearances and paperwork can amount to almost as much as the freight charges. This subject is covered in detail in reference [18].

Equipment

Most single-operator DXpeditioners will choose the smallest and lightest transceivers available. At the time of writing (early 2000) the single-operator DXpeditioners' rigs of choice are the Kenwood TS-50, Alinco DX-70, Icom IC-706 and its later variants, and the Yaesu FT-100.

In the case of a multi-operator DXpedition, when two or more stations will be operating simultaneously, it pays dividends to check the transceivers' performance in the presence of strong out-of-band signals. A rig which may appear to be perfectly satisfactory (even on 40m, at night, using a 40m beam, during the *CQ* World Wide DX Contest – surely the most severe test of any receiver's front-end performance!) may still 'fall over' if you try to operate a second station on another band from the same location.

It is not just receive performance which can be a problem: RF from one station can get into the other rig's microprocessor, causing all sorts of unexpected operational difficulties. If you are planning to use two stations simultaneously (and this applies to non-DXpedition special-event stations too) consider investing in a set of bandpass filters such as the Dunestar range [22]. When the appropriate filter is inserted in the RF lead it will provide approximately 40dB additional rejection of out-of-band signals, both on receive and transmit. If one is used on each station, a combined 80dB rejection of signals between the two bands can be obtained. Ensure the filters used are rated for at least 150W and, if you are using a linear amplifier, insert the filter *between* the driver and amp!

The only sure way to tell whether inter-station interference will be a problem is to try out all the rigs, in as close a simulation as possible, before departure. The CDXC 9M0C Spratly DXpedition used Yaesu FT-1000MPs and FT-920s with no problem at all, and members of the Voodoo Contest Group, which has operated as 5V7A from Togo and 9G5AA from Ghana among other locations, have recommended Kenwood TS-930S and TS-570D transceivers in this regard.

If you are taking your valuable amateur radio equipment on an aircraft, you will probably want to carry it yourself and not entrust it to baggage handlers, who have, just occasionally, been known not to treat suitcases with the sort of respect their owners would like.

A small, lightweight transceiver, switch-mode power supply, microphone and/or key, headphones, and even a notebook PC for logging, plus the PC's PSU, can easily be accommodated in a single aircraft carry-on bag. Do be aware, however, that in addition to the maximum *size* requirement, the *weight* of such bags should be limited to *5kg per person*. Airlines are, of course, perfectly within their rights to weigh these bags at the check-in and to make excess baggage charges if the allowance is exceeded. If two people travel together and each have a 20kg checked-in allowance, it should be possible to ensure that the *combined* checked-in and carry-on weight for both does not exceed 50kg – even if amateur radio equipment is going along for the ride. If the excessive weight of the carry-on bag is queried when checking-in, providing the *total* weight is below the permitted maximum, most airlines will not object,

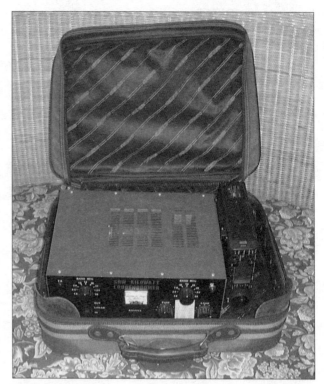

Icom IC-706MkII, Icom switch-mode PSU, microphone and/or key, miniature headphones, notebook PC, and the PC's PSU all fit neatly into an aircraft carry-on bag. Use 'bubble wrap' to ensure the equipment is not scratched in transit

An alternative DXpedition station: Icom IC-706MkII and SRW Kilowatt Loudenboomer amplifier packed in one carry-on bag. The transceiver's switch-mode PSU can be packed carefully in the checked-in luggage and a notebook PC for logging (if required) carried separately

especially if the fragile nature of the equipment is pointed out to them.

It is important to be 'loud' when on a DXpedition, so you should consider taking a linear amplifier. Before rejecting the idea out of hand because of the weight, do consider how useful a linear might be. If your destination is a long-haul one to an area without a large concentration of relatively local amateurs to work, you may find the extra power almost essential. Locations such as the Maldives, 8Q; the Gambia, C5, or any Pacific island are fairly remote places where 100W to a simple wire or vertical antenna may prove disappointing. Also, if you plan to use the low bands you may find the additional power makes all the difference between not being heard at all on 160 and 80m and having a nightly DX pile-up.

The problem, of course, is the weight. Given the phenomenal increase in DXpeditioning in recent years, it is perhaps surprising how few small, lightweight, linear amplifiers there are available. There are some, though: the FinnFet 1kW amplifier using power FETs and a switch-mode power supply is one, although at over £3000 it is rather expensive in the UK. It was developed especially for long-haul DXpeditions by Pekka Kolehmainen, OH1RY. The Icom IC-2KL (despite its name, capable of 500–600W output) is now a few years old and can sometimes be found second-hand at a reasonable price. It comes in two small boxes: the transistorised RF deck, which is light enough to be hand-carried, and a hefty PSU which could probably be packed in a Samsonite-type suitcase, if surrounded by plenty of bubble wrap and holiday T-shirts, and sent as checked-in luggage. Some years ago Steve Webb, G3TPW, marketed a small amplifier which he called the 'SRW Kilowatt Loudenboomer'. Using four PL519 TV sweep tubes, which have the advantage of being very cheap, it produces around 400W PEP output in a package measuring $14W \times 11D \times 5H$ in

and weighing just 7kg. Loudenboomers are now difficult to obtain, but are ideal for a one-man DXpedition requiring more than 100W of power.

Even if a lightweight amplifier is not available, with two or more members of a DXpedition team it could still be worth considering taking a 'standard' amplifier. They weigh in from 14.5kg (for the Ameritron AL-811X), though most are around 25kg or more. If packed well, preferably in the original box, they

A complete 400W DXpedition station – equipment in the bag and a Butternut HF6V antenna – can easily be carried by one person

should survive the rigours of an aircraft journey. Naturally the valves should be removed and hand-carried, wrapped in plenty of bubble wrap.

Antennas

Arguably the most-popular antenna for the one-man DXpedition is the Butternut HF6V. This is a six-band vertical, 26ft high, covering 80, 40, 30, 20, 15 and 10m (but not 12 and 17m) and which uses resonating circuits rather than traps. It requires radials in order to work properly. This antenna is popular with DXpeditioners as it packs into a box measuring just 4in square by 4.5ft long and weighing 4.5kg and is thus easy to check-in on commercial flights. It performs well, especially so if the DXpedition destination is an ocean-side location.

Other verticals, such as the Cushcraft R5000 and R7000, are also popular DXpedition antennas and also work well. They have two advantages over the Butternut: 12 and 17m are included (although 80m is not included on the R7000, and 30, 40 and 80m are not on the R5000) and, being half-wave design antennas, they do not need long wire radials (short 'spoke' radials are part of the design). Their disadvantages are that they do not pack into such a small box and that they are somewhat heavier, as they use traps.

Single-band dipoles work well too, and have the advantage of being extremely lightweight and easy to transport. Most DXpeditions, even single-man ones, will want to operate on several bands though, and, depending on the location, it may prove difficult to put up several different single-band antennas. Multiband wire antennas, such as the G5RV or Carolina Windom, tend not to be as popular as multiband verticals on DXpeditions. In most cases an ATU will be required, making it necessary to transport one additional box (plus an additional coax patch lead).

For multi-operator DXpeditions, a triband Yagi should be considered. An antenna such as the Cushcraft A3S weighs under 13kg and so is feasible to transport if there are two or three operators to pool weight allowances. There's no question that a beam will make a real difference to the final QSO count. Probably more difficult than transporting a beam, though, is taking something to mount it upon. Telescopic 'push-up' masts are ideal, though they can be as bulky and heavy to transport as the antenna itself. If the DXpedition is to a location right on the sea, eg a low-lying island, a beam will often work well at a height of just 20ft or so, whereas inland, and particularly if there are structures such as buildings in close proximity, the minimum height for a beam to perform reasonably well is about 30ft. If the DXpedition is operating from a high-rise hotel and you can gain access to the roof, beams mounted on short poles of 8–10ft should work well, providing the antenna is mounted close to the edge and 'looking over' the roof. Here, though, you will also need rotators, an additional complication and weight to consider.

Logging

These days almost all DXpeditions log by computer. The exception is the single-operator DXpedition on a 'holiday' type operation which does not expect to make a large number of QSOs. If you *are* using paper, a standard RSGB Log Book [23] will probably be the method of choice, although if you are really operating in 'DXpedition mode' and working a pile-up quickly you will probably not be stopping for a chat and will be giving 59 or 599 reports to allcomers. If that's the case you could log in an exercise book, with the date, band and mode at the top of the page, and the list of callsigns worked in a series

A vertical mounted close to the sea can be a surprisingly effective antenna. This is a Butternut HF6V at P29DX/P on Yule Island in Papua New Guinea (IOTA OC-153)

Yagis work well even when only about 20ft high, when mounted close to the sea. This 20m 3-ele monobander is at VK9MM on Mellish Reef

A multi-operator DXpedition should always consider taking at least one Yagi: it will pay dividends in the QSO count. Here members of the 9M0C DXpedition team put up a Cushcraft 3-ele 20m beam in the Spratly Islands. In the background is a Titanex vertical for 160 and 80m

of columns down the page. If the pile-up is thick and fast it is only necessary to note the time every five or 10 QSOs, ie every two or three minutes. The well-known German DXpeditioner Baldur Drobnica, DJ6SI, used this method in the days before computer logging.

There's no doubt, though, that computer logging makes log checking and QSLing much quicker and easier. Most of the popular station logging programs such as TurboLog [24] and SHACKLOG [25] work fine for DXpeditions. Some contest logging programs, including K1EA's CT [26], also have a 'DXpedition mode' which allows you to log QSOs made on the WARC bands, for example.

The CDXC 9M0C Spratly DXpedition took PC logging a step further and sent each QSO as it was logged (from each of up to six stations operating simultaneously) to a central server, using modified PacketCluster software and 70cm handhelds. In this way all the logs were immediately backed up on to a second PC; it was possible to make back-ups of the combined master log easily without disturbing the DXpedition operators; there was a central clock on the server PC to ensure that the logs did not get out of synch; and it was possible to upload the complete log to the Internet on a daily basis. The DXpedition operators could interrogate the server to determine which frequencies the other stations were on, how many QSOs had been made on each band and mode, how many with each continent and how many by each operator. The use of the DXpedition server is covered in detail in reference [18].

FURTHER INFORMATION

[1] 'Going HF mobile – some experiments in vehicle suppression methods', R V Heaton, G3JIS, *Radio Communication* October 1981, p920.

[2] 'Suppression of vehicle interference for mobile radio operation', D Morris, G3AYJ, *Radio Communication* May 1976, p336.

[3] *The Highway Code*, HMSO, 1999.

[4] 'A transmit control for mobile operation', J M Bryant, G4CLF, *Practical Wireless* July 1989, p24.

[5] *Radio Communication Handbook*, 7th edn, ed Dick Biddulph, M0CGN, RSGB, 1999.

[6] Amateur Radio Insurance Services Ltd, Freepost, 10 Philpot Lane, London, EC3B 3PA. Tel: 020 7335 1647. E-mail: aris@stuartalexander.co.uk.

[7] *RSGB Yearbook*, ed M Dennison, G3XDV, RSGB.

[8] *UK Repeater List*, RSGB (updated frequently).

[9] RSGB web site: www.rsgb.org.

[10] *A Guide to Small Boat Radio*, Mike Harris, G0HOC, available from RSGB.

[11] 'Aerials for portable operation', J E Hodgkins, G3EJF, *Radio Communication* May 1971, p319.

[12] *HF Antennas for All Locations*, 2nd edn, Les Moxon, G6XN, RSGB, 1993.

[13] 'Erecting portable masts', Terry Robinson, G3WUX, *RadCom* June 1999.

[14] 'Portable operation with ground planes', J B Roscoe, GM4QK, *Radio Communication* March 1972, p150.

[15] 'The G8ENN instant beam', D A Tong, G4GMQ, *Radio Communication* April 1976, p270.

[16] *HF Antenna Collection*, ed Erwin David, G4LQI, RSGB, 1991.

[17] 'Take to the hills', I J Kyle, GI8AYZ, *Radio Communication* July 1972, p436.

[18] *DXpeditioning Behind the Scenes*, eds Neville Cheadle, G3NUG, and Steve Telenius-Lowe, G4JVG, Radio Active Publications, 2000.

[19] CDXC (Chiltern DX Club) – The UK DX Foundation. Details from the Chairman (1999–2000), Neville Cheadle, G3NUG, Further Felden, Longcroft Lane, Felden, Hemel Hempstead HP3 0BN.

[20] The RSGB *IOTA Directory 2000*, RSGB, 2000.

[21] *Operation Under CEPT* (RA247), available free of charge from the Radiocommunications Agency, Wyndham House, 189 Marsh Wall, London E14 9SX; tel: 020 7211 0160.

[22] Dunestar Systems, PO Box 37, St Helens, OR 97051, USA; dunestar@QTH.com; www.dunestar.com/dunestar.

[23] RSGB Log Book (Transmitting), RSGB Sales.

[24] TurboLog, TurboLog Communications, Brinkweg 5, D-27321 Morsum, Germany; tel/fax: +49 4204 5321.

[25] SHACKLOG, current (Jan 2000) version 5.1.01, Alan Jubb, G3PMR, 30 West Street, Great Gransden, Sandy SG19 3AU; www.shacklog.co.uk; info@shacklog.co.uk.

[26] CT, current version 9.42, K1EA Software, c/o XX Towers Inc, 814 Hurricane Hill Road, Mason, NH 03048, USA; www.k1ea.com.

7 Amateur satellite and space communications

AMATEUR satellites have come a long way since the relatively primitive devices launched in the 'sixties, and those currently in orbit include sophisticated communication and research satellites which would have cost millions of dollars if constructed on a commercial basis.

The main organisation involved with the launch of amateur satellites is the Radio Amateur Satellite Corporation (AMSAT-NA) with its headquarters in the USA [1]. There are affiliated groups across the globe, including AMSAT-UK [2]. Many of these AMSAT groups have been actively engaged in the design, building and launch operations with almost 40 launches to date.

Russia (and the former USSR) has also launched several amateur satellites, co-ordinated by the Radio Sport Federation in that country.

The satellites constructed by AMSAT Groups in USA, UK, Germany, Mexico, Israel, South Africa, and Japan, loosely known as *AMSAT-International*, are generically known as *OSCARs* (Orbital Satellites Carrying Amateur Radio). These OSCARs will have their own name in their own countries and organisations, but in general all amateur satellites now have an OSCAR numbering, eg Oscar 17 is called also *Dove* (Peace) by the Brazilian owners. Oscar 16 is called *Pacsat*, and so on. These two satellites were part of a clutch of micro-satellites launched in a batch of six from the same spacecraft in 1993. To date we are up to number 36, UoSat 12, launched in April 1999 from Kazakhstan on a converted Russian ICBM SS18 rocket. These, added to the Russian series, make the above total. The Russian amateur satellites are designated RS for Radio Sport, eg RS15.

These have been classified into three *phases*:

- Phase 1 – experimental (Oscar 1–5), long since dormant or decayed.
- Phase 2 – long-life, medium-orbit, satellites with transponders (linear or inverting) allowing two-way communication (Oscars 16, 18, 19, 20, 22, 23 etc)
- Phase 3 – high-altitude (DX) elliptical orbit satellites with sophisticated control systems (Oscar 10).

PHASE 2 SATELLITES

Oscars 11 and 17 are Phase 2 research/educational satellites and do not contain an amateur-usage transponder.

The USSR/Russian satellites (RS1–RS15) have all been of the Phase 2 type, allowing two-way analogue communication. Of these, now only RS12/13 and RS15 provide a service. RS15 was launched on 29 December 1994 – this had some defects during the first few months in orbit, and is now giving a measure of service with Mode A analogue communications (uplink on 2m with a 10m downlink). It has been suggested that the on-board batteries were not fully charged prior to launch in Russia. This may well be the last Russian satellite to be built

Top: UoSat 12 (Oscar 36), launched in April 1999. *Bottom:* **example of the excellent ground images received from it – this is Detroit Metro International airport**

for some time to come, although Russia does provide a launch service for other nations. The very successful UoSat 12 (Oscar 36) was launched using this service.

However, during April 1995 two satellite launches failed. These were the Technion (Israel) and University of Mexico satellites. Both satellites were lost when the rocket launch vehicle failed to ignite on the fifth stage of flight over Eastern Russia. These satellites would have been numbers 29 and 30 if placed into Earth orbit. Subsequently the two satellites were replaced and successfully deployed in orbit from Baikonur on 10 July 1998. The Israel satellite, Techsat 1B, is now Oscar 32 but regrettably the Mexican satellite failed in orbit (it was designated Oscar 30).

In the Phase 2 designated satellite classification there is also the Japanese JAS-1 and 2 series.

JAS-1 is now defunct, although still in orbit. Reports have been received of intermittent operation but this is not substantiated. JAS-1B, named Fuji (Oscar 20), is currently in operation with the usual beacon on 437.795MHz CW and with communications taking place in analogue mode permanently. JAS-2 (Oscar 29) was launched on 17 August 1996 and is operational with an added Digitalker mode (digitally synthesised speech).

Also using Phase 2 type orbits are the purely digital satellites: Oscars 22 (UoSat 5), 23 (Kitsat A) and 25 (Kitsat B). These three satellites are used in 9600 baud mode for message handling with a computer (high-speed packet). Specific frequencies with uplinks on 2m and downlink on 70cm are available with orbits mostly available early morning and evening over the UK. Oscar 22 is now used almost exclusively as a *satgate* store-and-forward device. This means that an amateur who has no satellite equipment can place a message to his local packet station for onward transmission to GB7LAN or GB7LDI (in the UK) for onward transmission via Oscar 22 to any other packet station on the system. Full information is well beyond the scope of this chapter and the reader is referred to the many books written on all satellites, modes, and frequencies, obtainable from AMSAT organisations and the RSGB. AMSAT-UK [2] carries a wide range of publications. Oscar 36 (UoSat 12) has recently been added to the trio with even higher data rates up to 1Mb/s for high-definition image transmission.

Phase 2 and 3 satellites require a different approach in their use, and are considered separately in this chapter. Since Phase 2 satellites are low-orbit devices, their communication range is restricted and they move across the sky quite quickly (each orbit being completed in approximately 100 minutes). Therefore considerable attention to tracking them is required as they are in view of a given station for only some 15 minutes, depending upon satellite altitude. Phase 3 satellites are in a higher elliptical or Molnya orbit and appear to move very slowly when at apogee (the highest point of the orbit). Tracking is not then a particular problem but, due to the increased range, more attention then needs to be given to the efficiency of the user's station, in particular the receiving antenna and the receiver. High-efficiency front-ends should be the aim of the satellite user.

UoSat-Oscars 11, 22, 23, 25 and 36

Designed and constructed at the University of Surrey, UoSat 2 (Oscar 11) is a research/educational satellite which also transmits in the amateur bands. Oscar 11 does not offer two-way amateur communications, but is often used for updates of news and information on the whole satellite hobby. Full telemetry, which can be studied and processed by educational or radio amateur stations is also transmitted. Software to decode the data and information is available from AMSAT-UK.

Oscars 22, 23 and 25 were also designed and built by the University of Surrey Space Department, Surrey Satellite

Technology Ltd. Oscar 22 is owned by the University, while Oscars 23 and 25 are owned by the Korean Advanced Institute of Technology, KAIST. These satellites have store-and-forward amateur packet with 9600 baud transponders on board, and have given excellent service since launch. Oscar 36 is also part of the Surrey stable and is still undergoing commissioning. It carries a number of experiments as well as digital store and forward communications. Improved imaging payloads are carried. The results from these have been so good with 10m resolution that some American amateurs have referred to the satellite as a 'spy' satellite! UO-36 has L-band and S-band transponders. The images are downloaded at rates between 128kb and 1Mb on the S-band downlink.

UoSats are research tools that the radio amateur and the layman can use, with the professionals also interested because of the several experiments they carry. Full details of the satellites are available in book form from AMSAT-UK [4]. To whet one's appetite, here are a few details: Using Oscar-11, 145MHz enthusiasts can tell, by interpretation of their radiation counter readings, if an aurora is likely. UHF/microwave enthusiasts have the 2.4GHz beacon to use in propagation studies. Scientists are using the radiation detectors and the three-axis magnetometer (the first time one of these has flown in near-Earth space) and computer buffs can feed the telemetry, in ASCII code, straight into their machines. The telemetry, (and other data), can be received by voice, (Digitalker) as well as in binary code. The primary beacon frequencies are 145.825 and 435.025MHz.

RS10–RS15

Unfortunately, for various reasons, less detail was made available for the Soviet amateur satellites than for their western brothers, although more information is forthcoming now for RS15 since the Cold War ended. It is known that RS satellites are partly pressurised and heated in order to maintain frequency stability. The satellite 145MHz antennas, from their pictures, look like canted turnstiles, although no requirement for circular polarisation on the uplink has been publicised. Power output is in the 500mW to 1W PEP range. The previous satellites, RS1 and RS2, are known to have been about 2m by 1m in dimension.

A very useful device carried on RS satellites is known as the *robot*. One can actually have a contact with, rather than through, the satellite when the correct procedure is used, employing CW (Morse code) only. The robot communicator calls CQ, expects to be called back with its own callsign (and yours) and then responds with a sequential serial number; to use RS13 as an example:

CQ CQ CQ DE RS13 QSU ON 29454 KCS

RS13 DE G3AAJ $\overline{AR}$ *(not K)*

DE RS13 QSO NR 1645 G3AAJ DE RS13 QSO NR 1645 OP
    ROBOT TU FR QSO 73 $\overline{SK}$

Morse code speed can vary between about 12 and 40wpm, and the calling station can be heard coming back from the robot, making it sound a little like a CW repeater; in fact some stations have used it as such. The robot also has a limited vocabulary when it is not sure; it can send:

QRM, QRZ, QRS, QRQ and RPT

as dictated by circumstances. Later on the callsigns of stations worked are dumped by ground command, and a QSL card can be expected for the contact from the QSL bureau in Moscow.

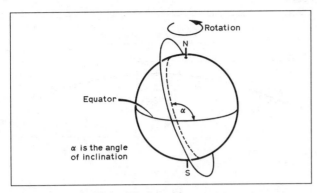

Fig 1. Phase 2 satellite orbit viewed from space

Locating Phase 2 satellites

Since radio communication via an amateur satellite is essentially line-of-sight, it follows that it can only take place when the satellite is above the user's horizon, and it is therefore necessary to determine the periods when this occurs. In addition, if beam antennas are in use it will be necessary to ascertain the beam headings of the satellites during these periods. Furthermore, many satellite stations use antennas adjustable in elevation as well as azimuth rotation (az-el antennas), and in this case it is also desirable to know the approximate elevation of the satellite above the horizon at any time.

All this information can be derived quite simply from orbital data or Keplerian elements which can be obtained from a number of sources. It helps to have a basic understanding of the terms involved. Some of these will be clarified by the following description – much more complete orbital information concerning amateur satellites is given in references [4], [6] and [9].

Orbital data

The simplest way to understand satellite orbits is to imagine one is out in space observing the Earth rotating on its axis from west to east with the north pole at the top (Fig 1). If an object as small as a satellite could be detected it would be seen coming from behind the Earth over the Antarctic region, proceeding northwards across the equator, then disappearing over the north polar region. This part of the orbit is known as the *ascending node*. The remaining half of the orbit, when the satellite travels from north to south is called the *descending node*.

It will be noticed that the orbit is tilted anti-clockwise from the north-south line, or is *retrograde* in space terminology. The satellite never goes right over the poles. The time which a satellite takes to complete one orbit is known as the *period*,

during which the Earth will have rotated eastwards. The satellite will therefore cross the equator at a different longitude, the difference being the *longitude increment*.

An orbit is said to commence when a satellite crosses the equator travelling north. All orbit predictions consist of this time, given in Co-ordinated Universal Time (UTC) and degrees west of the Greenwich meridian where the crossing occurs. This method of stating longitude often confuses the uninitiated who are more used to degrees east and west of Greenwich. It is really quite simple; up to 180°W there is no problem, after that 170°E is equivalent to 190°W, 160°E equates with 200°W, and so on round to the Greenwich meridian again; hence 5°E is given as 355°W.

The first orbit each day is called the *reference orbit*; these orbits are usually out of range of the UK. Future orbits can, however, be calculated with fair accuracy, given any reference orbit information. This may be obtained from the calendar available from AMSAT-UK [2] (see Fig 2).

Alternatively predictions are available on news bulletins such as GB2RS, the UK packet BBS, and many AMSAT and AMSAT-UK nets. Latest Keplerian elements from NASA are normally posted within a day of them being released by NASA. For those with internet access, AMSAT-UK operates a web page at http://www.uk.amsat.org/ which carries a lot of very useful information with all the latest news, links to other AMSAT organisations, as well as the latest Keplerian elements. Keplerian elements can of course be fed into the many satellite prediction programs now available for most computers [2].

Using the predictions

Having obtained the time of equatorial crossing and corresponding longitude for a suitable orbit, usually referred to as "EQX's" from one of the above sources, acquisition-of-signal (AOS) and loss-of-signal (LOS) times for your QTH need to be found, and possibly beam headings and elevation if directional antennas are in use.

Numerous ingenious ways of deriving this information have been published. These fall broadly into three types: (a) the use of a special map with a rotating cursor which represents the satellite track; (b) the use of 'look-up' tables or charts; (c) computer programs.

Method (a) was developed by G2AOX and K2ZRO but is now less popular as computers become universal. It uses a device which is now universally known as the *Oscarlator*. It consists of a polar projection map of the northern hemisphere, a small section of which is shown in Fig 3, and a transparent acetate overlay (Fig 4). The small circle is cut out of the latter

OSCAR11	20-08-95	03:06:52	143.0<	09:39:07	241.1<	19:27:29	28.2>
OSCAR16	20-08-95	00:13:30	14.1>	15:20:21	240.8<	23:44:09	6.8>
OSCAR17	20-08-95	09:52:22	158.3<	14:54:37	233.9<	23:18:22	359.8>
OSCAR18	20-08-95	01:04:56	26.5>	14:30:57	228.0<	22:54:43	353.9>
OSCAR19	20-08-95	09:54:55	158.6<	14:57:09	234.2<	23:20:52	0.1>
OSCAR20	20-08-95	00:45:08	30.5>	11.58.46	199.0<	23:12:24	7.4>
OSCAR26	20-08-95	00:52:03	33.2>	12:38:27	209.8<	22:43:56	1.2>
OSCAR27	20-08-95	00:24:29	26.4>	12:10:55	203.0<	23:57:22	19.6>
RS10-11	20-08-95	01:34:12	339.7>	10:19:09	111.5<	15:34:07	190.6<
RS12-13	20-08-95	01:34:17	298.1>	12:03:26	96.2<	22:32:35	254.2<
RS15	20-08-95	00:10:05	178.6<	12:56:24	11.6>	23:35:00	172.4<

Fig 2. Satellite predictions for 20 August 1995 as listed in the orbital calendar produced by AMSAT-UK. Only three orbits are given for each day – these will normally be the first, last and mid-orbits within range of some point in the UK, but there will be exceptions. Subsequent or previous orbits can be calculated using data shown in the calendar. The times given are those that the satellite crosses the equator (EQX). At the time the satellite crosses the equator it will either be moving north (ascending node >) or south (descending node <)

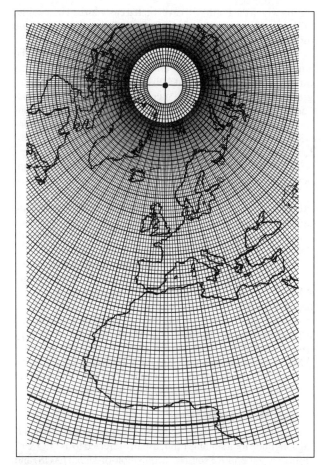

Fig 3. Part of a polar projection map of the northern hemisphere. The line near the foot of the map is the Equator, which is a complete circle on the whole map

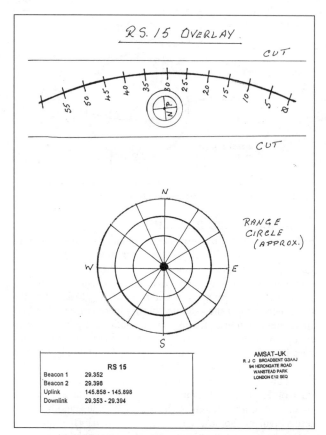

Fig 4. Typical Oscarlator for RS15 satellite, much reduced from actual size. The large circle corresponds to the equator on the polar projection map; the small circle to maximum satellite range when pasted on to the map

and stuck on the map with its centre at the user's location and its straight line pointing towards the North Pole. The large circle is cut out and placed over the map so that its centre corresponds with the north pole, and the 'O' mark corresponds with the point at which the satellite crosses the equator going north.

The curved line across the circle denotes the satellite track, and the point where this intersects the small circle is the AOS. The time after EQX (equatorial crossing) that this occurs can be found by reference to a time scale printed on the curved line. The approximate direction can also be ascertained from the map. The details for LOS and in fact any intermediate point can be determined in the same way.

The map and transparent overlays are available from AMSAT-UK at a small charge [2].

Method (b) utilises a table of satellite paths calculated for every 2° of EQX longitude (ascending node). Strictly speaking, the tables are only applicable for the location for which they were calculated but in practice the error is small at other locations up to 200km away. The time is given in minutes after ascending node and beam headings are true bearings.

Computer programs for method (c) are available from AMSAT-UK in PC format. The most popular suites of programs are Wisp by Chris, G7UPN/ZL2TPO, and Station by Paul, VP9MU. Fully automatic satellite station operation is possible using these programmes and suitable antenna turning control gear. Broadly, Wisp is used for the digital data (packet) satellites and the Station program for the analogue (voice) satellites.

Equatorial crossing times for most satellites are broadcast on the various AMSAT-UK nets every Sunday, Wednesday and on the local VHF nets, which are usually on a Thursday.

Receiving satellite signals

Reception in the 29MHz band from RS satellites differs from normal reception from distant ground stations in that it is basically line-of-sight, and it is also usually necessary to be able to receive the satellite's signals while the user's transmitter is on the air. Furthermore, as the satellite's power has to be shared

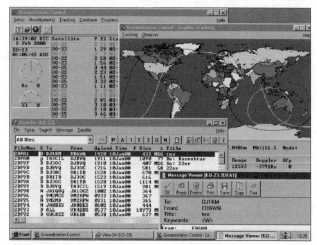

The Wisp satellite tracking program in action

Table 1. Transponder mode designations

Mode	Uplink	Downlink	Satellites
A	145MHz	29MHz	RS10–13
B	435MHz	145MHz	Oscar 10, 13, 21
J	145MHz	435MHz	Oscar 20, 22, 23, 25–28
JA	145MHz	435MHz	Oscar 20 analogue
JD	145MHz	435MHz	Oscar 20 digital
JL	1.2GHz/145MHz	145MHz	Oscar 13
K	21MHz	29MHz	RS10–13
KA	21/145MHz	29MHz	RS10–13
KT	21MHz	29/145MHz	RS10–13
L	1.2GHz	435MHz	Oscar 13
S	1.2GHz	2.4GHz	Oscar 13
T	21MHz	145MHz	RS10–13

between all the relayed signals, it is likely that only a few milliwatts is being radiated by the satellite for each signal. The performance of many amateur-band receivers and transceivers falls off at 29MHz, particularly at the satellite end of the band. Consequently a simple low-noise preamplifier to provide about 10dB gain will often bring about a dramatic improvement in reception.

A simple dipole or ground-plane antenna will enable good signals to be heard from the nearer satellite passes but for serious DX work at maximum range, a low-angle beam antenna of some kind should be considered, mounted in the clear as far as possible from local noise sources. Simple and compact beams with good gain for their size are the HB9CV and ZL-Special designs. To cope with the nearly overhead passes, a fixed horizontal dipole (16ft overall) running east-west is sufficient since the satellite will be north-south or south-north on such orbits.

Transmitting to the satellite

The uplink passbands for all modes of the Phase 2 low-orbit satellites are given in Table 1.

In the case of the RS satellites the maximum recommended ERP at horizon ranges is 50W. However, it has been frequently proved quite possible to access with less than 1W to a dipole, provided the other users are keeping to low power.

Band plans

As with other amateur allocations, there are band plans for the orderly use of the space sections of the amateur frequency allocations. These are as detailed by international agreement at ITU and IARU conferences and are shown on this page for Phase 2 satellites. It should be noted that where the transponder passband is inverted then HF becomes LF, USB becomes LSB etc. This plan applies to all future spacecraft, at all times, except when special trans-satellite tests are being held. It should be remembered that all band plans only work on a gentlemens' agreement basis, albeit to mutual advantage. However, if you do transmit outside recommended band limits, don't be surprised at failure to get any response from another satellite user.

Range of communication

For a station at sea level in the British Isles the range to Oscar 11 as it clears the horizon is about 2900km. It follows that the satellite has a 'footprint' of twice this value, or approximately 5800km. However, the effective 2m range can be exceeded considerably when there is anomalous propagation on 145MHz. For example, E-layer ionisation can enable a VHF or UHF signal to be received by or from a satellite when it's below your horizon. This mechanism allows a few enthusiasts to obtain great DX contacts by using the outer edges of satellite 'footprints', notably those of the RS satellites, and by choosing the

SATELLITE BAND PLANS

Band plans have been adopted by all IARU countries as the recommended operational usage of all Oscar satellite communications. It conforms to the accepted 'one-third' rule in terrestrial amateur band plans, ie lower end CW, middle phone/CW, upper SSB. The plan, which has been agreed since 1979, allocates a percentage of the available radio frequency spectrum as seen on the downlink to different modes of communication. The relative amount of spectrum for each mode is thus the same for any transponder in any satellite. Irrespective of the current use of channels on some satellites the thirds rule has been a success in the majority of cases. The allocations are as follows:

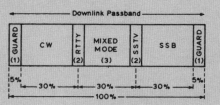

Notes:
1. Guard area to avoid interference to beacons.
These frequencies are available for emergency and bulletin stations.
2. RTTY and SSTV, if permitted and included within the capability of the satellite. are placed at the edge of the CW and SSB passbands, conforming to their usage at HF where RTTY is present within the CW space and SSTV is transmitted in the SSB sub-band.
3. Mixed mode area. This is recommended for crystal controlled stations, or dxpedition stations, or anyone wishing to work both CW and SSB stations.
4. Mode A guard channels 5kHz non-inverting pass band. Mode B guard channels 2–5kHz inverting passband Mode J guard channels 5kHz inverting passband.

The plan is always based on percentages of the downlink passband. It applies to both inverting and non-inverting transponders. The allocations of frequency for Oscar satellites are as follows:

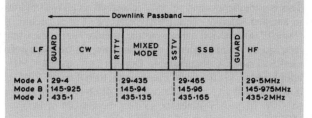

appropriate orbit. The footprint will obviously depend upon the altitude of the satellite in question. In the case of Oscar-11, the height is about 700km.

Polarisation

Imagine a satellite in orbit without a sophisticated stabilisation system. That satellite is likely to roll and tumble according to external influences of solar and terrestrial origin, and so the orientation of the satellite's antennas, as seen from the ground, can (and does) change in direction. The polarisation to cater for all orientations and combat fading (except when the antenna is end-on to the observer) is circular. Satisfactory results can be obtained with horizontal or vertical polarisation alone for most of the time but fading will occasionally be troublesome; better results can be obtained by having both polarisations available and switching between them to select the strongest signal.

Huge antenna arrays are not required for satellite operation, although beams with az-el rotation are desirable. This photo shows a typical small installation in a garden

Some satellites' antenna systems are themselves circularly polarised (on VHF and upwards) and some advantage can be gained by ground stations using the same sense of circular polarisation, but only when the satellite antenna system is pointing towards the observer. At other times selected linear polarisation will be better. The best all-round results will be achieved by the station which can switch between horizontal, slant, vertical, left-hand circular and right-hand circular polarisations; a means of obtaining these options from a crossed Yagi array is detailed in reference [5].

Doppler shift

This is a frequency- and velocity-related effect all too obvious in amateur satellite operation (typical speed of a satellite is about 17,000mph). It appears, as the satellite approaches, that its frequency starts high and is drifting lower in frequency. To use RS15 mode A as an example: Doppler shift on the 145MHz uplink added, via the linear transponder, to that on the 29MHz downlink will give a cumulative 'drift' of up to 5 or 6kHz per pass. As frequency increases, so does the amount of Doppler shift for a given velocity. So use of higher frequencies, such as 435 and 1270MHz, results in a total amount that can get extremely large. Remember that the shift of both links has to be taken into account and, in the case of Mode A, the 4 or 5kHz of 145MHz added to the 1kHz of 29MHz to give the total amount.

Designers of transponders, however, can in the mixer stage, extract a negative component rather than a positive component (or turn it upside down) such that one Doppler shift is subtracted from another. As an example, Doppler shift on the 145MHz uplink (4–5kHz) is subtracted from that on the 435MHz. downlink (about 12kHz) resulting in only 6 or 7kHz instead of the 18kHz which would result if the positive component were used. This is why the transponder is made *inverting* – frequencies at the top of the uplink band come out at the bottom of the downlink band and vice versa; consequently the convention of transmitting lower sideband results, through the translation process being inverted, in upper sideband being received.

Another satellite array mounted on an outbuilding

Telemetry

It is hoped that the foregoing will have given a newcomer to amateur satellite communication some insight on the subject. Sufficient at least to find out when and where to listen for amateur radio satellites. Short-wave listeners will be able to hear many countries via satellite and it is interesting to copy the telemetry data sent down continuously on the beacon frequencies, especially as this does not require a transmitting license. *Oscar News*, the official organ of AMSAT-UK, gives a list of 'Elmers' who will be only too happy to talk the newcomer through the opening stages of setting up a station, and 'working the birds.'

The telemetry from Oscar 11 is transmitted in ASCII format by tone frequency modulation of the beacon. Engineering data is transmitted in binary code by the same method. RS12–15 telemetry is made up of letters and figures in a variable format; it is transmitted in Morse code and speeds can be between 12 and 20wpm. Telemetry values, when decoded, give details of such parameters as battery voltage, current and temperature, power outputs etc, ie the 'housekeeping' data. Telemetry is an essential feature of satellites as it enables the ground control stations to take appropriate action if, for example, excessive battery current is being drawn or if the PA stage of the transmitter is becoming overheated.

RTTY and ASCII telemetry carry similar parameters to the CW telemetry and, with up to 60 channels available, much information can be transmitted. Speeds and formats are detailed in the handbook mentioned previously. The tone frequencies transmitted by frequency modulation from Oscar 11 conform to the once-popular Kansas City CUTS system and can be received on a home computer with a suitable interface.

In addition to telemetry formats requiring some skill or machinery, Oscars 11 and 17 (Dove) contain a speech synthesiser (Digitalker), which can be used to 'speak' short messages which have been uploaded from the command station.

Practical communication

Undoubtedly the most common abuse of a satellite is the use of far too much power. The operating schedules for all satellites allow them to be used whenever heard except that if you use the satellite while an experiment is in progress you could upset the whole experiment.

The use of excessive power seems to be due to the rather inferior receiving systems employed by some operators and this causes these stations to run a lot more power than necessary in order to find their own signal; they are known as 'alligators' – all mouth and no ears! This causes problems to other users. It is 'undue interference' according to the terms of the UK amateur licence, and cases have occurred where stations have lost their licence because of this. The receiver sensitivity required is $0.1\mu V$ for 5dB over the noise and this will enable weak signals to be heard. There are few amateurs fortunate enough not to be troubled with RF interference from switched-mode power supplies and line time-base oscillators in TV sets, faulty thermostats and sundry industrial and domestic noises. Much of this interference can be dealt with by a good noise blanker as opposed to a noise limiter. The rule of thumb is to monitor the satellite beacon; your own downlink signal should not be any stronger than the beacon.

To illustrate how to communicate through an amateur satellite, let us take a step-by-step example for RS13 in Mode A. It is assumed that you have a separate receiver and transmitter.

1. From the orbital calendar, or other published orbital data, check when a convenient orbit will occur and calculate AOS/LOS times and any beam pointing information needed. There are approximately 12 orbits per day on the LEO (Low Earth Orbiter) type satellites.
2. Tune up your transmitter in the 145.96–146.00MHz segment before the calculated AOS time, using a dummy load (do not tune up 'on-air').
3. If your transmit and/or receive antennas are rotatable, point them in the general direction from whence the satellite will come over the horizon. Keep the beam-pointing information where it can be seen at a glance.
4. Set up your receiver to cover the 29.4–29.5MHz section of the band, and make sure that the receiver is not muted or desensitised when the transmitter is used, and also that the squelch control, if fitted, is fully open.
5. Listen for the telemetry on 29.458MHz but remember that the satellite is travelling towards you at an orbital velocity of about 4.5 miles per second, causing a Doppler shift upwards in frequency of about 700Hz at the start of the orbit.
6. Having established that the satellite is within range and functioning, you should hear a number of mainly SSB and CW signals as you tune across the passband, so the next goal is to find out if your signal is being relayed by the satellite. What you must not do is to transmit a powerful carrier, 'swishing' your VFO across the whole passband trying to hear yourself. This is anti-social and unfortunately all too

prevalent with newcomers who have not paid attention to getting the best possible *receive* system from the antenna through to the loudspeaker.

You can quickly calculate where your signal ought to be within a few kilohertz on 29MHz. For example, if you transmit on 145.975 MHz on the uplink, your 29MHz signal would be in the centre of the downlink passband, ie 29.475MHz. Fig 5 should make this relationship clear. Again, due to the cumulative effects of Doppler shift on

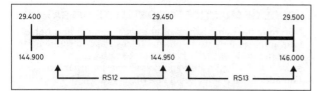

Fig 5. Relationship between uplink and downlink bands for RS12/13 (see text concerning effects of Doppler shift)

both 145 and 29MHz, you should tune up 4–5kHz away from the calculated frequency, higher in frequency if the satellite is approaching you or lower if it is receding from you. It is then permissible to 'swish' the VFO for fine adjustment of frequency but better to swing the receiver tuning. (Now read this again!)

It is also suggested that newcomers to the satellite scene spend at least a few sessions of shack time in listening mode only. This time will be well spent, and will help you to get a feel for the band. More importantly, you will not be known as a 'lid' to fellow enthusiasts who will be pleased to help you if you ask. Most satellite radio amateurs have had to learn the new techniques and are ready to help you get the most out of this hobby.

7. If you put out a CQ call, monitoring your own signal on 29MHz, be prepared for someone breaking in; that is the beauty of satellite working – it is the ultimate 'break-in' system of operation. Some CW stations are crystal controlled so you should tune around in case anyone is answering on other than your own frequency. Once contact is established a QSO would proceed in the usual way but, as the maximum 'visible' time of the satellite is between 15 and 35 minutes, most contacts are 'contest-style'. Maximum visibility occurs on overhead passes.

8. If you generate the required ERP by a low-power transmitter feeding a high-gain beam, do not forget in your initial enthusiasm to turn the antenna from time to time. Accurate time-keeping is necessary to have the antenna pointing in the right direction at the right time. A convenient approach is to turn the array in increments equal to the half-power beamwidth, say 45° for an eight-element Yagi, although you may find in times of good tropospheric and ionisation propagation that the 'true' path is not necessarily the best.

9. After you have made contacts, if you wish to send a QSL card, mark the card with "145/29MHz via RS13 orbit No xxx etc" in the frequency space on the card.

PHASE 3 SATELLITES
Oscar 10 (Phase 3B)

Launched in June 1983, it was the first Phase 3 satellite be placed in orbit (an earlier attempt was made in 1980 but the launch vehicle malfunctioned). Unlike the Phase 2 satellites, it is in a highly elliptical orbit with a much longer period, allowing worldwide VHF/UHF communication for hours at a time (Fig 6).

Oscar 10 contains two transponders, the U transponder (Mode B) (input 435MHz, output 144MHz) and the L transponder (Mode L) (input 1269MHz, output 436MHz). Currently, it is still in operation on Mode B only. It also has two beacons, the engineering and general beacons, active on both transponders. The general beacon transmits CW and 400 baud PSK – a special decoder is required for the latter mode. A circuit and PCB is available from G3RUH.

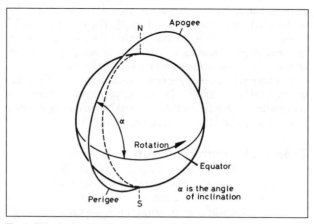

Fig 6. Elliptical orbit viewed from space

Ground-station requirements

All ground-station antennas should have right-hand circular polarisation (linear polarisation will result in signal flutter and loss of at least 3dB). The recommended receive antenna gains are 13.5dBi for the L transponder (3dB receiver noise figure), while the corresponding transmit antenna EIRP is 28.8dBW.

Normal SSB receivers having the above noise figures will be suitable – accurate frequency readout is a distinct advantage.

For the U transponder (Mode B), ordinary SSB transmitters should prove satisfactory, as will a 28 to 435MHz transverter. A power amplifier is unlikely to be required, with the high antenna gains possible today, unless feeder losses are significant.

As an example, a typical station using SSB through the U transponder (Mode B) could consist of the following:

- 435MHz 'black box' transceiver, providing 20W PEP SSB.
- 10-element crossed Yagi, for 435MHz, right-hand circular polarised.
- 145MHz converter, feeding 28–30MHz HF receiver.
- Six-element crossed Yagi, right-hand circular polarised for 146MHz.
- Manual az/el adjustment. A Trackbox or Kansas City Tracker [2] can of course be used for automatic tracking with a computer interface. You will also find an orbital calendar useful.

This station should have no difficulty hearing itself for 75% of the satellite's visible time. CW and SSB communication should present few problems provided that both transmitter and receiver are accurately calibrated. The limiting factor is the requirement to aim the antennas at the satellite, and the absence of high-power transmissions on the uplink.

Oscar 13 (P3C)

Oscar 13 was launched in June 1988 from Kourou by ESA. It has since re-entered the Earth's atmosphere on 5 December 1996. The spacecraft's orbit was highly elliptical but due to the differential pull of Sun and Moon the orbit eccentricity increased, making the perigee lower. Atmospheric drag took effect and resulted in a premature re-entry. However, Oscar 13 was a highly popular and successful satellite.

Tracking

Because Oscar 10 is in an elliptical orbit, tracking is a little difficult, and much benefits from automated assistance as the satellite approaches perigee. However, at apogee the satellite has a very slow motion relative to the ground, and will appear to hang in the sky at the same position for hours on end, preceded and followed by a much faster motion. Tracking methods available are:

1. The very easy way. Listen to the AMSAT nets which will probably provide local AOS and LOS (acquisition and loss of signal) predictions. If you are very lucky they will also provide antenna aiming (azimuth and elevation) information. (AMSAT-UK will always provide this information.)
2. A more reasonable way. Use the orbital calendar that is produced by AMSAT-UK. (This is a bimonthly calendar, available at low cost to members.)
3. Use a computer tracking program, and AMSAT-provided Keplerian orbital elements. Keplerian orbital elements are published in *Oscar News* or can be found on the AMSAT web page (http://www.uk.amsat.org).

Transmitting through the satellite

A variety of communication modes can be used – CW, SSB (LSB, received by ground stations as USB), and DPSK/PSK (data). FM and AM telephony, RTTY, SSTV and other high-duty cycle modes of transmission should *not* be used. The relatively new mode PSK31 has been tried with some success and seems at first sight to be admirably suited to satellite communication with the provision that transmission levels are kept low. It remains to be seen how popular the mode becomes.

The transmitter power and bandwidth should not be greater than necessary to achieve reliable communication – no SSB telephony transponded through the satellite should return a peak strength stronger than that of the general beacon (transponded CW signals should be not less than 9dB below that of the general beacon).

The Doppler shift experienced by ground stations varies according to the frequency in use, the position of the satellite in its orbit, and the position of the ground station. Maximum figures for stations located at latitude 50°N are ±2kHz (145MHz), ±6kHz (435MHz). As well as Doppler shift, there is a small time delay due to the distance involved – the round trip can take up to 250ms at apogee, and the receiver audio should then be turned down while transmitting to avoid 'echo'.

In general, operating through the satellite is similar to using Phase 2 satellites and the reader is referred to advice given earlier. Further information and technical data is also available from AMSAT-UK [7] (stamped, addressed envelope or three IRCs, please).

MANNED SPACECRAFT

One exciting possibility is that of communication with a manned spacecraft. Russian cosmonauts on Mir have been encouraged to make use of amateur radio in their off-duty periods. As a result many amateurs have talked to Mir on FM simplex and packet radio. The crew transmitted some good SSTV pictures from the spacecraft until they departed in August 1999. The future of Mir is somewhat indeterminate at the time of writing (January 2000). As indicated it is currently unmanned and under ground control. The Russian Space Agency is looking for funding to support a further mission of some 45 days on the ageing spacecraft. Should this not be forthcoming it is expected that a crew will fly a brief mission to consign Mir to a fiery re-entry, followed by a watery grave in the Pacific Ocean.

The crews of the American space shuttles are also encouraged to operate amateur radio gear when their work schedules permit. To this end most astronauts are licensed amateurs but they do have a heavy workload so be prepared. Simplex FM, packet and SSTV are the main modes employed. Information is usually posted on the AMSAT Bulletin Board System prior to every shuttle flight, giving full details on any amateur radio activity likely to occur during a mission.

A very exciting development is the building of the International Space Station (ISS). This is currently under construction and it will have a module carrying an amateur radio station. It is hoped to have a working set-up on VHF and UHF by mid-2000. The equipment for this has already been space qualified. The ISS programme is subject to some delays so it will possibly be 2004 or 2005 before the station is in full operation. When complete, the amateur station should be active on most bands, including several of the HF bands. The callsign of the station is likely to cause a problem or two. It is usual to base the call on the sovereign state owning the station but this one is international – perhaps it will have a special prefix like the Vatican! Whatever it is, this station holds out a very exciting prospect for the future.

FUNDING THE SATELLITE PROGRAMME

Membership of AMSAT-UK is open to anyone, including SWLs; there is a minimum donation and extra donations are always welcomed. Members receive *Oscar News* approximately six times per year which keeps them in touch concerning past achievements, present activities and future plans.

AMSAT-UK also runs several nets (which are open to non-members) where news is disseminated and questions can be asked. The most regular of these is on 3780kHz (± QRM) every Sunday morning at 1015 UK local time (G0AUK). Other, more local, nets take place from time to time; usually weekly and mostly on 144.28MHz. There is also a half-hour information net on 3780kHz on Wednesday evenings at 7pm. The Sunday and Thursday nets are run on 144.280 at 7pm local time.

AMSAT-NA also runs nets on Sunday evenings on 14.282 and 21.280MHz (usually at 1800 and 1900 GMT respectively). These nets, from the USA, are more of a worldwide news service than a place for questions but there is occasionally time for these. Further details of AMSAT-UK activities can be obtained from reference [2] for the courtesy of a self-addressed and stamped envelope (9in × 6in) or three IRCs.

THE FUTURE

Now prepared for launch sometime in 2000 is the heaviest (at half a tonne), the most complicated, and the most sophisticated satellite ever built by radio amateurs. Currently known as Phase 3D (Oscar numbers are not allocated until the spacecraft is in orbit), this satellite was built jointly by teams in Germany and USA by AMSAT International. With equipment designed and built by groups across the world including AMSAT-UK, this satellite should satisfy every kind of mode of amateur operation for a long time. It now sits at Kourou in air-conditioned quarters, batteries uncharged, waiting to be mated with an Ariane 5 rocket. Assuming a successful launch, its expected life span is 15 years at least.

Unlike previous amateur radio satellites, there are no preset transponder modes but a series of transmitters and receivers arranged in a matrix. Thus practically any transmitter can be linked with any receiver and the information in Table 1 becomes inadequate. A new mode designation has been devised where letters designate the frequency bands in use. 'U' is UHF (435MHz), 'V' is VHF (144MHz), 'S' is 10cm (2.4GHz), 'L'

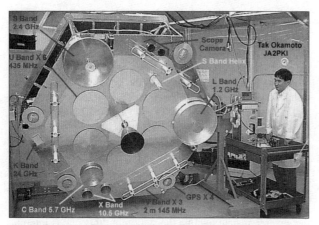

The underside of Phase 3D, showing the various antennas

Close-up of units partially installed

is 1.2GHz and so on. Hence Mode B would now become Mode UV and Mode L would be Mode LU. This uplink/downlink designation is very logical. All transponder arrangements are inverting to minimise the doppler effect.

Transmitter powers are much higher than normal with the power budget permitted by the large solar panels and satellite operation using very simple ground stations is envisaged. Full details of P3D operations are beyond the scope of this chapter, if only because the satellite is not in orbit at this date. Full information will be the subject of technical books from AMSAT groups after lift-off.

Financing the building, integration and transport of this massive (for us) satellite has strained the resources of AMSAT International to such an extent that it is unlikely that another project of this magnitude will be undertaken. The trend now is for small satellites, which are well within the budget of university departments, to be built. This is an excellent scheme in that it ensures that there is always a pool of satellite engineers to meet the challenges of the future and keep this aspect of amateur radio alive. Will the dream of an amateur radio, geostationary, satellite cluster ever be realised? Who knows?

FURTHER INFORMATION

[1] AMSAT-NA, 850 Sligo Avenue, Ste 600, Silver Spring, MD 20910-4730, USA.

[2] AMSAT-UK (Fred Southwell, G6ZRU), 40 Downsview, Small Dole, Henfield, West Sussex, BN5 9YB.

[3] AMSAT-UK – a series of A4 technical, PVC-laminated, information sheets.

[4] *The New Guide to Amateur Satellites*, G3RWL, AMSAT-UK.

[5] *VHF/UHF Handbook*, ed Dick Biddulph, G8DPS, RSGB, 1997.

[6] *Space Radio Handbook*, J Branegan, GM4IHJ, RSGB, 1991.

[7] *How To Use The Amateur Radio Satellites,* AMSAT-NA (available from AMSAT-UK).

[8] *AMSAT-UK Satellite Frequency Guide* showing all current satellites. PVC laminated on A4 card.

[9] *The Satellite Experimenter's Handbook,* Martin Davidoff, K2UBC (available from AMSAT-UK).

8 | Data communications

DATA communication is one of the fastest growing operating modes in amateur radio. No longer confined to sending simple text messages, amateurs are hooking into worldwide DX information networks using complex software and hardware. Most data communications take place on VHF and UHF using packet radio but, in spite of slower data rates, HF digital communications are also popular and have led to some sophisticated new modes being developed to cope with typical HF band conditions.

The present chapter reflects this emphasis, with an outline of the various HF modes followed by a more detailed discussion of how to get going on VHF packet radio and use its network facilities. The sections on message writing and operating guidelines which complete the chapter are of course broadly relevant whichever mode is in use.

HF DATA COMMUNICATIONS

At first, the world of HF data modes could be thought of as a 'jungle', with strange-sounding abbreviations for the different modes. But the undisputed future of *all* radio communication is digital, and right now there are plenty of amateurs around the world having great fun in communicating with each other using digital techniques. So let's take a look at how to get in on the action.

It could be argued that Morse code is the simplest of HF data modes, and besides manual operation it's easy to automatically generate and decode Morse using electronic-based systems. These transmit 'perfect' Morse and often add DSP bandwidth filtering and noise reduction to help in reception. But, after CW, the 'simplest' commonly used HF data mode is RTTY (from 'radio teletype'). Following RTTY comes AMTOR (from 'amateur teleprinter over radio'), which brings a degree of automatic error checking, then packet which adds automatic correction but is admittedly more suited to VHF and UHF use. Then comes PacTOR which is a 'combination' of AMTOR and packet and is well suited to HF, as are Clover and GTOR which also have their merits on HF. Finally, PSK31 and Hellschriber are almost 'back-to-basics' modes and are popular because of their ability to get through noise and interference in a narrow bandwidth using current DSP techniques. More of each mode a little later. The future possibly lies in spread-spectrum modulation of multimedia (ie speech and video) HF communications and amateurs have already used this to good effect on the bands, but that's another subject!

Getting equipped

To get started, you'll need something between your transceiver and your 'user terminal' (PC, keyboard, microphone, camera or whatever) to handle the data translation. A stand-alone solution is a multimode terminal node controller (TNC), which is simply connected between your transceiver's rear-panel data

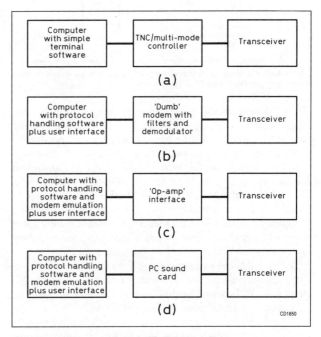

Fig 1. Equipment needed for HF digital modes

modes interface connector (or the TX/RX audio and PTT connections on earlier radios) and your terminal (Fig 1(a)). Typical popular multimode TNCs are the Kantronics KAM and KAM Plus, the AEA/Timewave PK-232, DSP-2232 and PK900, and the PacComm PacTOR controller, each giving data facilities including CW, RTTY, AMTOR, Packet and PacTOR. Although such multimode stand-alone controllers can be relatively expensive, you might often find bargains on the second-hand market. All you then need is a simple 'dumb terminal' program running on your PC (eg HyperTerminal) for instant two-way QSOs. Or you can run specific software on your PC to add facilities such as automatic file transfers, split-screen transmit/receive operation, and so on. An alternative is a plug-in multimode TNC card for your PC such as the HAL interface, which gives these modes as well as Clover, and comes with ready-to-run driver software.

Another method is to use the power of your PC together with an external modem, such as the BARTG (British Amateur Radio Teledata Group) 'Multyterm' terminal unit, which gives PacTOR, AMTOR, RTTY, CW, fax and SSTV capabilities. This uses specifically written software such as 'BMK Multy' to handle the data processing, with the external modem converting the 'ones' and 'zeros' to audio tones and vice-versa (Fig 1(b)).

An even-lower-cost method is to use a simple add-on 'data slicer', typically a single op-amp circuit which you can easily build yourself, between your PC's serial port and your radio

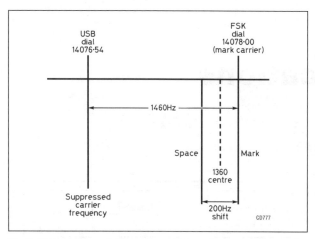

Fig 2. Relationship of transceiver dial readings to mark carrier frequency for USB and FSK operation

A typical multimode data controller, the Kantronics KAM

(Fig 1(c)). Here, the PC goes one step further and, besides handling the data processing, it also handles the timing to give the required bit rates and tone frequencies. Again, specifically written software is used, and there are a variety of shareware programs available for this type of modem, such as HamComm and Mscan, to give multi-mode operation on modes such as CW, RTTY and AMTOR. These and other programs can be downloaded from a multiplicity of sources on the Internet. Complete circuit details of the simple interface are usually given in the software documentation; ready-built types are also commercially available for around £10 to £20.

But probably the easiest way to get started on data is to use what you may already have in your house, a multimedia PC. Here, your PC's sound card is used as the modem, and you only need a couple of leads, for RX and TX audio, between the PC sound card and your transceiver. Transmit/receive switching can often be done using the VOX facility on your transmitter in SSB mode, the transmitter automatically detecting when transmit audio is present from the sound card.

RTTY

Much on-air data activity has been through the use of VHF/UHF packet mailbox systems, ie 'message boards', where amateurs don't communicate directly with each other on a one-to-one basis, or indeed in a 'net'. Also, the error-resistant protocols used in the TOR handshaking modes described later don't invite quick-fire 'break-in' type communication over the airwaves, nor roundtable 'nets' between a group of amateurs. RTTY allows others to 'listen in', to have quick 'one-line' overs and is especially suitable (and widely used) in HF contests, yet it's based on technology invented many, many years ago. Like Morse, its usefulness and simplicity have kept it going in the face of other more advanced data modes which, technically, are far superior in 'getting through when all else fails'. RTTY can use only upper-case letters, figures, and limited punctuation, ie like CW it's a basic 'text-only' mode.

Amateur RTTY uses FSK (frequency shift keying) with a shift of 170Hz, although some terminal units use a 200Hz shift. Commercial RTTY typically uses wider shifts such as 450Hz and 750Hz. Whether you use a stand-alone multimode TNC, an external interface or modem, or a sound card, the software (or hardware in the case of a TNC) will always have some form of 'tuning indicator'. You'll need to carefully tune your transceiver, in either the appropriate SSB or dedicated 'RTTY' mode, for accurate reception of the alternating 'mark and 'space'

RTTY tones. An SSB transceiver is typically used in 'AFSK' mode, where your terminal unit generates two alternating audio tones which, when fed to a correctly adjusted SSB transmitter (ie don't overload it with audio level, keep the ALC within limits!), gives two alternating carrier frequencies. You may hear of 'low' tones and 'high' tones; low tones are 1275Hz and 1445Hz (a 170Hz difference), 'high' tones as favoured by US amateurs are 2125Hz and 2295Hz (again a 170Hz difference). It's the frequency shift that's important – with each being the same they're quite compatible on the bands, so it makes no difference which you use. But be aware that if a frequency is given as "xx.xxxxx MHz Mark", the 'mark' tone refers to the higher-pitched tone; in Europe the convention is to use low tones on USB but US hams typically use high tones on LSB. So there will be a frequency difference with US high-tone USB signals being 1.46kHz higher than European LSB low-tone ones (Fig 2). In general tuning around there's no difference, but when arranging a 'sked' it's important to define which you're using!

Handshaking

Some other modes like AMTOR, Packet, GTOR, PacTOR etc go one further than RTTY in being 'handshaking modes'. Here, in QSO the Information Sending Station (ISS) sends a small group, or 'packet', of characters at a time, and the Information Receiving Station (IRS) receives these and performs a parity check to make sure the information 'adds up', ie 'were the characters received correctly?' If so, the IRS sends a brief 'OK' and the ISS then sends the next 'packet', and so on. When the time comes for the partner station to have its 'over', a 'changeover' occurs (your TNC or software handles all this for you) and the ISS and IRS change between the two stations.

AMTOR

On air, AMTOR sounds like a 'chirp, chirp' when receiving the ISS with just over two chirps a second, with briefer 'chirps' in between the gaps being transmitted back from the IRS. As such, AMTOR, like all TOR handshaking modes, is a 'one-to-one' QSO system. For CQ calls, a 'Forward Error Correction' system is used, ie a 'broadcast', almost like RTTY but where each character is transmitted twice and the receiving station checks to see if a correct character has been received before displaying it on your PC screen. AMTOR will, like RTTY, handle only text and figures, although upper and lower case plus punctuation symbols can be transmitted using the more recent 'mixed-case' AMTOR. Communication with AMTOR makes use of a 'selcal', which is a four-letter combination. This is typically the first letter and the final three letters of your callsign, ie the selcal for 'G4HCL' would be 'GHCL', and when you're transmitting a 'CQ' you should also include your selcall in your CQ text as well as your callsign. Once in QSO, normal

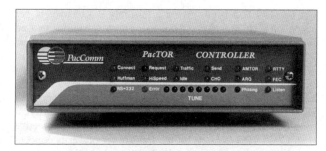

PacTOR controllers from PacComm *(top)* and SCS *(bottom)*

callsigns are exchanged as usual within the QSO text. For listening, all AMTOR terminal units and software also have a 'listen' mode where, providing signals levels are satisfactory, you can 'listen into' existing contacts just like on RTTY, so you can call one of the stations at the end of a QSO if you wish.

HF packet

Virtually all dual-port (HF/VHF) packet TNCs as well as multimode TNCs include 300 baud packet for HF operation. BBS mail-forwarding systems, especially in earlier days, used this and it can be reasonably effective on co-ordinated frequencies with no other users. But for the individual amateur seeking contacts, modes specifically designed for HF use such as PacTOR are notably better at getting through. One significant use for HF packet, however, is for APRS (Automatic Position Reporting System) 'beacons' from fixed and mobile stations, primarily on 30m.

PacTOR and PacTOR II

PacTOR combines the 8-bit data capability of packet, ie it can handle data files including images and software programs as well as plain text, with robustness on a typical HF path. It uses error detection and correction by 'building up' a pattern of received packets, repeated by the ISS as required and storing these in an internal memory, gradually adding the data components and reducing the noise components. Through these repeated packets it can often give 100% copy where the signal is literally buried in the noise and totally inaudible to the human ear. The transmission rate automatically adjusts itself to the conditions, and data compression is again automatically introduced for higher throughput. PacTOR II goes further by using DSP techniques for even better error correction of signals 'buried in the noise' as well as faster data throughput, PacTOR II TNCs are 'backward compatible' with PacTOR. There are a number of PacTOR and PacTOR II Bulletin Board Stations

around the world, often with a 'gateway' to local VHF/UHF packet radio networks in their country. Like AMTOR, terminal units and software have a PacTOR 'Listen' mode, where you can listen-in to on-going contacts.

Clover and Clover 2000

Clover is a proprietary commercial rival to PacTOR, and has found popularity primarily in the US. If your multimode TNC includes Clover capabilities, you'll find it useful for QSOs in demanding signal conditions, although the cost of Clover-specific TNCs is currently relatively high.

GTOR

GTOR, 'Golay TOR', was developed by Kantronics and is included in the later versions of their multimode TNCs such as the KAM Plus. Under good signal level conditions with little or no interference it's reportedly faster than PacTOR for information transfer, with speeds of 100, 200 and 300 baud. You might like to try it for one-to-one contacts with other similarly equipped stations.

PSK31

RTTY benefits from a 'broadcast' protocol, where others can listen in and join in as required, and Morse aficionados know that, the narrower the bandwidth of a mode, with 'human' audio filtering and preferably narrow receiver filtering as well, both in the IF and at AF, the better you can reject other signals on the band to receive the information you want.

This is what PSK31 does, whilst retaining the benefits of RTTY operation. PSK is 'phase shift keying', where the phase of your transmitted carrier signal changes in line with the binary '1' and '0' changes in information. The '31' comes from the rate of change, approximately 31 baud, and with PSK31 you get a communication speed of around 50 words per minute. For better performance in reasonable signal conditions PSK31 can also be switched under software control to QPSK (quadrature phase shift keying). Unlike TOR modes, it doesn't use any handshaking between stations, thus allowing 'quick-fire' message exchanges and multiple-station contacts just like RTTY, but with a much narrower bandwidth used, hence a better signal-to-noise ratio and more resilience against unwanted interference. PSK31 is often inaudible to the human ear, you'll typically need to use your PC software's tuning display as well as a very frequency-stable transceiver to 'tune in' and get in on the activity.

PSK31 software

There's plenty of software available to get you going on PSK31, for DOS, Windows, Linux and Macintosh. PSK31 for Windows on the Sound Card by Peter Martinez, G3PLX, is very easy to use on air – it even has a useful tuning scope and 'waterfall' display to help you get spot-on frequency. A 'once-only' set-up involves an audio-frequency check, and if you're feeding your sound card output to the mic socket of your rig rather to a data audio input socket on the rear panel, you'll just need to make up and fit a simple two-resistor attenuator in line with the lead to your rig to prevent overdriving the SSB transmit stages. All this information is given in the excellent 'help' file that comes with the program. You can get a fuller technical description of PSK31, and download several software offerings, via the Internet at http://aintel.bi.ehu.es/psk31.html.

PSK31 activity frequencies

Since PSK31 started, the plan has been to concentrate activity starting from the lower edge of the IARU RTTY band plan,

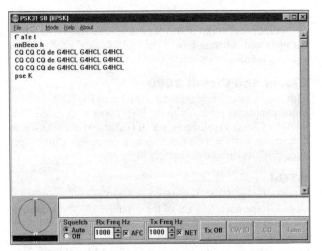

PSK31 for Windows using a sound card, by G3PLX

moving upwards as activity increases. In fact, activity is usually centred 150Hz higher than these, and remember you'll only need about 100Hz frequency separation between QSOs. You'll find most PSK31 activity on 3.58015MHz and 14.07015MHz. 40m and 15m are also active bands for PSK – look around 7.03515MHz and 21.08015MHz. You'll need to tune very carefully, and your receiver or transceiver should ideally tune in 1Hz steps. Once 'netted', you can invoke the AFC (automatic frequency control) function of your PSK31 program to 'fine-tune' the received audio if you or your partner's frequency is drifting slightly, but remember to keep the signal well within the IF passband of your receiver if you're using narrow filtering, eg a 200Hz or 300Hz CW filter and/or the receiver's DSP audio filtering. Current synthesised rigs are usually sufficiently stable, but you might have a hard time with earlier analogue VFO-controlled sets.

Hellschreiber

Hellschreiber was inverted by the German Dr Rudolf Hell in 1929, hence the name. It was essentially an early 'facsimile' mode, yet nowadays you can use your sound-card-equipped PC to make getting onto the mode, 'Hell' for short, very easy. A common type of Hell used on the amateur bands at the time of writing is Feld Hell (from the word 'Feldfernschreiber', ie field teleprinter, used during the Second World War), which transmits text in a 'dot matrix' type form, like early PC printers. The

dot rate is 122.5 baud, giving a leisurely 2.5 characters per second in a 7×7 matrix, and due to the narrow bandwidth used it can be sensitive and good at rejecting adjacent interference.

An improvement is concurrent multi-tone Hell (C/MT Hell) which uses 16 tones and an FFT receiver which works well on the lower HF bands – it's less sensitive but remarkably immune to interference. Another is sequential multi-tone Hell (S/MT Hell) where each row of dots is sent with a different tone frequency. S/MT Hell can sometimes be more effective than C/MT Hell because all the transmitter power is concentrated in one tone at a time; it's also the only MT Hell version suitable for CW and QRP rigs. Hellschreiber is a synchronous mode, and the text is printed twice to negate the effects of phase and small timing errors even though it's not transmitted twice. You can get more information, and download PC sound card software, from http://www.qsl.net/zl1bpu/.

Most Hellschreiber activity is concentrated on 20m at weekends, on 14.0635MHz LSB (try tuning in at 1300 UTC using Feld Hell). Hell activity also occurs on 7.037MHz and 3.580MHz LSB (Tuesdays 2000 UTC) around Europe. DX Hell frequencies include 10.140MHz, 14.0635MHz and 21.0635MHz LSB.

Get in on the action

As you'll see, you probably already have most if not all the equipment you need to get active on HF data modes – just load the software on your PC and get in on the action! Take care when interfacing your PC's sound card to your transceiver, especially when setting both input and output levels, as this is the most common cause of problems for beginners. In the 'Image Communications' chapter in this handbook you'll see that you can also use exactly the same system for HF SSTV and fax. Read the usually excellent help files in the software you're using, and have fun on the air!

PACKET RADIO

Most VHF/UHF data communications are carried out these days using packet radio. Like RTTY, packet radio is a means of sending messages from somewhere to elsewhere in the form of text data. However, it includes a number of important advantages, such as error correction and the ability of one frequency to handle more than one contact at one time. There are also advantages in the type of information which may be transmitted, which includes pictures and text messages. Messages, like the e-mail seen on the Internet, can be posted to or taken from a local packet bulletin board (BBS). The network of stations currently in use also provides other facilities such as conference (even worldwide) and DX information nets.

In cases where the distance between stations is too high for normal point-to-point working, a network of stations, called *nodes*, exists for the purposes of relaying these messages to their destination. In this respect, they are akin to the FM repeaters used by mobile operators to extend the range of operation.

How it works

Packet radio has an affinity with RTTY, in that the messages are transmitted and received by the normal amateur radio station equipment and, as with other digital modes mentioned earlier, the signal comprises two tones, each

Feld Hell QSO on 14.0635MHz (from Nino Porcino, IZ8BLY)

representing mark or space. For most purposes, FM is used on VHF and UHF, while SSB may be used on HF. The data to be transmitted is assembled into packets which one can think of as being similar to a letter in the post, using a *protocol* (method) called *AX25*. The message is transmitted in sequential parts together with address information. In this way it might be said to resemble the envelope of the letter.

The typical station

A typical installation for operating VHF/UHF packet radio comprises a number of parts:

1. A VHF/UHF FM radio.
2. A terminal node controller (TNC) which converts the computer data into a suitable form for transmission (or a computer add-on which makes it appear as TNC signals).
3. A computer of some sort – maybe the shack PC or Mac, together with some suitable software. You could also use a computer terminal unit (VDU) if you can find one.
4. A suitable power supply (usually 12V DC) for the TNC.

The radio

The choice of radio may require a little thought. Several makers of amateur equipment now incorporate a connector for data communications at 1200 baud and 9600 baud. Furthermore, various suppliers are able to supply ready-made cables for many of the amateur radios on the market.

However, most amateurs use a standard mobile radio or even a hand-held, but use of your favourite radio can tie up what amounts to a considerable investment. Many operators, particularly those who leave their packet stations on for long periods of time, have purchased surplus (ex-PMR) equipment. These radios can represent a sound investment.

One thing is important: deviation. Most amateur FM radios are set to too wide a deviation. Many nodes and BBSs are run with ex-PMR equipment which often features narrow filters for the commercial standard 12.5kHz channel separation, and 5kHz and more deviation is simply too much. Reliable communication, particularly at 1200 baud, is satisfactory with a lower deviation than one might use on voice. So, when you set up the radio, set the packet deviation to about 3kHz on the high tone and no more.

It is not usually necessary to have a couple of hundred watts into a whacking great beam antenna – useful though that is in some cases. This is because there are many stations out there and a lot of nodes which will act as a means of relaying your signals onwards to the final destination. These days, even satellites can be accessed on the power of a hand-held radio and a very modest antenna.

The TNC

The device which converts data from the computer (or terminal/VDU), is called a *terminal node controller* (TNC). This assembles the packets which are transmitted to the other station and converts the received packets for the computer.

The basic standard type of TNC is called the *TNC-2*, or TNC-2 clone, and sends signals at 1200 baud. Many TNC makers use this as the basis of their own design, including the built-in software. Some TNCs incorporate methods of generating the 300 baud signals required for HF operation. Which TNC you choose will depend a lot upon what you want to do and whether there is one for sale on the club notice board or the selling skills of your local shop.

There are different types of TNC, some of which are dedicated to some facet of packet radio. For example, there are:

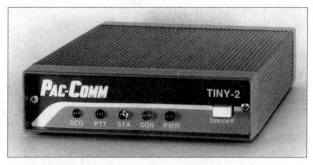

An example of a basic TNC-2 clone: the Tiny 2

* TNCs which also handle other modes of transmission.
* TNCs which have twin radio ports, (HF/VHF operation, for example, or 1200/9600 baud speeds)
* TNC on a plug-in card fitted inside the PC. Some are dual powered, so the TNC is active even if the PC is switched off.

Actually, a TNC can be dispensed with altogether – a low-cost system called *BayCom* uses a PC to do the conversion work normally handled in the TNC and uses a small hardware unit connected to the serial port to generate and convert the transmitted tones.

It might be unwise to use the same PSU that you use for your radio to power the TNC, at least in the long term. It is, however, important to ensure that the earths (frame, ground) are common. Remember that most of the chips used in the TNC and the computer are sensitive to static electricity. If the PSU is variable, set it to about 12.6V (it can ease the heat stresses). Some TNCs can run from a +5V supply.

Terminal or computer?

Some stations use a terminal unit or VDU which has an RS-232, D-type serial communications connector to connect it to the TNC. The disadvantage of a terminal is often its inability to keep a record or transfer files to or from your local bulletin board system (BBS). However, as a cheap way of getting on air and holding a QSO, it has much to commend it.

A more flexible solution is to use the computer found in most shacks, although in some cases some form of serial interface card may be required. It really is vital that it is a true RS-232 serial port, although some TNCs can handle TTL-level signals. A PC-type computer is usually fitted with at least one, and usually two, serial ports in addition to a printer port.

Most computers have software available to enable them to emulate a terminal, and many have a complement of facilities such as binary transfer and even compression. The more comprehensive the computer and its software, the better your chances of access to the files and software available on many BBSs. For many of us, this means a PC.

The ability to transfer files, whilst not being very important initially, may well be a useful consideration at a later stage. Many bulletin boards make available public domain software for a wide variety of computers.

The best advice that can be given is probably to choose a simple software terminal emulation first; then, when you've a grasp of what's going on in the TNC, you can entertain thoughts of installing something better suited to your requirements as they develop.

The basic installation

See Fig 3, which shows a simple diagram of a packet radio installation.

The terminal (or computer) communicates with the TNC at a speed no slower than the TNC sends data for transmission by the radio. Many VHF operators set up the serial link at a baud rate of 4800 and some at 9600 for a transmitted baud rate of 1200. The manuals for your TNC will often detail both the ideal and minimum requirements for this serial link.

Many older TNCs are equipped with an industry standard 25-pin, D-type connector. Most of the 25 connections, you will be pleased to learn, can be ignored. Exactly what connector is fitted depends upon the type of TNC, as several feature a nine-way D-type connector. Some of the miniature TNCs have non-standard connectors, such as (American) telephone jacks and so matching cables, unless supplied, could prove to be an expensive (if necessary) extra to your budget.

To send data back to other stations and communicate, the TNC needs control of your radio, usually via a three-core screened cable of suitable length.

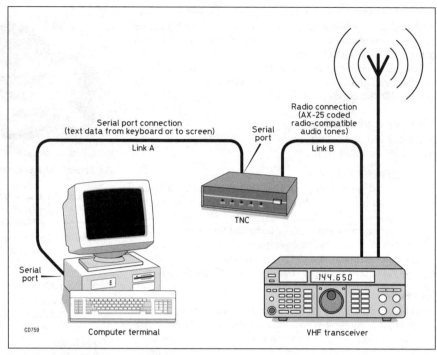

Fig 3. Packet radio station

It needs to listen to the traffic on the channel (so the volume might need adjustment), send a suitable signal to the microphone input (so the microphone sensitivity may need adjustment) and be able to activate the PTT line. If it is unsuccessful, it will stop automatically – in other words, it will time out. Connections at the TNC end are usually a five-pin, 180° DIN plug, or a nine-pin D-type. These days, it is a lot easier to buy the necessary cables than to make them.

Layout
The beauty of packet radio is that many shacks already have a computer. The addition of a TNC is a minor matter, particularly the smaller types. As there is no adjustment of a TNC by any other methods than using the computer, there is no reason for it to be at the forefront of the shack in the manner of the radio or ATU.

Interference
Many computers, including their monitors, can develop a lot of radiated interference. Some computer makers screen their circuit boards and many PC-type computers are in metal cases which naturally cuts down the level of radiation. To minimise noise, screened leads should be used for all connections. Bonding the screen of the microphone connection to the case of the plug will also reduce interference in sensitive audio circuitry. This screen should not be connected to anything at the other end. Naturally, the common (earthy) lead should be connected. If the screen is connected at only at the radio end, made off to the actual case, you should find a marked improvement.

Attention should be given to the actual layout of both cables and main units in the installation. For instance, it would be folly to place the computer too near a sensitive preamplifier. Really difficult cases of radiation may be mitigated by placing the computer in some sort of earthed metal framework, which will act as a Faraday cage, but first check the mains earthing on a PC, most of which have metal cases. It may be possible to screen the interior of a computer, although this is not recommended without a good knowledge of the interior workings or the approval of the maker.

It might be worth buying or making a simple mains filter unit if your local mains is subject to interference. Filter units for mains (usually 3 or 6A rating), can often be purchased at your local rally for a modest sum. This has the advantage of separating the supplies to the radio and computer, thus ensuring that there will be no mains-borne interference.

Typical signal levels
Much heartache can be caused by overloading either the TNC (volume turned up too high) or the radio (excessive deviation or distortion). The audio level into the TNC should lie in the range 100–600mV (it works best at the lower end of the range). Signals to the radio from the TNC should be sufficient to give a deviation of about 3kHz on the high tone. This is typically about 8–15mV RMS depending on the impedance presented at the microphone socket and the sensitivity of the microphone amplifier/s. A higher level of signal will be required for those radios having preamplifiers built into the microphone units.

Setting up
In the same way that an ATU should be carefully tuned to get the best from your radio, there are a number of adjustments which should be made to the TNC and the terminal software before going on air, because going on the air is the easy part. Most parameters are set from the keyboard.

The TNC
Many TNCs have the settings for both serial and radio rates to be set by switches, although in some it is inside the case or by use of commands from the keyboard.

If you are starting out at VHF or UHF, set the radio rate to 1200 baud (on HF, to 300 baud). As mentioned before, the serial link speed should be at least 1200 baud, although 4800 is not unusual. 9600 is possible but, to a certain degree, it will depend on the speed of the TNC's internal clock.

Setting up the terminal software configuration

You may need to set up your terminal's parameters correctly before switching it all on. If you are using a computer program, check the configuration file. For the time being, set the computer software or terminal to give 1200 baud, 7-bit, even parity, 1 or 2 stop bits, which is what most TNCs default to. Check the manual for any particular default settings. For example, some TNCs default to 8-bit so leave it at that.

If you are able to set the TNC baud rate by the switches, set them up to 1200 baud for now; later you can set it to whatever speed you want the computer to communicate to the TNC. If your TNC has software settings for speed, set the computer up to communicate at that speed.

It is vital that both the computer and TNC are OFF when you plug in the serial cable. You might get away with it once but the ICs inside are highly sensitive to voltage spikes and static and you don't always know until some time later that you've 'crisped' something.

Plug the serial cable in. Make sure the PSU for the TNC is fully serviceable and switch on the computer. When it's all settled down, load the terminal program. If you are using a terminal or VDU, it should light up with little problem. Check that it is set up as mentioned above.

Switch on the TNC. If you have managed to get both the terminal and the TNC talking properly, you should see the 'sign on' screen followed by the prompt:

```
cmd:
```

Pressing the Enter (Return) key should get you several more. It's a Very Good Sign.

Setting up the radio

At this point you should have your computer or terminal communicating with the TNC. We advise you not to try transmitting at this stage; there are several things within the TNC to set up first.

To receive 'normal' 1200 baud transmissions, all you have to do now is switch on and set up the radio to the appropriate frequency. On 2m, 144.950MHz is a good place to start or, on 70cm, 432.650MHz. You should hear the sound of packet radio traffic (bursts of a raucous, almost rasping, tone). Note: 144.625MHz is usually the place for TCP/IP transmissions, so you may hear traffic but not be able to decode it without the appropriate software. Local conditions may also enable you to find additional activity on the 4m, 6m and 23cm bands. Modulation is usually FM for 1200 baud signals.

Adjust the squelch so it is just on, as for normal FM operation. Turn the volume well down to near zero. Connect the TNC to the radio. Slowly adjust the volume control up from quiet and make sure that the DCD light on the TNC turns ON when a packet burst is detected. If your radio has a BUSY light, the DCD and BUSY lights should light up at roughly the same time. Stop when you have reliable flashes corresponding to received packet signal, as there is no point in deafening and de-sensitising the TNC. If it does not, check the cable connections between the TNC and the radio. Your screen should show packets, in the form of part-complete texts, whizzing up like something demented, unless you've found a quiet channel or your monitor command (MON) is set to OFF.

At this point, you might like to watch the traffic (don't send anything yet!), or get a friend to send something out. You can sometimes learn a lot by watching.

You should now be in a condition to make a few necessary adjustments. Don't try them unless you are confident that your serial settings are set correctly and that you have the cmd: prompt displayed. If, however, you are confident that all is well with your hardware and wiring, the next thing is to set up the TNC for your own particular installation by putting in the right commands.

Setting basic parameters

It is a very good idea to spend a while just looking at the traffic on a channel before trying a session yourself. Without a doubt, the one thing which will tell all and sundry that you are a newcomer to packet is to connect up the station and start sending packets willy-nilly.

It will be assumed, for the purposes of this present text, that you are using a PC and a TNC-2 clone. The principles should be the same for other combinations of computer and TNC.

With the exception of those commands using Alt or Ctrl, all commands entered to the TNC are implemented by pressing the Enter key. A TNC operates in what are known as *modes of operation*, each having a particular application to the job in hand. Fortunately, most terminal emulation software (ie packet programs) and the TNC will do the necessary changes automatically. The important one is called *command mode*, which has the prompt 'cmd:' at which you may type in a command to the TNC. The others are:

- *Converse*, which is the mode used when connected. Both text and commands (for example, to a BBS) entered in this mode will be transmitted.
- *Transparent*, which is used when moving certain files; and
- *KISS*, which is used only with particular program software, such as TCP/IP.

To get to command mode enter Ctrl-C, which is often written as '^C'. The '^' indicates the control function of the PC. In case you are not familiar with this, it means:

1. Press and hold the Ctrl key;
2. press the C key;
3. release both.

By the same token, any Alt commands are done the same way, but using the Alt key in place of the Ctrl key. Other computer software may have alternative methods, so if you don't use a PC, look in your manual. Now you can turn to the TNC itself and set it up. The reply will be a sign-on display and a prompt, 'cmd:'.

The TNC possesses some 'intelligence' of its own and does not always require the whole word of command to be typed: you can use a bit of short-hand. For example, 'MYCALL' is the same as 'MY' or even 'my'. They aren't all as simple as this but you should get the idea. To make life even easier the TNC will say what value the parameter was prior to any changes you may make. Note that all commands are executed after the Enter key has been pressed.

For example:

```
cmd:MY G8UYZ <Enter>    [enter my callsign & then
                         press the 'Enter' key]
MYCALL was NOCALL       [this is the old setting,]
cmd:                     [and returns the prompt]
```

Note that CAPITAL letters are not always needed and that many commands can be shortened to a few letters. This is an example of a direct command sent to the TNC for immediate execution.

If all you see on the screen is a load of gibberish, you can set up the terminal or the TNC (sometimes both), so that they talk in the same manner. Check out your TNC manual and the reference notes; then try a few simple commands. A command badly typed or not understood by the TNC will evoke a reply of "Eh?". You should try again.

It is possible to obtain the current state of a particular parameter by just typing in the command with no change, eg:

```
cmd:MY
MYCALL G8UYZ
cmd:
```

As might be expected, this is what tells the TNC your station's callsign. Typing "MY g9baa" will put the callsign G9BAA into the TNC's memory. Make sure that you put your own callsign in properly.

Parameters put in will stay in until changed or you issue a RESET command. The internal battery will maintain the parameters in the memory for as long as it lasts. Any changes that you do make, however, will affect the way that your TNC operates, so in these early stages you might like to make a note of what each parameter was before the change. That way, if you suddenly think "I wish I hadn't done that", you can change back to the factory settings without resorting to the big reset. Note: RESET is a software command and results in all settings being reset to the factory default, loss of whatever messages were stored in the PMS etc.

A few more commands need to be set prior to operating. These are 3rdparty, which you should make sure is OFF (the licensing requirements are different for a BBS) and the CW identification. The UK licence requires you to transmit a CW ident. Consult your TNC manual for full details but make sure you do set it. If you have a TNC with an early-issue ROM which does not have a CWident, you will need to modify it. A new ROM will usually sort it out. Talk to your TNC dealer or local AX.25 user group; it is usually neither difficult nor expensive.

Moving on a bit

As you get more experience and wish to do more than pass simple text messages you will almost certainly find it necessary to make a change to 8-bit. Some of the features and benefits of packet radio are simply not possible without setting your whole installation to 8-bit use. These include file transfer and handling 7-PLUS files. To set your system to 8-bit, set the following TNC parameters:

```
AWLEN 8
PAR 0
RESTART
8bit ON
```

. . . and in that order! If you can manage to do it early on, it will probably save you a fair bit of confusion later. When doing it, it is important to set the TNC up first; then do the terminal or software on the computer.

A simple ASCII text file with these settings should be kept for just this problem. Provided that the RESTART command is included, the TNC will respond and you should have no problem. Don't forget to set the terminal (or program) after you've set up the TNC.

If anything goes wrong, make sure that the program is set up and send the text file to the TNC (ASCII Send). Applications for setting it all up for 8-bit include binary transfer and use of 7PLUS files.

If when turning on the TNC you don't get the 'cmd:' prompt, check your serial cable and settings at both ends. Don't go further forward until you do.

The connections to the radio can be made with fewer problems. The TNC needs an audio feed from the radio and access to the microphone and PTT inputs. A screened lead is of great benefit in helping to control interference. You can buy a suitable cable from your dealer. Consult the TNC and radio manuals if you wish to make your own.

Screen width and packet length

The default setting for Screenln is usually 0, which defines that you have a screen 256 characters wide. This can make for a cluttered appearance on the screen which is very difficult to read, bearing in mind that most screens are of 80 columns. Far better to set it to 80, which gives a screen width of 80 characters. Some circumstances may require this to be set to slightly less.

The other command sets up how much data in a packet is sent and it is called PAClen, which is set to 128 as a default. To some extent, how you set this is determined by how experienced you are in using a keyboard or the traffic density in your area but if you set it to 77 you will find that you send a packet at the end of every line you type.

Even if you have a machine with a display of 40 characters, you would be advised to use these settings, since those who have a wider screen will find it tricky to read. Note that if you pay no attention to your screen as you type, you may still get words cut in half, but this may be a function of your terminal software. Some have 'word wrap'; if so, set it to ON.

Timing parameters

Getting these settings wrong can cause havoc to a lot of local users. Experience has shown that these settings are the best values. Set up your TNC to the following:

DWait	0
PPersist	ON
Frack	7
RESPTIME	30
MAXframe	2
SLottime	30
PERSIST	38

Ideally PERSIST should be 25 but some TNC software has a bug which means that any setting of less than 38 will ensure it never transmits.

TXDelay

This timing will be different from station to station. It is the time the TNC waits (in tens of milliseconds) between keying the transmitter and actually starting to send the data. The short delay allows your transmitter to lock up prior to sending the packet data. The default setting is 30.

To find your ideal setting, slowly decrease the setting of TXD, one at a time until you stop connecting reliably. Then add two or three to this figure and it will be about the best. Note: Don't set TXDelay too low, even if you can connect reliably with most stations. It's possible that other stations may need a comparatively long preamble to open the squelch on their receivers. If you have problems with a TXD setting of less than 25 on a fairly modern radio, your deviation could be too high or you could be off frequency a little.

Appearance on your screen

You can do a lot to make sure that what is printed on your screen is the way you like it. When monitoring packets (with MON ON), it will be seen that the header, showing the stations involved, is in the same line:

```
G8ZWU > G8UYZ : Hi Dave, how are you today?
```

Setting the HEaderline command to ON will produce:

```
G8ZWU > G8UYZ :
Hi Dave, how are you today?
```

You may wish to include information on which station actually sent the signal. The command is MRpt (Monitor Repeater). With MR set to ON, the greeting message about might look like this:

```
G8ZWU > G8UYZ, G4RVK* :
Hi Dave, how are you today ?
```

The '*' indicates the callsign of the station actually being received.

You can 'date-stamp' your contacts with the date/time, using the MStamp command. Setting it to ON will produce:

```
G8ZWU > G8UYZ, G4RVK* [27-06-94 21:30] :
Hi Dave, how are you today?
```

The date and time may be displayed in two forms, USA or European. This uses the DAYUsa command, which defaults to ON so the date is shown in the American form, mmddyy, eg 06/27/94. Setting it to OFF gives you 27-06-94.

Of course if you'd rather see the month in letters, not numbers, use the AMonth command: AMonth On/Off. If ON, the month in dates will be shown as Jan, Feb etc. If OFF, they will be 01, 02 etc. It is possible to set the date and time held in the TNC by hand, although many terminal programs include a method of taking the data from the PC system's internal clock (it's also likely to be more accurate). The command is DA:

```
DA yymmddhhmm    [year, month, day, hour, minute]
                 [2 digits for each]
```

Connecting to another station

C (Connect) is the primary command telling the TNC to go out and find someone:

```
cmd:C G8ZWU
```

The TNC will then send a sequence of packets to the stated station, in this case, G8ZWU. If he is not otherwise engaged his screen will show:

```
** CONNECTED TO G8UYZ **
```

Suppose we are in a rural area and that the nearest station is G3KQJ. If he leaves his 'digipeater' set to ON, G8UYZ can connect to G8NZU via this digipeater. In this case, the transmitting station will determine the route to be taken:

```
cmd:C G8NZU via G3KQJ
```

This will cause the TNC to transmit a packet to G3KQJ and request the digi to pass on the packet.

The contact can then continue as usual. You can use a digipeater to get into a BBS but it can take a bit of patience. Although up to eight digipeating stations can be nominated, the network of nodes has made long-distance digipeating unnecessary and even undesirable. A node behaves like a digi but does it better.

When using a node, the calling station must first connect to the node, then instruct it to connect to the next station:

```
cmd:C GB9PPP
```

Once you are connected, you will then get some form of acknowledgement that you are connected and a prompt at which you can continue:

```
C G8NZU
```

When the node connects to the distant station, you get confirmation and you can continue the contact as normal. There are several variations to this theme and the complexity and development of the whole network is continually undergoing change, so we will restrict ourselves to the simple things for the moment.

Now might be a good time to arrange with a local packet station to attempt a connect to him/her by way of a little practice.

As the message you send is sent in small bits, sometimes only part of a line of text, it is advisable to use some symbol to indicate that your transmissions are complete and you await the reply. It is pointless to put your callsign in as you do on a local FM net; that information is already transmitted. So use the chevron symbol; it is as good as any: >>.

To conclude a contact, use ^C to gain the command prompt (cmd:) and type the letter "D", "DIS" or even "disconnect"; then press the Enter key. You should shortly see:

```
***DISCONNECTED
```

Note: Some nodes may be dedicated to some function such as a DX cluster, or may simply include facilities such as 'Chat' or 'Conference' mode. An enquiry on the local BBS or to the system operator (sysop) should get you the information.

The personal mailbox system (PMS)

Several types of TNC are fitted with a system which makes it is possible to leave a message for the recipient in the manner of using a full-sized BBS. It's rather like an automatic telephone-answering machine but looks a bit like a bulletin board.

The commands for the PMS in the TNC are very similar in operation, if not in name. Many users of these handy gadgets leave them on for long periods, thus ensuring that they can get any messages from, or even sending to, the local BBS.

Setting up your PMS

To set one up takes a little patience, the co-operation of your local sysop, and certain commands set in the TNC. Note that most PMSs have a limited memory in which to store the messages. You should clear them out quickly.

If someone tries to connect as if you were in and paying attention he would call you in the usual way. However, if he wanted to leave a message, there has to be some way of showing that there is a difference between your usual callsign and your letter box. This dictates the callsign to be used by callers wishing to leave a message. It is set by using the following command:

MYPcall n

where n is the callsign secondary ident (SSID) in the range of 1–15, and is the callsign used by the HOM BBS to connect to your PMS. In this author's case, it is set to G8UYZ-2. On 70cm, -7 might be more appropriate, but it is a matter of personal taste. You should NOT set it to be the same as MYCall. It can cause no end of confusion!

PMS on/off

Default is OFF. If it is OFF, callers will see:

```
*** G8UYZ-2 BUSY
*** DISCONNECTED
```

on their screens. Messages are stored numerically and this is set back to 1 on the issue of a RESET command). If it is set to ON, users will see the PMS sign-on screen.

3rdparty

This should be set OFF, since there are separate licences for operators of bulletin boards. Setting it to ON will allow your TNC to act in the same way as a BBS. As the owner of the MYPcall, you can send any message to anyone but messages IN may only be left for you.

HOMebbs

Type in HOM followed by the callsign of your local BBS.

AUTO

Auto-forward. This marks any message with a callsign in the @ field to be marked for collection/forwarding.

KILONFWD

When ON, messages are deleted after forwarding. If OFF, they aren't. Note: A PMS has about 15kb of storage space, so if you are going to be away from your system for a while, and you expect to do several messages a day, leave it ON (default).

So much for the basic set-up. Given that your sysop has approved and done his bit, you may now have mail collected from your PMS. All you have to do first is put it in! The command to send mail (for collection) is the SEND command:

Send xxxx @ yyyy

where xxxx is the callsign and @ yyyy is the BBS address. Depending on the model or version, you may also be able to use SB or SP in place of Send. The 'Subject' has a maximum length of 20 characters. Text is terminated by either ^Z or /EX on a clear line. If successful, you will see "Message Saved as no. _" displayed.

You would be well advised to use an ASCII text editor to write any messages and then load them into the PMS.

So much for your acting as your own sysop. You'll also need to be able to manipulate the messages sent to you:

List

Lists your messages in order: Number, Month/Day, From, To & Subject. The short-form command entry is L.

Read n

This will display of message number n.

Kill n

This will delete the number n message from the memory. If successfully issued, you will see "Message Erased". Short-form command is KI or sometimes K.

Of course, it helps to know if someone has dared to use your PMS. When a message is on your box, the STA light will flash slowly, which indicates a message waiting. Reading the message/s will reset the flasher, but not necessarily wipe them from the memory.

Note: A change made in 1994 to the UK licensing regulations means that you need permission to run your station 'unattended'. Contact the RSGB or your local Radio Investigation Service (RIS) office for details. It is usually a question of a simple letter (or a form) stating who is available to switch the system off in your absence, and details of your particular set-up, such as frequency, power output etc. If approved, you will receive a confirming letter.

Packet radio and satellites

There are satellites transmitting the normal terrestrial 1200 baud signals which can be worked on very modest equipment. The NASA Shuttle missions also sometimes carry a packet transponder which can be worked with a bit of effort. To access these you will need satellite tracking software in addition to your basic installation. However, you do not require lots of power or special antennas; a simple pair of crossed dipoles and modest power is all that's needed, although good results have been obtained by many stations using a simple vertical or Yagi beam. Some satellites act as a 'flying BBS' and relay information about the globe. However, working these satellites is an art in itself and requires more comprehensive equipment. Contact AMSAT-UK for further details (see Chapter 7).

Bulletin boards

Your local bulletin board system is operated by a sysop (system operator). It usually comprises a computer and a number of TNCs and radios which enable a station to access on a number of frequencies. Typically these are 144.650 and 432.650MHz, although there are some in the 4m and 23cm bands. Most BBSs operate on 1200 baud, but many forward mail to other BBSs on special RF links running at 9600 baud.

Messages from one person to another in a more distant location are generally passed via a series of BBSs, which behave to the packet operator in the same way as those on the telephone system. There are different ways to send a message from a BBS:

> SB – Send Bulletin   A message sent to everyone
> SP – Send Private   A message to a person

There are a couple of variations. Suppose you've read a message you can send a reply without typing in the full address:

> SR – Send Reply

If you want a friend to see a copy of a message, you can send him a copy:

> SC <number> <callsign> [@ address]

This ends a copy of the message of <number> to the owner of the <callsign> at the address given.

Consult your sysop for full details of the facilities available on your local BBS. The 'help' file is usually obtained with a question mark '/?' or '/help'.

There is much that could be said on the contents of the message. It is better that you try and liken your packet messages to a letter written to a friend; polite and pertinent. If you are asking for help, it might be better if you gave as much information as possible. No facts, no help! See later for some tips on message content.

Addressing

In the same way you address an ordinary letter properly, with the postcode and so on, packet messages should be similarly addressed correctly. This is particularly true of mail to foreign parts. This is a typical UK address:

> G8UYZ @ GB7IPT.#28.GBR.EU

This shows that G8UYZ has GB7IPT as his normal home BBS but the rest is a bit more tricky. Reading from the back, the address is broken into parts. 'EU' is the particular part of the world. For example, America is broken into two: 'NOAM' and 'SOAM', while Australia is in OC (Oceania). Countries also have a code: 'GBR' is obviously Great Britain, whilst France is 'F'. Countries are broken down in turn to smaller areas, usually indicated by a '#' symbol. '#28' shows that the station is located in region this and county that.

If you send mail to a person, you do not always need to put in anything other than, say, @ GB7IPT. However, if you are send mail outside the UK, you must put the whole thing in clearly and properly. For example:

QW1AAA @ WB7ZZZ.#SVA.VA.USA.NOAM

The DX cluster system

If you are a keen DX enthusiast, the DX cluster node system and the right software is just waiting for your attentions. When you have logged into the DX cluster node, the software keeps you from being disconnected (it's called *timing out*). The screen displays a list of one-line entries sent from those stations also logged on who have heard something of interest:

```
DX de G4BUO:    7005.2  W1KM    Greg. Very early indeed! 1558Z
DX de G3PSM-7: 10101.1  OY2H                             1601Z
DX de G3VMW:    7003.4  9N1KC                            1602Z
DX de G3MCS:   14260.0  VP29EI                           1602Z
DX de G0TOG:   21287.5  KQ4PL   JUST CLEARING ID G4JCZ   1603Z
DX de G0JHC:   50110.0  OM3ID   599 800Hz TX DRIFT!      1603Z
DX de G3VMW:    7003.4  9N1CC   C/S sorry - QRM heavy!   1607Z
DX de G0JHC:   50107.0  OK2PPP  59                       1605Z
DX de G0JHC:   50115.0  OM3ID   ssb                      1607Z
```

As you can see, in a few moments there was much of interest to the DX enthusiast. An enquiry via your local BBS should elicit the details of the local DX cluster node. You should join the local group (it shouldn't cost a lot but the software is usually commercial), and you'll get a full manual with details of all the commands. Some operating guidelines for DX cluster operation are given in the panel on this page.

File transfer

Given a good path, text and text files are not the only thing that can be sent between stations using packet. It is possible to send binary files from one to another, given that the computer software and the TNC wiring are appropriate.

As the packet network is an essentially text-based message system, anything other than text distributed on the network needs to be converted into a form which can be sent as if it is text. Such utilities are called *ASCII code converters* and examples are 7-PLUS and UUCODE.

You are reminded not to send commercial software or material unrelated to amateur radio. See the panel on the next page for operating guidelines.

Local groups

If nobody in your local club knows anything, watching the screen for your local BBS should give you a good clue as to any local packet groups. Many are connected with a particular BBS, rather in the same way as repeater groups.

Writing packet messages

The tools

The art of writing clear messages is not as easy as you might first think. It is best to write the message as you would write a letter to a friend. If the quality of your writing is not quite up to standard then use a separate ASCII text editor with a spelling checker, if possible, and work it out on that.

The use of a simple ASCII text editor cannot be underlined enough and, although most of us possess one, we infrequently use it. Note that it is important that it is an ASCII text editor and not a real word processor, because these often use their own particular brand of storing the 'embedded format codes', which is not what we need. It has to be a simple ASCII output; even WordStar in 'non-doc' mode can include undesirable characters.

DOS AND DON'TS FOR PACKETCLUSTER OPERATION

DON'T use ANNOUNCE/FULL unless the information is 'perishable' (ie time-sensitive and *really* needs to go to *all* users immediately.

DO ask yourself before you do an ANNOUNCE/FULL whether the info wouldn't be better in a message to ALL where a lot more people than those logged on at the time will see it.

DO try to keep ANNOUNCEments and mail to DX related matters. If you want to send out general mail, please use the BBS network.

DO send replies/comments to an ANNOUNCE back to the originator using TALK, *not* ANNOUNCE.

DO try the W6GO QSL database before putting out an ANNOUNCE/FULL to find a QSL manager.

DO update QSLNEW when you find a new QSL manager – then it would be available to the rest of the folks using the cluster.

DON'T rise to the bait if someone is obviously abusing the cluster facilities.

DO hit Enter or Return after 68 characters or so when you are sending mail on the cluster.

DON'T fill your mail messages with lines of spaces – they all take air time to transmit; the object of the exercise is to convey information, not to make a message look pretty!

DO use your nearest PacketCluster. Don't DX through the node network, as this will probably slow the inter-cluster links down considerably.

DO use L4 connects to your cluster; if it appears in the node list on your local access node, use: C (Cluster callsign).

DON'T try to tailor a route through the network by connecting to intermediate nodes if it isn't necessary.

DO pick a cluster and stay with it – don't flit from one-to-another. Any mail for you will follow from cluster-to-cluster and this congests the network.

DO kill your mail when you've read it – *please*!

DO use the UPLOAD/USERCMD utility to tailor your own configuration, and especially the filters.

DON'T complain about VHF spots if you are an HF DXer, or HF spots if you are a VHF DXer. Use the SET/FILTER command inside a USERCMD file to tailor what you get to your own tastes.

DO if you have any problems, please send your sysop a mail message or ring him up. Please don't use the ANN/FULL facility to ask for advice which might well be in the user manual.

DO enjoy the facilities and if you have any suggestions for us that you think will improve the system, please let us know.

DO consider the effects of your actions on other users of the network; remember there is a rush hour and if possible, avoid heavy usage at that time.

DO SH/CLUS in preference to SH/CON whenever possible. Then, if you see a whole load of clusters connected, you can do a SH/CON GB7xxx for whichever cluster you want, rather than getting the whole lot.

DO send B or Q to disconnect rather than a hard disconnect which leaves the network cluttered with packets that have nowhere to go.

DON'T use ANNOUNCE to let folks know you have received QSL cards; use a message to ALL.

Issued by the UK Cluster Working Group in the interests of better Clustering for all.

RSGB GUIDELINES FOR SENDING 7PLUS FILES

During the RSGB Data Communications Committee meeting held on 6 November 1993, it was decided to issue a set of guidelines for the use of 7PLUS:

1. The files should be relevant to amateur radio.
2. Local tests should be carried out to prove successful sending and dealing with error requests.
3. The total file content should be no longer than 10 parts of 5k each, unless absolutely necessary.
4. All parts should *not* be sent out at the same time, but sent out over a period of a few days. Sysops should consider holding all locally generated mail.
5. An introductory message should be sent containing the details of the file contents.
6. Offer to make files available on disk.
7. It would be better to offer the distribution of files to other countries by post, for example EU and WWW.
8. Be careful not to breach copyright regulations when sending programs by 7PLUS.
9. Try to include '7+' in the subject line as this helps automatic servers; eg Subject: 7+ DCC.Zip P01/2

Examples for the PC include EDIT (free with MS-DOS 5 and later), ZED and BREEZE. Notepad may also work. If you do normally use a word processor, make sure it will export the file in proper ASCII format.

Some text editors always make out a full page, filling up the unused lines with blanks, which can leave a whacking great hole in your log file. This has another disadvantage in that those operators with simple terminals may actually miss the message. Set it up so it only takes up the minimum needed to do the job right. Make sure that you use single line spacing. However, don't crush it all up into one big block of text. Use a line between paragraphs (yes, use them as well!).

The use of a text editor has the advantage that you can take your time to get it right, which is particularly important if the subject is controversial. There's nothing more embarrassing than having to issue an apology or correction because you failed to check what you were sending. Then you can upload the message with confidence, knowing that it says exactly what you want and that it can be understood by all.

If you really are going to write an opus, do it in several parts, if you please. With the increase in traffic, anything longer than 5k, particularly if sent to foreign parts, could mean truncation or even loss of a message.

Layout and contents

Let's face it, if you want a coherent reply, it's good practice to send an understandable message the first time out. Remember always that, whilst you may enjoy the benefits of e-mail and have the necessary software for HTML files, not everyone else has and the extra characters in an HTML file can clutter up the network all too easily with unnecessary characters so please use a simple ASCII text editor.

Do use capitals and lower case; it can make it easier to read. Excessive use of lower case can look really strange, to say nothing of taking away something of the message. Use upper case for abbreviations (RSGB, DTI, IBM etc). You really do not need to be an expert typist; most of use one or two fingers of each hand and still manage to type at a respectable speed.

Off-line typing practice is a great teacher. If excessive use of lower case looks odd, the use of capitals and nothing else looks at best very impersonal. Such messages appear like a telegram or even a Telex message from head office. Such practice often stems from a familiarity with RTTY and you should

practice using all the keys at your disposal. An all-capitals message can even look insulting; do you like being shouted at? Capitals are a good way of emphasising some point you are trying to make, so using only capitals denies you a useful tool to help make your point.

Don't use special characters, even the pound sign. There's been a growth of peculiar boxes (some quite pretty, even clever) which contain all manner of 'sign-off' text. Everything from shadows to blocked-out '73' and so on, to great logos of what the operator uses, his sports, affiliations and an Internet address, all taking up half a page. Fine if you're proud of it, but not much good in a short text to say "thanks". Such extravagances really are not necessary and can spoil an otherwise good and simple layout. In fact, sometimes they can even look odd, particularly where the graphics characters used, for example, in the box, come out on the screen with a line like

MMMMMMMMMMMMMMMMMMMMMMMMMMMM

instead of a neat box line. Just leave the sign-off as a simple name, callsign and home BBS, please. You may think it looks a little lonely, even boring, but it's easier to read, especially for those using simple software or a VDU.

Spelling, grammar and punctuation are also matters which rise in the subject "DEBATE @ GBR" from time to time. A dictionary or a spelling checker, and careful use of the apostrophe, can avoid some spectacular howlers worthy of inclusion in the Hall of Fame. Remember the famous advert: "For Sale; cabinet, by lady with large legs". We all know what is meant, but there is often a clearer way of saying it.

A good, simple and clear message is preferable to one which may have a better standard of English but don't take it to extremes. Take care and do try to spell simple words correctly. For instance, 'amatuer' is a common error, as is 'Epsom' for 'Epson'. A little care and thought can make your message clearer to all readers!

UK readers are reminded to note BR68 clause 1.4.a on the subject of what can and cannot be transmitted, and the RSGB/RA approved guidelines (see panel). And don't think that no one will see; you'd be surprised what can be examined.

Etiquette

Don't use the medium of packet radio (or any other mode) to attack a person. If you absolutely must air your view about what someone else has said, do it in a direct (SP) message to the station concerned. There are those who do but, to many operators' minds, vociferous SBs just clog up an already busy network with trivialities or inflammation. Don't do it as an SB to all, particularly if you are not quite sure about the details of the subject. Libel can have quite serious consequences, not least for what further restrictions might be imposed on the rest of us.

Don't use the packet network to air your grievances about some trader who cannot reply in the same manner. It's unfair and often rude. If you have a complaint with a supplier of any sort, take it up with him first. You can always tell the tale later, when you have, hopefully, received some satisfaction. As the sign in a shop once put it: "If you're happy, tell your friends. If not, tell us". In the case of a rally, remember the legal maxim: "Caveat emptor" (Let the buyer beware).

On getting it out

General broadcast (SB) messages can be fraught with problems. Some mailboxes refuse to accept messages to 'ALL'. Known and used words like ICOM, YAESU, EPSON, DEBATE, and even HELP can make the purpose clearer.

There have been many examples of some stations which send

RSGB GUIDELINES FOR THE USE OF THE PACKET RADIO NETWORK

The packet radio network in the UK and throughout the world is an immensely useful tool for the dissemination of information, the seeking of help and advice and the publication of amateur radio related news. It is not uncommon to find messages giving information on AMSAT, RAYNET or other similar amateur radio related activities. The GB2RS news is also available on the network, as is local club news in the area of a particular mailbox. This use of the network is what was in many operators' minds when they spent large amounts of time and money in developing it.

With the advent of high-speed modems and dedicated links, some in the microwave bands, the packet radio network is developing, and hopefully will continue to do so for many years to come.

The RSGB Data Communications Committee (DCC), in consultation with the Radiocommunications Agency, has devised the following guidelines with which all operators are urged to comply. These guidelines have been split into five sections in order to reflect:

1. The need for messages to be within the terms of the licence conditions and the implications if they are not.
2. Messages which could result in legal action being taken by other amateurs or outside bodies.
3. Actions to be taken when amateurs identify cases of abuse.
4. Unattended operation.
5. General advice.

Section 1. Types of message
(a) All messages should reflect the purposes of the amateur licence, in particular 'self training in the use of communications by wireless telegraphy'.
(b) Any messages which clearly infringe licence conditions could result in prosecution, or revocation, or variation of a licence. The Secretary of State has the power to vary or revoke licences if an amateur's actions call into question whether he is a fit and proper person to hold an amateur licence.
(c) The Radiocommunications Agency has advised that the Amateur Radio Licence prohibits any form of advertising, whether money is involved or not.
(d) Messages broadcast to ALL are considered acceptable but should only be used when of real value to other radio amateurs, in order to avoid overloading the network.
(e) Do not send anything which could be interpreted as being for the purpose of business or propaganda. This includes messages of, or on behalf of any social, political, religious or commercial organisation. However, our licence specifically allows news of activities of non-profit making organisations formed for the furtherance of amateur radio.
(f) Do not send messages that are deliberately designed to provoke an adverse response. Debate is healthy but can sometimes lead to personal attacks and animosity which have no place on the packet network.
(g) Unfortunately the very success of the network has resulted in messages appearing which are of doubtful legality under the terms of the UK licence. The use of 7+ and other like programs to pass text and binary-based material in compressed form via the network has become commonplace. Users must always be aware of the licence conditions in BR68 (also copyright, and illegal use of software, or software which when decoded and used may contravene the licence) when entering such messages into the network via their local BBS. If in doubt consult your local sysop or your local RSGB DCC representative.

Section 2. Legal consequences
(a) Do not send any message which is libellous, defamatory, racist or abusive.
(b) Do not infringe any copyright or contravene the Data Protection Act.
(c) Do not publish any information which infringes personal or corporate privacy, eg ex-directory telephone numbers or addresses withheld from the call book.

Section 3. Action in cases of abuse
(a) Any cases of abuse noted should be referred in the first instance to the RSGB DCC chairman, care of RSGB HQ.
(b) It is worth noting that any transmissions which are considered grossly offensive, indecent or obscene, or contain threatening language, may contravene the Wireless Telegraphy (Content of Transmission) Regulations 1988 and should be dealt with by the police. This action should also be coordinated by the RSGB DCC initially.
(c) Mailbox sysops have been reminded by the Radiocommunications Agency that they should review messages and that they should not hesitate to delete those that they believe to contravene the terms of the licence or these guidelines. It is worth remembering that their licence is also at risk as well as your own.

Section 4. Unattended operation
As of July 1994 unattended operation of digital communications cannot be carried out without giving seven days' notice in writing of operation to the manager of the local Radio Investigation Service (RIS) – see BR68 para 2 (5). The manager may, before the commencement of operation, prohibit the unattended operation or allow the operation on compliance with the conditions which he may specify.

The RSGB DCC recommends supplying the following information when applying to your local RIS office for permission to operate unattended digital operations:

(i) An external close down switch or other means of closing down the station, which is separate from the rest of the premises.
(ii) A list of four persons, including telephone numbers, who can close down the station. (Not all need to be amateurs.)
(iii) Travelling times and availability times (ie 24 hours) of close down operators of the station.
(iv) Frequencies of operation (within BR68 cluase 2 4(c)), antennas and power used.
(v) Use of station for digital operation (ie PMS, node etc).
(vi) Only the licensee can reactivate the station after permission from the RIS.

Section 5. General advice
(a) With the advance in software writing it is now possible for packet users to set up intelligent software nodes (G8BPQ and similar applications) via a PC and radio. The RSGB DCC strongly recommends that users contact their local packet group and local BBS sysop before starting operation of such nodes. Appearance of such nodes without co-ordination causes problems within the local and inter BBS/DX cluster/TCP/IP network routing tables. Network sysops work closely with each other to determine route qualities and node tables, to aid the fast movement of traffic via the national trunk system. The appearance of unco-ordinated ('rogue') nodes causes in some cases severe problems in traffic routing. Packet users can experiment with software nodes, without affecting the node tables of local network nodes, by setting the software parameters of the node to stop propagation of the node into the network. Advice can be sought on the setting of software parameters from local node sysops or BBS sysops.
(b) Do not send 'open bulletins' to individuals.
(c) Do not write in the heat of the moment. Word process your bulletin first, then re-read it. You may feel differently after a few minutes.
(d) Stop to think before sending GIF images and the like which are sometimes in large multi-part files. Do you really need them, are they amateur radio related, and would they be better sent on disc in the post?
(e) Please try to show some consideration for your local sysop. Remember that you are using in most cases his own equipment, which is in his home. Try to comply with any requests he makes of you.
(f) When accessing your local mailbox at busy times and having problems holding the link, try not to turn your power up just to maintain the link – try later when it is not as busy.
(g) Obey the Golden Rule: "If you would not say it on voice; do not send it on packet".

GLOSSARY OF DATA COMMUNICATIONS TERMS

Address A character or sequence of characters designating the origin or destination of the data being transmitted and, to some extent, routing information. It is also used to denote a particular location in memory in a computer.

Alias An alternative to the callsign of a station. Sometimes, notably with nodes, this may be a mnemonic of the location. With personal stations, this may be the same as the Secondary Station ID (SSID), as set in the MYAlias command in your TNC.

amprnet (AMateur Packet Radio NETwork). The designator for TCP/IP.

ampr.org The high-level 'domain' recognised on Internet for amateur packet radio TCP/IP.

ASCII American Standard Code for Information Interchange. The USA version of the ISO seven-bit data code with 128 letters and numbers. It is usually transmitted as eight-bit characters, with the addition of a 'parity' bit. The 'extended' ASCII set, usually found on most computers, is of 256 characters and is transmitted as eight bits.

AMTOR AMateur Teleprinter Over Radio. An error-correcting data transmission system sending seven-bit data, devised by Peter Martinez, G3PLX, based on the maritime SITOR system. It is usually used on the HF bands.

Analogue An analogue signal is one which can vary continuously between defined limits. The human voice is an analogue signal, varying in frequency, volume and timbre.

ARQ Automatic Repeat reQuest. A method of error correction whereby a transmitting station is requested to send again certain blocks of text found by the receiving station to be in error. This is the principal mode used by AMTOR stations.

Asynchronous transmission A method of transmission that allows data to be sent at irregular intervals by preceding each character with a 'start' bit and following it with a 'stop' bit. A simple example would be teleprinter signals. See also 'Synchronous transmission'.

BayCom A packet system of truly minimum configuration which relies on the PC to do all the functions of the PAD and terminal. The hardware is restricted to a few ICs in a small unit, forming the modem function of a TNC, connected directly to the serial port. Devices are available for both 1200 baud on the serial port and 9600 baud on the parallel port.

Baud A unit of measurement that indicates the number of discrete 'signal elements' (bits) transmitted per second. It is named after the French telegraph engineer J M E Baudot (1845–1903).

Binary file A number of sequential bytes, held in memory or stored on a disk, which may make up a complete program instruction sequence: They are the 'actual programs' which are in a form which may be executed or run. For the purposes of packet radio, this might also include image files. The transfer by packet of binary files requires an eight-bit transmission system.

Bit A bit is the smallest unit of storage within a computer. It is a contraction of 'Binary digIT, and can take a value of 1 or 0. In the International Telegraph Alphabet (ITA) No 5, seven bits are required to make up a character. ITA 5 is similar to the lower ASCII set. A bit is represented in packet radio by a mark or space tone. It compares equally with the 'signalling element', the smallest unit of transmitted data, like a dot in Morse or five in a group to make up a teleprinter character. See also 'Byte'.

bps or bit/s Bits per second. A measure of the transfer rate of information on a channel, usually used at higher rates of transmission.

Byte A group of consecutive bits, normally eight, forming a unit of storage within the computer, which represent one data character, such as a letter or number of the alphabet. It is also used to denote the storage capacity of floppy and hard disks, although generally seen in thousands (kilobytes) or millions (megabytes). Some hard disks can store several gigabytes.

Case Letters are described as being either upper or lower case. Upper case is CAPITALS (usually accessed by pressing the 'shift' key). Lower case is 'normal' like this text. The term has its roots in the printing industry. Sentences in messages should be made up of both cases. They are then easier to read.

Check sum See 'Parity checking' and 'CRC'.

Clock A commonly used name for any source of timing signals. Examples are those used in computer equipment to ensure that all processes are synchronised with respect to one another. This word is also used as a comparative reference to indicate the speed of a PC. Clock speeds in excess of 20MHz are common. Most TNCs are 2–10MHz. See also 'System clock'.

Clover An HF data mode where the simple two-tone mark and space modulation is abandoned, and modern DSP techniques are used to provide a choice of more complex modulation schemes to make efficient use of narrow HF channels.

Configuration/Config file A short file used by a program to determine initial settings prior to use. Most can be edited by use of a simple ASCII text editor and/or from within the program itself. Many have the extension '.CFG'. Some terminal emulation (comms) software will store changes automatically and adopt this setting when next switched on; one example is paKet.

Compression See 'File compression'.

CRC Cyclic redundancy check. A method of deriving a number, by the transmission of additional data, in order to ascertain whether an error has been received. It is also used on disk files in the computer.

CSMA Carrier sense multiple access. The system allowing many stations to use the same radio frequency simultaneously for packet communications.

DCE Data Circuit terminating Equipment. The equipment usually provides the functions required to establish maintain or terminate a connection, and signal conversions. Examples are a modem or TNC.

DSP Digital signal processing. The representation and manipulation of signals using numerical techniques and digital processors as opposed to analogue electronics.

DTE Data Terminal Equipment. Equipment comprising the data source, data sink or both. Examples include the computer serial port.

Digipeater DIGItal rePEATER. An ordinary packet radio station used for repeating packets, enabled by a command in the ROM. See also 'Node'.

Digi Short name for a digipeater.

Downlink The signals or data sent to you (downloaded); these are the signals you listen to or the file you take. Alternatively, the data you receive from the BBS. See also 'Uplink'.

Emulator A program running on one computer that mimics the action of a different computer or device. For example, the PC emulator for an Atari or Archimedes machines should enable the owner of one of these machines to run some PC software.

EOF End Of File. A marker or other indicator to show that it is the end of the file. It is usually transparent to an operator. Examples used in packet radio are ^Z (Ctrl-Z, ASCII 26) or /EX.

EOT End of Text. A marker to show that it is the end of the (transmitted) text. It is character 04 in the ASCII character set.

EPROM Erasable programmable read-only memory. A type of ROM which may be re-programmed. It is more usually seen as having a small quartz 'window' above the active elements through which UV light is passed to erase the contents prior to loading a new program. Data is not lost when power is removed; it is said to be 'non-volatile'. There are also 'electrically erasable' PROMs but they are not as common. Warning: these devices are sensitive to both static electricity and UV light, so keep the cover on and take the necessary precautions.

File A number of sequential bytes, held in memory or stored on a disk, which may make up some entity understood by the particular application program. It could be a program or part thereof, or a simple letter or even a picture.

File compression A process where a file is reduced in size for storage or transmission. It originated in a requirement to pack more information on a single disk, or to 'archive' sections of a hard disk. It may now be used as part of a security system by use of passwords. See also 'Self-expanding file'.

Firmware The name usually given to software supplied in a ROM, usually for a TNC. This is as opposed to other supply methods, such as files on a disk.

Frame An arbitrary unit making up a packet, consisting of different types, sent over the packet network.

G-TOR An HF data mode originating from Kantronics and available in the KAM controller. It features faster throughput than Pactor under good conditions. The letter 'G' refers to Golay forward error correction, which provides sufficient redundancy in the transmitted data to enable correction of simple errors without re-transmission. G-TOR can use speeds of 100, 200 and 300 baud.

Handshaking The process whereby the TNC and its computer exchange acknowledgement of each others' existence and functions. This may take the form of control codes sent with the data, called 'software' handshaking, or on control lines separate from those carrying the data, called 'hardware' handshaking. Hardware handshaking is used by some software in order that neither the binary file nor the transfer process are corrupted by control codes. Software handshaking may be used for most applications like normal messages and the transfer of text files which are ASCII.

Hex, Hexadecimal A base-16 numbering system used in a computer, where memory addresses are often quoted in hex numbers. The numbers run:

Hex	0	1	2	3	4	5	6	7	8	9	A	B	C	D	E	F
Decimal	0	1	2	3	4	5	6	7	8	9	10	11	12	13	14	15

The numbers are indicated as being hexadecimal in a number of ways. The simplest is hex or h (14h). However, for historical reasons you will often find the dollar sign, $, used ($14), particularly in American manuals. You may also occasionally see &H used to precede the hex number (&H14), a notation often used in programming.

KA-Node A simple node networking scheme where acknowledgements are passed between connected stations, not end-to-end. It was developed by TNC maker Kantronics and is similar in operation to a digipeater.

KA9Q NOS KA9Q Network Operating System. A TCP/IP program originally developed by Phil Karn, KA9Q. There are many different versions available for a variety of computers. The program (also called 'NET') is one of the most commonly used version of TCP/IP in packet radio. NOS was originally written for the PC compatible, but it has been ported to many different computers such as the Amiga, Macintosh and others systems like Unix.

Keps Satellite-tracking computer software use Keplerian elements (also known as 'orbital' or 'tracking' elements or 'Keps') to pinpoint the location of a satellite (or shuttle) at any given time. Keplerian elements, named after astronomer and mathematician Johannes Kepler, provide the software with a snapshot of a satellite's orbital track, which the computer uses to calculate the future whereabouts of the satellite. Using such a computer tracking program allows an observer to determine when a satellite is to appear above his or her horizon. Note: This information may also be on your local BBS as a simple message (TO KEPS @ GBR, for example). Some BBS software, such as the FBB, has the capability to display the predictions of a chosen satellite; yours may also include Shuttle data. Ask your sysop for details.

KISS "Keep it simple, stupid." Originally an admonition to servicemen, but now denotes (in the packet radio field) a simple interface developed for communications between the TNC and its host computer. Commonly used with TCP/IP, but increasingly with more conventional AX.25 packet programs (GP, SP etc), particularly in Germany.

Modem Short for 'MODulator/DEModulator'. In the packet radio application, it is usually part of the TNC. A device which converts digital data into audio tones for transmission (called 'base band' technique). A unit of similar nature at the 'other' end reverses the process. On its own it is used on the telephone network for access to the usual BBS facilities thereon. Modern high-speed devices do not use 'base band' techniques but employ complex modulation techniques, which are like those seen on some 9600 baud AX.25 transmissions. This has been described as 'modulated noise', but it sounds like a 'hiss'.

Multiplexing A technique whereby one circuit can carry several messages simultaneously. It was first developed by Emile Baudot in 1872.

Multi-tasking The process whereby several 'jobs' being executed by the CPU are apparently done simultaneously. Real multi-tasking would require multiple processors. It usually relies on a fast CPU executing instructions so quickly that there are periods of time where there is a pause (for example, something external is expected, like data from the serial port or a keystroke), and the CPU can get on with something else in that time. Software for the PC exists which purports to make it a multi-tasking system. Under these circumstances, a fast CPU and lots of RAM are vital.

NET/ROM A commercial software for packet radio networking nodes.

Node The digital equivalent of a voice repeater. It is a device that links two or more stations together, which enables messages to be passed over long distances between connecting stations. The packet network is made up of a multiplicity of nodes. Often a network node runs NET/ROM or similar software (TheNet etc but there are other types of node).

PacTOR This data mode was designed specifically for use at HF with a faster throughput than AMTOR, with a full 8-bit character set, and with more reliable error detection. It features built-in data compression which activates automatically when advantageous, and dynamic speed changes between 100 and 200 baud depending on conditions.

PAD Packet assembler/disassembler. A device in a TNC which converts the terminal's flow of data to and from packet frames and handles call set-up and addressing.

Parity A technique of error-checking in which one extra (parity) bit is added to the bits that make up the character, so the number of '1' bits per character is always even (or, in some cases, odd). See also 'CRC'.

Port The name given to a facility on a computer where signals may pass, such as 'serial' or 'parallel'. It is under the control of the I/O controller within the computer. The serial port passes bi-directional data bit-by-bit down one wire at one time although there may actually be several other wire connections involved at any one time (see also 'Hand-shaking'). The parallel port passes bi-directional data byte-by-byte on a 'bus' of eight wires simultaneously; for example, to the printer.

Protocol A set of rules governing information flow in a communication system. In practice, it includes electrical, electronic, software and hardware specifications.

Public domain Publication or software which is not subject to copyright and which may be freely copied, modified or re-written. Not the same as 'freeware' which may still be copyright even if the author allows it to be copied.

RAM Random access memory. Also known as 'read/write memory', because it can be written to and read from. The memory in a computer (or TNC) which stores or supplies the program or data in use. This data is lost when power is removed, ie it is 'volatile'.

RESET An immediate command sent to the TNC to put all the parameters back to their default values and re-initialise it. This can be made easier by uploading to the TNC a simple ASCII text file containing all your settings.

RESTART A command sent to the TNC to re-initialise it as if the power supply had been turned on. Previously entered or stored parameters are maintained.

ROM Read only memory. An array of memory cells organised in such a way as to present a program to the CPU. This memory cannot be written to, only read from. It is usually non-volatile (data is not lost on 'power off') and comes in different types and capacities. See also 'EPROM'. Not to be confused with CD-ROM, which is a compact disc holding up to 600Mb of read-only binary data.

RS-232 (RS-323C, EIA-232-D) The standard for interconnection of serial peripherals to all computer systems. In packet radio, RS-232 is the most common physical interface between TNCs and the computer or terminal.

Self-expanding file Many programs are now supplied as a single 'self-expanding' file. This usually involves placing all the files of a program (and there can be many) into a single, compressed-format file. When the file is run for the first time, it will 'un-compress' itself into its constituent parts, sometimes into the appropriate directory. Methods include LHArc and PKZip.

Shareware Commercial software not generally available via retailers. You usually get an early version which should be registered with the author. You may then get the most recent version together with full manuals etc, although this may depend on the source. Costs are markedly lower than retail software and some of it is very good.

Software The instructions or program routines which cause a desired action in the hardware. It is usually found in ROM or is loaded from some exterior source (eg disks or ports) and put into RAM, from where it is executed. Some is in the 'public domain', which see.

Start/stop A serial data transmission system which contains control information in the data stream. In packet practice, this may be described as 'software' handshaking. The term owes its origin to asynchronous transmission, where there is no additional timing information.

Synchronous transmission A method of transmission that allows data to be sent at regular, timed, intervals: one end makes a request and waits a finite period of time, then reads the reply. The system usually relies on an external clock for timing. See also 'Asynchronous transmission'.

Sysop SYStem OPerator. The person who runs your local BBS. Treated with normal respect and politeness, your local sysop can help you a lot, but remember he/she does it for the love of it.

System clock PCs and TNCs have a clock chip which keeps real time and date data which is maintained by a small rechargeable battery. This data is fed automatically to a TNC to update the time and date data by some software. See also 'Clock'.

TCP/IP Transmission Control Protocol/Internet Protocol. Two main network communication protocols, part of the commercially used Internet Protocol suite (used on, for example, the Internet). Another method of sending data used by amateurs which relies on a completely different system technique to AX.25. The TNC used *must* have KISS facilities, because all the hard work is done in the computer. The full TCP/IP suite contains different transmission facilities such as FTP (File Transfer Protocol), SMTP (Simple Mail Transport Protocol), Telnet (remote terminal protocol), and NNTP (Net News Transfer Protocol).

Terminal A device for sending and receiving data. Also sometimes known as a 'VDU' (visual display unit).

Terminal emulation A program within the computer to enable the computer to act as a terminal.

Time out Several processes in packet radio are controlled by internal timers which are reset by some expected event or other, such as the reception of a packet. When the expected event does not happen within the stated time, it is said to 'time out' and some other action is taken, such as automatic disconnection.

TSR Terminate and stay resident. A type of program running on the PC which remains active despite being 'switched off' from immediate use. Only a small 'hot key' section of the program is retained in memory which calls the main programme when activated.

Uplink The transmission of signals to the remote station, eg you 'upload' data to your BBS or even a satellite.

WA8DED firmware A program, usually in ROM, which enables a different method of achieving AX.25 data transmission. It has nothing to do with TCP/IP. It may be found in the standard ROM supplied with some PacComm TNCs (for instance, Tiny-2 Mk2) and it is accessed by use of a simple push switch as an alternative to the TAPR command set. It uses a different method of controlling the TNC. Used by such software as Graphic Packet.

YAPP Yet Another Packet Program. A protocol devised by Jeff Jacobsen, WA7MBL, to permit the transmission of binary and ASCII files between packet stations. It uses the TNC in 'Transparent' mode, and requires hardware handshaking to control the functions of the TNC. It is also the name of a program which acts as a terminal emulator. Many packet programs incorporate the YAPP protocol in some form (enhanced or not).

44 net The Class A network designator for TCP/IP amateur packet radio. All numerical TCP/IP addresses are in the format of: 44.131.xxx.yyy. The '44' represents the amateur network, and '131' the country (or state in the USA), 'xxx' is the region code and 'yyy' the user number.

7PLUS A program for encoding and decoding eight-bit (binary) files to seven-bit text so that they can be sent around the packet network. This widely-used program was devised and written by Axel Bauda, DG1BBQ.

---

out the same bulletin to different TO fields, which is a bit of overkill best left for dead. Examples include one to each of PC, IBM, HELP, and DISK, all with the same text content, seeking assistance with a disk problem on a PC. The network has enough traffic without cluttering it up unnecessarily. Pick the one you think best suits the problem and use that. If you don't get any satisfaction, try again later. Again, it is inadvisable to use 'special' characters not normally visible, such as the smiling face, which is Ctrl-C, since this is a command and some software may try to accept it.

You don't have to put an '@' in at all, in which case the message will stay on your local box for local users to read. Avoid @ EU unless important and you know that the answer is likely to come from Europe. As for @WWW (note that it is 'WWW' not 'WW'), it must be really vital for that to be put on. There are, however, good examples including QSL information and HUMOUR @ WWW, which, by the way, is not usually appreciated in France. Even with the appearance of satellites which carry packet traffic, there's still a finite limit to the volume of traffic which can be borne by the global network. Think carefully before using it. Some mailboxes have a server facility for '@ REGION' addressing, which is of use if you want to locate something in a particular area. Ask your sysop for details.

The subject field (title) is another trap for the unwary. Ideally, both TO and SUBJECT should work together to illustrate the nature of the message. Obscure titles are more often ignored than read, so think carefully. You know what you want, but does your subject field make it clear? The maximum length is 29 characters, but don't go on typing as if you are in the message part already. Words may be shortened: Please = Pse, HELP = HLP etc.

Another well-known horror is to see SB ALL @ GBR (or similar) in the subject field. A little thought can go a long way to getting a useful reply. What can be more aggravating to see than:

```
HELP WANTED: TRANSISTOR TYPE
```

or

```
Hi, I am looking for help wi
```

It is also worth noting that BBS software refuses a message without something in the SUBJECT field.

The actual layout of the message is important. It can be surprisingly difficult to read something upon which little or no visible care has been taken and it's frustrating when you're reading it in an effort to see if you can help him! Set your line width to something less than 76; anything longer may cause the reader to see a short extra line printed below each transmitted line, which can make it

difficult to read.

If your text editor features wordwrap, turn it ON.

TCP/IP

The AX25 standard (which is what 'normal' packet radio is all about) can be used to transmit data of a type more often seen in big networks or even the Internet. The protocol is called *TCP/IP* and the software most seen for its use by amateurs is usually based on the KA9Q Network Operating System (NOS). Compatibility is ensured by placing an AX25 header before the TCP/IP data. As it is an 8-bit system, use of software such as UUENCODE or 7PLUS is not usually required. Hardware handshaking is a MUST.

A station wishing to make a start on TCP/IP will need to contact the local IP coordinator to get a number which is unique to that station. The number, which is the a form known as a 'dotted quad' looks exactly like the IP addresses one sometimes finds on the Internet, eg [44.131.xxx.yyy] where x and y are the area and personal numbers of the station; (the 44 means is is an AMATEUR address and 131 is for UK operation). Convention has it that the square brackets [ ] are used when quoting the number.

Armed with this and suitable software such as JNOS or TNOS, the TNC is put into KISS mode and the PC software does all the work. It is a steep but satisfying learning curve which features short-hand command entries more often seen in UNIX or LINUX but it does have a great many uses and features. By using this method, connection to a wide variety of stations is possible, including direct keyboard-to-keyboard contacts to many stations around the world. Some overseas BBSs have access via the Internet and it is possible to access one and leave a TELNET message to be picked up by an amateur station via his packet system.

There are an increasing number of BBS and nodes dedicated to its use and it is well worth enquiring about it after you are reasonably familiar with the usual AX25. TCP/IP might be fairly seen as 'the next step' between normal AX25 packet and other Internet complexities.

THE BRITISH AMATEUR RADIO TELEDATA GROUP

This society, affiliated to the RSGB, is a specialist group for the data enthusiast. It offers a quarterly journal, *Datacom*, organises contests, an annual rally, an awards scheme, publishes a range of books and does lots more. For more details, contact the membership secretary at the address given in the current edition of the *RSGB Yearbook*.

FURTHER INFORMATION

[1] *The BARTG Guides to RTTY/Amtor/Pactor/AX25*, British Amateur Radio Teledata Group.
[2] 'Data Comms' column in *RadCom*.
[3] *Packet Radio Primer*, Dave Coomber, G8UYZ, and Martyn Croft, G8NZU, RSGB.
[4] *Practical Guide to Packet Operation in the UK*, Mike Mansfield, Compaid Graphics.
[5] *Your Gateway to Packet Radio*, Stan Horzepa, ARRL.
[6] *AX.25 Amateur Packet-Radio Link-Layer Protocol, Version 2, October 1984*, ARRL.
[7] *NOSintro, TCP/IP over Packet Radio*, Ian Wade, G3NRW, Dowermain Ltd, 1992, ISBN 1-897649-00-2.

9 Image techniques

THE term *image techniques* as applied to amateur radio encompasses quite a wide selection of differing modes, which include in the main fast-scan television as in commercial broadcast television (FSTV), slow-scan television (SSTV), facsimile (fax) and remote imaging, which is essentially the reception of weather satellite images. However, this latter mode is more to do with the reception of extremely low-level signals, rather than with the technicalities of television or image techniques themselves.

SSTV and fax are narrow-band modes, which means that signals in these modes can be transmitted on normal voice channels. You can thus reach with SSTV and fax signals wherever you can reach by voice mode; in other words, on appropriate bands they are worldwide communication modes.

FSTV, on the other hand, is a wide-band mode, and as such is essentially a 'local' communication system although, when propagation conditions permit, relatively long-range contacts can be established over several hundreds of kilometres. However, it must be said that the quality of received pictures may not always be up to broadcast TV standards.

An exciting new mode of image transmission for the radio amateur is digital amateur television (DATV) and a basic introduction into this state-of-the-art mode will be discussed here.

SLOW-SCAN TELEVISION

Slow-scan television has changed greatly since the idea was developed by Copthorne Macdonald and first used by radio amateurs many years ago. Nowadays, we have a plethora of systems within the mode and a variety of equipment that can be used to transmit and receive SSTV pictures. The original concept behind SSTV was to find a method by which a normal wide-band television picture could have its bandwidth reduced, so as to allow its transmission over a single-channel voice communication system. This means that a typical 6MHz or so wide colour television signal has to be reduced to around 3kHz – around a 2000 to 1 reduction in bandwidth! Because of this severe narrowing of the bandwidth, the system is only suitable for the transmission of still pictures – moving images are still the preserve of FSTV, but only until digital television becomes the norm.

To reduce the bandwidth of a television signal, both the horizontal (line) and the vertical (frame or field) scanning rates must be reduced to as low a frequency as possible. A convenient stable source from which to obtain these low scanning rates is the domestic mains supply, which is 50Hz in the UK and 60Hz in the USA and elsewhere.

The basic parameters of a slow-scan signal are shown in Table 1, from which it can be seen that the line rate of 16.6MHz is easily obtained by dividing the mains frequency by three and the frame rate by diving the mains by 360. This system resulted in a picture made up of 120 lines. However, with the

Table 1. SSTV standards

Parameter	50Hz mains	60Hz mains
Line speed	120 or 128	120 or 128
Frame speed	7.2s or 7.68s	8s or 8.53s
Aspect ratio	1 to 1	1 to 1
Scanning direction:		
Horizontal	Left to right	Left to right
Vertical	Top to bottom	Top to bottom
Sync pulse duration:		
Horizontal	5ms	5ms
Vertical	30ms	30ms
Subcarrier frequency:		
Sync	1200Hz	1200Hz
Black	1500Hz	1500Hz
White	2300Hz	2300Hz
Required transmission bandwidth	1.0–2.5kHz	1.0–2.5kHz

wide use of computers in SSTV use it is more common to have pictures made up of 128 lines, which translates to a nicer binary number, ie 10000000.

The basics of a slow-scan picture

The composition of a single picture using the original specifications in Table 1 is shown in Fig 1. A complete picture line is 65ms long, with a 5ms line synchronisation (sync) pulse and 60ms of video information. The line sync pulse is the method by which the receiving system determines when to start 'writing' a new line. The frame pulses, which appear at the start of every 120 (or 128) lines of video information, are 30ms long, in order for the receiving system to recognise them apart from the line sync pulses. The frame pulses tell the receiving system when to start 'writing' a new picture from scratch. Unlike conventional TV the aspect ratio was set at 1:1, which results in a square picture. The reason for adopting this standard originally was the use then of round ex-radar display cathode ray tubes in the receiving system display monitor.

The base-band (audio) video signal is modulated onto a subcarrier, which helps to overcome phase shift and drift problems in SSTV demodulators. Obviously, this subcarrier had to be within the audio band, otherwise it could not be transmitted over the audio speech channel. With reference to Fig 1 it can

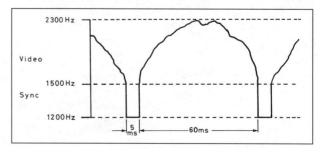

Fig 1. Frequency composition of a single slow-scan line

be seen that the basic frequency representing the black level (no video information) is 1500Hz, which rises to 2300Hz representing peak white. The various tones, hues and grey levels are translated into frequencies within this spectrum.

The sync pulse frequency is 1200Hz, which is 'blacker-than-black', and thus the visible raster is blanked out when the spot moves to the beginning of the next video line or picture frame, which is normally termed the *retrace*.

Receiving slow-scan television

SSTV signals can be received using an ordinary communications receiver or transceiver covering the amateur bands. Table 2 shows the amateur bands where SSTV is to be found, the IARU recommended working frequencies and the popular operating spots. The mode of transmission may be either FM or SSB. No modifications are required to the receiver, although the internal IF filter should have a bandwidth not less than 2.5kHz and ideally 3kHz. The SSTV baseband signal is extracted either from the headphone or line-output jack of the receiver/transceiver.

The SSTV base-band signal must now be further processed in a purpose-designed demodulator system. Essentially, over the years, there have been three systems for demodulating and viewing SSTV signals:

1. A monitor containing an integral long-persistence cathode-ray-tube (CRT), together with signal processing and deflection circuits.
2. A digital scan-converter in which the received signal is digitised and stored in a memory. This memory is then scanned at fast-scan (625 lines in the UK) rate for displaying on a conventional TV or monitor.
3. A computer which processes the picture either directly or via a hardware interface, depending on the software system in use.

This latter method is the most widely used now, particularly with the advent of full-featured, cost-effective software packages for the IBM PC, such as the JV-FAX package.

The SSTV monitor

The SSTV monitor, as outlined in the block diagram in Fig 2, is now only probably used by the SSTV purist, as it requires a great deal of home construction and the resulting pictures are perhaps a little disappointing! In simple terms the incoming SSTV signal is FM detected, resulting in an AM signal, but still consisting of a subcarrier centred around 1500Hz. The two sync signals are recovered by using a tuned-circuit sync discriminator.

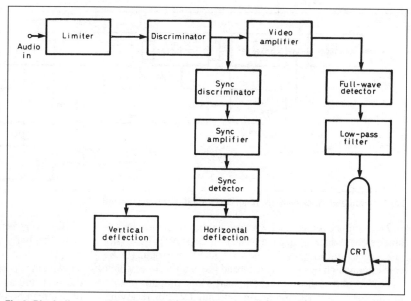

Fig 2. Block diagram of a slow-scan TV monitor

The AM subcarrier signals are detected, often by using simple full-wave rectification.

The monitor deflection circuits are controlled by the sync pulses using similar methods to those employed in broadcast TV receivers. The nature of the long-persistence CRT used in an SSTV monitor requires that attention is paid that the scanning spot is not bright enough to burn the phosphor. In more sophisticated circuits spot suppression is included to blank out the raster, and thus the scanning spot, in the absence of received signals. The detected base-band video information is used to bright up the CRT as with a conventional TV, the principles of which are not being discussed here.

The results obtained from a slow-scan monitor are rarely comparable to those received by more modern systems. Because of the reliance on the long-persistence properties of the CRT often as not the top half of the picture fades out before the second half has been scanned in. Consequently, it is often necessary to view the received pictures in a darkened room. Also, there is no direct way to received colour pictures on a slow-scan monitor.

The digital scan converter

For many years the most popular system for decoding SSTV pictures was the digital scan converter. The popularity of this type of equipment has only been surpassed in latter years with the advent of versatile computer-based systems. A simplified block diagram of a digital scan converter is shown in Fig 3. A distinct advantage of digital scan converters is the ability to use a standard domestic TV set or video monitor to view the received pictures.

The incoming audio signal from the receiver is firstly limited and then passed through a sync detection circuit. The remaining picture information signal is fed to an analogue-to-digital (A-D) converter, where each line of information is converted into 128 4-bit binary words. Two 512-bit shift registers act as odd and even line buffers, the necessary switching being carried out by a clock signal generated from the sync pulses. While a line is being read into a buffer the preceding line held in the other buffer is read out into memory. Without doing all the maths here, suffice it to say that a 16k memory is required to store a 128-line by 128-pixel picture.

Table 2. SSTV frequencies

Band (MHz)	IARU recommended frequency (MHz)	Popular frequency (MHz)
3.5	3.735	3.730
7	7.040	7.040
14	14.230	14.230
21	21.340	21.340
28	28.680	28.680
144	144.500	144.500

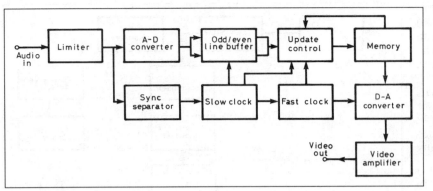

Fig 3. Simplified block diagram of an SSTV digital scan converter

The memory is continually scanned at fast scan rates by another clock, which is also controlled by the sync pulses. This fast-scanned information is fed to a digital-to-analogue (D-A) converter, which results in a base-band fast-scan picture, which can either be fed direct to a composite video monitor or to a modulator for viewing on a conventional TV. For receiving colour pictures three 16k memory banks are required and also further circuits for detecting the extra control signals present in a colour SSTV signal that are used to inform the receiving system which of the three colours are being transmitted (a colour SSTV signal is transmitted as three separate signals, red, green and blue).

An immediate advantage of this type of system is that there is always a picture on the screen. As each new picture is received it is slowly scanned in over the preceding one. Consequently, a picture once received can be viewed on the screen for as long as you wish, without it slowly fading into the west!

As mentioned earlier, the aspect ratio of SSTV pictures is 1:1, whereas with FSTV it is currently 4:3. If no correction was carried out on the received picture then it would appear distorted on the screen. To overcome this problem the converter is configured such that the first and last eighths of each line are blanked out during the fast-scan memory read cycle, resulting in a square picture of the TV/monitor screen.

Computerised SSTV

There have been computerised SSTV systems for just about all makes of computer, but they are now mainly developed for the PC. Most systems are full SSTV transceivers, although there have been one or two receive-only or transmit-only packages. Some of the earlier systems require quite complex signal-processing circuitry that plugs into an expansion port on the computer, some of the PC-based ones utilise a purpose-built modem that plugs into an internal slot in the computer and some have very sophisticated software routines that require only a very simple A-D converter that plugs into a serial port for example. Further information on computers and SSTV will be examined in the following section dealing with transmitting SSTV.

Transmitting SSTV

Sending SSTV pictures over the air requires nothing special as far as the radio equipment is concerned – the video information on its 1500Hz subcarrier is simply fed into the microphone socket of the transmitter. The band on which you are operating largely dictates the mode of emission, although both FM and SSB are used on VHF.

No modifications to the transmitting equipment are required. However it must be borne in mind that if using an SSB mode of emission, unlike normal speech which has a low duty cycle, the SSTV signal has a 100% duty cycle. Therefore, the transmitter must be capable of producing full power continuously on SSB when transmitting SSTV.

Picture sources

Nowadays, the most common picture sources are computers, and it is not hard to see why with the graphics capabilities of modern machines. Also, with relatively inexpensive image digitisers now available, camera pictures can be brought into the computer as SSTV pictures.

However, there have been and are still other picture sources available. Early SSTV stations built and employed SSTV keyboards, which were essentially electronic typewriters which allowed the operator to type messages which were output from the keyboard as a standard SSTV signal. The main drawback of using an SSTV keyboard, however, is that it is a text-only mode, and television is of course a visual medium.

Digital scan converters have been the most widely used method of generating SSTV pictures for many years. The principle is the same as for the receive converter but in reverse. In actuality, of course, only one digital scan converter is used, it being in fact a transverter. The advantage here is that circuitry is incorporated which enables a video camera to be plugged in, allowing the user to 'snatch' a picture for transmission.

Other equipments have been used over the years for generating SSTV pictures including:

1. Slow-scan and sampling cameras, where the camera vidicon is scanned at slow-scan rates to produce the SSTV signal from the fast-scan picture the camera is looking at.

2. Flying-spot scanners – where a photographic slide of the required picture is held in front of a CRT being swept at slow-scan rates, causing a spot to move over the CRT screen exactly as if it was scanning the picture in. In front of the CRT and slide is located a photomultiplier tube, which reacts to the varying light intensity transmitted through the slide from the moving spot on the CRT. Further circuitry processes the output from the photomultiplier into the standard SSTV signal.

Commercial systems

Much of the equipment discussed here bears a great deal of home-brew construction. There are several commercially built slow-scan systems available, both new and second hand. The major makers, both current and past, are Wraase Electronik, Davtrend Ltd and Robot Research. Sadly, both the latter two are no longer in production, although examples of Robot digital scan converters are often seen advertised in *RadCom* and elsewhere. However, with the recent advent of extremely versatile computer-based slow-scan systems, it is clear that the era of the hardware-based digital scan converter is coming to an end.

Slow-scan television, as it has developed over the years, has collected along the way a vast array of variations on the original theme, with many differing modes, speeds of transmission and internal protocols. This in itself has been good for slow-scan, because it has helped to maintain the interest in this specialist mode. However, it has also produced many problems, for example Robot Research with their large range of scan converters developed their equipment using strict protocols – the Robot Modes. Similarly, so did Wraase Electronik and all the other mainstays of SSTV over the years. This presents the SSTVer with a problem – generally speaking, for example, you can only receive Robot mode pictures using Robot equipment, or equipment specifically built and configured for Robot Modes. Thus there has developed over the years fractional elements within the SSTV side of the hobby that may have done a disservice to the mode.

Never fear though, all is not lost. With the advent of the PC in particular, SSTV systems have been developed for the computer, using its mode-independent versatility and immense processing power, which allow the user to receive just about any mode, speed and protocol of SSTV picture being transmitted, and what is more transmit pictures similarly in any format.

The age of the affordable high-specification computer has seen an upsurge in the interest in SSTV around the world. Whilst it is true to say that systems are available, and have been for many years, for other computers such as the Sinclair Spectrum, the Dragon, the Commodore series, the Amiga etc, it has only been the arrival of the PC in the amateur shack that has revolutionised SSTV as well as other amateur radio modes.

The two main systems that have been available worldwide are the software based JV-FAX package and the Pasokon TV system, although this system is no longer readily available other than on the second-hand market.

The JV-FAX package is at present the most widely used system in SSTV. It is essentially a powerful software package that requires only a very simple interface between a serial port of the PC and the microphone and headphone sockets of the transceiver. The software does not take long to master and allows a variety of slow-scan modes, speeds and protocols to be received and transmitted. The package is virtually shareware and as such is obtained at very low cost, complete with all information and a pre-built interface if required. A UK source for JV-FAX is given in reference [1].

Pasokon TV is also a very versatile PC-based SSTV system, but differs from the latter in that it has a dedicated modem which requires to be plugged into a spare 16-bit slot in the computer, connection to and from the transceiver being made to a nine-way plug at the rear of the card. The software package is extremely versatile and allows transmission and reception in all the current modes. The software has built-in selectable error correction and colour processing modes, and a graphic user interface for mouse operation. Whilst the package is excellent in operation and hard to fault, it is much more expensive than the JV-FAX system and perhaps a little less versatile in future expansion than a wholly software-based system.

Further details of the various SSTV modes, speeds, formats etc, plus full construction details on various units for operating SSTV can be found in the British Amateur Television Club (BATC) publication *Slow Scan Television Explained*. The BATC is an RSGB affiliated society which publishes its own quarterly journal *CQ-TV*; membership information etc is given in reference [3] at the end of this chapter.

How to start an SSTV contact

To start an SSTV contact simply call "CQ SSTV [your callsign] CQ SSTV" on one of the frequencies shown in Table 1 and then, once a contact has been established, change frequency to a mutually agreed one in the SSTV portion of the band. Decide who is to transmit first and establish the speed and mode to be used. Remember that it is not only good operating practice, but also plain good manners, to call "Is this frequency in use?"

A SSTV picture received using the JV-FAX software

An attractive SSTV station identification picture

SSTV picture received off-air from IT9MRW

before calling CQ and moving there after a contact has been made.

Often you will come across existing contacts or nets taking place. If this is the case, await a gap in the transmissions, announce your callsign on the frequency and await to be called in.

FACSIMILE

Facsimile (fax) is the process by which graphic information is converted into electrical signals which are transmitted by cable or radio and then reproduced exactly as the original. Until a few years ago the only method of operating fax available to the radio amateur was by converting ex-commercial equipment, which was essentially electromechanical in nature. Whilst it is still possible to obtain, modify and use this equipment today, most amateur fax operation is conducted using computer-based systems. Software packages are available for most of the popular makes of personal computer, including the JV-FAX package for the PC as described earlier.

Much of the amateur activity with fax is actually in the reception of the various weather and news fax services around the world, rather than by reception and transmission of this mode between amateur stations.

FAST-SCAN TELEVISION

Amateur fast-scan television (ATV) is perhaps one of the more fascinating and self-construction oriented modes in amateur radio today. Whilst there is an amount of commercially made equipment for ATV, this is mainly in the form of kits or units for the microwave bands. The starting place for all ATVers is 70cm, where relatively simple equipment can be built and contacts can be easily made. It must be mentioned here that it is not possible to use existing radio equipment for fast-scan television. Transmitters and receivers must be specially commissioned for this mode, whether it be for AM television on 70cm or FM television on 24cm or the microwave bands.

Fast-scan television is essentially a short DX mode, in that the average contact distance on 70cm is probably of the order of 30 to 50 miles or so for good-quality received pictures. It must be noted here that for a received picture to be of good, relatively noise-free quality, the received signal strength must be in excess of 20dB over S9 – much higher signal strength than for a noise-free phone contact!

Fast-scan ATV is, to all intents and purposes, identical to standard broadcast television, with a few relaxations of some of the more-stringent specifications. In the UK, amateur television transmissions may be found on several of the UHF and microwave bands, the principal one traditionally being the amateur 70cm allocation (amplitude modulation around 436MHz – approximately channel 17). The 23cm band is where the ATV repeaters are to be found. Table 3 shows the various bands and frequencies where ATV activity may be found.

AM is used on the 70cm band, but on the higher bands FM is preferred because of the higher quality of pictures that can be sent and received and because of the better attributes of the signal-to-noise characteristics of FM transmissions. On 70cm only the video content is transmitted, with the audio channel being usually on 2m, nominally 144.750 which is the UK ATV calling channel. However, on 24cm and above, because of the much greater bandwidth available, ATV signals also carry the sound channel, just as with broadcast TV. So an ATV contact on 24cm can be, and usually is, conducted totally on the one band, without the use of the 2m talkback channel.

If portable operation appeals to you, then try 10GHz ATV, where by converting simple doppler microwave burglar alarm units, or commercial satellite TV receivers, a simple, but very effective, FM ATV station can be built.

An amateur TV station need be only as complex as the individual desires. Many stations employ no more equipment than a camera, transmitter and a receiver. This basic system is

Table 3. Popular ATV frequency allocations

Band	Centre frequency (MHz)	Mode
70cm	435.4	AM
23cm	1249.0–1265.0	FM
	1316.5	
13cm	Not allocated	FM
3cm	10,000.150	FM

adequate for normal communication and has the advantage of being easy to use and maintain. The block diagram of such a basic amateur TV station is shown in Fig 4.

Amateurs who do not wish to put together a complete TV station often set up a simple receiving system in order to 'eavesdrop' on ATV pictures in their area. This can be very rewarding, especially if they live in a high-activity area, and particularly when lifts in propagation conditions occur. Looking-in on a local ATV repeater is a common pastime for many stations.

Transmitters and receivers

Full information on constructing simple ATV transmitters and receivers can be found in the RSGB *Radio Communication Handbook* (7th edition) and BATC publications *An Introduction to Amateur Television* and the *ATV Compendium* [3].

The simplest ATV station is a receive-only one. All that is required to receive 70cm fast-scan TV (FSTV) is a receive converter and a normal television set. The receive or up-converter converts the received signals around 436MHz to usually channel 36 on the TV set (approximately 700MHz). Assuming that you already own a TV set then all you need is the converter and a 70cm beam.

It may be advantageous to use a preamplifier as well but it is not essential for starting out. With some of the Japanese TV sets in particular even the receive converter may not be required, as these sets often tune down to below 436MHz and still have enough RF gain to be useful, although if using this method considerable advantage can be obtained by using a preamplifier. The same is true for a variety of video cassette recorders, whose tuners often tune to the ATV operating area on 70cm.

To transmit FSTV then a purpose-built transmitter is required. Commercially built equipment is generally no longer available new in the UK, but several types can often be found at rallies or in the Members' Advertisements in *RadCom* or *CQ-TV*. The most popular commercial units made were probably the Microwave Modules series. This company produced a range of transmitters for 70cm and 24cm and a matching set of receive

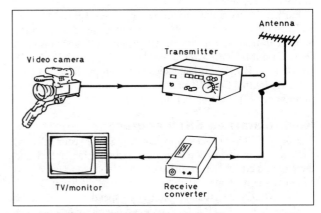

Fig 4. Block diagram of a basic fast-scan station

converters. They also produced a range of suitable linear amplifiers as well, which can often be found at rallies etc.

Another company that produced ATV equipment over the years was Wood & Douglas. Although they mainly marketed kits, much of their equipment can be found built and working offered for sale. A correctly built W&D unit will be as good as anything commercially built.

A specialist in 24cm component units in the past was Camtech Electronics. When new these modules were priced at the high end of the market, reflecting their quality, but nowadays they can often be found for sale and are well worth investigating.

If you are into home construction then Mainline Electronics [4] offer a range of kits, many based on BATC designs, from 70cm converters through to 3cm transmitters. They use well-made quality printed-circuit boards and are easy to construct and align, using the minimum of equipment. You do not need to be an electronic genius or a master craftsman to build these kits. Another source of kits and pre-built 24cm ATV units is the Worthing & District Video Repeater Group [5].

Video sources

The most obvious video source is a camera. At one time the only cameras that were readily available to the amateur were ex-surveillance security black-and-white cameras, that more often than not had an imprint after the fashion of 'Fred Bloggs – Grocer to the Rich and Famous' permanently burnt into the vidicon. However, nowadays many households own camcorders, which have standard video and audio outputs and what is more are capable of colour. Also readily available now are small, compact, LCD colour cameras, with superb specifications and found often at quite low prices.

Please note here that if a colour camera or colour source of any description is going to be used with a 70cm system, then a low-pass filter, cutting off at around 3MHz, should be incorporated between the source and the transmitter. If this is not done then the spectrum of the transmitted signal will cause severe interference with other users of 70cm and will probably also result in out-of-band transmissions.

Another video source easily obtained at rallies, or even from the lounge, is the video cassette recorder – useful for transmitting those wonderfully interesting videos of your holidays etc!

Computers can also be used as video sources, although certain models are difficult to modify so as to extract the necessary composite video and blanking signal (CVBS) – this unfortunately includes the PC. However, plug-in cards are available for the PC which will output the necessary signals from the computer, certain models also having a camera input allowing the camera pictures to be overlaid with the PC pictures and/or text to produce a combined output.

Computers such as the BBC 'B' and the Atari series are easily modified (some models already have a CVBS output) but it should be remembered that because computers generate their pictures digitally, large transient signals may appear at odd frequencies. Thus, if using these older computers, a low-pass filter cutting off at around 5MHz should be used and ideally also an amplitude limited/clipper as well. (PCs using plug-in cards should cause no problems, as the outputs from these purpose-built cards should meet the requirements of correctly filtered outputs etc.)

Test signals provide useful video sources, enabling the operator to correctly set up his camera, monitor, transmitter or receiver. The most commonly used test signal is a test card, of which there are two main types, those printed onto card and

held in front of a camera and those which are digitally generated. Printed card types are the only type that can be used for setting up cameras and are available from the British Amateur Television Club. Digitally generated test cards are available in kit form from Maplin Electronics [6] and provide high-specification colour test cards for alignment or general transmission, the operator's callsign being encoded into the data.

Setting up a simple ATV studio

The term *television studio* tends to conjure up pictures of what we see and imagine is used by the broadcast TV programme makers. However, for ATV use we need nothing so complex, although if your interest in the hobby is the actual production of programmes then you may need something more sophisticated than the following.

A simple ATV station requires only one camera and perhaps a video cassette recorder. Alternatively the family camcorder can be pressed into service, which can double as both and will almost certainly also be colour. If a camcorder is being used then no particular attention needs to be paid to shack lighting. The author can see the purists cringing at that statement, but unless you wish to produce BBC-quality programmes then simple lighting will suffice. As long as the subject area is sufficiently lit to allow the camera to produce a well-contrasted and, if in colour, a nicely colour-balanced picture, then the use of fluorescent tubes is often more than adequate.

The content of the pictures is entirely your choice. However, try to avoid long transmissions with the camera staring back at you. Compose camera shots so that you are off-centre, with a nice shot of the rest of the shack taking centre stage. The rest is up to you.

ATV repeaters

The list in Appendix 9 shows the callsigns, locations and operating channels of the operational ATV repeaters in the UK, and as can be seen they now cover most of the major areas of the country. Apart from the Bath repeater GB3UT and the Hastings repeater GB3VI, which are AM or FM input and AM output, the mode is FM with vertical polarisation.

Essentially, ATV repeaters are much the same as their phone counterparts. However, there are three very important differences:

1. to access an ATV repeater all you need is a fast-scan FM TV signal with the video conforming to the CCIR specifications for 625-line television (ie exactly what you get from your camera, camcorder, VCR etc);
2. there are no time-outs operating on ATV repeaters; and
3. perhaps most importantly for the constructor, when a repeater is not repeating a received picture it broadcasts its own pictures, from test cards, colour test screens, textual information screens, outside masthead cameras shots, etc, all scrolling round in an endless loop. When in repeat mode many ATV repeaters have large selections of options available, from relaying weather pictures to giving signal and picture reports, and in the case of GB3ZZ in Bristol to being able to select which of several receive antennas you are being picked up on.

In other words, ATV repeaters are operational 24 hours a day, either radiating on-board generated pictures or repeating incoming received pictures.

Table 4 gives the channel frequencies of UK ATV repeaters. Repeaters operating with their inputs on RT1-1 will accept either AM or FM signals, whilst inputs on RT1-2 and RT1-3

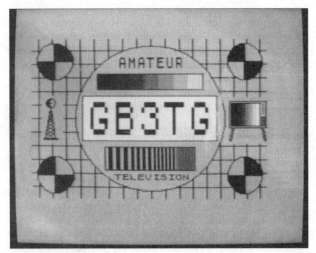

A typical test card – this is one received off-air from the GB3TG TV repeater on 10GHz

Off-air picture recived from G7SEC via the GB3LO TV repeater. Note the picture-in-picture feature

A picture received from G8XTW on 10GHz

Here Belgian and Dutch amateurs are exchanging TV pictures via GB3LO

are FM only. Repeater output on RT1-1 is AM only and the outputs on RT1-2 and RT1-3 are FM only. A Morse code identification giving callsign and location is present on the 6MHz audio carrier. FM transmissions can employ standard CCIR video pre-emphasis as an option.

Although the actual coverage of each repeater will depend on local geography, and in some cases antenna directivity, the average area covered is of the order of a 30km radius from the repeater. Apart from one or two exceptions, antennas are omnidirectional. *All* ATV antennas operate horizontally, regardless of band. Details of ATV repeaters are given in Appendix 9.

The widest frequency difference between an input and an output signal is 67.5MHz. A check on the specification of

commercial antennas will show that most of them lack the necessary bandwidth necessary for single-antenna ATV operation. Fortunately, there are one or two which are suitable for ATV applications: the helix, or helical beam, and the quad-loop Yagi are two of the more readily available designs. An excellent Yagi antenna for 24cm ATV work is available from the Severnside Television Group [7].

How to conduct an ATV contact

So, how do you start an ATV contact? The UK and international calling channel is 144.750MHz FM. If the frequency is clear, then simply put out a CQ call such as "CQ ATV, CQ ATV, [your callsign], CQ ATV, CQ 70cm ATV. [your callsign] listening". Alternatively, if you know an ATV station who may be listening then, as with any other mode, call the station direct. Once a contact has been established it is generally good practice, unless attempting to generate further activity, to QSY up or down 25kHz away fromn the calling channel.

Having established phone contact and lined the antennas up on each other, you then decide who is to transmit ATV first. The transmit station starts sending an ATV picture on 70cm (nominal centre frequency 432.4MHz) and the receive station starts tuning his/her ATV receiver until hopefully a picture is established.

Table 4. UK ATV repeater channel frequencies

Channel	Input (MHz)	Output (MHz)	Mode
RT1-1	1276	1311.5	AM
RT1-2	1249	1316.5	FM
RT1-2R	1249	1316	FM
RT1-3	1248	1308	FM
RT10-1	10,200	10,040	FM
RT10-2	10,255	10,150	FM
RT10-3	10,250	10,150	FM

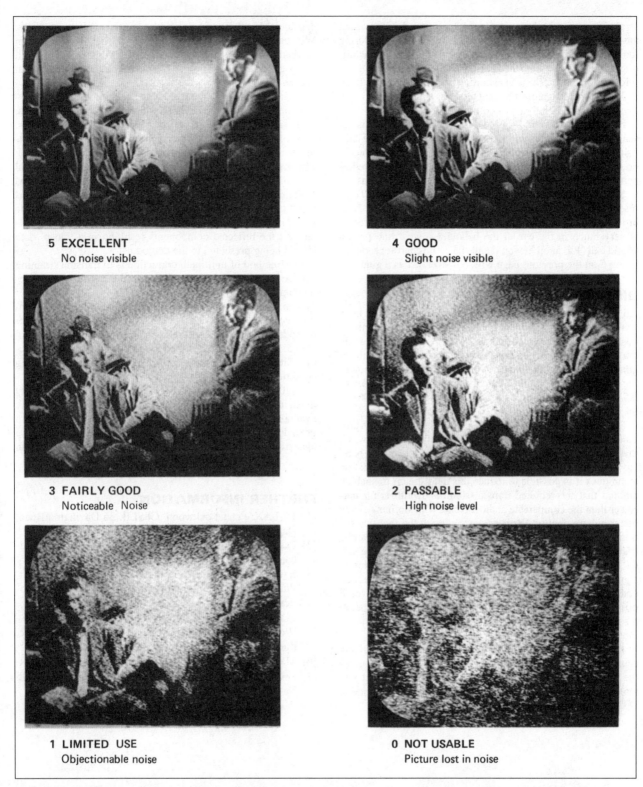

5 EXCELLENT
No noise visible

4 GOOD
Slight noise visible

3 FAIRLY GOOD
Noticeable Noise

2 PASSABLE
High noise level

1 LIMITED USE
Objectionable noise

0 NOT USABLE
Picture lost in noise

Fig 5. The British Amateur Television Club ATV reporting chart for P0–P5 (Reproduced with acknowledgement to the BATC)

It will often be necessary to fine tune the beam headings until the best picture is received. The best method for achieving this is for the transmit station to start swinging his/her antenna first in one direction and then the other, announcing on the talkback channel, say, "Going north, going south", and dropping the carrier on 2m between changes. The receiving amateur watches the TV screen and announces over the talkback channel "Stop" when the received picture disappears for both antenna swings. The transmit station then splits the difference between the two extremes and points the antenna in that direction. The receive station then optimises the antenna direction by peaking for strongest picture.

Having thus established the best path headings, the two stations can enjoy a 70cm ATV QSO with talkback on 2m, swapping pictures, the content of which may be just about anything which complies with the licence conditions. During long periods of ATV transmission, the callsign of the transmitting station must be announced on the talkback channel with the information that he/she is transmitting fast-scan ATV on 70cm, and also the callsign must be sent either as a picture on its own or superimposed on to the transmitted TV picture.

For fast-scan TV on 24cm and above, the procedure is essentially the same, except that once a QSO has been established via 2m then the entire ATV contact may be conducted on the TV band, with the audio talkback being transmitted as part of the TV signal, just as with broadcast trasnmissions.

Signal reports on ATV are given using the P code – P0 representing no picture, to P5 representing a perfect, noise-free, broadcast-quality one. As with RST reports, the exact P report is very much in the eye of the beholder – what one person would call 'P4' another would call 'P5'. The chart reproduced in Fig 5 on the previous page will, however, act as a guide.

DIGITAL AMATEUR TELEVISION

All the modes of amateur television described so far have been analogue systems, in that the image is modulated as a continuously varying voltage and frequency onto a carrier. The main drawback with analogue image systems is the bandwidth required to represent the image with full dynamic range, equating to the full spectrum of colour, nominally of the order of 6.5MHz in conventional systems on 70cm and up to 20MHz on 23cm.

Straightforward digitising of a colour signal initially provides a data stream that requires considerably more bandwidth than the comparable analogue signal, so no advantage here. However, due to modern digital compression and reduction techniques it is possible to reduce this data stream to such an amount that the required transmission bandwidth is far narrower than the comparable analogue one, with similar or even better picture quality. Furthermore, owing to the fact that the transmitted signal is digital as opposed to analogue it can be transmitted over greater distances with good results. The signal is far less susceptible to picture strength variation due to signal reduction and digital technology offers error correction coding that is not possible with analogue systems. The overall possibilities are larger transmission distances with much better resolution of received pictures and virtual elimination of noise, interference and fading.

Rather than reinvent the wheel, modern video digitising standards are adopted: MPEG-1 and MPEG-2. MPEG-1 gives the well-known Video-CD quality with a 1.5Mbit/second data stream, which is more than sufficient for DATV and a

considerable saving of bandwidth over the analogue systems. With the error-correction data added to the MPEG-1 data stream this bandwidth increases to around 2 to 2.5Mbits/s, which is still a very great saving and is independent of which band the system is used on.

A large problem encountered with analogue television transmission is *multipath reception*, which is the reception of the same signal from various reflecting sources (telegraph poles, hills, office blocks, etc.) which causes distortion of the resulting demodulated image giving rise to ghosting (double images). Multipath distortion is even more of a problem with digital transmissions as the resulting summed data stream presented to the demodulation stage would be at best severely distorted resulting in a virtually unrecognisable image. However, as the signal is digital, correction techniques can be employed to 'filter' out the reflected components resulting in a 'clean' data stream being presented to the demodulator.

One method of multipath correction is to transmit 'training data sequences' to the data stream at the transmitter the demodulated results of which are already 'known' by the receiver system. Consequently, any distortion to the transmitted training sequence causes a different output from the receiver demodulator than the expected one and an error signal is generated. This error signal is then used to shape a digital correction filter to optimise the received signal and react to varying conditions.

At the time of publication of this edition DATV is still very much in the experimentation stage and no hardware is currently available. Amateurs in the UK and abroad are carrying out a great deal of experimentation and development and articles are appearing in the BATC publication *CQ-TV* and the German ATV Club (AGAF) publication *TV-Amateur*.

FURTHER INFORMATION

[1] JV-FAX: Peter Lockwood, G8SLB, 36 Davington Road, Dagenham, RM8 2LR.

[2] Pasokon TV: now only readily available on the second-hand market.

[3] BATC Membership enquiries: BATC, Grenehurst, Pinewood Road, High Wycombe, HP12 4DD. BATC Publications: BATC, 14 Lilac Avenue, Leicester, LE5 1FN.

[4] Mainline Electronics, PO Box 235, Leicester, LE2 9SH.

[5] Worthing Video Repeater Group, 21 St James Avenue, Lancing, West Sussex, BN15 0NN.

[6] Maplin Electronics, PO Box 3, Rayleigh, Essex, SS6 8LR.

[7] Severnside Television Group, 15 Witney Close, Saltford, Bristol, BS18 3DX.

10 Special-event stations

FROM time to time groups of radio amateurs organise and operate special radio stations at exhibitions, conferences and other events. These fall mainly into two groups: demonstration stations and talk-in stations.

In the first case the object is to demonstrate amateur radio to members of the general public and the whole operation must be planned as a public-relations exercise. In the second case the purpose is to guide mobile amateur radio visitors to the event and efficient communication procedure and navigational expertise is more important.

GENERAL POINTS

Although quite different in their function, demonstration and talk-in stations do share some common ground in respect of planning matters, and these will be outlined briefly before discussing each type of station in more detail.

Personnel

One of the first steps is for a station manager to be appointed. In turn the manager can then appoint assistants to be in charge of various aspects such as equipment, power supplies, antennas etc. At the same time the identities of possible operators can be established.

Budget

Planning, establishing and operating a special-event station will almost invariably incur some expense, no matter how generous club members prove to be with support in terms of equipment and time. Therefore at the very earliest stage the club honorary treasurer should be asked to draw up a budget. This should include the cost of hire of any equipment (antennas, masts, tables, chairs and motor generators, and tents if the event is being held under canvas). Do not overlook coaxial cable and plugs, fuel for generators (very expensive if long periods of operation are undertaken) and the cost of refreshments.

If possible it is better for the operators to cater for themselves as they can then arrange meals to suit operating times. Sometimes exhibition organisers supply meals free of charge as an expression of gratitude to the amateurs concerned, but it is obvious that where a number of operators are involved for a period of more than a couple of days or so the cost becomes so restrictive that they have to ask for financial assistance.

Other items include postage and telephone charges incurred during the planning period, and gratuities to the venue staff in order to make sure that your antenna is placed at the very highest point possible. QSL cards may need to be printed and maps, felt-tipped pens, pencils and paper, logbooks and notice boards will all have to be purchased if they cannot be begged.

Having drawn up the budget make certain that you have sufficient funds to cover all items.

Security

Amateur equipment is expensive and much of it is readily portable. At the event make sure that all equipment is out of reach of light-fingered visitors. "Out of sight – out of mind" is a good motto to follow, and spare transceivers and accessories can be stored under tables or behind screens. Owners of equipment must make adequate insurance arrangements, and the station manager should discuss this subject with the event organiser.

While on the subject of insurance, cover should be obtained for personal injury to those participating in setting up and operating the station as well to members of the public. Antennas can blow down, electricity can be dangerous and claims could be substantial.

Licensing

Do you really need a GB callsign?

From 1 June 1990, all club stations have been able to pass greetings messages sent by a non-licensed third party. This means that applying for a GB callsign no longer holds any advantages!

Provided the club uses the prefix letters (see below), the club station is allowed to pass greetings messages and to operate simultaneously on more than one band. The club prefixes are very distinctive and create interest through their rarity. If a club regularly operates a special event station using the club callsign, this will help increase the club's identity. It will also benefit by being able to print its QSL cards in larger, more economic quantities.

Old Club Prefix	New Club Prefix
G/M	GX/MX
GD/MD	GT/MT
GI/MI	GN/MN
GJ/MJ	GH/MH
GM/MM	GS/MS
GU/MU	GP/MP
GW/MW	GC/MC

Another advantage is that any suitably licensed and authorised club member may operate the club station. This gives greater flexibility over a GB callsign which is just a variation to an individual's licence. Best of all, you don't have to fill in any forms or wait 28 days!

Remember when operating away from the Main Address, if you give prior written notice to the RIS office covering your area, you do not have to sign /P or give your location to within 5km.

GB callsigns

Since 1983 the RSGB has despatched on behalf of the Radiocommunications Agency the Notices of Variation issued by the Secretary of State for Trade & Industry authorising special event stations. Consequently, all enquiries and correspondence should be addressed to the Society and not to the RA.

The RA has stated that it ". . . requires the Society when distributing Notices of Variation allowing radio amateurs to set up special event stations, to ensure that the request is for an event which is of special significance and therefore is generally accepted as one requiring celebration, and that the event is open to viewing by members of the public".

Applying for a GB callsign

No charge is made by the RSGB for a special-event callsign, but application forms must be returned at least 28 days prior to the start of the event!

Applications are normally processed shortly after receipt. If nothing has been received 14 days prior to the event, please contact the Amateur Radio Dept at RSGB HQ immediately. *Please note that no authority exists until this notice has been received.*

A letter of variation will only be issued to individuals who hold a current full UK A or B licence (ie not Novices). It will be valid for a maximum of 28 consecutive days. Please note that a notice of variation will only be issued for an individual's licence; variations will not be issued on a club licence.

The station may only be established and operated at one specified location. This must be the address stated on the application form which must be detailed enough for anyone to find easily. Operation of a special event station from a licensee's home address is not normally permitted.

Only the person responsible for the station need sign the form, as the authorisation is by notice of variation to that individual's licence. This person is required to be present to supervise the correct operation of the station. Additional operators need only sign and write their callsigns in the logbook.

If you have not used the callsign before, you can avoid last-minute disappointment by first contacting RSGB. We can then check that it is available and reserve it for you. A GB callsign may be reserved for up to six months in advance. When a GB callsign has been used it will not normally be re-issued to another amateur for use at a different event for a period of 24 months.

If operation of the special event station on the HF bands is desired, a Class A licence holder must apply for a Class A GB callsign.

Subject to availability special event callsigns are available in the following formats:

Class A	Class B
GB0 + 2 or 3 letters	GB1 + 2 or 3 letters
GB2 + 2 or 3 letters	GB5 + 3 letters
GB4 + 2 or 3 letters	GB6 + 3 letters
GB5 + 2 letters	GB8 + 3 letters
GB6 + 2 letters	
GB8 + 2 letters	

Other callsign formats

The RSGB is not authorised to issue callsigns in formats different from the above. Applications for special anniversaries which require the numbers such as 25, 40, 50, 60 or 75 must be specially supported by the RSGB to the RA. The Radiocommunications Agency will only consider applications via the RSGB from the headquarters of a nationally based organisation. Applications cannot be entertained from individuals, local clubs or branches of nationally based organisations.

Greetings messages

The guidelines agreed with the RA are:

1. Each greetings message should not exceed two minutes.
2. Each person may pass only one message to each station with which the originating station is in contact.
3. A non-licensed person may speak into the microphone but the licensed radio amateur must identify the station and operate the transmitter controls at all times.
4. Greetings messages by third parties may only be sent from and received by stations within the UK or the USA, Canada, Falkland Is and Pitcairn Is. The licensee may exchange greetings as in any QSO, with any station.

Charitable events

It is recognised that some special event stations will be established at certain charitable events where a major concern will be the raising of funds.

The RA has agreed that the charity (if one is involved) or the reason for establishing the special-event station may be mentioned 'on-air' provided that under *no* circumstances may a donation be requested during the contact, and sending of QSL cards must *not* be conditional upon the pledge of a donation. It is in the interests of everyone who holds a special event station licence that operators keep within the spirit of this by not asking for any money over the air.

The station may be sponsored per contact, ie the licensee may in advance of the event seek from his/her friends and relatives sponsorship assurances under the usual arrangements for sponsorship. You must *not* seek sponsors 'on-air' at any time.

QSL information

Special event stations generate many QSL cards; it is therefore important that you use the QSL Bureau correctly, so please follow the guidelines given in Chapter 3. GB stations have their own QSL Bureau submanagers – see the *RSGB Yearbook* for their addresses.

DEMONSTRATION STATIONS

The object of a demonstration station is to present amateur radio in an attractive and informative manner to the general public. Clearly, a poorly organised station with a confusing and unwelcoming display hardly serves this purpose – visitors will probably just pass straight on to the next stand!

Attention to the points outlined in the previous section well before the event is important and should provide the basis for a successful operation. At the same time the display should be planned. A few radio maps, some posters explaining the use of Q-signals etc and some colourful QSL cards will do much to brighten up the look of the station. It is also a good idea to have a few radio magazines handy and in addition it is possible to obtain posters and leaflets from the RSGB for display purposes. Contact the Amateur Radio Department for further details.

If you know a local amateur who is able to construct simple equipment, even if only for CW use, get him to display and operate it. The fact that an onlooker can see something working that he feels he could construct himself is a spur to his interest.

Publicity

If a GB licence is applied for, send details of the event to *RadCom* and all the other UK amateur radio magazines as soon as possible. If you are involved with the Press at any stage make sure they do not print any misleading information. In fact it is advisable to provide them with an information handout to ensure the details are correct. The use of such terms as 'broadcasting' and 'message passing' should be avoided at all costs. Guidance on this topic is given in the *RSGB Yearbook*.

Power supplies

Unless the equipment will operate from batteries perhaps the most important single factor affecting the station is the electricity supply. If possible try to operate from a mains source as this is normally reliable and noiseless in comparison with generators. On the other hand, if the radio station is only one of many electrically operated displays and is near the end of the supply line it is quite possible for the voltage to drop well below 200V and render radio equipment inoperative. Also, the other displays may cause electrical interference and it is as well to check these points before operating commences.

If a generator is used, diesel fuel or petrol will be required and cost alone might easily restrict the number of operating hours. Generators must be tested before the event despite any verbal assurances that they are suitable. Check the voltage on full load (including lights) and your receiver on all frequencies for the amount of hash produced. The earth wire should be connected to a ground stake at both the generator and the equipment end of the supply line as this reduces hash.

Where more than one generator is in use it is an extremely important safety factor to ensure that the supplies are not only kept separate but are connected to entirely separate groups of equipment. Sharing can result in twice the supply voltage appearing across some distribution terminals – a very dangerous situation. See Chapter 5 for more safety information.

You will need lights for the evenings or for operation in dark locations and the use of spotlights will enhance the overall effect. Avoid fluorescent lights and keep as far away as possible from other displays using them as they can cause considerable interference. The fact that you are operating at an exhibition does not mean you can neglect the normal safety principles. Make sure the supply is correctly fused and that there is one main and clearly labelled switch to disconnect it in an emergency.

Accommodation and the station

When operating in the open try to obtain the use of the largest tent or marquee possible as this must not only house the station but a fair number of spectators too. The next most important decision will be where to actually locate the station on the site and this will usually be a compromise. The station will obviously have to be located fairly near the mains source unless a generator is to be used. If late-evening operation is envisaged or a generator is to be used it is not advisable to be close to any sleeping areas.

Should a public address system be in operation, try to ensure that the station is sited as far away from the input end as possible, especially if they intend to use a wandering microphone with a long lead. If the station does cause interference carry out tests on all bands as you will normally find that only one or two are affected. Unless you can clear the interference it will be necessary to liaise with the PA operator to keep off the offending bands while his equipment is in use.

Unless the station has some definite objective (eg to make contacts in certain parts of the world involving directive antenna arrays) it is advisable to keep the antenna system as simple as possible. Complicated antennas and masses of equipment create their own problems and often need days of installation work in order to operate satisfactorily.

A few friendly faces, some posters and a couple of radios being operated in full view of passers-by was the welcoming scene at GB2STF

Once again remember that it is important to observe the normal safety principles and provide an efficient earthing system for the antenna arrays. Long cables carrying mains voltages should not be trailed over the ground and are better sited up in the air well out of reach.

If sleeping on site, it is as well to screen off a portion of the operating position as living accommodation. This has the added benefit that you will provide your own guard for the station equipment during the silent hours.

Try to make the display as attractive as possible – do not use so much material that it looks cluttered and 'bitty'. Battered old tables can be improved beyond recognition by covering them with cloths or crepe paper. If the stand is located against a wall this will probably be dull and uninspiring so cover it with sheets or curtains. Such action helps to highlight the equipment and posters etc, all of which are your 'selling' points to the public.

It is advisable to erect some form of barrier between the visitors and the operators to avoid crushing. Any really interested onlooker can always be invited 'in' to sit with the operators.

Running the station

When operating a demonstration station it should be remembered that one is dealing with varying degrees of interest. Some members of the public will be interested in becoming amateurs or short-wave listeners but the majority will be showing just a general interest. It is important to cater for this second group as it is an advantage for as many members of the public as possible to realise that radio amateurs are serious and responsible members of the community and that the hobby is controlled by examination, licensing and official inspection.

Unless the station is part of a technical exhibition the most friendly and able speakers should be in the front of your stand to explain what is going on. Remember the majority of the public will not be technically minded and what is required is a good 'salesman' and not a technical expert. It is important that the onlookers should receive a friendly welcome and are not faced with the backs of operators all far too busy to talk to them.

With the present state of telecommunications, where it is possible to pick up a pocket-sized mobile phone and dial numbers on the other side of the world, speaking over great distances by radio no longer holds the mystery and excitement it did a few decades ago. It is therefore far more impressive to make clear audible local contacts than almost unreadable ones over thousands of kilometres.

Most of your audience will not have heard SSB signals previously so do remember that it takes some time for their ears to become accustomed to the sounds and for them to understand what is being said.

Also remember the many varied facets of amateur radio: short-wave listening, mobile working, packet radio, SSTV etc. All of these should be mentioned even if they are not demonstrated.

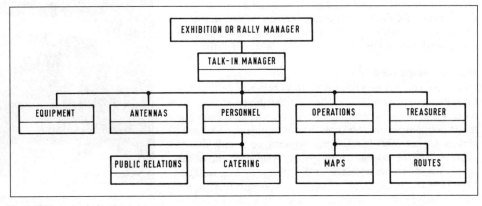

Fig 1. Talk-in organisation

Do not make your display too complicated as there is nothing to be gained by confusing the onlookers. The fixation in the minds of many amateurs that they must work as many stations as possible during these events is quite unfounded as the main aim is to arouse interest in the hobby.

TALK-IN STATIONS

"GB2VHF – This is G3AFT/mobile on the M1 near Watford – request talk-in – over."

This crisp message was sufficient to bring into action the talk-in station at a major RSGB event held in London some years ago. Half an hour later the relaxed visitor to London was personally thanking the talk-in operator for making the drive through the capital's northern suburbs so simple.

Introduction

Talk-in facilities have become a familiar feature at amateur radio mobile rallies and most exhibitions. "Talk-in on 2m and 70cm" appears on the notices for such events with a regularity which belies the considerable effort involved in setting up and operating a talk-in station. The purpose of this section is to set out the main planning points to be considered in order to ensure that mobile stations are helped, in a safe, orderly and expeditious manner, to their destination.

A little bit of history

In an earlier age cars arriving at mobile rallies would have been equipped for top band. This often meant installing a seriously large antenna, guyed to the four corners of the vehicle. A little later, with the move to VHF, transmitters were crystal controlled, but the associated receiver had to be tuned 'high to low' when seeking a contact. The 2m halo atop a large pole was as daring a structure as its 160m counterpart. No two installations were alike, and there was never a shortage of technical discussion amongst those attending mobile rallies. The mobile operator would be expected to change frequency, fine tune both TX and RX for best results, fill in the log book for each contact and at the same time drive what often was an older car. But in those days the pace of life was leisurely, traffic moved slowly and nor was there much of it to spoil happy motoring and QSOing. Long overs were the order of the day.

Today the scene is very different; the accent is on safety, conformity and anonymity. Hands-free microphones are essential and mobile antennas are discreet so as not to draw attention to expensive equipment within. The operation of a talk-in station must reflect these changes by delivering brief, clear and well-thought-out navigation information. This information is provided in the form of advice rather than instruction and the car driver remains responsible in the final analysis for arriving safely at the destination.

Organisation

Any event, be it a major exhibition, a mobile rally or a local club junk sale, will invariably be organised by a committee. This committee will have to decide such matters as the dates, times and location of the event, how to attract trade support, and which guest speakers to invite if lectures are to be held. Further subjects to come before the organising committee will include catering, car parking, publicity, family attractions and of course talk-in facilities. It is therefore important that the person in charge of the talk-in service, the talk-in manager, is a member of this organising committee.

In turn the talk-in manager will gather round him a team of specialists to help set up and run the talk-in station. Formal reporting lines are indicated in Fig 1.

It is never too early to start planning, and as soon as the date and place of the event have been decided the talk-in manager must choose and bring together his team. If the event is not a major international meeting then one person can do more than one of the jobs identified in Fig 1. In the case of a very minor occasion such as a club social evening the whole operation could be undertaken single-handed. Even so, all the factors discussed in this section must be taken into account regardless of the number of people involved.

Frequencies

Frequency modulation in the 145–146MHz band is the mode of choice by the great majority of mobile operators. Furthermore, the constraints of band planning and the limitation in the number of fixed channels available has meant in practice only a handful of frequencies need to be considered for a talk-in station.

145.500MHz is by convention the general mobile calling channel and should not be monopolised by any one station. 145.550MHz is likely to be installed in even the oldest FM rig and is therefore usually selected as the primary talk-in frequency. 70cm FM is also very popular and a talk-in station should be capable of operating on, for example, 433.55MHz.

For the more elaborate talk-in operation it will be necessary to choose additional frequencies for approach, the handling of lost mobiles, and for a parking information channel – see later.

Equipment

Whether a variety of frequencies is to be used or, as in the case of a more modest station, only 145.55MHz FM, it is vital that the talk-in operator can hear (and be heard loud and clear) over a wide area.

The merits or otherwise of individual commercial transceivers cannot be discussed here but in all cases reliability is of paramount importance. Transmitter power outputs in excess of 10W are usually unnecessary at VHF provided that the antenna is well sited. This often means that a 'black-box' transceiver running off a car battery is the only equipment needed. A standby battery together with a charger should be available as a sensible precaution, however. If two channels in the same band are to be used simultaneously it may be necessary to install cavity filters to provide sufficient separation – such a technical matter must be investigated well in advance of the event, as it can be a great disappointment having made plans for two transceivers only to find on the day that they cause mutual interference. This mode of operation can be very difficult if the frequencies are not far apart. There may also be adjacent-channel interference to mobiles in the close vicinity.

A desk microphone is more convenient than a handheld one, and a boom microphone attached to headphones is better still. If headphones are used it may be desirable from a public-relations point of view to connect extension loudspeakers for the benefit of visitors.

The quality of transmission must always be of the highest order – decent punchy audio but without a high level of background noise (where did all those Second World War throat mikes go?) It is especially important to check performance and quality when the gear is running from (unknown) generators.

Antennas

At VHF the basic rule is to mount the antenna as high as possible. However, the choices open for the siting of antennas are often limited, and will be dictated by the type of building housing the event and by the position of the stand made available to the talk-in station.

The type of antenna can influence the efficiency of the station and advantage should be taken of the gain of a stacked array, collinear or $5\lambda/8$ ground plane. An ideal arrangement would be a collinear mounted on a tall mast attached to the highest point of the building with the station in a room as close as possible. Use of low-loss coaxial cable together with attention to accurate matching will further help effective radiation and clear reception. Another possibility, particularly if the event is being held in a single-storey building or under canvas, is to hire or borrow a telescopic mast.

A few words of warning to the person in charge of antennas (see also Chapter 5):

- take safety seriously;
- wear protective headgear;
- use correct clamps and other hardware. A good antenna system will almost certainly ensure a good talk-in station.

Operating procedures

Having installed the equipment and erected an efficient antenna system the next item to be considered is how best to provide talk-in advice. Talk-in information is divided into two parts; the approach phase and the final run-in.

Approach phase

With the aid of a map having a scale of about 500,000:1 (6km/cm or 10 miles to the inch) decide on the major routes from surrounding towns to the scene of the event. Observe if these roads converge on to a ring road, bridge over a river or some other unmistakable feature about 3km (two miles) from

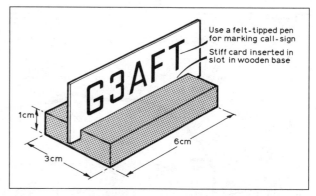

Fig 2. Wooden identification 'chuck'. These are placed on the operational map and show the position of all mobiles being worked; their position is updated as the talk-in proceeds. Use a fine felt-tipped pen for writing callsigns on cards, selecting contrasting colours if more than one talk-in station is to be used. Operators can then tell at a glance which mobiles are under their care

the destination. Try to choose features common to roads coming in from more than one outlying town. Two, or at most three, such features should be selected as 'reporting points'. The journey from a distant place to the reporting point can be considered as the approach phase and quite often only one instruction is needed for this part of the talk-in, eg "Continue towards London and turn left onto the North Circular Road" in the case where the North Circular Road is the reporting point.

Final phase

This stage of the talk-in begins at the reporting point and consists of detailed instructions to get the mobile station to the destination. If only two reporting points are chosen, then only two sets of detailed instructions need be planned for the final phase. This will greatly simplify the job of the operators.

No doubt there will be endless debate as to the best routes to choose and in the event of a dispute the talk-in manager must make the final decision. To help in arriving at the right choice it will probably be necessary to survey all likely roads and make careful notes of important road signs and other useful features. Familiarity with the locality is imperative and before the event all operators should tour the selected routes.

Having decided on routes it will next be necessary to describe each, briefly and through the eyes of a driver. The examples quoted in this chapter may give some guidance on this key part of the planning of the talk-in station.

Maps

Although Ordnance Survey and road maps will be needed at the planning stage it may be better to construct a special map for actual use by the talk-in operators. This would show only major roads and would highlight only those features of interest to a driver being talked-in. In this respect garages are very prominent, as are roundabouts, railway bridges and ring roads.

Another reason for constructing an operational map is to take advantage of the flexibility it can provide at a major event. At any moment perhaps a dozen or more mobile stations may be being talked-in. In order to keep track of this activity it will be helpful to use identification 'chucks'; see Fig 2. These are moved around the map as mobile stations report in, and will show the location of all approaching traffic.

Normal maps, being of fixed scale, will show too much detail in the outlying areas and not enough of the streets close in. A map with a variable scale will be much more useful, but it will take time and effort to draw.

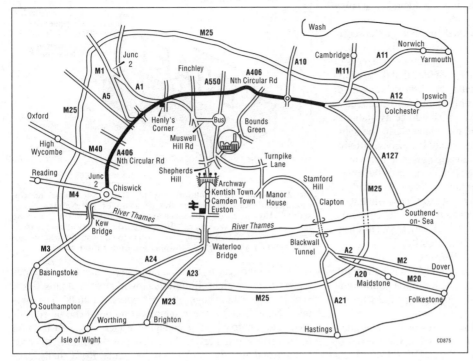

Fig 3. Typical talk-in map

Fig 3 shows an example of such a map of the south-east England which was drawn on 25 sheets of A2 paper (equivalent to 100 sheets of A4 paper). The main roads from Norwich, Dover and Southampton are shown in broad outline whereas the minor roads in the immediate vicinity of the exhibition, in this case at Alexandra Palace in North London, are shown in considerable detail.

The smaller scale at a greater distance is compatible with the frequency with which a mobile station needs assistance. For example a mobile station calling in from Maidstone, 50km (30 miles) out, would be advised: "Continue on M20 and A20 towards Blackwall Tunnel – call again when approaching Blackwall Tunnel". This same mobile, when at Manor House and only 5km (3 miles) from his destination, would be told, "Turn right at Manor House – after 1½ miles turn left at Turnpike Lane and call again". The variable scale map will have been able to accommodate the mobile station both when 50km away near Maidstone and 5km away at Manor House. The talk-in operator will have known his whereabouts with an

appropriate degree of accuracy in both circumstances. Incidentally the talk-in station will have transmitted only three messages from the Blackwall Tunnel to Alexandra Palace, on what many would describe as a nightmare journey across London.

A schematic map, produced on an Apple Macintosh computer for a large exhibition, is shown in Fig 4. This has the merit of being easily updated with last-minute news on road works and it also includes distances to run and other useful information.

Lost procedure

Should a mobile station become lost, and this will usually occur in a large city, it is vital that the talk-in operator establishes an accurate 'fix' before trying to give further advice. First the lost mobile should be asked to check the name of the road on which he is driving together with the name of a side road. The operator then calmly looks up this reference on his town map and decides on a route back to a main road. If the whereabouts of the mobile cannot be readily established he should be asked to stop when safe to do so. (He will not mind this as he is relying on the talk-in operator for correct information to get him back on course.)

CD-ROM maps now provide the almost perfect solution to finding a 'lost' mobile; enter the street name and post code and in a fraction of a second the appropriate section of map appears. In moments the visitor can once again be brought back on track.

Parking

Up-to-date parking information is a frequently asked question by approaching mobiles. A useful facility is a dedicated parking channel operated by a person with a handheld transceiver and located in the car park area. This can be especially helpful for disabled visitors, directing them to reserved parking spaces adjacent to eg ramp access for wheel chairs.

Words

Having established contact with a mobile station needing advice the next question which arises is which words should be used? Brevity with clarity is the key and the following example may be helpful. In the case of the mobile who called in at the beginning of this section the assistance might go as follows (callsigns omitted):

"Your present location please."

The mobile gives his location which is pin-pointed on the map with a 'chuck'.

"Continue on M1 for eight miles – leave at junction 2 and call again at North Circular Road."

Later, when the mobile reports at the North Circular Road (and his 'chuck' having been moved accordingly) he is advised:

A talk-in map similar to that shown in Fig 3

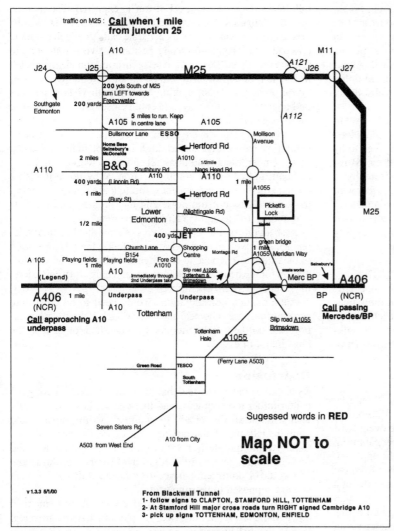

Fig 4. Computer-generated schematic map showing the significant landmarks, distances and suggested words

"Continue on North Circular Road for three miles – leave on B550 for Muswell Hill and call again."

The final instruction would be:

"Continue to Muswell Hill Roundabout (with bus depot in middle) – go in at 6 o'clock and out at 12 o'clock [see Fig 5], down steep hill and turn hard left left at bottom of hill into grounds of Alexandra Palace."

Other examples of suggested words are shown on the map illustrated in Fig 4, and suitable words for *talk-out* after the event should not be overlooked.

Courtesy at all times

The rationale for selecting frequencies for a talk-in service has already been explained and it is worth mentioning that no exclusive right can be claimed for them. However, a word or two on how to establish and maintain a clear channel may not be out of place.

Assuming a well-sited antenna and satisfactory equipment the talk-in station is likely to be heard over a wide area. Therefore if upon switching on another contact is in progress, wait until it is finished. Then call, say, "GB2ZZZ listening" once only. If you receive a reply from a mobile coming to the event,

you are in business! If the reply comes from a fixed station, you should explain that you are listening out for mobiles who might request directions to the event.

Should there be no reply, then the channel is almost yours; call again, "GB2ZZZ listening".

A procedure such as that suggested above will allow other amateurs to speak up if they feel that you are pushing your way in a little too hard. Radio amateurs will respond to a polite, brief explanation of what you are doing, but few will relish being steam-rolled off a frequency which they may have regarded as being 'theirs'.

Having gently eased your way onto a frequency the next problem becomes how best to hold on to it. 14MHz operators will know how easy it is to lose their place on a frequency by remaining silent for too long at the end of a contact. Somebody, somewhere is always waiting on the side, ready to take over the channel. Therefore you must make sure that other stations are aware that the frequency is in use. The simplest way, and the one which is least likely to aggravate other amateurs, is to announce "GB2ZZZ listening" every 20 to 30 seconds.

Do *not* put out a long call stating that you are the talk-in station to the such-and-such rally being held at so-and-so, and that you can provide a talk-in service for those needing navigation advice and that you are listening out for any mobiles etc etc etc! All that you need to do in order to hold on to the frequency is to say every 20 to 30 seconds "GB2ZZZ listening".

Personnel

How many people are needed to operate a talk-in service? The answer will depend mainly on the number of simultaneous frequencies in use and also upon the length of the event.

Each channel will need an operator who must have a thorough knowledge of the area as well as a good operating manner, and a logkeeper who should have a tidy hand. A third member of the team will be needed to take care of the maps, particularly during the period when a mobile reports being lost. He will also keep the operator up to date on the whereabouts of all stations being worked by moving the 'chucks' across the map.

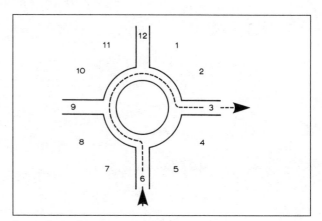

Fig 5. Giving directions at roundabouts

A team of three can work well for two hours without a break, but if three-hour or four-hour shifts are decided upon, then tea, sandwiches and biscuits should be available. In the case of an elaborate talk-in station with perhaps a dozen or more people on each shift a catering manager should be appointed. Attention in the area of refreshments can make amateur radio that much more civilised. Note, however, that alcohol is absolutely prohibited prior to or during operating.

Next, having identified the jobs to be done, it is important to ensure that the workload does not fall upon too few shoulders. A personnel roster should be drawn up and when agreed to by all concerned it should be prominently displayed. If more than 12 or so people are involved it is as well to write formally to each enclosing a statement of their times on duty, together with a return slip on which they should signify their agreement to these times. Such a written commitment should ensure that personnel attend on time. The talk-in manager or his deputy must be present throughout the period of the event.

Public relations

An ambitious talk-in station can, if well presented and efficiently operated, be an excellent way of demonstrating amateur radio to the public. Nominate one person on each shift to answer questions about amateur radio in general and the talk-in station in particular. Do not overlook this opportunity to promote the usefulness of the hobby.

Some thoughts for the future

Earlier it was suggested that the talk-in operation should be considered in two parts: the initial approach phase followed by the final run-in. Mobiles at some distance from the event have little chance of being heard due to those much closer capturing the attention of the talk-in operator. Why not site the approach station on really high ground, perhaps some distance from the final destination and establish *initial* contact with mobiles at a much greater range, using a frequency adjacent to the main talk-in channel? General directions only need be given at this stage, for example ". . . call GB2LON on 145.55 when passing junction 24 . . ." (see Fig 4). The callsign, present position and name of the mobile operator would then be passed back to the main event station by packet radio. This information, displayed as a time-stamped listing on a screen, would alert the main

talk-in operators to the level of traffic expected. The mobile, having changed to the local frequency of 145.55MHz, will be greeted by name as well as by accurate callsign. Such a remote approach station would ideally be run by two people, one to operate the rig and the other to relay information via the packet link. The remote station would be on air well before the exhibition or rally opened, making contact with visitors perhaps half an hour or more away, and could close down when traffic levels began to fall off.

Another remote device might be considered for use at an exhibition hall which suffers high levels of radio frequency noise. Under such conditions reception of weaker signals is difficult and can spoil communication. Why not establish a receiver on the talk-in frequency, locate it at a remote corner of the site well away from noise and interference, and relay its output back to the hall via a microwave link?

Restrictions on the use by drivers of hand-held cellular telephones may well extend to radio amateurs; the talk-in station of the future might need to consider 'broadcasting' basic information for the benefit of solo drivers, and only work cars carrying a separate radio operator.

Authorisation considerations would need to be taken into account for these schemes, and clearly such elaborate preparations are unlikely to be thought worthwhile at other than the most major of events.

Conclusion

By now the reader may feel that so much planning is taking the matter rather too seriously, and that one could get by equally well without all this detail. Be warned! An ill-prepared and rambling talk-in is embarrassing to the casual listener, and frustrating to the mobile in need of clear navigation advice who finds he cannot get a word in edgeways. Tempers will fray and you would be doing a disservice to both your club and the event you are supposed to be supporting.

On the other hand a well-managed talk-in, carefully planned down to the last word of advice, is a pleasure to listen to, a real help to the distant visitor and can only bring goodwill to your club as well as a great deal of pleasure to all concerned.

"GB2VHF – this is G3AFT mobile – I've arrived in the car park – many thanks for your assistance – out."

1 Continental and regional maps

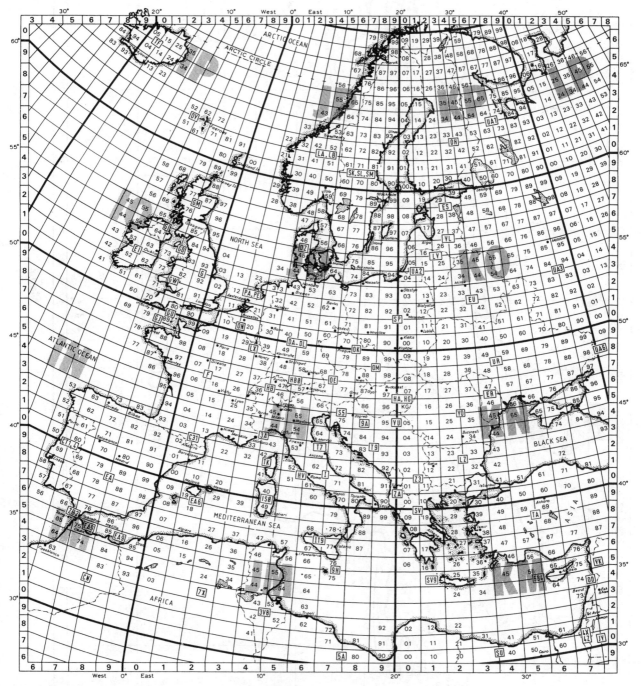

Map 1. IARU locator map of Europe

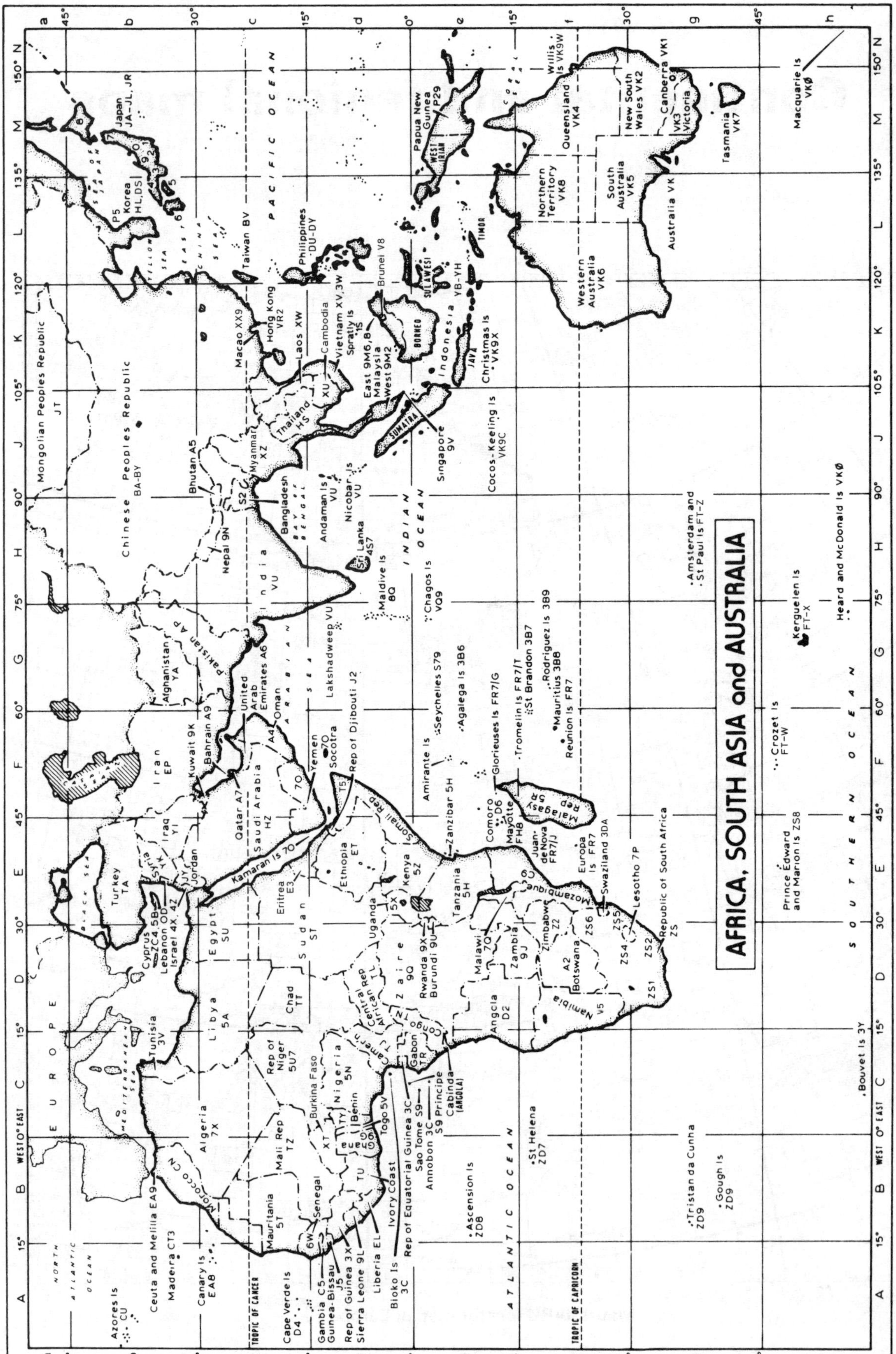

AFRICA, SOUTH ASIA and AUSTRALIA

Map 2

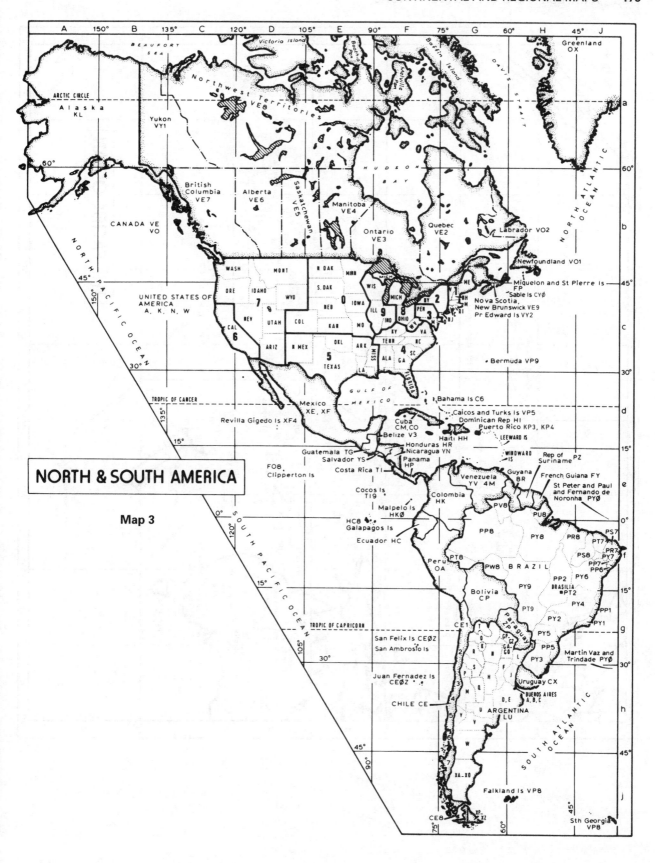

NORTH & SOUTH AMERICA

Map 3

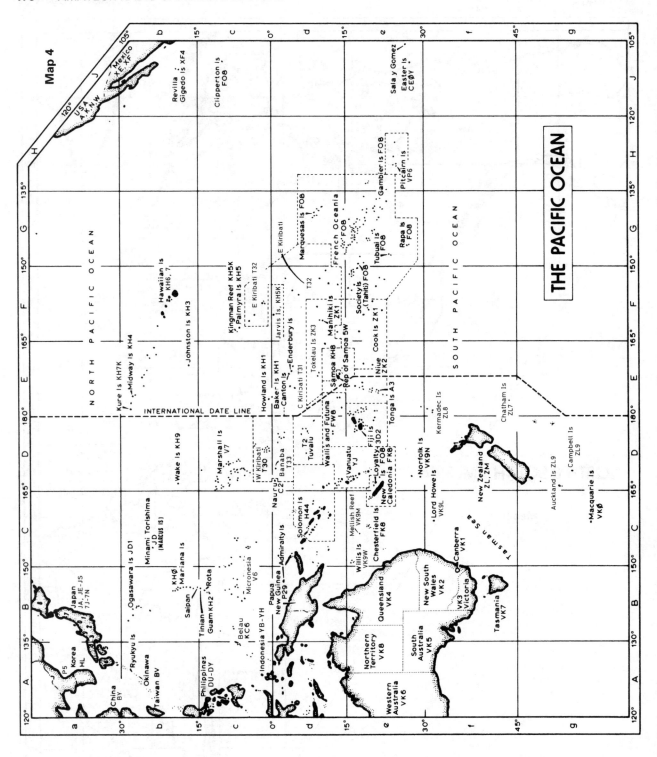

Map 4

THE PACIFIC OCEAN

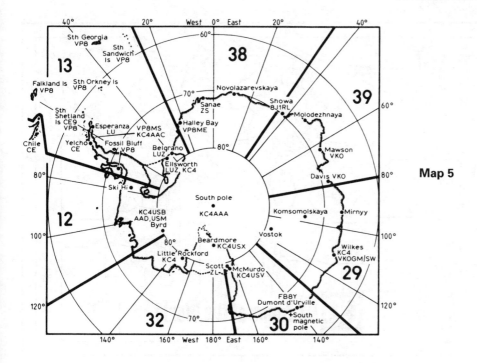

Map 5

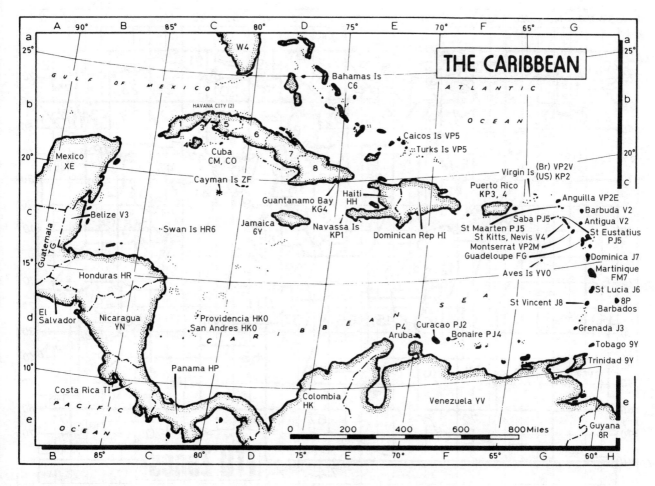

Map 6

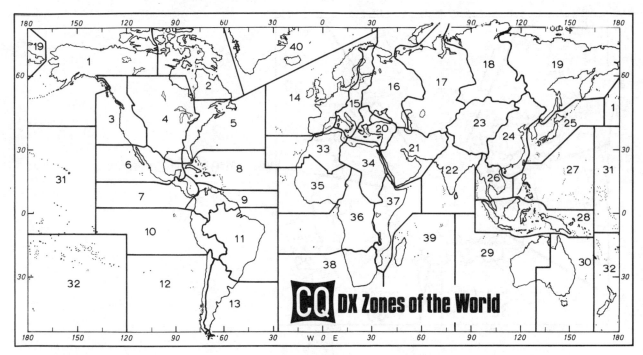

Map 7

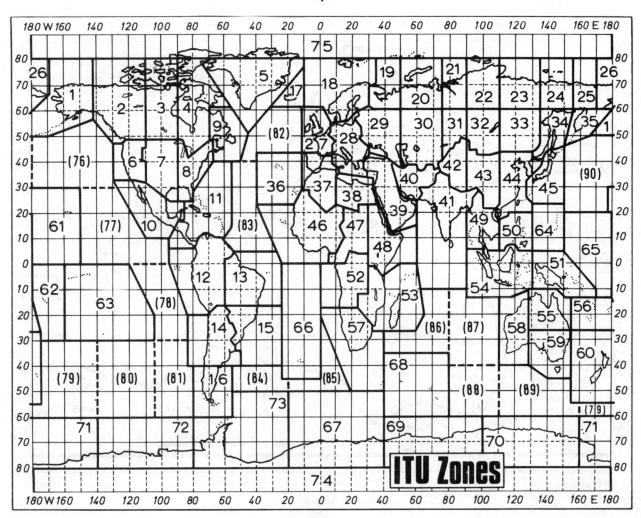

Map 8

T HIS appendix gives information of amateur radio interest for each organisation, country (or group of countries) which has been assigned one or more ITU international callsign series.

The callsign system information is supplemented by call area maps where possible; this information for larger countries may instead appear in Appendix 1. In some cases maps are not provided for countries with call area systems; here a city or town from each province is given in a table so that the reader can still locate the approximate area corresponding to a callsign in an atlas.

A note to the effect that third-party traffic is permitted does not necessarily mean that such communication is possible with all other countries permitting it. Such traffic is by individual agreement between the countries concerned (see Chapter 1).

If foreign amateurs from certain countries are permitted to operate inside a particular country on a reciprocal basis, some indication of these other countries is given where known but this may well be incomplete due to political change or fresh reciprocity agreements. It is therefore advisable to check the up-to-date position with the relevant national society or licensing administration. The amount of notice required for processing foreign amateur operation licences where given is in respect of completed applications on the correct forms with all necessary enclosures, including any necessary fee. Therefore allowance should be made for initial correspondence and answering of queries (see Chapter 6). The licensing administration addresses given are the correspondence addresses for foreign amateur licensing and may not be correct for other telecommunication matters. Such addresses are not given where application must be made via the national society which is the case in several countries.

All telephone numbers given are in the international form $+x$ (y) z where x is the country code, y is the area code (if applicable) and z is the number. They will have to be prefaced by the appropriate international dialling prefix (00 in the UK). If dialling from within the country, the area code may need to be prefaced by a zero.

AFGHANISTAN (Democratic Republic of)
ITU allocation: T6A–T6Z, YAA–YAZ.
Callsign system: Prefix: YA.
Licensing administration: Ministry of Communications, Kabul.

ALBANIA (Socialist People's Republic of)
ITU allocation: ZAA–ZAZ.
National society: Albanian Amateur Radio Association, PO Box 1501, Tirana.
Licensing administration: Direction générale des postes et télécommunications, 42 rue Myslym Shyri, Tirane.

ALGERIA (Algerian Democratic and Popular Republic)
ITU allocation: 7RA–7RZ, 7TA–7YZ.
Callsign system:

Prefix	City/town and area
7X2	Algiers (El Djezair)
7X3	Ghardaia
7X4	Oran
7X5	Constantine

Licence notes: Single licence class, 200W PEP.
Foreign amateur operation: 7X0 prefix. All countries except Israel.
National society: Amateurs Radio Algeriens, PO Box 1, 16000 Alger-Gare. Tel: +213 (2) 644432. E-mail: 7x2ara@caramail.com.
Licensing administration: Ministère des postes et télécommunications, Secretariat générale, 4 boulevard Salah Bouakouir, Algiers.

ANDORRA (Principality of)
ITU allocation: C3A–C3Z.
Callsign system: Prefix: C3.
Licence notes: Single licence class: 1kW PEP.
National society: Unio de Radioaficionats Andorrans, PO Box 1150, La Vella. Tel/fax: +376 825380. E-mail: ura@andorra.ad. Web: www.sta.ad/ura.

ANGOLA (People's Republic of)
ITU allocation: D2A–D3Z.
Callsign system: Prefix: D2.
Licensing administration: Service dos correios, telegrafos e telefones, Luanda.

ANTIGUA and BARBUDA
ITU allocation: V2A–V2Z.
Callsign system: Prefix: V2.
Licence notes: Single licence class: 1kW DC input.
Foreign amateur operation: Countries include: British Commonwealth, Canada and USA.
National society: Antigua and Barbuda Amateur Radio Society, PO Box 1111, St John's.
Licensing administration: Telecommunications Officer, Ministry of Public Works and Communications, St John St, St John's.

ARGENTINA (Argentine Republic)
ITU allocation: L2A–L9Z, AYA–AZZ, LOA–LWZ.
Callsign system: Prefixes: LU, LW. For call areas see map (letters shown are the first serial letter(s)).
Licence notes: Five licence classes: Special: all bands, 1kW DC input. Superior: all bands, 1kW DC input. General: all bands, 1kW DC input. Intermediate: VHF, restricted HF, CW only, 100W DC input. Novice: 3.5MHz and VHF, CW only, 50W.
Foreign amateur operation: Call is own call/LU. Countries include: USA, Venezuela, Costa Rica, Paraguay, Peru, Chile, Uruguay.
National society: Radio Club Argentino, Carlos Calvo 1420/24, 1102 Buenos Aires. Tel: +54 (11) 4305-0505. Fax: +54 (11) 4304-0555. E-mail: lu4aa@lu4aa.org. Web: www.lu4aa.org.
Licensing administration: Dirección National de Telecomunicaciónes, Sección Radioaficionados, Sarmiento 141, 1000 Buenos Aires.

ARMENIA
ITU allocation: EKA–EKZ.
Callsign system: Prefix: EK.

ARUBA
ITU allocation: P4A–P4Z.
Callsign system: Prefix: P4.
National society: Aruba Amateur Radio Club, PO Box 2273, San Nicolas. Tel: +297 (8) 52773. Fax: +297 (8) 31545. E-mail: aarc@qsl.net. Web: www.qsl.net/aarc.

AUSTRALIA
ITU allocation: AXA–AXZ, VHA–VNZ, VZA–VZZ.

ARGENTINA – LU

VP8
Falkland Islands

0   500   1000
kilometres

Callsign system: See Map 2 for prefixes. Licence class is denoted by serial letters as follows:

Suffix	Licence class
AA–ZZ	Unrestricted
AAA–GZZ	Unrestricted
IAA–IZZ	Full
JAA–KZZ	Limited plus novice (licensees holding both licences)
LAA–NZZ	Novice
PAA–PZZ	Novice
RAA–RZZ	Beacons and repeaters
SAA–SZZ	Unrestricted
TAA–UZZ	Limited
VAA–VZZ	Novice
WAA–WZZ	Unrestricted
XAA–ZZZ	Limited

Operation outside own call area uses CALL/VK followed by number denoting state.
Licence notes: Three classes of licence. Unrestricted call: 120W mean or 400W PEP SSB. Novice: crystal-controlled 3.525–3.565MHz, 21.125–21.200MHz, 28.100–28.600MHz, 146–148MHz phone/CW, 10W mean, 30W PEP. Limited: VHF phone only, 120W mean, 400W PEP. Third-party traffic permitted.
Foreign amateur operation: Three types of licensing. All require a fee and form. Three months' notice by post but can obtained over the counter at the local Spectrum Management Agency office in each state. *Category A:* Reciprocal. Countries include: Canada, Denmark, France (incl New Caledonia), Germany, India, Israel, Japan, Malaysia, New Zealand, Papua New Guinea, Poland, Solomon Is, Spain, Switzerland, UK, USA, Singapore. *Category B:* Unilateral. Countries include: Argentina, Falkland Is, Greece, Hong Kong, Indonesia, Irish Rep, Italy, Luxembourg, Malta, Nauru, Netherlands, Norway, Phillipines, Singapore, South Africa, Sri Lanka, Sweden, Vanuatu and West Indies. *Category C:* 10W 144MHz licence. All other countries.
National society: Wireless Institute of Australia, PO Box 2175, Caulfield Junction, VIC 3161. Tel: +61 (3) 9528 5962. Fax: +61 (3) 9523 8191. Web: www.wia.org.au.

Licensing administration: Spectrum Management Agency, Purple Building, Benjamin Offices, Chan Street, Belconnen ACT 2617. Tel +61 (6) 256 5555. Fax: +61 (6) 256 5200. Web: www.sma.gov.au.

AUSTRIA

ITU allocation: OEA–OEZ.
Callsign system: See map for prefixes. Serial letters in the XAA–XZZ series denote special stations, beacons or repeaters.
Licence notes: Four licence classes: 25, 50, 100 and 250W anode dissipation. 250W class is VHF phone-only for clubs only.
Foreign amateur operation: *CEPT:* A + B using OE/own call. *Non-CEPT:* A + B. Fee. Six weeks' notice. Call is OE series. Countries include: Australia, Belgium, Costa Rica, Germany, Finland, France, UK, Israel, Italy, Yugoslavia, Canada, Liechtenstein, Luxembourg, Monaco, Netherlands, Norway, Poland, Romania, Sweden, Switzerland, USA. Applications should be made to the local branch of the licensing administration. Further information can be obtained from the telecomm contact officer at ÖVSV.
National society: Österreichischer Versuchsenderverband, Theresiengasse 11, A-1180 Vienna. Tel: +43 (1) 408 5535 (Wednesdays, 4–8pm). Fax: +43 (1) 403 1830. E-mail: oevsv@oevsv.at. Web: www.oevsv.at.
Licensing administration: BMöWV – Sektion IV – Rechtsabteilung, Kelsenstr 7, A-1030 Vienna. Tel: +43 (1) 79731-4100. Fax: +43 (1) 79731-4109. Local offices are: *OE1, OE3, OE4* – Fernmeldebüro für Wien, NÖ und Bgld, Nordbergstrasse 15, A-1091 Vienna. *OE2, OE5* – Fernmeldebüro für OÖ und Salzburg, Domgasse 1, A-4010 Linz. *OE6, OE8* – Fernmeldebüro für Kärnten und Steiemark, Neutorgasse 46, A-8011 Graz. *OE7, OE9* – Fernmeldebüro für Tirol und Vorarlberg, Maximillianstrasse 7, A-6010 Innsbruck.

AZERBAIJAN

ITU allocation: 4JA–4KZ.
Callsign system: Prefix: 4J, 4K followed by area number (see below).

Area no	Area
2	Nakhichevan
3	Azerbaijan (except Nakhichevan and Baku)
4–9	Baku

BAHAMAS (Commonwealth of the)

ITU allocation: C6A–C6Z.
Callsign system: Prefix: C6.
Licence notes: Similar to USA General class. Maximum power: 100W PEP.
Foreign amateur operation: Unilateral. A only. Countries include UK, USA.
National society: Bahamas Amateur Radio Society, PO Box SS-6004, Nassau.
Licensing administration: Bahamas Telecomms Corp, PO Box N3048, Nassau. Tel: +1 (809) 323-4911.

BAHRAIN (State of)

ITU allocation: A9A–A9Z.
Callsign system: Prefix: A9.
Licence notes: Single licence class. Maximum power: 500W PEP.
Foreign amateur operation: Must have resident permit.
National society: Amateur Radio Association Bahrain, PO Box 22371, Muharraq. Web: members.tripod.com/a92c.
Licensing administration: Secretary General, Permanent Telecommunications Committee, PO Box 831, Manama.

BANGLADESH (People's Republic of)

ITU allocation: S2A–S3Z.
Callsign system: Prefix: S2.
Foreign amateur operation: Call is own call/S2.
National society: Bangladesh Amateur Radio League, GPO Box 3512, Dacca 1000. E-mail: barl@pradeshta.net.

Licensing administration: Telegraph and Telephone Board of Bangladesh, Capital Exchange Building, Dacca 15.

BARBADOS

ITU allocation: 8PA–8PZ.
Callsign system: Prefix: 8P.
Licence notes: Two licence classes: CW: all bands, CW only, 500W. General: all bands, all modes, 500W input. There is also an advanced class.
Foreign amateur operation: Unilateral. A only. Call is 8P9 series. Countries include: British Commonwealth countries, USA.
National society: Amateur Radio Society of Barbados, PO Box 814E, Bridgetown. Tel: +1 (809) 406-2502.
Licensing administration: Senior Telecomms Officer, Ministry of Public Works, Communications and Transportation, Herbert House, Fontabelle, St Michael. Tel: +1 (809) 426-2669.

BELARUS

ITU allocation: EUA–EWZ.
Callsign system: Prefixes: EU, EW followed by area number.

Area no	Area	Capital
1	Minsk city	
2	Minsk Obl	Minsk
3	Brest Obl	Brest
4	Grodno Obl	Grodno
6	Vitebsk Obl	Vitebsk
7	Mogilev Obl	Mogilev
8	Gomel Obl	Gomel

National society: Belarus Federation of Radio Amateurs and Radio-sportsmen, Ul Kazintca 48, 220099 Minsk. Tel/fax: +375 (17) 278 3016.

BELGIUM

ITU allocation: ONA–OTZ.
Callsign system:

Prefix	Area	Licence class
ON0	Belgium	Repeaters
ON1	Belgium	Class BI, BII
ON4, 5, 6, 7	Belgium	Class A, BI, BII, C
ON9	Belgium	Foreign amateurs
OR4	Overseas incl Antarctica (Belg)	

Licence notes: Four classes of licence. Class A: 125W all bands, all modes. Class BI: 250W VHF bands only. Class BII: 125W VHF bands only. Class C: 500W all bands, all modes.
Foreign amateur operation: *CEPT:* A + B using ON/own call. *Non-CEPT:* Reciprocal. Two months' notice. Fee. Form. Countries include UK, USA.
National society: Union Belge des Amateurs-Emetteurs, rue de la Presse 4, B-1000 Brussels. Web: www.uba.be.
Licensing administration: IBPT, Av de l'Astronomie 14, B-1030 Brussels. Tel: +3 (2) 207 77 77. Fax: +32 (2) 207 78 88.

BELIZE

ITU allocation: V3A–V3Z.
Callsign system: Prefixes: V31 (Class 1), V32 (Class 2). For call areas see below.

Serial letters	Area
AA–BZ	Corozel
CA–DZ	Orange Walk
FA–KZ	Belize
LA–MZ	Stann Creek
NZ–OZ	Cayo
PA–QZ	Toledo

National society: Belize Amateur Radio Club, PO Box 296, Belize.
Foreign amateur operation: Countries include USA.
Licensing administration: Belize Telecommunication Authority, PO Box 603, Belize City.

BENIN (People's Republic of)

ITU allocation: TYA–TYZ.
Licensing administration: M le Directeur général de l'Office des postes et télécommunications, Cotonou.

BHUTAN (Kingdom of)

ITU allocation: A5A–A5Z.
Callsign system: Prefix: A5.

BOLIVIA (Republic of)

ITU allocation: CPA–CPZ.
Callsign system: See map for prefixes. CP0 is a special station prefix.
Licence notes: Two licence classes: Second class: 3.5, 7MHz SSB/CW,

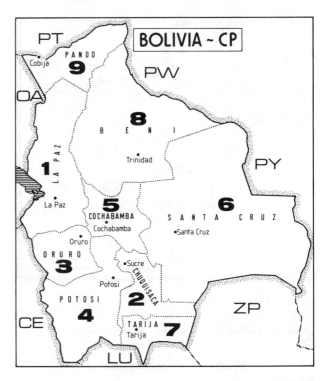

200W PEP. First class: all bands, SSB/CW, 1kW PEP. Third-party traffic permitted.
Foreign amateur operation: Call is own call/CP. Countries include: Colombia, Germany, Peru, Switzerland, Venezuela, USA.
National society: Radio Club Boliviano, PO Box 2111, La Paz. Tel/fax: +591 (2) 224921.

BOSNIA-HERCEGOVINA

ITU allocation: T9A–T9Z.
Callsign system: Prefix: T9.
Foreign amateur operation: *CEPT:* A + B using T9/own call.
National society: Asocijacija Radioamatera Bosne i Hercegovine, PO Box 61, 71001 Sarajevo. Tel/fax: +387 (71) 663414.

BOTSWANA (Republic of)

ITU allocation: A2A–A2Z, 8OA–8OZ.
Callsign system: Prefixes: A22, A24 (novice).
Licence notes: Maximum power 150W DC input. Also novice class.
Foreign amateur operation: Reciprocal (A only). Countries include UK. Call is A25/own call.
National society: Botswana Amateur Radio Society, Box 1873, Gaborone.
Licensing administration: Botswana Telecoms Corporation, Radio Spectrum Management Division, PO Box 700, Gaborone. Tel: +267 358000.

BRAZIL (Federative Republic of)

ITU allocation: PPA–PYZ, ZVA–ZZZ.
Callsign system: See map for prefixes. Abrolhos Is also has PY0 prefix. WAA–WZZ serial letters: Class C or licensee aged 14–18 years. ZAA–ZZZ serial letters: foreign citizens. State prefix is added when operating outside own state, eg PY1ZZ/PT2.
Licence notes: Three licence classes. Class A: all bands, 1kW PEP. Class B: 1.8, 3.5MHz, 1kW PEP. Class C: 7MHz, 100W PEP. Third-party traffic permitted.
Foreign amateur operation: Reciprocal (A only). Temporary calls are own call/PY. Six months' notice. Countries include: Bolivia, Canada, Chile, Colombia, Guatemala, Denmark, Dominica, Germany, UK, Paraguay, Portugal, Sweden, USA, Venezuela, Uruguay.
National society: Liga de Amadores Brasileiros de Radio Emissao, PO Box 00004, 70359-970 Brasilia, DF. Tel: +55 (61) 223 1157. Fax: +55 (61) 223 1161. E-mail: labre@labre.org. Web: www.labre.org.
Licensing administration: Departamento Nacional de Telecomunicações, Dentel 4° Andar, Ministerio das Comunicações, 70000 Brasilia, DF.

BRUNEI DARUSSALAM

ITU allocation: V8A–V8Z.
Callsign system: Prefix: V8.

BRAZIL

Foreign amateur operation: Unilateral (A + B).
National society: Negara Brunei Darussalam ARA, PO Box 73, Gadong, Bandar Seri Begawan 3100.
Licensing administration: Director of Telecommunications, Telecom Dept, Ministry of Communications, Negara. Tel: +673 42324.

BULGARIA (People's Republic of)

ITU allocation: LZA–LZZ.
Callsign system: Prefix: LZ. KAA–KZZ serial letters denote club call.
Licence notes: Maximum power 1kW DC input.
Foreign amateur operation: CEPT: A + B. Call is LZ/own call. Non-CEPT: Unilateral (A only). Apply to BFRA.
National society: Bulgarian Federation of Radio Amateurs, PO Box 830, 1000 Sofia. Tel: +359 (2) 623022 ext 258. Fax: +359 (2) 9801458. E-mail: secretary@bfra.org. Web: www.bfra.org.
Licensing administration: Commitee of Post and Telecomms, Radioregulatory Dept, Gurko Street 6, 1000 Sofia. Tel: +359 (2) 889511.

BURKINA FASO (Republic of) *(previously Upper Volta)*

ITU allocation: XTA–XTZ.
Callsign system: Prefix: XT.
National society: Association des Radioamateurs du Burkino Faso, c/o Youssouf Kaba, ONATEL, PO Box 01, Ouagadougou 10000.
Licensing administration: Office des postes et télécommunications, Ouagádougou.

BURUNDI (Republic of)

ITU allocation: 9UA–9UZ.
Callsign system: Prefix: 9U.
Licensing administration: Ministère des Postes et Télécommunications, BP2000, Bujumbura.

CAMBODIA

ITU allocation: XUA–XUZ.
Callsign system: Prefix: XU.
Licensing administration: Ministère des postes et télécomunications, Phnom-Penh.

CAMEROON (United Republic of)

ITU allocation: TJA–TJZ.
Callsign system: Prefix: TJ.
Licensing administration: Ministère des postes et télécommunications, Yaounde.

CANADA

ITU allocation: CFA–CKZ, CYA–CZZ, VAA–VGZ, VOA–VOZ, VXA–VYZ, XJA–XOZ.
Callsign system:

Prefix	Province or island
CY0	Sable Is
CY9	St Paul Is
VE1	Nova Scotia
VA2, VE2	Quebec
VA3, VE3	Ontario

VE4	Manitoba
VE5	Saskatchewan
VE6	Alberta
VE7	British Columbia
VE8	NW Territories
VE9	New Brunswick
VO1	Newfoundland
VO2	Labrador
VY0	Nunavut (Inuit homeland)
VY1	Yukon Territories
VY2	Prince Edward Is

Licence notes: Three licence classes. Amateur: all bands (CW on HF) 1kW. Advanced: all bands, all modes, 1kW. Digital: 220MHz and above, 1kW. Third-party traffic permitted.
Foreign amateur operation: CEPT: A + B using VE/owncall. Non-CEPT: Reciprocal (A + B). No notice required. Call is own call/Canadian prefix. All countries.
National society: Radio Amateurs of Canada, 720 Belfast Road, Suite 217, Ottawa, Ontario K1G 0Z5. Tel: +1 (613) 244 4367. Fax: +1 (613) 244 4369. E-mail: rachq@rac.ca. Web: www.rac.ca.
Licensing administration: The Industry Canada, Amateur Radio Dept, 300 Slater Street, Ottawa, Ontario K1A 0C8. Tel: +1 (613) 998 3693 1700.

CAPE VERDE (Republic of)

ITU allocation: D4A–D4Z.
Callsign system: Prefix: D4.
Licensing administration: Servicios dos Correios, Telegrafos e Telefones, Praia.

CENTRAL AFRICAN REPUBLIC

ITU allocation: TLA–TLZ.
Callsign system: Prefix: TL.
Licensing administration: Direction générale de l'office centrafricain des postes et télécommunications, Bangui.

CHAD (Republic of the)

ITU allocation: TTA–TTZ.
Callsign system: Prefix: TT.
Licensing administration: Ministère des postes et télécomunications, N'Djamena.

CHILE

ITU allocation: CAA–CEZ, XQA–XRZ, 3GA–3GZ.
Callsign system: Prefixes: CE, XQ. See map for mainland prefixes CE1–8.

Prefix		Area
CE9	AA–AM	Antarctica
	AN–AZ	S Shetland Is
CE0Y		Easter Is
CE0Z		San Felix Is, San Ambrosio Is
CE0Z		Juan Fernandez Is

Licence notes: Four licence classes. 'Aspirante': 3.5, 7MHz (in presence of licensed operator only). Novice: 3.5, 7MHz, SSB/CW, 200W. General, Superior: all bands, 1kW.
Foreign amateur operation: Unilateral (A only). All countries. Apply to RCC.
National society: Radio Club de Chile, PO Box 13630, Santiago 21. Tel: +56 (2) 696-4707. Fax: +56 (2) 672-2623. E-mail: rcch@itn.cl.

CHINA

ITU allocation: BAA–BZZ, VRA–VRZ, XSA–XSZ, 3HA–3UZ.

PEOPLE'S REPUBLIC OF CHINA

Callsign system:

Prefix	Class
B	VHF and UHF stations and contest stations
BA–BC	Class 1
BD–BF	Class 2
BG–BI	Class 3
BR	Repeaters
BS7	Scarborough Reef
BT	Special events
BY	Club stations
BZ	Club operators
VR2	Hong Kong

Licence notes: *Mainland:* Three licence classes for individual operators. Class 1: all bands, 500W HF, 50W VHF. Class 2: all bands

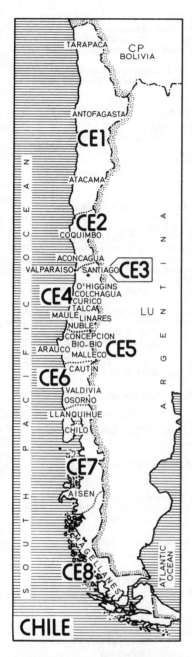

CHILE

COLOMBIA
HK

(restricted), 100W HF, 25W VHF. Class 3: 3.5, 7, 21, 28 and VHF and above, 10W HF, 3W VHF. Three licence classes for club operators. Class 1: all bands, full power of club station. Class 2: all bands, 100W HF, 10W VHF. Class 3: all bands except 14MHz, 100W HF, 10W VHF.
Foreign amateur operation: *Mainland:* Unilateral. A only. Fee. Club operator licence only. Call is own call/club call or just club call. No individual operation permitted. Apply to CRSA. *Hong Kong:* Reciprocal (A + B). Call is in VR2 series.
National society: *Mainland:* Liaison Department, Chinese Radio Sports Association, PO Box 6106, Beijing 100061. Tel: +86 (10) 6702-5488. Fax: +86 (10) 6701-6974. E-mail: crsa@public.bta.net.cn. *Hong Kong:* HARTS, PO Box 541. E-mail: hartscom@harts.org.hk. Web: www.harts.org.hk.

TAIWAN
Callsign system:

Prefix	Area
BO	Quemoy, Matsu Is
BV	Taiwan
BV9P	Pratas Is
BV9S	Spratly Archipelago

National society: Chinese Taipei Amateur Radio League, PO Box 1039, Chuanghua 500. Tel: +886 (4) 738-8746. Fax: +886 (4) 738-5441. E-mail: hq@ctarl.org.tw. Web: www.ctarl.org.tw.

COLOMBIA (Republic of)
ITU allocation: HJA–HKZ, 5JA–5KZ.
Callsign system: See map for mainland prefixes HK1–9. HK0 prefix is used for Bajo Nuevo, Malpelo Is, San Andres Is, Providencia Is and Serrana Bank.
Licence notes: Three licence classes: Class 1, 2, all bands, 2kW PEP. Class 3: 1.8, 3.5, 7MHz and 144MHz, 50W. Third-party traffic permitted.
Foreign amateur operation: Unilateral (A only). Call is HK7/own call. Countries include: Peru, Ecuador, Brazil, USA, Spain, Argentina, Bolivia, Dominican Rep, Canada. All others: 30 days only.
National society: Liga Colombiana de Radioaficionados, PO Box 584, Santafe de Bogotá. Tel: +57 (1) 610 8499. Fax: +57 (1) 610 9877. E-mail: hk31r@yahoo.com. E-mail: www.lcra.org.co.
Licensing administration: Ministerio do comunicaciónes, Bogotá, DE1.

COMOROS (Federal and Islamic Republic of the)
ITU allocation: D6A–D6Z.
Callsign system: Prefix: D68.
Licensing administration: Direction des télécommunications, BP 348, Moroni.

CONGO
ITU allocation: TNA–TNZ.
Callsign system: Prefix: TN.

CONGO (Democratic Republic of)
ITU allocation: 9OA–9TZ.
Callsign system: Prefix: 9Q.
Licence notes: Single licence class: 100W DC input maximum.
National society: Association des Radio Amateurs du Congo, PO Box 1459, Kinshasa 1. E-mail: aracongo@altavista.net. Web: www.multimania.com/aracongo.
Licensing administration: Département des postes, télégraphes et téléphones, PO Box 800, Kinshasa 1.

COSTA RICA
ITU allocation: TEA–TEZ, TIA–TIZ.

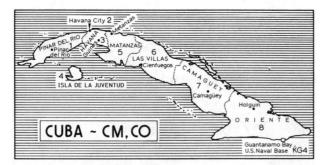

CUBA ~ CM,CO

Callsign system:

Prefix	Area
TI2	San José
TI3	Cartago
TI4	Heredia
TI5	Alajuela
TI6	Limón
TI7	Peninsula de Nicoya
TI8	Puntarenas, Golfito
TI9	Cocos Is

Licence notes: Three licence classes: Class A: all bands, 1kW. Class B: 3.5, 7, 14MHz, 100W. Class C: 7, 14MHz, 100W. Third-party traffic permitted.
Foreign amateur operation: Unilateral (A only). Call is own call/TI. Most countries, including USA. Apply to RCCR.
National society: Radio Club de Costa Rica, PO Box 2412-1000, San José 1000. Tel: +506 280-7855. E-mail: ti0rc@qsl.net. Web: www.qsl.net/ti0rc.

CROATIA
ITU allocation: 9AA–9AZ.
Callsign system: Prefix: 9A.
Foreign amateur operation: CEPT: A + B. Call is 9A/own call.
National society: Hrvatski Radioamaterski Savez, Dalmatiska 12, HR-10000 Zagreb. Tel: +385 (1) 484 8759. Fax: +385 (1) 484 8763. E-mail: hrs@hztk.tel.hr. Web: ham2.irb.hr/hrs.

CUBA
ITU allocation: CLA–CMZ, COA–COZ, T4A–T4Z.
Callsign system: Prefixes: CO (Class A), CM (Class B), CL (Novice Class). See map for call areas.
Licence notes: Three licence classes. Class A: all bands, 450W CW 1kW PEP SSB. Class B: 3.5, 7MHz, 200W CW, 500W PEP SSB. There is also a Novice Class.
National society: Federacion de Radioaficionados de Cuba, PO Box 1, Havana 10100. Tel: +53 (7) 34811. Fax: +53 (7) 335365. E-mail: frcuba@mail.infocom.etecsa.cu. Web: www.infocom.etecsa.cu/cgi-bin/frc/frcuba.htm.
Licensing administration: Ministerio de communicaciones, Plaza de la Revolucion "Jose Marti", Havana.

CYPRUS (Republic of)
ITU allocation: C4A–C4Z, H2A–H2Z, P3A–P3Z, 5BA–5BZ.
Callsign system: Prefix: 5B4. For UK bases, see UK Overseas Territories.
Licence notes: Single licence class, 150W DC input.
Foreign amateur operation: CEPT: A + B. Call is 5B4/own call. Non-CEPT: Reciprocal (A + B). One month's notice. No fee. All EU countries, USA, British Commonwealth.
National society: Cyprus Amateur Radio Society, PO Box 51267, 3053 Limassol. Web: www.spidernet.net/cars.
Licensing administration: Chief Comms Officer, Ministry of Communications and Works, 1424 Nicosia. Tel: +357 (2) 302268. Fax: +357 (2) 360578.

CZECH REPUBLIC
ITU allocation: OKA–OLZ.
Callsign system: See map for call areas.

Prefix	Type of station
OK1, 2	Ordinary stations (KAA–KZZ and RAA–RZZ serial letters denote club calls)
OK4, 5, 6, 7	Special stations
OK8	Foreign amateurs
OK0	Repeaters
OL	Novice

Licence notes: Four licence classes. Class A, B: all-band, 750W, 300W respectively. Class C: 1.8, 3.5, 10, 21, 28MHz plus VHF, 100W. Class D, 3.5MHz, 144MHz, 100W.

Foreign amateur operation: CEPT: A + B. Call is OK/own call. Non-CEPT: Reciprocal (A + B). Countries include UK.
National society: Czech Radio Club, PO Box 69, 11327 Prague 1. Tel: +42 (2) 8722 240. Fax: +42 (2) 8722 242. E-mail: crklub@mbox.vol.cz. Web: crk.mlp.cz.
Licensing administration: Czech Communication Office, Att Mrs Bocková, Klimentská 27, 12502 Prague 1. Tel: +420 (2) 2491 1605. Fax: +420 (2) 2491 1658.

DENMARK
ITU allocation: OUA–OZZ, XPA–XPZ, 5PA–5QZ.
Callsign system: REA–REZ serial letters denote a repeater station.

Prefix	Area
OX3	Greenland (nationals)
OX5	Greenland (USA personnel)
OY	Faroe Is
OZ	Denmark

Licence notes: Four licence classes: Classes A, B: 500,100W DC input respectively, all bands. Class C: 10W DC input, HF bands CW only, plus 100W on VHF. Class D: 100W DC input, VHF phone-only.
Foreign amateur operation: CEPT: A + B. Call is OZ/own call (Denmark), OY/own call (Faroe Is), OX/own call (Greenland). Non-CEPT: Reciprocal (A + B). Call is OZ/own call (Denmark), OY/own call (Faroe Is), OX/own call (Greenland). One month's notice. Form. Fee. Countries include: Australia, Belgium, Brazil, Canada, Germany, Finland, France, Iceland, Luxembourg, Netherlands, Norway, Portugal, Spain, Sweden, Switzerland, UK, USA.
National societies: Denmark and Greenland: Eksperimenterende Danske Radioamatorer, Klokkestoebervej 11, DK-5230 Odense. Tel: +45 66 156511. Fax: +45 66 156598. E-mail: hb@edr.dk. Web: www.edr.dk. Faroe Is: Foroyskir Radioamatorar, PO Box 343, FR-110 Torshavn.
Licensing administration: National Telecom Agency, Holsteingarde 63, DK-2100 Copenhagen Ø. Tel: +45 35 43 03 33.

DJIBOUTI (Republic of)
ITU allocation: J2A–J2Z.
Callsign system: Prefix: J2.
National society: Association des Radioamateurs de Djibouti, BP 1076, Djibouti.

DOMINICA (Commonwealth of)
ITU allocation: J7A–J7Z.
Callsign system: Prefix: J7.
Foreign amateur operation: Countries include USA.
National society: Dominica Amateur Radio Club, PO Box 389, Roseau. Tel: +1 (767) 448 8533. Fax: +1 (767) 448 7708.

DOMINICAN REPUBLIC
ITU allocation: HIA–HIZ.
Callsign system:

Prefix	Area	Town/city
HI1	Beata Is	
HI2	Saona Is	
HI3	North	Santiago
HI4	North-west	Montecristi
HI5	South-west	Barahona
HI6	West	San Juan
HI7	East	La Romana
HI8	South	Santo Domingo
HI9	North-east	Nagua

Licence notes: Four licence classes: Novice: 3.5, 7MHz, CW only 50W. Technician: all bands, 300W. General and Limited: all bands 1kW. Third-party traffic permitted.
Foreign amateur operation: Countries include: Canada, Brazil, El Salvador, UK, USA, Venezuela.
National society: Radio Club Dominicano Inc, Apartado Postal 1157, Santo Domingo. Tel: +1 (809) 533-2211.
Licensing administration: Dirección General de Telecomunicaciónes, Santo Domingo.

ECUADOR
ITU allocation: HCA–HDZ.
Callsign system:

Prefix	City/town and region
HC1	Quito
HC2	Guayaquil
HC3	Loja
HC4	Esmeraldas, Portoviejo
HC5	Cuenca
HC6	Amabato
HC7	East Ecuador

| HC8 | Galapagos Is |
| HC9 | Maritime mobile |

Licence notes: 1kW DC input maximum. Third-party traffic permitted.
Foreign amateur operation: Call is own call/HC. Countries include: USA, Venezuela. Apply to GRC.
National society: Guayaquil Radio Club, PO Box 09-01-5757, Guayaquil. Tel: +593 (4) 294671. Fax: +593 (4) 690241. E-mail: hc2grc@mail.com. Web: www.ecuared.com/hc2grc.

EGYPT (Arab Republic of)

ITU allocation: SSA–SSM, SUA–SUZ, 6AA–6BZ.
Callsign system: Prefix: SU.
Licence notes: Four licence classes. Class A: all modes, 100W DC input. Class B: all modes 50W DC input. Class C: CW only, 25W DC input. Class D: CW only, 10W DC input. Separate VHF licence.
National society: Egypt Radio Amateurs Assembly, PO Box 78, Heliopolis, Cairo 11341. Tel: +20 (2) 574 4841. E-mail: radioass@idsc.gov.eg. Web: www.qsl.net/su0era.
Licensing administration: Telecommunications Organisation, PO Box 2271, Cairo.

EL SALVADOR (Republic of)

ITU allocation: HUA–HUZ, YSA–YSZ.
Callsign system:

Prefix	City/town and area
YS1	San Salvador
YS2	Santa Ana
YS3	San Miguel
YS9	Foreign amateurs

Licence notes: Two licence classes: First class: all bands, 1kW. Second class: all bands, 250W. Third-party traffic permitted.
Foreign amateur operation: Callsign is own call/YS. Countries include: Canada, USA, UK, Dominican Rep, Guatemala, Nicaragua, Costa Rica.
National society: Club de Radio Aficionados de El Salvador, PO Box 517, San Salvador. Tel: +503 260 5882. Fax: +503 260 5874. Web: www.milian.com/faf/cras.shtml.
Licensing administration: Administración Nacional de Telecomunicaciónes, Centro de Gobierno, San Salvador.

EQUATORIAL GUINEA (Republic of)

ITU allocation: 3CA–3CZ.
Callsign system: Prefixes: 3C (mainland), 3C0 (Annobon).
Licensing administration: Dirección General de Telecomunicaciónes, Malabo.

ERITREA

ITU allocation: E3A–E3Z.
Callsign system: Prefix: E3.

ESTONIA

ITU allocation: ESA–ESZ.
Callsign system: Prefix: ES.
Licence notes: Four licence classes: 200W, 50W, 20W and novice.
Foreign amateur operation: *CEPT:* A + B, call is ESx/own call. *Non-CEPT:* Reciprocal (A + B). Countries include UK.
National society: Eesti Raadioamatooride Uhing, PO Box 125, EE-10502 Tallinn. Tel/fax: +372 (6) 570774. Web: www.erau.ee.
Licensing administration: Inspection of Telecommunications Republic of Estonia, Adala 4D, E-0006 Tallinn. Tel: +372 6399075.

ETHIOPIA

ITU allocation: ETA–ETZ, 9EA–9FZ.
Callsign system: Prefix: ET3.
Foreign amateur operation: Call is own call/ET3.
National society: Ethiopian Amateur Radio Society, PO Box 60258, Addis Ababa.
Licensing administration: Ethiopian Telecommunications Authority, PO Box 1047, Addis Ababa.

FIJI

ITU allocation: 3DN–3DZ.
Callsign system: Prefix: 3D2.
Licence notes: Two licence classes. General: all bands, 150W DC input. Technician: 144MHz only, 50W DC input.
Foreign amateur operation: Unilateral (A + B). Fee. Several weeks' notice. Local licence issued. All countries.
National society: Fiji Association of Radio Amateurs, PO Box 184, Suva.

FINLAND **OH**

Licensing administration: The Director, Regulatory Unit, Ministry of Information, Broadcasting, Television and Telecommunication, Government Buildings, Suva. Tel: +679 211257. Fax: +679 300766.

FINLAND

ITU allocation: OFA–OJZ.
Callsign system: See map for prefixes. Market Reef is also OH0M.
Licence notes: Three licence classes: General: all bands, 600W output. Technician: VHF, all modes, 150W output. Novice: 3.5, 7, 21MHz CW only, 28MHz SSB and 144MHz all modes, 15W output.
Foreign amateur operation: *CEPT:* A + B. Call is OH/own call. *Non-CEPT:* Reciprocal (A + B). Four weeks' notice. Fee. Form. All countries.
National society: Suomen Radioamatööriliitto RY, PO Box 44, SF-00441 Helsinki. Tel: +358 (9) 562 5973. Fax: +358 (9) 562 3987. E-mail: hq@sral.fi. Web: www.sral.fi.
Licensing administration: Telehallintokeskus, PO Box 53, SF-00211 Helsinki. Tel: +358 (0) 69661. Fax: +358 (0) 6966410.

FRANCE

ITU allocation: FAA–FZZ, HWA–HYZ, THA–THZ, TKA–TKZ, TOA–TQZ, TVA–TXZ.
Callsign system (France): New callsigns have prefix which indicates licence class A–E, ie FA–FE.

Prefix	Country
F, FA, FB	France
FF	Clubs (France)
FG	Guadeloupe
FH	Mayotte
FJ	St Barthelemy
FK	New Caledonia
FM	Martinique
FO	French Oceania: Clipperton Is, Gambier Is, Marquesas Is,

	Society Is (Tahiti), Tubuai Is
FP	St Pierre et Miquelon
FR	Reunion
FR/E	Europa Is
FR/G	Glorieuses Is
FR/J	Juan de Nova Is
FR/T	Tromelin Is
FS	St Martin
FT-W	Crozet Is
FT-X	Kerguelen Is
FT-Y	Terre Adelie
FT-Z	Amsterdam and St Paul Is
FW	Wallis and Futuna Is
FX	beacons
FY	French Guiana
FZ	repeaters
TK	Corsica

Licence notes (France): Five licence classes. Class A: 144MHz only. Class B: 144MHz plus some CW segments in HF bands and 28MHz phone. Class C: above 30MHz. Class D: all bands. Class E: issued after three years in Class D.
Foreign amateur operation: *CEPT*: A + B. Call is prefix letters/ own call. *Non-CEPT (France)*: Reciprocal (A + B). Call is F series. Three months' notice. Form. Fee. Countries include: Andorra, Belgium, Canada, Germany, Luxembourg, Monaco, Netherlands, Spain, Switzerland, UK, USA.
National society: *France*: REF-Union, BP7429, F-37074 Tours Cedex 2. Tel: +33 (2) 4741 8873. Fax: +33 (2) 4741 8888. E-mail: ref@ref.tm.fr. Web: www.ref-union.org, www.ref.tm.fr. *French Polynesia*: Club Oceanien de Radio et d'Astronomie, PO Box 5006, Pirae 98716, Tahiti.
Licensing administration (France): Direction Génerale des Télécommunications, Direction des Affairs Industrielles et Internationales, Services des Affairs Internationales, 7 boulevard Romain Rolland, F-92128 Montrouge. Tel: +33 (1) 45 642222.

GABON REPUBLIC

ITU allocation: TRA–TRZ.
Callsign system: Prefix: TR.
National society: Association Gabonaise des Radio-Amateurs, BP 1826, Libreville.
Licensing administration: M le Directeur général de l'Office des postes et télécommunications de la République Gabonais, Libreville.

GAMBIA (Republic of the)

ITU allocation: C5A–C5Z.
Callsign system: Prefix: C5.
Licence notes: Single class of licence. 3.5–28MHz bands only but endorsements available for 18 and 144MHz.
Foreign amateur operation: Unilateral (A + B). Call is C53/own call. Countries include UK.
National society: Radio Society of the Gambia, c/o Jean-Michel Voinot, C53GB, PMB 120, Banjul.
Licensing administration: The Managing Director, Gambia Telecommunications Co Ltd, PO Box 387, Banjul. Tel: +220 29999.

GEORGIA

ITU allocation: 4LA–4LZ.
Callsign system: Prefix: 4L.

First serial letters	Area	Capital
Q, S	Adzharia	Batumi
V, X	Abkhazia	Sukhumi
Other letters	Georgia	Tbilisi

GERMANY (Federal Republic of)

ITU allocation: DAA–DRZ, Y2A–Y9Z.
Callsign system:

Callsign series	Licence class	Notes
DA1AA–DA2ZZ	B	Military stations
DA4AA–DA4ZZ	C	Military stations
DB0AA–DB0ZZ	C	Clubs and repeaters
DB1AA–DB9ZZ	C	
DC0AA–DC0EZ	C	
DC0FA–DC0JZ	C	Reciprocal licence
DC0KA–DC0ZZ	C	
DC1AA–DC9ZZ	C	
DD0AA–DD4ZZ	C	
DD5AA–DD5ZZ	C	Reciprocal licence
DD6AAA–DD6ZZZ	C	
DD7AA–DD9ZZ	C	
DF0AAA–DF0ZZZ	B	Club station
DF1AA–DF9ZZ	B	

Greece

DG1AAA–DG9ZZZ	C	
DH0AAA–DH9ZZZ	A	
DJ0AAA–DJ0ZZZ	B/A	Reciprocal licence
DJ1AA–DJQZ	B	
DK0AA–DK9ZZ	B	
DL0AAA–DL9ZZZ	B	

Licence notes: Three licence classes. Class A: 3.5 and 21MHz (CW, rtty only), 28MHz plus VHF (all modes), 150W output. Class B: all bands, all modes, 750W output. Class C: VHF phone only, 75 output.
Foreign amateur operation: *CEPT*: A + B. Call is DL/own call (CEPT Class 1), DC/own call (CEPT Class 2). *Non-CEPT*: Reciprocal. Six weeks' notice. Fee. Call is DL/own call or DC/own call. Licence valid for three months only. Countries include UK, USA. Apply to DARC.
National society: Deutscher Amateur Radio Club, Postfach 1155, D-34216 Baunatal. Tel: +49 (561) 949 880. Fax: +49 (561) 949 8850. E-mail: darchq@t-online.de. Web: www.darc.de.

GHANA

ITU allocation: 9GA–9GZ.
Callsign system: Prefix: 9G.
Licence notes: Maximum power 400W PEP.
National society: Ghana Amateur Radio Society, Box 3936, Accra.
Licensing administration: Director-General, Posts and Telecommunications Corporation, Accra.

GREECE

ITU allocation: SVA–SZZ, J4A–J4Z.
Callsign system: See map for call areas.
Licence notes: Three licence classes. Classes A, B, C: all bands, 300, 150, 50W PEP respectively.
Foreign amateur operation: *CEPT*: Call is SVx/own call. *Non-CEPT*: Two months' notice. A + B. Call is SVx/own call for temporary licence. All EU countries, otherwise reciprocal (including USA).
National society: Radio Amateur Association of Greece, PO Box 3564, GR-10210 Athens. Tel: +30 (1) 522 6516. Fax: +30 (1) 522 6505. E-mail: raag@athena.domi.gr. Web: www.domi.gr/eer.
Licensing administration: Ministry of Transport and Telecommunications, Administration of Posts & Telecommunications, 49 avenue Sygrou, GR-11780 Athens. Tel: +30 (1) 923-2906.

GRENADA

ITU allocation: J3A–J3Z.
Callsign system: Prefix: J3
Licence notes: Maximum power 2kW PEP.
Foreign amateur operation: All countries.
National society: Grenada Amateur Radio Club, PO Box 737, St George's.
Licensing administration: Wireless Officer, Ministry of Communications and Works, St George's.

GUATEMALA (Republic of)

ITU allocation: TDA–TDZ, TGA–TGZ.
Callsign system:

Prefix	City/town area
TG4	Escuintla
TG5	Huehuetenango
TG6	Chiquimula
TG7	Izabal
TG8	Quezaltenango
TG9	Guatemala City
TG0	Special stations

Licence notes: Two licence classes. Novice: 7MHz CW/AM only. General: all bands, all modes, 2kW PEP. Third-party traffic permitted.
Foreign amateur operation: Call is own call/TG. Countries include USA.
National society: Club de Radioficionados de Guatemala, PO Box 115, Guatemala City 01901. Tel: +502 597 2027. E-mail: crag@micro.com.gt. Web: members.xoom.com/CRAG.
Licensing administration: Dirección General de Radio y Television Nacional, Guatemala City.

GUINEA (People's Revolutionary Republic of)

ITU allocation: 3XA–3XZ.
Callsign system: Prefix: 3X.
Licensing administration: Ministère des postes et télécommunications, Conakry.

GUINEA-BISSAU (Republic of)

ITU allocation: J5A–J5Z.
Callsign system: Prefix: J5.
Licensing administration: Commissariado de Estado dos Correios e Telecomunicações, Bissau.

GUYANA

ITU allocation: 8RA–8RZ.
Callsign system: Prefix: 8R.
Foreign amateur operation: Call is own call/8R. Countries include USA.
National society: Guyana Amateur Radio Association, PO Box 101122, Georgetown.
Licensing administration: The General Manager, Guyana Telecommunication Corporation, 55 Brickdam, Georgetown.

HAITI (Republic of)

ITU allocation: HHA–HHZ, 4VA–4VZ.
Callsign system: Prefix: HH.
Licence notes: Four licence classes: First, second and third class: all bands, 2kW DC input. Third class (B): all bands, 350W DC input. Third class (C): 1.8, 3.5, 7, 21MHz CW/phone, 100W DC input. Third-party traffic permitted.
Foreign amateur operation: Countries include: Canada, France, Switzerland, USA, Dominican Rep.
National society: Radio Club d'Haiti, PO Box 1484, Port-au-Prince. Tel: +509 22 1336. Fax: +509 57 4925.
Licensing administration: CONATEL, Ave Marie Jeanne, Port-au-Prince.

HONDURAS (Republic of)

ITU allocation: HQA–HRZ.
Callsign system:

Prefix	Area	City/town
HR1	—	Tegucigalpa, DC
HR2	Cortes	San Pedro Sula
HR3	Atlantida	La Ceiba
HR4	Valle	Amapala
HR5	Copan	Santa Rosa
HR6	Bay Is	
HR0	(mobile)	

Licence notes: 1kW DC input maximum. Third-party traffic permitted.
Foreign amateur operation: Call is own call/HR. Countries include USA.
National society: Radio Club de Honduras, Apartado 273, San Pedro Sula. Tel/fax: +504 556 6173. E-mail: radioclub@globalnet.hn.
Licensing administration: Dirección de Radio Nacional, Hondutel, Tegucigalpa.

HUNGARY

ITU allocation: HAA–HAZ, HGA–HGZ.
Callsign system: See map. The letters shown in the map are the first letters after the call number. K is reserved for club stations and Y for youth club stations.

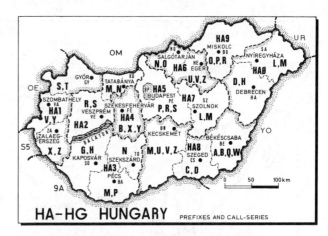

HA–HG HUNGARY   PREFIXES AND CALL-SERIES

Licence notes: Five licence classes. Young class: 3.5MHz, 10W DC input. Class A: 3.5MHz, 7MHz, 25W DC input. Class B: all bands, 100W DC input. Class C: all bands, 250W DC input. Class C (contests): all bands, 500W DC input.
Foreign amateur operation: *CEPT:* A + B. Call is HA/own call (CEPT Class 1), HG/own call (CEPT Class 2). *Non-CEPT:* A only. All countries. Apply to MRASZ.
National society: Magyar Radioamator Szovetseg, PO Box 11, H-1400 Budapest. Tel: +36 (1) 312-1616. Fax: +36 (1) 311 6204. E-mail: mrasz@elender.hu. Web: www.mrasz.hu.
Licensing administration: Hungarian Telecommunications Corporation, Krisztina Krt 6-8, H-1541 Budapest. Tel: +36 (1) 175-6885. Fax: +36 (1) 175-3093.

ICELAND (Republic of)

ITU allocation: TFA–TFZ.
Callsign system: See map for call areas. Novice and Technician callsigns have N or T respectively as the last of three serial letters, eg TF1AN. Foreign amateurs who have passed the Icelandic amateur radio examination have X as the first of three serial letters, eg TF1XAA. Icelanders who have been licensed for 25 years or longer and have a Class C licence can obtain single-serial-letter callsigns, eg TF1A.
Licence notes: Five licence classes. Novice: 3.5, 7, 21MHz, crystal-controlled CW only, 5W DC input. Technician: VHF phone only, 50W DC input. Class A: all bands, CW only, 50W DC input. Classes B and C: all bands, all modes, 200 and 500W DC input respectively.
Foreign amateur operation: *CEPT:* A + B. Call is TF/own call. *Non-CEPT.* Reciprocal (A + B). Three months' notice. Form. Countries include: Canada, Denmark, Luxembourg, Norway, Sweden, Switzerland, UK, USA, Germany.
National society: Islenzkir Radioamatorar, PO Box 1058, IS-121 Reykjavik.
Licensing administration: National Telecom Inspectorate, Malarhofda 2, IS-150, Reykjavik. Tel: +354 587 2424.

INDIA (Republic of)

ITU allocation: ATA–AWZ, VTA–VWZ, 8TA–8YZ.
Callsign system: Prefix: VU.

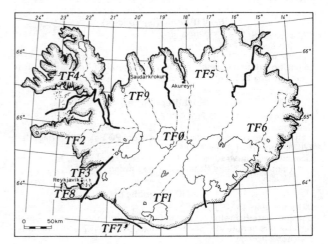

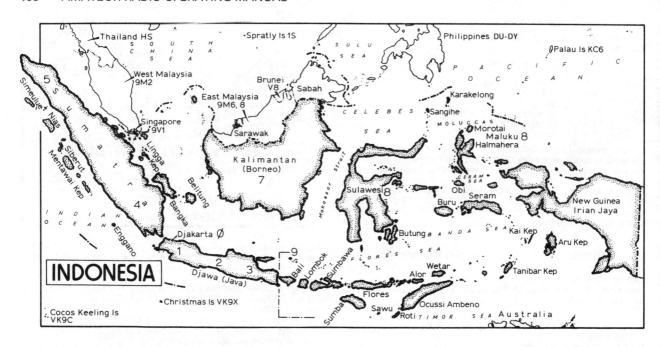

Licence notes: Three licence classes. Advanced, Grade 1, all bands, 150W DC input. Grade 2: 3.5, 7MHz CW only plus VHF, 25W DC input. Permitted powers on VHF are lower.
Foreign amateur operation: Reciprocal (A + B). Call is in VU series. Countries include: UK, USA, Canada, Brazil, Peru, Germany, Greece, Poland, Switzerland, Japan, Australia, New Zealand.
National society: Amateur Radio Society of India, 4 Kurla Industrial Estate, Ghatkopar, Mumbai, Bombay 400086. Tel: +91 (22) 514 7574. Fax: +91 (22) 511 5810.
Licensing administration: Ministry of Communications, Sanchar Bhavan, 20 Ashoka Road, New Delhi 110001. Tel: +91 (11) 335 5441. Fax: +91 (11) 371 6111.

INDONESIA (Republic of)

ITU allocation: JZA–JZZ, PKA–POZ, YBA–YHZ, 7AA–7IZ, 8AA–8IZ.
Callsign system: Prefixes: YB–YH. See map on next page for call areas.

Prefix	Licence class
YB	General
YC	Middle
YD	Novice

Licence notes: Three licence classes: Novice: 3.5MHz, CW/AM, 10W. Middle: all bands except 14MHz, all modes, 75W. General: all bands, all modes, 500W.
Foreign amateur operation: Unilateral (A only). Countries include: USA, Canada, Germany, UK, France, Sweden, Argentina, Australia. Apply to ORARI.
National society: Organisasi Amatir Radio Indonesia, Jalan Karang Tengah Raya 59B, Lebak Bulus, Jakarta 12440. Tel: +62 (21) 758 16884. Fax: +62 (21) 766 8726.

IRAN (Islamic Republic of)

ITU allocation: EPA–EQZ, 9BA–9DZ.
Callsign system: Prefix: EP.
Licence notes: 100W maximum output.
Licensing administration: Directorate General of Telecommunications, PO Box 931, Teheran.

IRAQ (Republic of)

ITU allocation: HNA–HNZ, YIA–YIZ.
Callsign system: Prefix: YI.
National society: Iraqi Radio Amateur Club, PO Box 55072, Baghdad 12001. Tel: +964 (1) 884 3521.
Licensing administration: The Director General of Posts, Telegraphs and Telephones, State Organisation, Baghdad.

IRELAND (Republic of)

ITU allocation: EIA–EJZ.
Callsign system: Prefixes: EI (mainland). EJ (offshore islands). Three serial letters denote VHF phone only except VAA–VZZ which denotes a visitor's licence.

Licence notes: Two licence classes. All bands, 150W DC input and VHF phone only.
Foreign amateur operation: CEPT: A + B. Call is EI/own call. Non-CEPT: Reciprocal (A + B). Call is in EI series (mainland) or EJ series (offshore islands). Countries include UK, USA.
National society: Irish Radio Transmitters Society, PO Box 462, Dublin 9. E-mail: secretary@irts.ie. Web: www.irts.ie.
Licensing administration: Office of Director of Telecommunications Regulation, Radio Section, Abbey Court Irish Life Centre, Lower Abbey Street, Dublin 1. Tel: +353 (1) 804 9600. Fax: +353 (1) 804 9680.

ISRAEL (State of)

ITU allocation: 4XA–4XZ, 4ZA–4ZZ.
Callsign system: Prefixes 4X, 4Z. Serial letters NAA–NZZ denote Class C licence.
Licence notes: Three licence classes. Class A, B all bands, and 250W DC input respectively. Class C: 7 and 21MHz CW only, 10W DC input. Third-party traffic permitted.
Foreign amateur operation: CEPT: A + B. Non-CEPT: Reciprocal (A + B). Call is 4X/own call (CEPT Class 1), 4Z/own call (CEPT Class 2). Countries include UK, USA.
National society: Israel Amateur Radio Club, PO Box 17600, Tel-Aviv 61176. Tel: +972 (3) 565 8203 (Friday mornings). E-mail: iarc@iarc.org. Web: www.iarc.org.
Licensing administration: Ministry of Communications, PO Box 29187, Tel-Aviv 61290. Tel: +972 (3) 519 8277.

ITALY

ITU allocation: IAA–IZZ.
Callsign system: See map for prefixes. IW is Technician class prefix.
Licence notes: Four licence classes. Classes 1, 2, 3: 75,150, 300W DC input respectively. Technician: VHF only, 10W.
Foreign amateur operation: CEPT: A + B. Call is IKx/own call (CEPT Class 1), IWx/own call (CEPT Class 2). Non-CEPT: Reciprocal (A + B). Fee. Form. Countries include: all EU, USA.
National society: Associazione Radioamatori Italiani, Via Scarlatti 31, I-20124 Milan. Tel: +39 (2) 669 2192. Fax: +39 (2) 667 14809. E-mail: ari@micronet.it. Web: www.ari.it.
Licensing administration: Ministero delle Poste e Telecomunicazioni, Direzione Centrale dei Servizi Radioelettrici, Divisione VI – Sezione IV, viale Europa 175, I-00144 Rome. Tel: +39 (6) 595 4894.

IVORY COAST (Republic of the)

ITU allocation: TUA–TUZ.
Callsign system: Prefix: TU.
Licence notes: 1kW DC input maximum.
National society: Association des Radio-Amateurs Ivoiriens, PO Box 2946, Abidjan 01.
Licensing administration: Office des postes et télécommunication de la République de Côte d'Ivoire, Direction générale des télécommunications, Abidjan.

JAMAICA

ITU allocation: 6YA–6YZ.
Callsign system: Prefix: 6Y.
Licence notes: 1kW DC input maximum. Third-party traffic permitted.
Foreign amateur operation: Unilateral (A only). Call is own call/6Y5. Countries include: British Commonwealth, USA, Netherlands.
National society: Jamaica Amateur Radio Association, 76 Arnold Road, Kingston 5. E-Mail: 6y5mm@toj.com. Web: www.buyjamaica.com/jara.
Licensing administration: Headquarters, Posts and Telegraphs Dept, South Camp Road, Kingston.

JAPAN

ITU allocation: JAA–JSZ, 7JA–7NZ, 8JA–8NZ.
Callsign system: Prefixes: JA, JD1 (Ogasawara Is, Minami Torishima), JE–JS, 7J–7N. See map for call areas.
Licence notes: Four licence classes. First class: 500W PEP. Second class: 100W PEP. Telegraphy class: 10W CW-only (except 14MHz) Telephony class: 10W PEP phone-only (except 1.9 and 14MHz).
Foreign amateur operation: Countries include: USA, Germany, Finland. Apply to JARL.
National society: Japan Amateur Radio League, Tokyo 170-8073. Tel: +81 (3) 5395-3106. Fax: +81 (3) 3943-8282. E-mail: hq@jarl.or.jp. Web: www.jarl.or.jp.

JORDAN (Hashemite Kingdom of)

ITU allocation: JYA–JYZ.
Callsign system: Prefixes: JY (individual stations), JY6 (club stations), JY9 (foreign amateurs).
Licence notes: 1kW DC input maximum. Third-party traffic permitted.
Foreign amateur operation: Unilateral. Call is in JY9 series. Countries include USA. Apply via RJARS.
National society: Royal Jordanian Radio Amateur Society, PO Box 2353, Amman. Tel: +962 (6) 666 235.

KAZAKHSTAN

ITU allocation: UNA–UQZ.
Callsign system: Prefix is UN.

1st serial letter	Area	Capital
A	Mangyshlak Obl	Shevchenko
B	Tselinograd Obl	Tselinograd
C	North Kazakh Obl	Petropavlovsk
D	Semipalatinsk Obl	Semipalatinsk
E	Kokchetav Obl	Kokchetav
F	Pavlodar Obl	Pavlodar
G	Alma-Ata city	—
I	Atyubinsk Obl	Atyubinsk
J	East Kazakh Obl	Ust-Kamenogorsk
K	Kzyl-Orda Obl	Kzyl-Orda
L	Kustanay Obl	Kustanay
M	Urals Obl	Uralsk
N	Chimkent Obl	Chimkent
O	Guryev Obl	Guryev
P	Karaganda Obl	Karaganda
Q	Alma-Ata Obl	Alma-Ata
R	Dzhezkazgan Obl	Dzhezkazgan
S	Kzyl-Ordinskaya	—
T	Dzhambul	Dzhambul
V	Taldy-Kurgan	Taldy-Kurgan
Y	Turgav Obl	Arkalyk

KENYA (Republic of)

ITU allocation: 5YA–5ZZ.
Callsign system: Prefix: 5Z4.
Licence notes: Single licence class: all bands, 150W DC input.

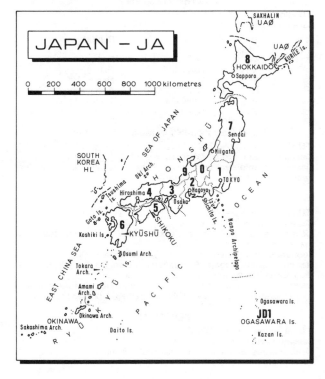

Foreign amateur operation: Unilateral (A only). Fee. Six months' notice or more. No short-term licences. Call is in 5Z4 series. Countries include UK.
National society: Radio Society of Kenya, PO Box 45681, Nairobi. Tel/fax: +254 (2) 449373.
Licensing administration: The Manager, Frequency Spectrum Management Branch, KP & TC, PO Box 30301, Nairobi. Tel: +254 (2) 449373.

KIRIBATI (Republic of)
ITU allocation: T3A–T3Z.
Callsign system:

Prefix	Area
T30	West Kiribati (Gilbert Is)
T31	Central Kiribati (formerly British/American Phoenix Is), including Kanton and Enderbury Is
T32	East Kiribati (Line Is)
T33	Banaba Is (Ocean Is)

Foreign amateur operation: Most countries, including USA.
Licensing administration: Controller of Telecommunications, Ministry of Communications & Works, PO Box 487, Betio, Tarawa Atoll.

KOREA (Democratic People's Republic of)
ITU allocation: HMA–HMZ, P5A–P9Z.
Callsign system: Prefix: P5.
Licensing administration: Ministry of Posts and Telecommunications, Pyongyang.

KOREA (Republic of)
ITU allocation: DSA–DTZ, D7A–D7Z, HLA–HLZ, 6KA–6NZ.
Callsign system: Prefixes: HL, DS.
Licence notes: Three licence classes: First class: all bands, all modes, 500W. Second class: all bands except 14MHz, 100W. Third class: 1.8, 3.5, 7MHz, 28MHz, VHF, 50W.
National society: Korean Amateur Radio League, CPO Box 162, Seoul 100-601. Tel: +82 (2) 575 9580. Fax: +82 (2) 576 8574. E-mail: karl@karl.or.kr. Web: www.karl.or.kr.
Licensing administration: Ministry of Communications, Taepyong-ro Choong-Ku, Seoul.

KUWAIT (State of)
ITU allocation: 9KA–9KZ.
Callsign system: Prefix: 9K.
Foreign amateur operation: Countries include USA.
National society: Kuwait Amateur Radio Society, PO Box 5240, Safat 13053. Tel: +965 533 3762. Fax: +965 531 1188. E-mail: kars@kuwait.net. Web: www.kars.org.

KYRGHYZSTAN
ITU allocation: EXA–EXZ.
Callsign system: Prefix: EX.

1st serial letter	Area	Capital
M	—	Frunze
N	Osh Obl	Osh
P	Naryn Obl	Naryn
Q	Issyk-Kul Obl	Przhevalsk
T	Talas Obl	Talas
V	Dzhalal-Abad Obl	

LAOS (Lao People's Democratic Republic)
ITU allocation: XWA–XWZ.
Callsign system: Prefix: XW.
Licensing administration: Ministère des postes et télécommunications, Avenue Lane Xang, Vientane.

LATVIA
ITU allocation: YLA–YLZ.
Callsign system: Prefix: YL.
Licence notes: Four licence classes: 200W, 50W restricted, 20W restricted, novice.
Foreign amateur operation: CEPT: Call is YL/own call.
National society: Latvijas Radioamatieru Liga, PO Box 164, LV-1010 Riga. E-mail: lral@ardi.lv. Web: www.lral.ardi.lv.
Licensing administration: Latvia Communication State Inspection, 41/43 Elizabetes Street, LV-1010 Riga. Tel: +371 (2) 333034.

LEBANON
ITU allocation: ODA–ODZ.
Callsign system: Prefix: OD.
Licence notes: Single licence class, 1kW DC input.
Foreign amateur operation: All countries.

National society: Association des Radio Amateurs Libanais, PO Box 11-8888. Beirut. Web: www.ral.org.lb.
Licensing administration: Ministère des PTT, Direction générale des télégraphes et téléphones, Bir Hassan, Beirut.

LESOTHO (Kingdom of)
ITU allocation: 7PA–7PZ.
Callsign system: Prefix: 7P.
National society: Lesotho Amateur Radio Society, PO Box 949, Maseru 100.
Licensing administration: The Director of Posts and Telecommunications, PO Box 413, Maseru.

LIBERIA (Republic of)
ITU allocation: A8A–A8Z, D5A–D5Z, ELA–ELZ, 5LA–5MZ, 6ZA–6ZZ.
Callsign system: See map for prefixes. 5L is special-event prefix.
Licence notes: Two licence classes. Basic: 3.5, 7 and 21MHz CW, 75W DC input. General: all bands, 1kW DC input.
Foreign amateur operation: All countries.
National society: Liberian Radio Amateur Association, PO Box 10-1477, 1000 Monrovia 10.
Licensing administration: Ministry of Post & Telecommunications, Monrovia.

LIBYA (Socialist People's Libyan Arab Jamahiriya)
ITU allocation: 5AA–5AZ.
Callsign system: Prefix: 5A.
Licensing administration: Posts and Telecommunication Corporation, Tripoli.

LIECHTENSTEIN – see Switzerland

LITHUANIA
ITU allocation: LYA–LYZ.
Callsign system: Prefix: LY.
Licence notes: Four licence classes: 160m only, 10W, 50W and 200W.
Foreign amateur operation: CEPT: A + B using LY/own call.
National society: Lietuvos Radijo Megeju Draugija, PO Box 1000, LT-2001 Vilnius. Tel: +370 (2) 221 836 (Tuesday evenings and Sunday mornings). Fax: +370 (2) 700 447. E-mail: hq@lrmd.ktl.mii.lt.

LUXEMBOURG (Duchy of)
ITU allocation: LXA–LXZ.
Callsign system:

Prefix	Licence category
LX1	Luxembourg nationals
LX2	Foreign nationals
LX9	Club stations
LX0	Beacons, repeaters, special

Licence notes: Power limit 100W or one year after issue of initial licence 1000W PEP on request.
Foreign amateur operation: CEPT: A + B. Call is LX/own call. Non-CEPT: Unilateral (A only). Call is in LX series. Countries include: Austria, Belgium, Canada, Denmark, Faroes, France, Germany, Netherlands, Norway, Switzerland, UK, USA.
National society: Reseau Luxembourgeois des Amateurs d'Ondes Courtes, c/o J Kirsch, LX1DK, 23 route de Noertzange, L-3530 Dudelange. Web: lx0.restena.lu/rl.

Licensing administration: Ministère des Communications, 18 Montée de la Pétrusse, L-2945 Luxembourg. Tel: +352 4781.

MACEDONIA

ITU allocation: Z3A–Z3Z.
Callsign system: Prefix: Z3.
National society: Radioamaterski Sojuz na Makedonija, PO Box 14, 91000 Skopje. Tel: +389 (1) 237371. Fax: +389 (1) 238257.

MADAGASCAR (Democratic Republic of)

ITU allocation: 5RA–5SZ, 6XA–6XZ.
Callsign system: Prefix: 5R8.
Licensing administration: Ministère des postes et télécommunications, Antaninarenina, Tananarive.

MALAWI (Republic of)

ITU allocation: 7QA–7QZ.
Callsign system: Prefix: 7Q.
Licensing administration: The Postmaster General, PO Box 537, Blantyre.

MALAYSIA (Federation of)

ITU allocation: 9MA–9MZ, 9WA–9WZ.
Callsign system:

Prefix	State
9M0	Spratly Is
9M2	West Malaysia
9M6	Sabah
9M8	Sarawak

Licence notes: Single licence class: 150W DC input.
Foreign amateur operation: Reciprocal (A only). Countries include: UK, USA, Philippines, Japan, Australia plus certain European countries.
National society: MARTS, PO Box 10777, 50724 Kuala Lumpur. Web: www.jaring.my/enrich/marts.
Licensing administration: IBU Pejabat, Jabatan Telekom Malaysia, (Kementerian Tenega Telekom Dan Pos Malaysia), Wisma Damansra, Jalan Sematan, 50668 Kuala Lumpur. Tel: +60 3255 6687.

MALDIVES (Republic of)

ITU allocation: 8QA–8QZ.
Callsign system: Prefix: 8Q.
Licensing administration: The Director, Telecommunication Department, Male.

MALI (Republic of)

ITU allocation: TZA–TZZ.
Callsign system: Prefix: TZ.
National society: Club des Radioamateurs et Affilies du Mali, PO Boz 2826. Bamako.
Licensing administration: Office des postes et télécommunications de la République du Mali, Bamako.

MALTA (Republic of)

ITU allocation: 9HA–9HZ.
Callsign system:

Prefix	Island
9H1	Malta
9H3	Malta (visitors and special)
9H4	Gozo
9H5	VHF licences
9H8	Comino
9H0	Special events

Licence notes: Two licence classes. Class A: all bands, 150W DC input. Class B: VHF phone only, 150W DC input.
Foreign amateur operation: *CEPT:* A + B using 9H/own call. *Non-CEPT:* Reciprocal (A + B). Call is in 9H3 series. Form. All countries.
National society: Malta Amateur Radio League, PO Box 575, Valletta.
Licensing administration: Wireless Telegraphy Dept, Evans Laboratory Buildings, Merchants Street, Valletta. Tel: +356 24-39-25. Fax: +356 23-44-94.

MARSHALL ISLANDS

ITU allocation: V7A–V7Z.
Callsign system: V7.
Foreign amateur operation: Countries include USA.

MAURITANIA (Islamic Republic of)

ITU allocation: 5TA–5TZ.
Callsign system: Prefix: 5T.

Licensing administration: Office des postes et télécommunications de la République Islamique du Mauritanie, Nouakchott.

MAURITIUS AND DEPENDENCIES

ITU allocation: 3BA–3BZ.
Callsign system:

Prefix	Island
3B6	Agalega Is
3B7	St Brandon Is
3B8	Mauritius
3B9	Rodriguez Is

Licence notes: 150W DC input maximum.
Foreign amateur operation: Reciprocal (A only). Countries include UK.
National society: Mauritius Amateur Radio Society, PO Box 104, Quatre Bornes.
Licensing administration: Mauritius Telecommunications Authority, 6th Floor, Blendax House, Dumas Street, Port Louis. Tel: +230 208 5623. Fax: +230 211 2871.

MEXICO

ITU allocation: XAA–XIZ, 4AA–4CZ, 6DA–6JZ.
Callsign system:

Prefix	Area
XE1	Central Mexico
XE2	North Mexico
XE3	South Mexico
XF1	Islands N of 20°N
XF2	Islands W of 90°W
XF3	Islands E of 90°W
XF4	Revilla Gigedo Is

Licence notes: Four licence classes: First class: all bands, 1kW. Second class: all bands: 250W. Third class: all bands, 150W, no mobile operation. VHF: VHF only, 100W.
Foreign amateur operation: Must be resident in Mexico. Countries include USA.
National society: Federación Mexicana de Radio Experimentadores, PO Box 907, 06000 Mexico, DF. Tel: +52 (5) 563-1405. Fax: +52 (5) 563-2264. E-mail: fmre@supernet.com.mx. Web: www.fmre.org.mx.
Licensing administration: Dirección General de Telecomunicaciónes, Av Lázaro Cardenas 567, Mexico 12, DF.

MICRONESIA (Federated States of)

ITU allocation: V6A–V6Z.
Callsign system: Prefix: V6.
Foreign amateur operation: Countries include USA.

MOLDOVA

ITU allocation: ERA–ERZ.
Callsign system: Prefix: ER.
National society: Asociatia Radioamatorilor din Republica Moldova, PO Box 9537, MD-2071 Kishinev. E-mail: arm@qsl.net.

MONACO (Principality of)

ITU allocation: 3AA–3AZ.
Callsign system:

Prefix	Licence class
3A1	VHF/UHF phone-only
3A2	All band

Licence notes: For licence classes see prefix table. All licences are for 100W anode dissipation.
Foreign amateur operation: *CEPT:* A + B. Call is 3A/own call. Note: Licensing administration must be informed of exact location of station prior to operation due to indistinct borders between Monaco and France. *Non-CEPT:* Reciprocal (A + B). Three months' notice. Call is 3A/own call. Countries include UK, USA.
National society: Association des Radio-Amateurs de Monaco, BP 2, MC-98001 Monaco Cedex. Tel: +33 93 254 727. Fax: +33 92 165 601.
Licensing administration: Direction Générale de Telecommunications, Service Radio-Amateurs, 25 boulevard de Suisse, MC98030 Monaco Cedex. Tel: +33 93 250 505.

MONGOLIA (Mongolian People's Republic)

ITU allocation: JTA–JVZ.
Callsign system:

Prefix	Area
JT0A–M	Bayan-Olgij
JT0N–Z	Khowd
JT1A–G	Ulan Bator
JT1H–Z	Nalajch

JT2A–M	Dornod
JT2N–Z	Suchbatar
JT3A–M	Chentij
JT3N–Z	Dornogobi
JT4A–M	Omnogobi
JT4N–Z	Dungobi
JT5A–M	Selenge
JT5N–Z	Tow
JT6A–M	Archangaj
JT6N–Z	Oworchangaj
JT7A–M	Chowsgol
JT7N–Z	Bulgan
JT8A–M	Gobi-Altai
JT8N–Z	Bayan-Chongor
JT9A–M	Uws
JT9–Z	Dzawchan

Foreign amateur operation: Unilateral (A only). Call is JT/own call. Apply to MRSF.
National society: Mongolian Radio Sports Federation, PO Box 639, Ulaanbaatar. Tel: +976 (1) 320058. E-mail: mrsf@magicnet.mn.

MOROCCO (Kingdom of)
ITU allocation: CNA–CNZ, 5CA–5GZ.
Callsign system: Prefix: CN.
Licence notes: 100W maximum.
Foreign amateur operation: Unilateral (A only). Call is own call/CN. Three months' notice. All countries except Israel but residents only.
National society: Association Royale des Radio-Amateurs du Maroc, BP 299, Rabat. Tel: +212 (7) 673703. Fax: +212 (7) 674757.
Licensing administration: Ministère des PTT, Division des télécommunications, Rabat.

MOZAMBIQUE (People's Republic of)
ITU allocation: C8A–C9Z.
Callsign system: Prefix: C9.
National society: Liga dos Radio Emissores de Mocambique, PO Box 25, Maputo.
Licensing administration: Ministerio dos Transportes e Communicações, Maputo.

MYANMAR
ITU allocation: XYA–XZZ.
Callsign system: Prefix: XY, XZ.
Licensing administration: Posts and Telecommunications Dept, Ministry of Transport and Communications, Ministers' Office, Yangon.

NAMIBIA
ITU allocation: V5A–V5Z.
Callsign system: Prefix: V5.
National society: Namibian Amateur Radio League, PO Box 1100, Windhoek 9000.

NAURU (Republic of)
ITU allocation: C2A–C2Z.
Callsign system: Prefix: C2.
Foreign amateur operation: Call is own call/C2.

NEPAL
ITU allocation: 9NA–9NZ.
Callsign system: Prefix: 9N.
Licensing administration: Ministry of communications, Singha Durbar, Kathmandu.

NETHERLANDS (Kingdom of the)
ITU allocation: PAA–PIZ.
Callsign system:

Prefix	Licence category
PA0	Class A
PA2	Class A, B
PA3	Class A, B
PA6	Special
PD0	Class D
PE0	Class C
PE1	Class C
PE2	Class C
PI1	Class A (schools)
PI2	Class A (schools)
PI3	Class C (repeaters, beacons)
PI4	Club stations

Licence notes: Four licence classes. Class A: all bands, 100W output. Class B: 3.5, 21, 28MHz (CW only), 100W output, plus VHF and above,

30W output. Class C: VHF, phone-only, 30W output. Class D: 144MHz, phone-only, crystal-controlled (S10, S11, S13, S14, S15, S16 channels only), 15W output.
Foreign amateur operation: CEPT: A + B. Call is PA/own call (CEPT Class 1), PE/own call (CEPT Class 2). Non-CEPT: Reciprocal (A + B). Call is PA/own call (UK Class A), PE/own call (UK Class B). Fee. Form. Two months' notice. Call is own call/PA or /PE. Countries include: Belgium, Botswana, Canada, Denmark, France, Jamaica, Luxembourg, Netherlands Antilles, Mexico, Norway, Panama, Portugal, Sierra Leone, Sweden, Switzerland, UK, USA, Germany.
National society: VERON Centraal Bureau, Postbus 1166, NL-6801 BD Arnhem. Tel: +31 (26) 442 6760. Web: www.veron.nl/veron.
Licensing administration: HDTP-02, PO Box 450, NL-9700 AL Groningen. Tel: +31 (50) 522 2214.

NETHERLANDS ANTILLES
ITU allocation: PJA–PJZ.
Callsign system:

Prefix	Area
PJ0, 1	Special stations
PJ2	Curaçao
PJ4	Bonaire
PJ5	St Eustatius
PJ6	Saba
PJ7	St Maarten
PJ8	St Eustatius, St Maarten, Saba (foreign nationals, special stations)
PJ9	Bonaire, Curaçao (foreign nationals, special stations)

Licence notes: Three licence classes. Class A: all bands, 1kW PEP. Class B: restricted bands, CW only, 150W. Class C: VHF phone-only, 150W PEP.
Foreign amateur operation: CEPT: A + B using PJx/own call. Non-CEPT: Six weeks' notice. Countries include: Netherlands, USA, Canada, UK, Australia, Indonesia, Jamaica.
National society: VERONA, PO Box 3383, Curaçao.
Licensing administration: Landsradio, PO Box 103, Willemstad, Curaçao.

NEW ZEALAND
ITU allocation: ZKA–ZMZ.
Callsign system: NAA–NZZ serial letters denote Novice licence, TAA–TZZ Grade 3.

Prefix	Area or licence category
ZK1	Cook Is, Manihiki Is
ZK2	Niue Is
ZK3	Tokelau Is
ZL1–ZL4	Mainland
ZL5	Antarctica
ZL6	Intruder watch and emergency
ZL7	Chatham Is
ZL8	Kermadec Is
ZL9	Auckland and Campbell Is

Licence notes: Four licence classes. Grade 1: full licence, 150W DC input. Grade 2: 1803–1813, 1875–1900, 3500–3900kHz and VHF, 150W DC input. Grade 3: VHF phone-only, 150W DC input. Novice: crystal-controlled 3525–3575kHz, 10W DC input.
Foreign amateur operation: CEPT: Call is ZL/own call. Non-CEPT: Reciprocal (A + B). Fee. Two weeks' notice. Countries include: British Commonwealth, USA, France, Switzerland.
National society: New Zealand Amateur Radio Transmitters, PO Box 40-525, Upper Hutt 6415. Tel/fax: +64 (4) 528-2170. E-mail: nzart@nzart.org.nz. Web: www.nzart.org.nz.
Licensing administration: Radio Operations, Communications Division, Ministry of Commerce, PO Box 1473, Wellington. Tel: +64 (4) 472-0030.

NICARAGUA
ITU allocation: HTA–HTZ, YNA–YNZ, H6A–H7Z.
Callsign system:

Prefix	City/town and area
YN1	Managua, DN
YN2	Granada
YN3	Leon
YN4	Bluefields
YN5	Jinotepe
YN6	Chinandega
YN7	Masaya, Rivas
YN8	Esteli
YN9	Matagalpa

Licence notes: Two licence classes. Novice: 3.5, 7MHz, 100W DC input. General: all bands, 1kW DC input.

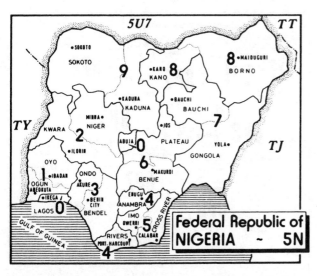

Foreign amateur operation: Countries include: USA, Canada, Guatemala, El Salvador, Honduras, Costa Rica, Panama; all others 90 days (in transit only).
National society: Club de Radio-Experimentadores de Nicaragua, Apartado de Correos 925, Managua, DN. Tel: +505 (2) 771274. Fax: +505 (2) 774973.
Licensing administration: Jefatura de Comunicaciónes, Managua, DN.

NIGER (Republic of the)
ITU allocation: 5UA–5UZ.
Callsign system: Prefix: 5U.
Foreign amateur operation: Call is own call/5U.
Licensing administration: Ministère des postes et télécommunications, Niamey.

NIGERIA (Federal Republic of)
ITU allocation: 5NA–5OZ.
Callsign system: Prefix: 5N. See map for call areas.
Licence notes: Single licence class, 150W DC input.
Foreign amateur operation: Fee. Two months' notice. All member countries of ITU and IARU. Apply to NARS.
National society: Nigerian Amateur Radio Society, PO Box 2873, GPO Marina, Lagos. Tel/fax: +234 (1) 884145.

NORWAY
ITU allocation: JWA–JXZ, LAA–LNZ, 3YA–3YZ.
Callsign system:

Prefix	Area/licence category
JW	Bear Is, Svalbard
JX	Jan Mayen Is
LA0	Norway (reciprocal)
LA1–9	Norway (Class A)
LB	Norway (Class B)
LC	NRRL Emergency Corps
LF	Norway (electronics companies)
LG	Morokulien (LG5LG)
LH	Norway (scientific stations)
LJ	Norway (schools)
3Y	Antarctica (Norway), Bouvet Is, St Peter Is

Licence notes: Two licence classes: Class A: all bands, 600W PEP input. Class B: all bands, CW only, 15W input. Electronic companies, scientific stations and schools permitted to operate within the amateur bands.
Foreign amateur operation: *CEPT:* A + B. Call is LA/own call (CEPT Class 1), LC/own call (CEPT Class 2). *Non-CEPT:* Reciprocal (A only). Fee. Form. One month's notice. For less than one year's operation, call is LA/own call. For longer periods, call is prefixed LA0, JW0 etc. Countries include: Austria, Canada, Denmark, Finland, France, Iceland, Spain, Sweden, Switzerland, UK, USA, Germany.
National society: Norsk Radio Relae Liga, PO Box 20, Haugenstua, N-0915 Oslo. Tel: +47 (22) 213790. Fax: +47 (22) 213791. E-mail: nrrl@online.no. Web: home.sn.no/~nrrl.
Licensing administration: NTRA, PO box 447 Sentrum, N-0104 Oslo. Tel: +47 (22) 824838.

OMAN (Sultanate of)
ITU allocation: A4A–A4Z
Callsign system: Prefix: A4.

Licence notes: Single licence class: 150W DC input.
Foreign amateur operation: Prefix is A45. All countries except Israel. Apply to ROARS.
National society: Royal Omani Amateur Radio Society, PO Box 981, Muscat. Tel: +968 600407. Fax: +968 698558. E-mail: roars@omantel.net.om. Web: www.roars.com.

PAKISTAN (Islamic Republic of)
ITU allocation: APA–ASZ, 6PA–6SZ.
Callsign system: Prefix: AP.
Licence notes: Single licence class: 100W DC input.
National society: Pakistan Amateur Radio Society, PO Box 1450, Islamabad 44000. Tel: +92 (51) 819077. Fax: +92 (51) 827581. Web: www.micro.net.pk/pars.
Licensing administration: The Director-General, Telegraph & Telephone Dept, Zero Point, Islamabad.

PALAU (Republic of)
ITU allocation: T8A–T8Z.

PALESTINE
ITU allocation: E4A–E4Z.
Callsign system: Prefix: E4.

PANAMA (Republic of)
ITU allocation: H3A–H3Z, H8A–H9Z, HOA–HPZ, 3EA–3FZ.
Callsign system:

Prefix	Area/town
HP0	Coiba Is
HP1	Panamá
HP2	Colon, St Blas
HP3	Chiriqui
HP4	Bocas del Toro, Cocle
HP5	Herrera, Los Santos
HP6	Veraguas
HP7	Darién
HP8	Perlas Archipelago

Licence notes: Three licence classes. Class A: all bands, 1kW DC input. Class B: all bands, 500W DC input. Class C: 3.5, 7, 144MHz, 200W. Third-party traffic permitted.
Foreign amateur operation: Unilateral. Call is own call/HP4X. Countries include: USA, Canada, Guatemala, Honduras, Costa Rica, Nicaragua, Ecuador, Peru, Chile, Argentina, Paraguay, Uruguay, Brazil, Bolivia, El Salvador. Applicant must be in Panama at time of application.
National society: Liga Panameña de Radioaficionados, Apartado Postal 175, Panamá 9A. Tel/fax: +507 226 3160. E-mail: ligaradio@hotmail.com. Web: www.qsl.net/lpr.
Licensing administration: Ministerio de Gobierno y Justicia, Dirección Nacional de Medio de Comunicación Social, Apartada Postal 1628, Zona 1, Panamá.

PAPUA NEW GUINEA
ITU allocation: P2A–P2Z.
Callsign system: Prefix: P2. NAA–NZZ serial letters denote Novice, ZAA–ZZZ Limited (VHF).
Licence notes: Three licence classes. Amateur: all bands, 400W PEP. Novice: 3.5, 21 and 28MHz, 30W PEP. Limited (VHF): VHF only, 400W PEP.
Foreign amateur operation: Reciprocal (A + B). Call is in P29V series. Countries include: UK, USA, Canada, Australia, France, New Zealand, Germany, Japan, and others.
National society: Papua New Guinea Amateur Radio Society, PO Box 204, Port Moresby.
Licensing administration: The Manager, Radio Branch, Posts and Telecomms Corp, PO Box 1783, Port Moresby. Tel: +675 274236.

PARAGUAY (Republic of)
ITU allocation: ZPA–ZPZ.
Callsign system:

Prefix	Area
ZP0	Special stations
ZP1	Boqueron
ZP2	Alto Paraguay, Presidente Hayes
ZP3	Amambay, Concepción
ZP4	Canendiyu, San Pedro
ZP5	Asunción
ZP6	Central, Cordillera, Paraguarí
ZP7	Caaguazu, Caazapá, Guaira
ZP8	Misiones, Neembucu
ZP9	Alto Parana, Itapuá

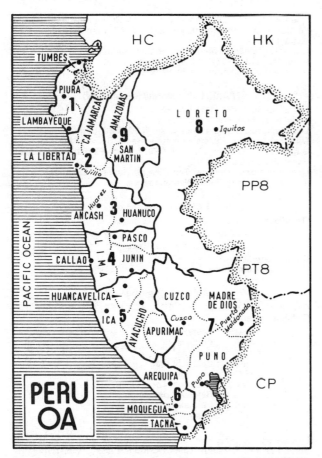

Licence notes: Three licence classes: Class A: all bands, 2kW PEP. Class B: 1.8, 3.5, 7, 14MHz, 2kW PEP. Class C: 3.5, 7MHz, 2kW PEP.
Foreign amateur operation: All countries. Apply to RCP.
National society: Radio Club Paraguayo, Casilla Postal 512, Asunción. Tel: +595 (21) 446124. Fax: +595 (21) 451410. E-mail: radioclu@conexion.com.py.

PERU

ITU allocation: OAA–OCZ, 4TA–4TZ.
Callsign system: See map for prefixes.
Licence notes: 2kW PEP maximum. Third-party traffic permitted.
Foreign amateur operation: *CEPT*: A + B. Call is own call/prefix. *Non-CEPT*: Unilateral (A only). Countries include: Colombia, Bolivia, Canada, Chile, Germany, Uruguay, USA, Netherlands, Venezuela, Argentina, Switzerland, Poland, Spain, Ecuador; other countries 90 days (in transit) only.
National society: Radio Club Peruano, Box 538, Lima 100. Tel: +51 (1) 441-4837. Fax: +51 (1) 440-8944. E-mail: oficina@oabbs.org.pe.
Licensing administration: Ministerio de Transportes, Comunicaciónes, Vivienda y Construccion, Av Wilson y 28 de Julio, Cercado de Lima, Lima. Tel: +51 (1) 433-7800.

PHILIPPINES (Republic of the)

ITU allocation: DUA–DZZ, 4DA–4IZ.
Callsign system: Prefixes: DU (Advanced), DV (General), DW (Novice), DY (VHF only).

Area no	Island
1–4	Luzon
5	Leyte
6, 7	Panay, Negros, Cebu
8, 9	Mindanao

Licence notes: Four licence classes: Advanced, General, Novice and VHF only. 2kW PEP maximum.
Foreign amateur operation: Reciprocal. Call is own call/DU1. Countries include: USA, Canada.
National society: Phillipine Amateur Radio Association, PO Box 4083, Manila Central 1080. Tel/fax: +63 (2) 681 6229. E-mail: para@semicon.net. Web: www.qsl.net/dx1par.
Licensing administration: Planning Division, Telecommunication Control Bureau, 5th Floor, Delos Santos Building, Quezon Avenue, Quezon City.

POLAND

ITU allocation: HFA–HFZ, SNA–SRZ, 3ZA–3ZZ.
Callsign system: Prefixes: SP, SQ. See map for call areas. Serial letters KAA–KZZ, PAA–PZZ, ZAA–ZZZ denote club stations.
Licence notes: Two main licence classes. Class 1: all bands, 20, 50, 250 or 750W. Class 2 VHF phone-only, 10W.
Foreign amateur operation: Reciprocal (A + B). Call is in SO series. Three months' notice. No mobile licences. Countries include UK.
National society: Polski Zwiazek Krotkofalowcow, PO Box 42, 64100 Leszno 7. Tel/fax: +48 (65) 5209529. E-mail: hqpzk@pzk.org.pl. Web: www.pzk.org.pl.
Licensing administration: Panstwowa Agencja Radiokomunikacyjna, Zarzad Krajowy, ul Kasprzata 18/20, 01-211 Warsaw.

PORTUGAL

ITU allocation: CQA–CUZ, XXA–XXZ.
Callsign system:

Prefix	Area
CS1, 4, 5, 6, 7	Portugal (special)
CT1	Portugal
CT3	Madeira Is
CT4	Portugal
CT5, 6, 7	Portugal (special)
CT0	(Repeaters)
CU	Azores Is (see below)
XX9	Macao

Azores prefix	Area
CU1	Santa Maria
CU2	São Miguel
CU3	Terceira
CU4	Graciosa
CU5	São Jorge
CU6	Pico
CU7	Faial
CU8	Flores
CU9	Corvo

Licence notes (Portugal): Three licence classes. Class A: VHF phone only. Class B: all bands. Class C: HF/VHF only.
Foreign amateur operation: *CEPT*: A + B. Call is prefix letters/own call. *Non-CEPT (Portugal)*: Reciprocal (A + B). Fee. Call is own call/area prefix. Licences can only be issued in Portugal. Countries include all EU countries, USA.
National society: *Portugal and Madeira Is*: Rede dos Emissores Portugueses, Rua D Pedro V-7-4°, 1250-092 Lisbon. Tel: +351 (1) 346 1186. Fax: +351 (1) 342 0448. E-mail: ct1rep@rep.pt. Web: www.rep.pt.
Licensing administration: *Portugal*: Instituto das Comunicações Portugal, Rua de S Jose 20, 1193 Lisbon Cedex. *Madeira Is*: Instituto das Comunicações Portugal, Delegação do ICP Madeira, Centro Fiscalizacao, Pico da Cruz, 9000 Funchal. Tel: +35191 762868.

QATAR (State of)

ITU allocation: A7A–A7Z.
Callsign system: Prefix: A7.

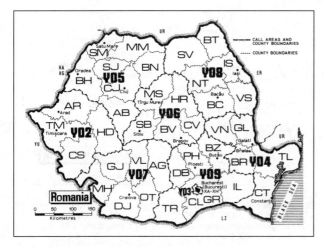

National society: Qatar Amateur Radio Society, PO Box 22122, Doha. Tel: +974 439191. Fax: +974 439595.
Licensing administration: Director of Telecommunications, Ministry of Communications and Transport, PO Aox 2633, Doha.

ROMANIA (Socialist Republic of)

ITU allocation: YOA–YRZ.
Callsign system: See map for prefixes.
Licence notes: 400W maximum DC input.
Foreign amateur operation: *CEPT:* A + B. YO/own call. *Non-CEPT:* Unilateral. No fee. All countries. Three months' notice.
National society: Federatia Romana de Radioamatorism, PO Box 22-50, R-71100 Bucharest. Tel/fax: +40 (1) 315 5575. E-mail: yo3kaa@pcnet.pcnet.ro.
Licensing administration: General Inspectorate of Radiocommunications, 202A Splaiul Independentei sector 6, R-77208 Bucharest. Tel: +40 (1) 386981.

RUSSIAN FEDERATION

ITU allocation: RAA–RZZ, UAA–UIZ.
Callsign system: Prefixes: R, RA, RK, RM, RN, RQ, RS, RU, RV, RW, RX, RZ, UA, 3T (Nishni Novgorod, formerly Gorkij). Current 1st, 2nd and 3rd class callsigns are made up of a two-letter prefix, a number which denotes the geographical area, a letter corresponding to the administrative region and two serial letters, eg RA-3-M-AA. The two serial letters are AA–VZ for individual stations and WA–ZZ for club stations. Note: older two-letter calls do not follow these rules.

GEOGRAPHICAL AREA/ADMINISTRATIVE REGION

The number given in brackets after the geographical area denotes the callsign number for stations in that area. AO = autonomous oblast, AOk = autonomous okrug, Obl = oblast, Rep = republic.

North European Russia (1)

1st serial letter	Area	Capital
A, B	St Petersburg city	
C	Leningrad Obl	St Petersburg
D	St Petersburg city	
F–J	St Petersburg city	
L, M	St Petersburg city	
N	Karelian Rep	Petrozavodsk
O	Arkhangelsk Obl	Arkhangelsk
P	Nenets AO	Naryan-Mar
Q, R, S	Vologda Obl	Vologda
T, U	Novgorod Obl	Novgorod
W, X	Pskov Obl	Pskov
Y, Z	Murmansk Obl	Murmansk

The Baltic (2)

1st serial letter	Area	Capital
F	Kaliningrad Obl	Kaliningrad

Central European Russia (3)

1st serial letter	Area	Capital
A, B, C	Moscow city	
D	Moscow Obl	Moscow
E	Orlov Obl	Orel
F	Moscow city	
G	Lipetsk Obl	Lipetsk
H	Moscow city	
I, J	Tvers Obl	Kalinin
L	Smolensk Obl	Smolensk
M	Yaroslavl Obl	Yaroslavl
N, O	Kostroma Obl	Kostroma
P	Tula Obl	Tula
Q	Voronezh Obl	Voronezh
R	Tambov Obl	Tambov
S	Ryazan Obl	Ryazan
T	Gorkiy Obl	Gorkiy
U	Ivanovo Obl	Ivanovo
V	Vladimir Obl	Vladimir
W	Kursk Obl	Kursk
X	Kaluga Obl	Kaluga
Y	Bryansk Obl	Bryansk
Z	Belgorod Obl	Belgorod

East European Russia (4)

1st serial letter	Area	Capital
A, B	Volgograd Obl	Volgograd
C, D	Saratov Obl	Saratov
F	Penza Obl	Penza
H, I	Samar Obl	
L, M	Ulyanovsk Obl	Ulyanovsk
N, O	Kirov Obl	Kirov
P, Q, R	Tatar Rep	Kazan
S, T	Mari Rep	Yoshkar-Ola
U	Mordov Rep	Saransk
W	Udmurt Rep	Izhevsk
Y, Z	Chuvash Rep	Cheboksary

South European Russia (6)

1st serial letter	Area	Capital
A–D	Krasnodar Kray	Krasnodar
E	Karacheyevo-Cherkessk Rep	Cherkessk
F, G, H	Stavropol Kray	Stavropol
I	Kalmyt Rep	Elista
J	North Osetia Rep	Ordzhonikdze
L–O	Rostov Obl	Rostov-na-Donu
P, Q, R	Chechen Rep, Ingush Rep	Grozny
U, V	Astrakhan Obl	Astrakhan
W	Dagestan Rep	Makhachkala
X	Kabardino-Balkar Rep	Nalchik
Y	Adigei Rep	Maykop

The Urals and West Siberia (9)

1st serial letter	Area	Capital
A, B	Chelyabinsk Obl	Chelyabinsk
C, D, E	Sverdlovsk Obl	Sverdlovsk
F	Perm Obl	Perm
G	Komi-Permiatsky AOk	Kudymkar
H, I	Tomsk Obl	Tomsk
J	Khanty-Mansiysky AOk	Khanty-Mansiysk
K	Jamalo-Nenets AOk	Salekhard
L	Tyumen Obl	Tyumen
M, N	Omsk Obl	Omsk
O, P	Novosibirsk Obl	Novosibirsk
Q, R	Kurgan Obl	Kurgan
S, T	Orenburg Obl	Orenburg
U, V	Kemerov Obl	Kemerov
W	Bashkir Rep	Ufa
X	Komi Rep	Syktyvkar
Y	Altay Kray	Barnaul
Z	Gorno-Altay AO (Altay Kray)	Gorno-Altaysk

East Siberia and the Far East (0, 8)

1st serial letter	Area	Capital
0A	Krasnoyarsk Kray	Krasnoyarsk
0B	Taymyr AOk	Dudinka
0C	Khabarovsk Kray	Khabarovsk
0D	Yevrev AO (Khabarovsk Kray)	Birobidzhan
0E–G	Sakhalin Obl	Yuzhno-Sakhalinsk
0H	Evenkiysk AOk	Tura
0I	Magadan Obl	Magadan
0J	Amur Obl	Blagoveshchensk
0K	Chukotsky AOk	Anadyr
0L, M, N	Primorsky Kray	Vladivostok

0O, P	Buryat Rep	Ulan-Ude
0Q, R	Sakha Rep	Yakutsk
0S, T	Irkutsk Obl	Irkutsk
0U, V	Chita Obl	Chita
0W	Khakass AO	Abakan
	(Krasnoyarsk Kray)	
0X	Koriatsky AOk	Palana
0Y	Tuvinsk Rep	Kyzyl
0Z	Kamchatsky Obl	Petropavlovsk-Kamchatsky
8T	Ust-Ordinsky Buriatsky AO	Ust-Ordinsky
8V	Aginsky-Buriatsky AOk	Aginskoye

Polar regions

Prefix	Area
R1A	Antarctica
R1F	Franz Josef Land
R1M	Malyji Vysotskij

Licence notes: Four licence classes. 1st class: all bands, 200W. 2nd class: all bands, 40W. 3rd class: 1.8, 3.5, 7, 14, 28MHz and VHF, 10W. 4th class: 1.8MHz, 5W.
Foreign amateur operation: Unilateral (A only). Apply via URR.
National society: Union of Radioamateurs of Russia, PO Box 88, Moscow 123459. Tel: +7 (095) 949 5302. Fax: +7 (095) 948 0604.

RWANDA (Republic of)

ITU allocation: 9XA–9XZ.
Callsign system: Prefix: 9X.
Licensing administration: Ministère des postes et télécommunications, BP 720, Kigali.

SAINT KITTS AND NEVIS

ITU allocation: V4A–V4Z.
Callsign system: Prefix: V4.

SAINT LUCIA

ITU allocation: J6A–J6Z.
Callsign system: Prefix: J6.
Foreign amateur operation: Application must be made in person. Countries include USA.
Licensing administration: Police HQ, Castries.

SAINT VINCENT (and Grenadines)

ITU allocation: J8A–J8Z.
Callsign system: Prefixes: J88, J80 (special stations).
Foreign amateur operation: Call uses J87 prefix. Countries include USA.

SAN MARINO (Republic of)

ITU allocations: T7A–T7Z.
Callsign system:

Prefix	Licence class
T70	Club station
T71	Special events
T72	Second class
T77	First class

Licence notes: Two licence classes. Second-class is VHF only.
National society: Associazione Radioamatori della Repubblica di San Marino, PO Box 77, 47031 San Marino. Tel/fax: +378 906790. E-mail: arrsm@inthenet.sm. Web: inthenet.sm./arrsm.
Licensing administration: Direzione Generale Poste e Telecommunicazioni, San Marino. Tel: +378 991349.

SAO TOME AND PRINCIPE (Democratic Republic of)

ITU allocation: S9A–S9Z.
Callsign system: Prefix: S9.
Licensing administration: Direcçâo dos Correios e Telecommunicaçôes da Republica Democratica de São Tomé e Principe, São Tomé.

SAUDI ARABIA (Kingdom of)

ITU allocation: HZA–HZZ, 7ZA–7ZZ, 8ZA–8ZZ.
Callsign system: Prefix: HZ, 7Z.
Licensing administration: The Deputy Minister of Posts, Telegraphs & Telephones, Riyadh.

SENEGAL (Republic of the)

ITU allocation: 6VA–6WZ.
Callsign system: Prefix: 6W.

See text for key to counties and towns

National society: Association des Radio-Amateurs du Senegal, BP 971, Dakar. Tel: +221 (8) 217034. Fax: +221 (8) 217032.
Licensing administration: Ministère de l'Information et des Télécommunications, Dakar.

SEYCHELLES (Republic of)

ITU allocation: S7A–S7Z.
Callsign system: Prefix: S7.
Foreign amateur operation: Unilateral (A only). Call is in S79 series. Countries include UK, USA.
Licensing administration: Seychelles Licensing Authority, PO Box 3, Victoria Mahe.

SIERRA LEONE

ITU allocation: 9LA–9LZ.
Callsign system:

9L1	Western Province
9L2	Northern Province
9L3	Southern Province
9L4	Eastern Province
9L7	Novice licences
9L8	VHF/UHF licences

Licence notes: Two licence classes, General and Novice.
Foreign amateur operation: Reciprocal. Call is own call/9L1. Countries include the UK, USA and most others but residence required.
National society: Sierra Leone Amateur Radio Society, PO Box 10, Freetown. Tel: +232 223335.
Licensing administration: Licensing Office, P & T Dept, GPO, Freetown.

SINGAPORE (Republic of)

ITU allocation: S6A–S6Z, 9VA–9VZ.
Callsign system: Prefix: 9V.
Licence notes: Single licence class: 150W DC input maximum.
Foreign amateur operation: Unilateral (A only). Call is in 9V1 series. All countries.
National society: SARTS, GPO Box 2728, Singapore 904728. Web: www.sarts.org.sg.
Licensing administration: Telecommunication Authority of Singapore, Radio Licensing Dept, 4–8 George Street #04-00, Singapore 0104. Tel: +65 322 1905.

SLOVAK REPUBLIC

ITU allocation: OMA–OMZ.
Callsign system: Prefix: OM. See map for call areas. Numbers in parenthesis below are keys to map.

Prefix	County
OM1	Bratislava metropolitan county (3)
OM2	Bratislava district county (4), Dunajska Streda (7), Galanta (8), Senica (25), Trnava (32)
OM3	All counties
OM4	Povazska Bystrica (20), Prievidza (22), Trencin (31)
OM5	Komarno (10), Levice (12), Nitra (17), Nove Zamky (18), Topolcany (29)
OM6	Cadca (5), Dolny Kubin (6), Liptovsky Mikulas (13), Martin (15), Zilina (37)
OM7	Banska Bystrica (1), Lucenec (14), Velky Krtis (33), Zvolen (35), Ziar nad Hronom (36)
OM8	Kosice (11), Poprad (19), Rimavska Sobota (23), Roznava (24), Spisska Nova Ves (26), Stara Lubovna (27)
OM9	All counties
OM0	Bardejov (2), Humenne (9), Michalovce (16), Presov (21), Svidnik (28), Trebisov (30), Vranov (34)

Callsign series	Type of station
OM1x–OM0x	Contest stations
OM1xx–OM8xx, OM0xx	Classes A and B
OM1xxx–OM8xxx, OM0AAA–JZZ	Classes C and D
OM3xx–OM3xxx	Existing callsigns from OK3 era. KAA–KZZ, RAA–RZZ serial letters used for club stations
OM9xx	Special event stations
OM9AAA–SZZ	Foreign visitors
OM0MAA–MZZ	Beacons
OM0NAA–NZZ	Packet
OM0PAA–PZZ	Packet BBS
OM0OAA–OZZ	Repeaters
OM0SAA–SZZ	Amtor, Pactor, Clover BBS

Licence notes: Four licence classes. Class A: all bands, 750W. Class B: all bands, 300W. Class C: 1.8, 3.5, 10, 21, 28MHz bands plus VHF and above, 100W. Class D: 3.5MHz band plus VHF and above, 100W.
Foreign amateur operation: *CEPT:* A + B. Call is OM/own call. *Non-CEPT:* Fee. Form. Call is in OM9AAA–SZZ series.
National society: Slovak Amateur Radio Association, Wolkrova 4, SK-85101 Bratislava. Tel: +421 (7) 6224 7501. Fax: +421 (7) 6224 5138. E-mail: sara@ba.sknet.sk. Web: www.hamradio.sk/szr.
Licensing administration: Telecomunikacný úrad SR, povol'ovanie rádiostanic, Jarošova 1, SK-83008 Bratislava. Tel: +42 (7) 279 2704.

SLOVENIA

ITU allocation: S5A–S5Z.
Callsign system: Prefix: S5.
Foreign amateur operation: *CEPT:* A + B. Call is S5/own call.
National society: Zveza Radioamaterjev Slovenije, Lepi Pot 6, SI-1000 Ljubljana. Tel/fax: +386 (61) 222459. E-mail: zrs-hq@hamradio.si. Web: www.hamradio.si.

SOLOMON ISLANDS

ITU allocation: H4A–H4Z.
Callsign system: Prefix: H4.
Foreign amateur operation: Unilateral (A + B). Call is in H44 series. All ITU member countries.
National society: Solomon Islands Radio Society, PO Box 418, Honiara.
Licensing administration: Director, Ministry of Posts and Communications, PO Box G25, Honiara.

SOMALI DEMOCRATIC REPUBLIC

ITU allocation: 6OA–6OZ, T5A–T5Z.
Callsign system: Prefix: T5.
Licensing administration: Ministry of Posts and Telecommunications, Mogadishu.

SOUTH AFRICA (Republic of)

ITU allocation: ZRA–ZUZ.
Callsign system: Prefixes: ZR (Class B licences), ZS (Class A licences), ZU (Novice licences). For call areas see map.
Licence notes: Two licence classes. Class A: all bands, 150W DC input or 400W PEP. Class B: VHF phone-only, 150W DC input or 400W PEP.
Foreign amateur operation: *CEPT:* A + B using ZS/own call (CEPT Class 1) and ZR/own call (CEPT Class 2). *Non-CEPT:* Reciprocal (A + B). Fee. Four weeks' notice. Countries include: UK, USA, Zimbabwe. Apply to SARL.
National society: South African Radio League, PO Box 1721, Strubensvallei 1735. Tel/fax: +27 (11) 675 2393. E-mail: sarl@intekom.co.za. Web: www.sarl.org.za.

SPAIN

ITU allocation: AMA–AOZ, EAA–EHZ.
Callsign system: See map for call areas. Area 8 is Canary Is.

Prefix	Licence class
EA	Class A
EB	Class B
EC	Class C

Licence notes: Three classes. Class A: all bands, 250W anode dissipation. Class B: VHF only, 50W anode dissipation. Class C: 3.550–3.575, 7.020–7.030, 21.000–21.150MHz CW only, 29.0–29.1MHz AM only, 20W anode dissipation.
Foreign amateur operation: *CEPT:* A + B. Call is EA/own call (UK Class A), EB/own call (UK Class B). *Non-CEPT:* Reciprocal (A + B). Call is EA/own call (UK Class A), EB/own call (UK Class B). Fee. Form. Countries include UK, USA.
National society: Unión de Radioaficionados Españoles, Apartado 220, E-28080 Madrid. Tel: +34 (91) 477 1413. Fax: +34 (1) 477 2071. E-mail: ure@ure.es. Web: www.ure.es.
Licensing administration: Dirección General de Telecomunicaciónes, Plaza Cibeles, Palacio de Comunicaciónes, E-28014 Madrid. Tel: +34 (1) 346 1500. Fax: +34 (1) 396 2229.

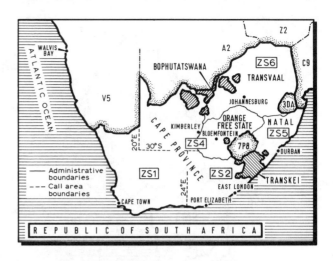

REPUBLIC OF SOUTH AFRICA

SRI LANKA (Democratic Socialist Republic of)

ITU allocation: 4PA–4SZ.
Callsign system: Prefix: 4S.
Licence notes: 100W DC input maximum.
Foreign amateur operation: Reciprocal (A + B). No specified countries. UK licence is usually accepted as a qualification for the issue of a licence.
National society: Radio Society of Sri Lanka, PO Box 907, Colombo. E-mail: rssl@mailexcite.com. Web: www.qsl.net/rssl.
Licensing administration: The Director General of Telecomms, Regulatory Authority, EH Cooray Building, 4th floor, 411 Galle Road, Colombo 3.

SUDAN (Democratic Republic of the)

ITU allocation: SSN–STZ, 6TA–6UZ.
Callsign system: Prefixes: ST, 6U.
Licensing administration: The Director, Department of Telecommunications, Ministry of Communications, Khartoum.

SURINAME (Republic of)

ITU allocation: PZA–PZZ.
Callsign system:

Prefix	City/town and area
PZ1	Paramaribo, Sipalwini, Wanica
PZ2	Nickerie
PZ3	Coronie
PZ4	Saramacca
PZ5	Foreign nationals
PZ6	Para
PZ7	Brokopondo
PZ8	Commewijne
PZ9	Marowijne
PZ0	Special stations

Spain—EA

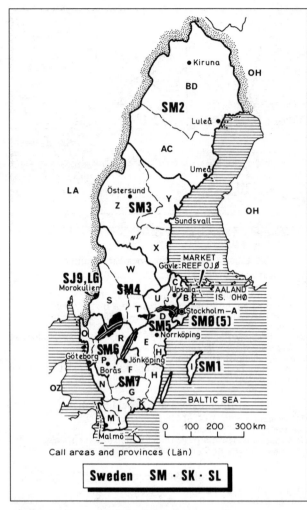

Call areas and provinces (Län)

Sweden SM · SK · SL

Licence notes: 150W DC input maximum.
Foreign amateur operation: Countries include: USA, Netherlands, Netherlands Antilles (reciprocal). All other countries on production of licence. Apply to VRAS.
National society: Vereniging van Radio Amateurs in Suriname, PO Box 1153, Paramaribo.

SWAZILAND (Kingdom of)
ITU allocation: 3DA–3DM.
Callsign system: Prefix: 3DA.
Licence notes: 500W DC input maximum.
Foreign amateur operation: Reciprocal (A + B). Countries include UK.
National society: Radio Society of Swaziland, PO Box 3744, Manzini.
Licensing administration: Engineer Frequency Management, Managing Director of Posts and Telecomms, PO Box 125, Mbabane.

SWEDEN (Kingdom of)
ITU allocation: SAA–SMZ, 7SA–7SZ, 8SA–8SZ.
Callsign system: For call areas see map and table. Serial letters RAA–RZZ in the SK prefix series are allocated to repeaters.

Prefix	Area/licence category
SH	Novice licences
SJ9	Morokulien
SK	Club stations, repeaters, beacons
SL	Military
SM	Individual
SM8	Maritime mobile

Licence notes: Four licence classes. Class A (advanced): all bands, 500W DC input. Class B (general): all bands CW, 28MHz and above phone, 75W DC input. Class C (basic): all bands CW, 28MHz and above phone, 10W DC input. Class T (technician): VHF phone-only, 75W DC input.
Foreign amateur operation: *CEPT*: A + B. Call is SM/own call. *Non-CEPT*: Unilateral (A + B). Call is SM/own call. Fee. Form. Two months' notice. All countries.

National society: Föreningen Sveriges Sändareamatörer, PO Box 45, SE-19121 Sollentuna. Tel: +46 (8) 585 702 73. Fax: +46 (8) 585 702 74. E-mail: hq@svessa.se. Web: www.svessa.se.
Licensing administration: Telestyrelsen, Frequency Division, Box 700, S-136 27 Haninge. Tel: +46 (8) 707-3500.

SWITZERLAND (Confederation of)
ITU allocation: HBA–HBZ, HEA–HEZ.

LIECHTENSTEIN
Callsign system:

Licence class	Call series
A	HB0A–Z
A	HB0AA–ZZ
A	HB0AAA–LZZZ
B	HB0MAA–ZZZ

Licence notes: Two licence classes. Class A: all bands, 1kW PEP. Class B: VHF, 1kW PEP.
Foreign amateur operation: *CEPT*: A + B. Call is HB0/own call. *Non-CEPT*. Reciprocal. Fee. Form. One month's notice. Countries include: Australia, Austria, Belgium, Brazil, Canada, Chile, Cyprus, Czech Rep, Denmark, Finland, France (incl overseas terr), Germany, Greece, Iceland, India, Irish Rep, Israel, Italy, Kuwait, Luxembourg, Malta, Monaco, New Zealand, Netherlands, Norway, Papua New Guinea, Peru, Portugal, Qatar, Slovakia, South Africa, Sweden, Thailand, UK, USA.
National society: Amateurfunk Verein Liechtenstein, Postfach 629, FL-9495 Triesen.
Licensing administration: Office pour Relations internationales, Regierung des Furstentums Liechtenstein, FL-9490 Vaduz. Tel: +41 (75) 61111.

SWITZERLAND
Callsign system:

Licence class	Call series
A	HB9A–Z
A	HB9AA–ZZ
A	HB9AAA–LZZ
A	HB4FA–FZ (military amateur radio clubs)
B	HB9MAA–ZZZ

Licence notes: Two licence classes. Class A: all bands, 1kW PEP. Class B: VHF, 1kW PEP.
Foreign amateur operation: *CEPT*: A + B. Call is HB9/own call. *Non-CEPT*. Reciprocal. Fee. Form. One month's notice. Countries include: Australia, Austria, Belgium, Brazil, Canada, Chile, Cyprus, Czech Rep, Denmark, Finland, France (incl overseas terr), Germany, Greece, Iceland, India, Irish Rep, Israel, Italy, Kuwait, Luxembourg, Malta, Monaco, New Zealand, Netherlands, Norway, Papua New Guinea, Peru, Portugal, Qatar, Slovakia, South Africa, Sweden, Thailand, UK, USA.
National society: Union Schweizerischer Kurzwellen-Amateure, PO Box 238, CH-4805 Brittnau. E-mail: hq@uska.ch. Web: www.uska.ch.
Licensing administration: General Directorate, Swiss Telecom, Radiocom and Broadcast Licensing, Speichergasse 6, CH-3030 Berne. Tel: +41 (31) 338 5191. Fax: +41 (31) 338 5191.

SYRIA (Syrian Arab Republic)
ITU allocation: YKA–YKZ, 6CA–6CZ.
Callsign system: Prefix: YK.
Licence notes: 500W DC input maximum.
Foreign amateur operation: Call is own call/YK.
National society: Technical Institute of Radio, PO Box 245, Damascus.
Licensing administration: Direction générale de l'établissement des postes et des télécommunications, Damascus.

TAIWAN – see China

TAJIKISTAN
ITU allocation: EYA–EYZ.
Callsign system: Prefix: EY.

Prefix	Area	Capital
EY4	Gorno-Badakstan AO	Khorog
EY5	Kulyab Obl	Kulyab
EY6	Kurgan-Tyube Obl	Kurgan-Tyube
EY7	Leninabad Obl	Khudzand
EY8	—	Dushanbe
EY9	Nureka	

National society: Tajik Amateur Radio League, c/o Nadir Tursoon-Zadeh, PO Box 203, Glavpochtamt, Dushanbe 734025.

TANZANIA (United Republic of)
ITU allocation: 5HA–5IZ.
Callsign system: Prefixes: 5H1 (Zanzibar), 5H3 (Tanzania).

National society: Tanzania Amateur Radio Club, PO Box 21497, Dar-es-Salaam.
Licensing administration: Director-General, East African P&T Corporation, PO Box 9070, Dar-es-Salaam.

THAILAND
ITU allocation: HSA–HSZ, E2A–E2Z.
Callsign system:

Prefix	City/town and area
HS0	(Club stations)
HS1	Bangkok
HS3	Korat
HS5	Chiang Mai

Licence notes: 1kW DC input maximum.
Foreign amateur operation: Countries include: USA, UK, Japan, Malaysia, Canada, Switzerland, Germany, Sweden, Denmark, Australia, New Zealand, Spain, and others. Apply to RAST.
National society: Radio Amateur Society of Thailand, GPO Box 2008, Bangkok 10501. Tel/fax: +66 (2) 618 4435. Web: www.qsl.net/rast.

TOGO (Togolese Republic)
ITU allocation: 5VA–5VZ.
Callsign system: Prefix: 5V7.
Licensing administration: Ministère des travaux publics, de la construction des postes et télécommunications, Lomé.

TONGA (Kingdom of)
ITU allocation: A3A–A3Z.
Callsign system: Prefix: A3.
Licence notes: Single licence class, 500W PEP.
Foreign amateur operation: Unilateral (A + B). Call is in A35 series. Most countries.
National society: Amateur Radio Club of Tonga, c/o Manfred Shuster, PO Box 1078, Nuku'alofa.
Licensing administration: Tonga Telecom, PO Box 46, Nuku'alofa. Tel: +676 24255. Fax: +676 22200.

TRINIDAD & TOBAGO
ITU allocation: 9YA–9ZZ.
Callsign system: Prefix: 9Y.
Licence notes: 1kW DC input maximum. Third-party traffic permitted.
Foreign amateur operation: Countries include: USA, Canada, UK, British Commonwealth. Apply in person to the Chief Wireless Officer.
National society: Trinidad and Tobago Amateur Radio Society, PO Box 1167, Port of Spain. E-mail: ttars@carib-link.net. Web: www2.carib-link.net/~ttars.
Licensing administration: Chief Wireless Officer, Ministry of Works, Transport and Communications, Wrightson Road, Port of Spain.

TUNISIA
ITU allocation: TSA–TSZ, 3VA–3VZ.
Callsign system: Prefix: 3V.
National society: Association Tunisienne des Radioamateurs, PO Box 2055, ISAJC, Bir El Bey.
Licensing administration: Ministère des transports et communications, 3 bis rue d'Angleterre, Tunis.

TURKEY
ITU allocation: TAA–TCZ, YMA–YMZ.
Callsign system: Prefix: TA. See map for call areas.
Foreign amateur operation: *CEPT*: A + B. Call is TA/own call. *Non-CEPT*: Unilateral.
National society: Telsiz Radyo Amatorieri Cemiyeti, PO Box 699, Karakoy 80005 Istanbul. Tel/fax: +90 (212) 245 3942. E-mail: hq@trac.org.tr. Web: www.trac.org.tr.
Licensing administration: Directorate General of PTT, Ankara.

TURKMENISTAN
ITU allocation: EZA–EZZ.
Callsign system: Prefix: EZ.

Prefix	Area
EZ3	Akhalsky veloyat
EZ4	Balkansky veloyat
EZ5	Maryisky veloyat
EZ6	Dashkhovuzsky veloyat
EZ7	Lebapsky veloyat
EZ8	Ashgabat

National society: Liga Radiolyubiteley Turkmenistana, PO Box 555, Ashgabat 744020.

TUVALU
ITU allocation: T2A–T2Z.
Callsign system: Prefix: T2.
Foreign amateur operation: Countries include USA.

UGANDA (Republic of)
ITU allocation: 5XA–5XZ.
Callsign system: Prefix: 5X.
National society: Uganda Amateur Radio Society, PO Box 22761, Kampala.
Licensing administration: The Director-General, East African P&T Corporation, PO Box 7106, Kampala.

UKRAINE
ITU allocation: EMA–EOZ, URA–UZZ.
Callsign system: Prefixes: UR–UU, UY, UX.

1st serial letter	Area	Capital
A	Sumy Obl	Sumy
B	Ternopol Obl	Ternopol
C	Cherkassy Obl	Cherkassy
D	Zakarpatsk Obl	Uzhgorod
E	Dnepropetrovsk Obl	Dnepropetrovsk
F	Odessa Obl	Odessa
G	Kherson Obl	Kherson
H	Poltava Obl	Poltava
I	Donetsk Obl	Donetsk
J (UU)	Krymsk Obl	Simferopol
J (UU9)	Sevastopol City	
K	Rovno Obl	Rovno
L	Kharkov Obl	Kharkov
M	Voroshilovgrad Obl	Lugansk
N	Vinnitsa Obl	Vinnitsa
P	Volynsk Obl	Lutsk
Q	Zaporozhe Obl	Zaporozhe
R	Chernigov Obl	Chernigov
S	Ivano-Frankovsk Obl	Ivano-Frankovsk
T	Khmelnitsky Obl	Khmelnitsky
U	Kiev city, Kiev Obl	Kiev
V	Kirovgrad Obl	Kirovgrad
W	Lvov Obl	Lvov
X	Zhitomir Obl	Zhitomir
Y	Chernovtsy Obl	Chernovtsy
Z	Nikolayev Obl	Nikolayev

The appropriate prefix is given in brackets.

National society: Ukrainian Amateur Radio League, PO Box 56, 01001 Kiev 1. Tel: +380 (44) 457 0972. Fax: +380 (44) 4577195.

UNITED ARAB EMIRATES
ITU allocation: A6A–A6Z.
Callsign system: Prefix: A6.
Licensing administration: The Director-General, Posts and Telecommunications, PO Box 902, Abu Dhabi.

UNITED KINGDOM (of Great Britain and Northern Ireland)
ITU allocation: GAA–GZZ, MAA–MZZ, VPA–VQZ, VSA–VSZ, ZBA–ZJZ, ZNA–ZOZ, ZQA–ZQZ, 2AA–2ZZ.
Callsign system (UK): See table for prefixes. M5AAA–M5ZZZ series are Class A/B licences. G1AAA–1ZZZ, G6AAA–6ZZZ, G7AAA–7ZZZ, G8AAA–8ZZZ and M1AAA–ZZZ series are Class B licences. Prefix varied when operating outside own area.

BRITISH ISLES

Prefix	Club prefix	Novice prefix	Area/licence category
G, M	GX, MX	2E	England
GD, MD	GT, MT	2D	Isle of Man

GI, MI	GN, MN	2I	Northern Ireland
GJ, MJ	GH, MH	2J	Jersey
GM, MM	GS, MS	2M	Scotland
GU, MU	GP, MP	2U	Guernsey and dependencies
GW, MW	GC, MC	2W	Wales
GB0, GB2, GB4	—	—	Special (all areas), Class A
GB3	—	—	Beacons and repeaters (all areas). Beacons have GB3+3 callsigns, repeaters GB3+2 callsigns.
GB1, GB6, GB8	—	—	Special (all areas), Class B
GB7	—	—	Data repeaters and mailboxes (all areas). Repeaters have GB7+2 callsigns, mailboxes GB7+3 callsigns.

UK OVERSEAS TERRITORIES

Prefix	Area
VP2E	Anguilla
VP2M	Montserrat
VP2V	British Virgin Is
VP5	Turks and Caicos Is
VP8	Falkland Is, Graham Land, S Orkney Is, S Sandwich Is, S Shetland Is, Antarctica (UK)
VP9	Bermuda
VQ9	Chagos Is
VR6	Pitcairn Is
VS6	Hong Kong
ZB2, 0	Gibraltar
ZC4	Cyprus (UK bases only)
ZD7	St Helena Is
ZD8	Ascension Is
ZD9	Tristan da Cunha
ZD9G	Gough Is
ZF1	Grand Cayman Is
ZF2	Foreign nationals
ZF8	Little Cayman Is
ZF9	Cayman Brac Is

Licence notes: *UK:* Three licence classes. Class A: all bands. Class A/B: all bands. Class B: above 30MHz only. Maximum power is 26dBW PEP (less for some bands). There are also two Novice classes. Novice Class A: 1950–2000kHz, 3550–3650kHz, 10,110–10,140kHz, 21,050–21,149kHz, 28,050–28,190 and 28,225–28,500kHz, plus VHF and above, CW and phone, 10W PEP. Novice Class B: VHF and above only, CW and phone, 10W PEP. See Appendix 9 for more details. *Overseas territories:* Usually similar to UK but without Class B licences.
Foreign amateur operation: *Bermuda:* Unilateral (A only). Call is own call/VP9. *British Virgin Is:* Unilateral (A + B). Call is VP2V/own call. *Cayman Is:* Unilateral. Call is in ZF1 series. *Falkland Is:* Unilateral (A + B). Call is in VP8 series. *Gibraltar:* Reciprocal (A + B). Call is ZB2/own call. *Montserrat:* Unilateral. Call is in VP2M series. *Pitcairn & Henderson Is:* Unilateral. *UK: CEPT:* A + B. Call is area prefix (M series)/own call. *Non-CEPT:* Reciprocal. Two types are available: Two month (no fixed address required) or 12 month (main station address required). Call is area prefix/own call. Fee. Form. 30 days' notice required. Countries include: Austria, Australia, Belgium, Bermuda, Botswana, Brazil, Canada, Cyprus, Denmark, Dominican Republic, Finland, France, Germany, Gibraltar, Hong Kong, Iceland, India, Irish Republic, Israel, Italy, Kenya, Malaysia, Malta, Monaco, Netherlands, New Zealand, Norway, Papua New Guinea, Poland, Portugal, Sierra Leone, South Africa, Spain, Sri Lanka, Sweden, Switzerland, Trinidad & Tobago, USA, Vanuatu, Zimbabwe.
National society: *Anguilla:* Anguilla Amateur Radio Society, PO Box 1, The Valley. *Bermuda:* Radio Society of Bermuda, PO Box HM 275, Hamilton HM AX. Web: www.bermuda-shorts.com/rsb. *British Virgin Is:* British Virgin Islands Radio League, PO Box 4, West End, Tortola. *Cayman Is:* Cayman Amateur Radio Society, PO Box 1029, Grand Cayman. E-mail: zf1a@candw.ky. *Gibraltar:* Gibraltar Amateur Radio Society, PO Box 292. Web: www.gibnet.com/gars. *Montserrat:* Montserrat Amateur Radio Society, PO Box 448, Plymouth. *Turks and Caicos Is:* Turks and Caicos Amateur Radio Society, PO Box 694800, Miami, FL 33269, USA. *UK:* Radio Society of Great Britain, Lambda House, Cranborne Road, Potters Bar, Herts EN6 3JE. Tel: +44 (1707) 659015. Fax: +44 (1707) 645105. E-mail: AR.Dept@rsgb.org.uk. Web: www.rsgb.org.
Licensing administrations: *Bermuda:* Dept of Telecomms; PO Box 101, Hamilton 5. Tel: +1 (809) 295-5151 ext 1120. *British Virgin Is:* Telecommunications (Radio) Officer, Ministry of Communications and Works, Government of the BVI, Road Town, Tortola. *Cayman Is:* The Postmaster, General Post Office, Grand Cayman. Tel: +1 (809) 949-2474. *Falkland Is:* Superintendent, Posts and Telecommunications, The Post Office, Port Stanley. Tel: +500 27135. *Gibraltar:* The Wireless Officer, General Post Office, 104 Main Street. Tel/fax: +350 75714. *Montserrat:* The Ministry of Communications and Works,

General Turning Road, Plymouth. *Pitcairn & Henderson Is:* Office of the Governor of Pitcairn, Henderson, Ducie and Oeno Is, c/o British Consulate General, Auckland, New Zealand. *UK:* Amateur and Citizens' Band Radio Unit, Radiocommunications Agency, South Quay Three, 189 Marsh Wall, London, E14 9SX. Tel: +44 (171) 211 0211.

UNITED NATIONS ORGANIZATION
ITU allocation: 4UA–4UZ.
Callsign system: Prefix: 4U1.
Licence notes: Permanent stations located at ITU HQ, Geneva; World Bank, Washington, DC; UN, Vienna; and UN HQ, New York.
Foreign amateur operation (4U1ITU, Geneva): Four weeks' notice. Apply to station manager at IARC.
Society: International Amateur Radio Club, Box 6, Place des Nations. CH-1211 Geneva 20. Switzerland.

UNITED STATES OF AMERICA
ITU allocation: AAA–ALZ, KAA–KZZ, NAA–NZZ, WAA–WZZ.
Callsign system: The current callsign system came into effect on 24 March 1978. Amateurs were, however, permitted to continue to hold existing callsigns.

GEOGRAPHICAL LOCATION
Contiguous 48 states and District of Columbia
The digit indicates the location at the time when the callsign was issued but in many cases not the current location.

Digit	Location
1	Maine, New Hampshire, Vermont, Massachusetts, Rhode Island, Connecticut
2	New York, New Jersey
3	Pennsylvania, Delaware, Maryland, District of Columbia
4	Virginia, North Carolina, South Carolina, Georgia, Florida, Alabama, Tennessee, Kentucky
5	Mississippi, Louisiana, Arkansas, Oklahoma, Texas, New Mexico
6	California
7	Oregon, Washington, Idaho, Montana, Wyoming, Arizona, Nevada, Utah
8	Michigan, Ohio, West Virginia
9	Illinois, Indiana, Wisconsin
0	Colorado, Nebraska, North Dakota, South Dakota, Minnesota, Iowa, Missouri, Kansas

Other locations (current system)

Prefix	Location
AH0, KH0, NH0, WH0	Northern Mariana Is
AH1, KH1, NH1, WH1	Baker, Howland Is
AH2, KH2, NH2, WH2	Guam
AH3, KH3, NH3, WH3	Johnston Is
AH4, KH4, NH4, WH4	Midway Is
AH5K, KH5K, NH5K, WH5K	Kingman Reef
AH5, KH5, NH5, WH5	(except K suffix) Palmyra, Jarvis Is
AH6, KH6, NH6, WH6	Hawaiian Is
AH7, KH7, NH7, WH7	Hawaiian Is
AH7K, KH7K, NH7K, WH7K	Kure Is
AH8, KH8, NH8, WH8	American Samoa
AH9, KH9, NH9, WH9	Wake, Wilkes, Peale Is
AL, KL, NL, WL	Alaska
KP1, NP1, WP1	Navassa Is
KP2, NP2, WP2	Virgin Is
KP3, NP3, WP3	Puerto Rico
KP4, NP4, WP4	Puerto Rico
KP5, NP5, WP5	Desecheo Is

Other locations (old system)

Prefix	Location
KA	Japan (USA military personnel)
KB6	Baker, Howland Is
KC4	Navassa Is, Antarctica (USA bases)
KC6*	Belau
KG4*	Guantanamo Bay
KG6	Mariana, Marcus and Guam Is
KH6	Hawaiian, Kure Is
KJ6	Johnston Is
KL7	Alaska
KM6	Midway Is
KP4	Puerto Rico
KP5	Desecheo Is
KP6	Palmyra and Jarvis Is
KS4	Swan Is, Roncador Cay and Serrana Bank
KS6	American Samoa
KV4	Virgin Is
KW6	Wake Is
KX6	Marshall Is

* Still current as these are not FCC administered.

USA LICENCE CLASSES (current system)

First letter/ No of serial letters	—	A	B	C	D	E	F	G	H	I	J	K	L	M	N	O	P	Q	R	S	T	U	V	W	X	Y	Z
A + 1			E	E	E	E	E	E	E	E	E	E	E														
A + 2	E	E	E	E	E	E	E	E	E	A	E	E	E	A													
A + 3	E																										
K + 1			E	E	E	E	E	E	E	E	E	E	E	E	E	E	E	E	E	E	E	E	E	E	E	E	E
K + 2	E	A^x	A	A^a	A	A	A	A^b	GT	A	A	A	GT	A	A	A	GT	A	A^c	A	A	A	A	A	A^d	A	A^e
K + 3	GT	N	N	N^f	N	N	N	N	N	N	N	N	N	N	N	N	N	N	N	N	N	N	N	N	N	N	N
N + 1			E	E	E	E	E	E	E	E	E	E	E	E	E	E	E	E	E	E	E	E	E	E	E	E	E
N + 2	E	A	A	A	A	A	A	A	GT	A	A	A	GT	A	A	A	GT	A	A	A	A	A	A	A	A	A	A
N + 3	GT																										
W + 1			E	E	E	E	E	E	E	E	E	E	E	E	E	E	E	E	E	E	E	E	E	E	E	E	E
W + 2	E	A	A	A	A	A	A	A	GT	A	A	A	GT	A	A	A	GT	A	A	A	A	A	A	A	A	A	A
W + 3	GT	N	Ra	N	N	N	N	N	N	N	N	N	N	C	N	M	N	N	R	N	Ty	N	N	N	N	N	N

Key: A Advanced; C Club; E Extra; G General; M Military recreation; N Novice; R Repeaters; Ra RACES; T Technician; Ty Temporary.
Notes: a Except KC6. b Except KG4, KG6R, KG6S. c Except KR6. d Except KX6. e Except KZ5. f Except KC4AAA–AAF, KC4USA–USZ. x KA1 only.

Licence classes: Five licence classes: Advanced, Extra, General, Novice, Technician. Maximum power is 1.5kW PEP output except in the Novice segments of the bands and on 10MHz. For permitted HF bands see below. Third-party traffic permitted.

CW/RTTY/DATA SEGMENTS

Extra	Advanced	General	Technician Plus and Novice
3500–3750	3525–3750	3525–3750	3675–3725*
7000–7150	7025–7150	7025–7150	7100–7150*
10,100–10,150	10,100–10,150	10,100–10,150	—
14,000–14,150	14,025–14,150	14,025–14,150	—
18,068–18,110	18,068–18,110	18,068–18,110	—
21,000–21,200	21,025–21,200	21,025–21,200	21,100–21,200*
24,890–24,930	24,890–24,930	24,890–24,930	—
28,000–28,300	28,000–28,300	28,000–28,300	28,100–28,300

* CW only

CW/PHONE/IMAGE SEGMENTS

Extra	Advanced	General	Technician Plus and Novice
3750–4000	3775–4000	3850–4000	—
7150–7300	7150–7300	7225–7300	—
14,150–14,350	14,175–14,350	14,225–14,350	—
18,110–18,168	18,110–18,168	18,110–18,168	—
21,200–21,450	21,225–21,450	21,300–21,450	—
24,930–24,990	24,930–24,990	24,930–24,990	—
28,300–29,700	28,300–29,700	28,300–29,700	28,300–28,500*

* SSB only

Foreign amateur operation: *CEPT*: A + B using prefix/own call. *Non-CEPT*: Reciprocal (A + B). Call is US prefix/own call. Fee. Form 610-A. 60 days' notice. Operating privileges do not exceed those of foreign amateurs' own licence. Countries include: Antigua, Argentina, Australia, Austria, Bahamas, Barbados, Belgium, Belize, Bolivia, Botswana, Brazil, Canada, Chile, Colombia, Costa Rica, Cyprus, Denmark, Dominica, Dominican Republic, Ecuador, El Salvador, Fiji, Finland, France, Germany, Greece, Grenada, Guatemala, Guyana, Haiti, Honduras, Hong Kong, Iceland, India, Indonesia, Irish Republic, Israel, Italy, Jamaica, Japan, Jordan, Kiribati, Kuwait, Liberia, Luxembourg, Marshall Is, Mexico, Micronesia, Monaco, Netherlands, Netherlands Antilles, New Zealand, Nicaragua, Norway, Panama, Paraguay, Peru, Philippines, Portugal, St Lucia, St Vincent, Seychelles, Solomon Is, South Africa, Spain, Sierra Leone, Suriname, Sweden, Switzerland, Thailand, Trinidad, Tuvalu, UK (including VP2M, VP2S, VP2V, VP5, VP8, VP9, ZB2, ZD7 and ZF), Uruguay, Venezuela and Yugoslavia.
National society: American Radio Relay League, 225 Main Street, Newington, Connecticut 06111-1494. Tel: +1 (860) 594-0200. Fax: +1 (860) 594-0259. E-mail: hq@arrl.org. Web: www.arrl.org.
Licensing administration: Federal Communications Commission, Gettysburg, PA 17326.

URUGUAY (Oriental Republic of)

ITU allocation: CVA–CXZ.
Callsign system: Prefix: CX. First serial letter denotes province (see map). Z indicates mobile/portable operation.
Licence notes: Four licence classes. 1st, 2nd and 3rd classes, all bands, 500, 100 and 30W respectively. Special: 1.8, 3.5, 7, 28MHz, 30W. Third-party traffic permitted.

Foreign amateur operation: Unilateral (A only). All countries. Apply to RCU.
National society: Radio Club Uruguayo, PO Box 37, Montevideo. Tel: +598 (2) 787879. Fax: +598 (2) 787523. E-mail: rcuhq@adinet.com.uy.

UZBEKISTAN

ITU allocation: UJA–UMZ.
Callsign system: Prefix: UK.

1st serial letter	Area	Capital
A	Tashkent city	
B	Tashkent Obl	Tashkent
C	Kashkadar'in Obl	Karshi
D	Syrdar'in Obl	Gulistan
F	Andizhan Obl	Andizhan
G	Fergana Obl	Fergana
I	Samarkand Obl	Samarkand
L	Bukhara Obl	Bukhara
O	Namangan Obl	Namangan
Q	Navoiy Obl	
T	Surkandar'in Obl	Termez
U	Khorezm Obl	Urgench
V	Dzhizak Obl	Dzhizak
Z	Kara-Kalpakstan	Nukus

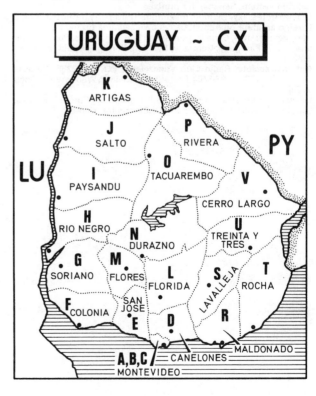

URUGUAY ~ CX

VANUATU

ITU allocation: YJA–YJZ.
Callsign system: Prefix: YJ.
Foreign amateur operation: Reciprocal (A only). Call is in YJ0A series. Countries include UK.
National society: Vanuatu Amateur Radio Society, PO Box 665, Port-Vila 3092. Tel: +678 3092.
Licensing administration: Director of Posts and Telecomms, Dept of Posts and Telecomms, Port-Vila.

VATICAN CITY STATE

ITU allocation: HVA–HVZ.
Callsign system: Prefix: HV.
Licensing administration: Governatorato – Secretariat General, Administration des PTT, Vatican City. Tel: +39 66982.

VENEZUELA (Republic of)

ITU allocation: YVA–YYZ, 4MA–4MZ.
Callsign system: See map for prefixes.
Licence notes: 1kW DC input maximum. Third-party traffic permitted.
Foreign amateur operation: Call is own call/area prefix/digit. Countries include: USA, Canada, Dominican Rep, Costa Rica, Peru, Argentina, Brazil, Paraguay, Germany, Sweden, Portugal, Spain, Ecuador, Bolivia.
National society: Radio Club Venezolano, PO Box 2285, Caracas 1010-A. Tel: +58 (2) 793-5404. Fax: +58 (2) 793-6883. Web: www.qsl.net/yv5aj.
Licensing administration: Ministerio de Transporte y Comunicaciónes, Division de Tramitación, Esquina de Carmelitas, Caracas 101.

VIETNAM (Socialist Republic of)

ITU allocation: XVA–XVZ, 3WA–3WZ.
Callsign system: Prefix: XV, 3W.
Licensing administration: Direction génerale des postes et télécommunications, 18 rue Nguyên Du, Hanoi.

WESTERN SAMOA

ITU allocation: 5WA–5WZ.
Callsign system: Prefix: 5W1.
Licence notes: Three licence classes, similar to New Zealand Grade, 1, 2 and 3. 1kW DC input maximum.
Foreign amateur operation: Unilateral. Fee. All countries.
National society: Western Samoa ARC, PO Box 2015, Apia.
Licensing administration: Director, Post Office and Radio, Chief Post Office, Apia.

YEMEN

ITU allocation: 7OA–7OZ.
Callsign system: Prefix: 7O.
Licensing administration: Director General of Posts and Telephones, General Post Office, PO Box 1000, Aden.

YUGOSLAVIA (Socialist Federal Republic of)

ITU allocation: YTA–YUZ, YZA–YZZ, 4NA–4OZ.
Callsign system: Prefixes: YU1/YT1 (Serbia). The 0 digit is used for special activities. Clubs are allocated serial letters AAA–MZZ.

Prefix	Category
YT	Class F
YU	Classes A–E
YZ	Special
4N	Special and repeaters

Licence notes: Six licence classes. Class A: all bands, 2kW input. Class B: all bands, 250W input. Class C: 3.5, 7MHz and VHF, 250W input. Class D: VHF only, 100W input. Class E: VHF phone only, 50W input. Class F: 3.565–3.575MHz, CW only, 10W input.
National society: Savez Radio-amatera Jugoslavije, PO Box 48, YU-11001 Belgrade. Tel/fax: +381 (11) 634437. E-mail: yu0srj@EUnet.yu.
Licensing administration: Community of Yugoslavia Posts, Telegraphs and Telephones, Palmoticeva 2, 11001 Belgrade. Tel: +381 (11) 338921.

ZAMBIA (Republic of)

ITU allocation: 9IA–9JZ.
Callsign system: Prefix: 9J2.
Licence notes: Single licence class: 150W DC input maximum.
National society: Radio Society of Zambia, PO Box 20332, Kitwe.
Licensing administration: Chief of Operations, Telecommunications Services, PTC HQ, Box 1660, Ndola.

ZIMBABWE

ITU allocation: Z2A–Z2Z.
Callsign system: Prefix: Z2.
Foreign amateur operation: Reciprocal (A + B). Call is own call/Z2. Countries include UK.
National society: Zimbabwe ARS, Postbox 2377, Harare.
Licensing administration: Manager, National Telecommunications Services, PO Box 8061, Causeway, Harare. Tel: +263 (4) 731989.

3 Callsign list

THIS list is intended to help in determining the country, continent, zones and true bearing (from London, England) of a station from its callsign. The prefixes given are those for standard stations; club ststions and special stations may use other prefixes although these should nevertheless conform with the ITU allocations given in the first column (see Chapter 1).

The country given on the same line as an ITU allocation is the holder of that allocation; further information on its callsign system may be found in Appendix 2. Certain countries encompass more than one zone; separate entries corresponding to each zone are therefore given on successive lines. Reference to Maps 7 and 8 in Appendix 1 should enable the correct zone to be ascertained in cases of doubt.

In 1998 the IARU approved exact definitions of ITU zones for amateur radio award purposes – for a complete list of these see the *RSGB Prefix Guide*.

ITU allocation	Prefix	Country	Continent	CQ	ITU	Deg
A2A–A2Z	A2	Botswana	AF	38	57	157
A3A–A3Z	A3	Tonga	OC	32	62	350
A4A–A4Z	A4	Oman	AS	21	39	104
A5A–A5Z	A5	Bhutan	AS	22	41	72
A6A–A6Z	A6	United Arab Emirates	AS	21	39	103
A7A–A7Z	A7	Qatar	AS	21	39	103
A8A–A8Z		Liberia				
A9A–A9Z	A9	Bahrain	AS	21	39	103
AAA–ALZ		USA				
	AA–AG	USA (see W)				
	AH0	Northern Mariana Is	OC	27	64	35
	AH1	Baker Is	OC	31	61	349
	AH1	Howland Is	OC	31	61	349
	AH2	Guam	OC	27	64	36
	AH3	Johnston Is	OC	31	61	11
	AH4	Midway Is	OC	31	61	358
	AH5J	Jarvis Is	OC	31	62	339
	AH5	Palmyra Is	OC	31	61	339
	AH5K	Kingman Reef	OC	31	61	339
	AH6, 7	Hawaii	OC	31	61	338
	AH7K	Kure Is	OC	31	61	338
	AH8	American Samoa	OC	32	62	346
	AH9	Peale Is	OC	31	65	13
	AH9	Wake Is	OC	31	65	13
	AH9	Wilkes Is	OC	31	65	13
	AI–AK	USA (see W)	NA			
	AL	Alaska	NA	1	1	348
	AL	Alaska	NA	1	2	348
AMA–AOZ		Spain				
APA–ASZ	AP	Pakistan	AS	21	41	87
ATA–AWZ		India				
AXA–AXZ		Australia				
AYA–AZZ		Argentina				
BAA–BZZ	BY	China	AS	23/24	33, 42–44	50
	BS	Scarborough Reef	AS	27	50	
	BV	Taiwan	AS	24	44	50
	BV9P	Pratas Is	AS	24	44	
C2A–C2Z	C2	Nauru	OC	31	65	17
C3A–C3Z	C3	Andorra	EU	14	27	172
C4A–C4Z		Cyprus				
C5A–C5Z	C5	Gambia	AF	35	46	207
C6A–C6Z	C6	Bahamas	NA	8	11	278
C7A–C7Z		WMO				
C8A–C9Z	C9	Mozambique	AF	37	53	148
CAA–CEZ		Chile				
	CE0Y	Easter Is	SA	12	63	266
	CE0Z	San Ambrosio Is	SA	12	14	247
	CE0Z	San Felix Is	SA	12	14	247
	CE0Z	Juan Fernandez Is	SA	12	14	241
	CE1–8	Chile	SA	12	14, 16	235
	CE9	Antarctica (Chile)	AN	13	73	209

ITU allocation	Prefix	Country	Continent	CQ	ITU	Deg
	CE9	S Shetland Is	SA	13	73	209
CFA–CKZ		Canada				
CLA–CMZ	CL, CM	Cuba	NA	8	11	279
CNA–CNZ	CN	Morocco	AF	33	37	196
COA–COZ	CO	Cuba	NA	8	11	279
CPA–CPZ	CP	Bolivia	SA	10	12–14	243
CQA–CUZ		Portugal				
	CT1, 4	Portugal	EU	14	37	208
	CT3	Madeira Is	AF	33	36	220
	CU	Azores	EU	14	36	248
CVA–CXZ	CX	Uruguay	SA	13	14	224
CYA–CZZ		Canada				
	CY0	Sable Is	NA	5	9	285
	CY9	St Paul Is	NA	5	9	287
D2A–D3Z	D2	Angola	AF	36	52	160
D4A–D4Z	D4	Cape Verde	AF	35	46	216
D5A–D5Z		Liberia				
D6A–D6Z	D6	Comoros	AF	39	53	171
D7A–D9Z		Korea (RK)				
DAA–DRZ	DA–DL	Germany	EU	14	28	96
DSA–DTZ	DS	Korea (RK)				
DUA–DZZ	DU–DY	Philippines	OC	27	50	57
E2A–E2Z		Thailand				
E3A–E3Z	E3	Eritrea	AF	37	48	120
E4A–E4Z	E4	Palestine	AS	20	39	
EAA–EHZ	EA–EC	Spain	EU	14	37	199
	EA6	Balearic Is	EU	14	37	167
	EA8	Canary Is	AF	33	36	213
	EA9	Ceuta	AF	33	37	193
	EA9	Melilla	AF	33	37	193
	ED	Antarctica (Spain)				
EIA–EJZ	EI, EJ	Eire	EU	14	27	303
EKA–EKZ	EK	Armenia	AS	21	29	91
ELA–ELZ	EL	Liberia	AF	35	46	195
EMA–EOZ		Ukraine				
EPA–EQZ	EP	Iran	AS	21	40	96
ERA–ERZ	ER	Moldova	EU	16	29	91
ESA–ESZ	ES	Estonia				
ETA–ETZ	ET	Ethiopia	AF	37	48	129
EUA–EWZ	EU, EW	Belarus	EU	16	29	71
EXA–EXZ	EX	Kyrghyzstan	AS	17	30	69
	EX	Kyrghyzstan	AS	17	31	69
EYA–EYZ	EY	Tajikistan	AS	17	30	77
EZA–EZZ	EZ	Turkmenistan	AS	17	30	81
FAA–FZZ	F	France	EU	14	27	–
	FG	Guadeloupe Is	NA	8	11	258
	FH	Mayotte	AF	39	53	171
	FJ	St Barthelemy				
	FK	Chesterfield Is	OC	30	56	45
	FK	Loyalty Is	OC	32	56	25
	FK	New Caledonia	OC	32	56	25
	FM	Martinique	NA	8	11	256
	FO	Austral Is	OC	32	63	
	FO	Clipperton Is	NA	7	10	291

ITU allocation	Prefix	Country	Continent	CQ	ITU	Deg
	FO	Gambier Is	OC	32	63	313
	FO	Marquesas Is	OC	31	63	313
	FO	Society Is	OC	32	63	313
	FO	Tubuai Is	OC	32	63	313
	FP	St Pierre et Miquelon	NA	5	9	286
	FR	Europa Is	AF	39	53	130
	FR	Glorieuses Is	AF	39	53	130
	FR	Reunion Is	AF	39	53	130
	FR	Tromelin Is	AF	39	53	130
	FR	Juan de Nova	AF	39	53	137
	FS	St Martin				
	FT-W	Crozet Is	AF	39	68	261
	FT-X	Kerguelen Is	AF	39	68	137
	FT-Y	Antarctica (France)	AN	30	70	165
	FT-Z	Amsterdam Is	AF	39	68	124
	FT-Z	St Paul Is	AF	39	68	124
	FW	Wallis Is	OC	32	62	356
	FW	Futuna Is	OC	32	62	356
	FY	French Guiana	SA	9	12	242
GAA–GZZ		UK				
	G	England	EU	14	27	–
	GD	Isle of Man	EU	14	27	321
	GI	Northern Ireland	EU	14	27	317
	GJ	Jersey	EU	14	27	215
	GM	Scotland	EU	14	27	345
	GU	Alderney	EU	14	27	224
	GU	Guernsey	EU	14	27	224
	GU	Sark	EU	14	27	224
	GW	Wales	EU	14	27	285
H2A–H2Z		Cyprus				
H3A–H3Z		Panama				
H4A–H4Z	H4	Solomon Is	OC	28	51	28
	H40	Temotu Province	OC	32	51	
H6A–H7Z		Nicaragua				
H8A–H9Z		Panama				
HAA–HAZ	HA	Hungary	EU	15	28	100
HBA–HBZ		Switzerland				
	HB0	Liechtenstein	EU	14	28	118
	HB	Switzerland	EU	14	28	125
HCA–HDZ	HC	Ecuador	SA	10	12	261
	HC8	Galapagos Is	SA	10	12	270
HEA–HEZ		Switzerland				
HFA–HFZ		Poland				
HGA–HGZ	HG	Hungary	EU	15	28	100
HHA–HHZ	HH	Haiti	NA	8	11	268
HIA–HIZ	HI	Dominican Rep	NA	8	11	267
HJA–HKZ	HJ, HK	Colombia	SA	9	12	260
	HK0	Malpelo Is	NA	9	12	266
	HK0	Providencia Is	NA	7	11	272
	HK0	San Andres Is	NA	7	11	272
HLA–HLZ	HL	Korea (RK)	AS	25	44	40
HMA–HMZ		Korea (DPRK)				
HNA–HNZ		Iraq				
HOA–HPZ	HP	Panama	NA	7	11	267
HQA–HRZ	HR	Honduras	NA	7	11	277
HSA–HSZ	HS	Thailand	AS	26	49	72
HTA–HTZ		Nicaragua				
HUA–HUZ		El Salvador				
HVA–HVZ	HV	Vatican	EU	15	28	132
HWA–HYZ		France				
HZA–HZZ	HZ	Saudi Arabia	AS	21	39	137
IAA–IZZ	I	Italy	EU	15	28	128
	IS0	Sardinia	EU	15	28	147
J2A–J2Z	J2	Djibouti	AF	37	48	123
J3A–J3Z	J3	Grenada	NA	8	11	255
J4A–J4Z		Greece				
J5A–J5Z	J5	Guinea-Bissau	AF	35	46	207
J6A–J6Z	J6	St Lucia	NA	8	11	260
J7A–J7Z	J7	Dominica	NA	8	11	260
J8A–J8Z	J8	St Vincent and the Grenadines	NA	8	11	260
JAA–JSZ	JA	Japan	AS	25	45	35
	JD	Minami Torishima	OC	27	—	34
	JE–JS	Japan	AS	25	45	35
JTA–JVZ	JT	Mongolia	AS	23	32	45
	JT	Mongolia	AS	23	33	45
JWA–JXZ		Norway				
	JW	Svalbard	EU	40	18	25
	JW	Svalbard	EU	40	75	25
	JX	Jan Mayen Is	EU	40	18	9

ITU allocation	Prefix	Country	Continent	CQ	ITU	Deg
JYA–JYZ	JY	Jordan	AS	20	39	113
JZA–JZZ		Indonesia				
KAA–KZZ		USA				
	K	USA (see W)				
	KA	USA (see W)				
	KB	USA (see W)				
	KC	USA (see W)				
KC4AAA–KC4AAF		Antarctica				
KC4USA–KC4USF		Antarctica				
	KC6	Belau	OC	27	64	49
KD–KG		USA (see W)				
	KG4	Guantanamo Bay	NA	8	11	272
	KH0	Northern Mariana Is	OC	27	64	35
	KH1	Baker Is	OC	31	61	349
	KH1	Howland Is	OC	31	61	349
	KH2	Guam	OC	27	64	36
	KH3	Johnston Is	OC	31	61	11
	KH4	Midway Is	OC	31	61	358
	KH5J	Jarvis Is	OC	31	62	339
	KH5	Palmyra Is	OC	31	61	339
	KH5K	Kingman Reef	OC	31	61	339
	KH6, 7	Hawaii	OC	31	61	338
	KH7K	Kure Is	OC	31	61	338
	KH8	American Samoa	OC	32	62	346
	KH9	Peale Is	OC	31	65	13
	KH9	Wake Is	OC	31	65	13
	KH9	Wilkes Is	OC	31	65	13
	KI, KJ	USA (see W)				
	KK	USA (see W)				
	KL	Alaska	NA	1	1	348
	KL	Alaska	NA	1	2	348
KM–KO		USA (see W)				
	KP1	Navassa Is	NA	8	11	270
	KP2	US Virgin Is	NA	8	11	262
	KP3, 4	Puerto Rico	NA	8	11	263
	KP5	Desecheo Is	NA	8	11	263
	KQ–KV	USA (see W)				
	KW	USA (see W)				
	KX	USA (see W)				
	KY, KZ	USA (see W)				
L2A–L9Z		Argentina				
LAA–LNZ	LA, LB	Norway	EU	14	18	23
LOA–LWZ	LU	Argentina	SA	13	16	229
	LU	Argentina	SA	13	16	229
	LU W, X	Argentina	SA	13	16	229
	LU Z	Antarctica (Argentina)	AN	13	73	209
LXA–LXZ	LX	Luxembourg	EU	14	27	108
LYA–LYZ	LY	Lithuania	EU	15	29	58
LZA–LZZ	LZ	Bulgaria	EU	20	28	108
MAA–MZZ		UK				
	M	England	EU	14	27	–
	MD	Isle of Man	EU	14	27	321
	MI	Northern Ireland	EU	14	27	317
	MJ	Jersey	EU	14	27	215
	MM	Scotland	EU	14	27	345
	MU	Alderney	EU	14	27	224
	MU	Guernsey	EU	14	27	224
	MU	Sark	EU	14	27	224
	MW	Wales	EU	14	27	285
NAA–NZZ		USA				
	N	USA (see W)				
NA–NG		USA (see W)	NA			
	NH0	Northern Mariana Is	OC	27	64	35
	NH1	Baker Is	OC	31	61	349
	NH1	Howland Is	OC	31	61	349
	NH2	Guam	OC	27	64	36
	NH3	Johnston Is	OC	31	61	11
	NH4	Midway Is	OC	31	61	358
	NH5J	Jarvis Is	OC	31	62	339
	NH5	Palmyra Is	OC	31	61	339
	NH5K	Kingman Reef	OC	31	61	339
	NH6, 7	Hawaii	OC	31	61	338
	NH7K	Kure Is	OC	31	61	338
	NH8	American Samoa	OC	32	62	346
	NH9	Peale Is	OC	31	65	13
	NH9	Wake Is	OC	31	65	13

ITU allocation	Prefix	Country	Continent	CQ	ITU	Deg
	NH9	Wilkes Is	OC	31	65	13
	NI–NK	USA (see W)				
	NL	Alaska	NA	1	1	348
	NL	Alaska	NA	1	2	348
	NM–NO	USA (see W)				
	NP1	Navassa Is	NA	8	11	270
	NP2	US Virgin Is	NA	8	11	262
	NP3, 4	Puerto Rico	NA	8	11	263
	NP5	Desecheo Is	NA	8	11	263
	NQ–NZ	USA (see W)				
OAA–OCZ	OA	Peru	SA	10	12	249
ODA–ODZ	OD	Lebanon	AS	20	39	109
OEA–OEZ	OE	Austria	EU	15	28	106
OFA–OJZ	OH	Finland	EU	15	18	42
	OH0	Aland Is	EU	15	18	42
	OJ0	Market Reef	EU	15	18	42
OKA–OLZ	OK, OL	Czech Rep	EU	15	28	94
OMA–OMZ	OM	Slovak Rep	EU	15	28	94
ONA–OTZ	ON	Belgium	EU	14	27	94
	OR	Antarctica (Belgium)	AN	–	67	–
OUA–OZZ		Denmark				
	OX	Greenland	NA	40	5	340
	OX	Greenland	NA	40	75	340
	OY	Faroe Is	EU	14	18	343
	OZ	Denmark	EU	14	18	48
P2A–P2Z	P2	Papua New Guinea	OC	28	51	47
P3A–P3Z		Cyprus				
P4A–P4Z	P4	Aruba	SA	9	11	262
P5A–P9Z	P5	Korea (DPRK)	AS	25	44	40
PAA–PIZ	PA, PB, PD, PE, PI	Netherlands	EU	14	27	73
PJA–PJZ		Netherlands Antilles				
	PJ2	Curaçao	SA	9	11	262
	PJ4	Bonaire	SA	9	11	262
	PJ5	St Eustatius	NA	8	11	262
	PJ6	Saba	NA	8	11	262
	PJ7	St Maarten	NA	8	11	261
PKA–POZ		Indonesia				
PPA–PYZ	PP1, 2	Brazil	SA	11	15	221
	PP5	Brazil	SA	11	15	221
	PP6–8	Brazil	SA	11	13	230
	PR7, 8	Brazil	SA	11	13	230
	PS7	Brazil	SA	11	13	230
	PT2	Brazil	SA	11	12, 15	229
	PT7	Brazil	SA	11	12, 13	230
	PT9	Brazil	SA	11	12, 15	230
	PU8	Brazil	SA	11	13	230
	PV8	Brazil	SA	11	12, 13	230
	PW8	Brazil	SA	11	12, 13	230
	PY0F	Fernando de Noronha	SA	11	13	218
	PY0S	St Paul Is	SA	11	13	215
	PY0S	St Peter Is	SA	11	13	215
	PY0T	Trindade Is	SA	11	15	208
	PY0	Martim Vaz Is	SA	11	15	208
	PY1–5	Brazil	SA	11	15	221
	PY6–8	Brazil	SA	11	13	230
	PY9	Brazil	SA	11	15	230
PZA–PZZ	PZ	Suriname	SA	9	12	245
RAA–RZZ		Russia (see UAA–UIZ)				
	R1A	Antarctica				
	R1F	Franz Josef Land	EU	40	20/75	10
	R1M	Malyj Vysotskij Is	EU	16	29	10
S2A–S3Z	S2	Bangladesh	AS	22	41	74
S5A–S5Z	S5	Slovenia	EU	15	28	113
S6A–S6Z		Singapore				
S7A–S7Z	S7	Seychelles	AF	39	53	121
S9A–S9Z	S9	São Tomé, Principe	AF	36	47	171
SAA–SMZ	SH, SK–SM	Sweden	EU	14	18	42
SNA–SRZ	SO–SQ	Poland	EU	15	28	79
SSA–SSM		Egypt				
SSN–STZ	ST	Sudan	AF	34	47	132
	ST	Sudan	AF	34	48	132
SUA–SUZ	SU	Egypt	AF	34	38	128
SVA–SZZ	SV	Greece	EU	20	28	119
	SV5	Rhodes	EU	20	28	115

ITU allocation	Prefix	Country	Continent	CQ	ITU	Deg
	SV9	Crete	EU	20	28	121
T2A–T2Z	T2	Tuvalu	OC	31	65	2
T3A–T3Z	T30	W Kiribati	OC	31	65	13
	T31	Central Kiribati	OC	31	62	332
	T32	E Kiribati	OC	31	61	350
	T32	E Kiribati	OC	31	63	350
	T33	Banaba	OC	31	65	13
T4A–T4Z		Cuba				
T5A–T5Z	T5	Somalia	AF	37	48	126
T6A–T6Z		Afghanistan				
T7A–T7Z	T7	San Marino	EU	15	28	125
T8A–T8Z		Palau				
T9A–T9Z	T9	Bosnia-Hercegovina	EU	15	28	113
TAA–TCZ	TA	Turkey	EU/AS	20	39	103
TDA–TDZ		Guatemala				
TEA–TEZ		Costa Rica				
TFA–TFZ	TF	Iceland	EU	40	17	331
TGA–TGZ	TG	Guatemala	NA	7	11	280
THA–THZ		France				
TIA–TIZ	TI	Costa Rica	NA	7	11	272
	TI9	Cocos Is	NA	7	11	271
TJA–TJZ	TJ	Cameroon	AF	36	47	166
TKA–TKZ		France				
	TK	Corsica	EU	15	28	141
TLA–TLZ	TL	Central African Rep	AF	36	47	152
TMA–TMZ		France				
TNA–TNZ	TN	Congo	AF	36	52	161
TOA–TQZ		France				
	TP	Council of Europe, Strasbourg				
TRA–TRZ	TR	Gabon	AF	36	52	165
TSA–TSZ		Tunisia				
TTA–TTZ	TT	Chad	AF	36	47	151
TUA–TUZ	TU	Ivory Coast	AF	35	46	187
TVA–TXZ		France				
TYA–TYZ	TY	Benin	AF	35	46	176
TZA–TZZ	TZ	Mali	AF	35	46	187
UAA–UIZ		Russia				
	UA0A	Krasnoyarsk	AS	18	22	24
	UA0B	Taymyr	AS	18	22	24
	UA0C	Khabarovsk	AS	19	34	24
	UA0D	Yevrev	AS	19	34	24
	UA0E–G	Sakhalin	AS	19	34	24
	UA0H	Evenkiysk	AS	18	22	24
	UA0I	Magadan	AS	19	25	24
	UA0I	Magadan	AS	19	26	24
	UA0J	Amur	AS	19	33	24
	UA0L–N	Primorsky	AS	19	34	24
	UA0O, P	Buryat Rep	AS	18	32	24
	UA0Q, R	Sakha	AS	19	23	24
	UA0S, T	Irkutsk	AS	18	32	24
	UA0U, V	Chita	AS	18	33	24
	UA0W	Khakass	AS	18	32	24
	UA0X	Koriatsky	AS	19	25	24
	UA0Y	Tuvinsk	AS	23	32	24
	UA0Z	Kamchatsky	AS	19	35	24
	UA1	Russia (N European)	EU	16	29	51
	UA1N	Karelian Rep	EU	16	19	51
	UA1O	Arhangelsk	EU	16	20	13
	UA1P	Nenets	EU	16	20	13
	UA1Y, Z	Murmansk	EU	16	19	51
	UA2	Kaliningrad	EU	15	29	66
	UA3	Russia (C European)	EU	16	29	63
	UA4	Russia (E European)	EU	16	29	67
	UA4N, O	Kirov	EU	16	30	67
	UA4P–R	Tatar Rep	EU	16	30	67
	UA4W	Udmurt Rep	EU	16	30	67
	UA6	Russia (S European)	EU	16	29	77
	UA8T	Ust-Ordinsky	AS	18	32	24
	UA8V	Aginsky-Buriatsky	AS	18	33	24
	UA9A, B	Chelyabinsk	AS	17	30	60
	UA9C–E	Sverdlovsk	AS	17	30	60
	UA9F	Perm	AS	17	30	60
	UA9G	Komi-Permiatsky	AS	17	30	60
	UA9H, I	Tomsk	AS	18	31	49
	UA9J	Khanty-Mansiysky	AS	17	20	30
	UA9K	Jamalo-Nenets	AS	17	20	30

ITU allocation	Prefix	Country	Continent	CQ	ITU	Deg
	UA9L	Tyumen	AS	17	21	30
	UA9M, N	Omsk	AS	17	30	60
	UA9O, P	Novosibirsk	AS	18	31	49
	UA9Q, R	Kurgan	AS	17	30	60
	UA9S, T	Orenburg	AS	16	30	62
	UA9U, V	Kemerov	AS	18	31	49
	UA9W	Bashkir	AS	16	30	62
	UA9X	Komi	AS	17	20	30
	UA9Y	Altay	AS	18	31	49
	UA9Z	Gorno-Altay	AS	18	31	49
UJA–UMZ	UK	Uzbekistan	AS	17	30	77
UNA–UQZ	UN–UQ	Kazakhstan	AS	17	29–31	67
URA–UZZ	UR–UZ	Ukraine	EU	16	29	82
V2A–V2Z	V2	Antigua, Barbuda	NA	8	11	260
V3A–V3Z	V3	Belize	NA	7	11	280
V4A–V4Z	V4	St Kitts & Nevis	NA	8	11	260
V5A–V5Z	V5	Namibia	AF	38	57	164
V6A–V6Z	V6	Micronesia	OC	27	64	32
	V6	Micronesia	OC	27	65	32
V7A–V7Z	V7	Marshall Is	OC	31	65	14
V8A–V8Z	V8	Brunei Darussalam	OC	28	54	67
VAA–VGZ		Canada				
	VA2	Canada	NA	2	9	293
	VA2	Canada	NA	5	9	293
	VA3	Canada	NA	4	4	308
	VE1	Canada	NA	5	9	290
	VE2	Canada	NA	2	9	293
	VE2	Canada	NA	5	9	293
	VE3	Canada	NA	4	4	308
	VE4	Canada	NA	4	3	315
	VE5	Canada	NA	4	3	319
	VE6	Canada	NA	4	2	323
	VE7	Canada	NA	3	2	329
	VE8	Canada	NA	1	2	328
	VE8	Canada	NA	1	3	328
	VE8	Canada	NA	2	3	328
	VE8	Canada	NA	2	4	328
	VE8	Canada	NA	2	5	328
	VE8	Canada	NA	2	75	328
VHA–VNZ		Australia				
	VK0	Antarctica				
	VK0	Heard Is	AF	39	68	138
	VK0	Macquarie Is	OC	30	60	113
	VK1	Australia	OC	30	59	63
	VK2	Australia	OC	30	59	65
	VK3	Australia	OC	30	59	76
	VK4	Australia	OC	30	55	75
	VK5	Australia	OC	30	59	75
	VK6	Australia	OC	29	58	79
	VK7	Australia	OC	30	59	83
	VK8	Australia	OC	29	55	65
	VK9C	Cocos Keeling Is	OC	29	54	92
	VK9L	Lord Howe Is	OC	30	60	63
	VK9M	Mellish Reef	OC	30	56	45
	VK9N	Norfolk Is	OC	32	60	27
	VK9W	Willis Is	OC	30	55	45
	VK9X	Christmas Is	OC	29	54	84
VOA–VOZ	VO	Canada	NA	2, 5	9	288
VPA–VQZ		UK				
	VP2E	Anguilla	NA	8	11	260
	VP2M	Montserrat	NA	8	11	260
	VP2V	British Virgin Is	NA	8	11	260
	VP5	Caicos Is	NA	8	11	270
	VP5	Turks Is	NA	8	11	270
	VP6	Pitcairn Is	OC	32	63	285
	VP8	Antarctica (UK)				
	VP8	Falkland Is	SA	13	16	216
	VP8	S Georgia	SA	13	73	202
	VP8	S Orkney Is	SA	13	73	202
	VP8	S Sandwich Is	SA	13	73	195
	VP8	S Shetland Is	SA	13	73	208
	VP9	Bermuda	NA	5	11	274
	VQ9	Chagos Is	AF	39	41	108
VRA–VRZ		China				
	VR2	Hong Kong	AS	24	44	58
VSA–VSZ		UK				
VTA–VWZ		India				
	VU	Andaman Is	AS	26	49	81
	VU	India	AS	22	41	85
	VU	Lakshadweep	AS	22	41	97
	VU	Nicobar Is	AS	26	49	81

ITU allocation	Prefix	Country	Continent	CQ	ITU	Deg
VXA–VYZ	VY1, 2		Canada			
VZA–VZZ		Australia				
WAA–WZZ	W0	USA	NA	4	7	305
	W1	USA	NA	5	8	290
	W2	USA	NA	5	8	290
	W3	USA	NA	5	8	289
	W4	USA	NA	4	8	288
	W4	USA	NA	5	8	288
	W5	USA	NA	4	7	292
	W6	USA	NA	3	6	317
	W7	USA	NA	3	6	316
	W7	USA	NA	4	6	316
	W8	USA	NA	4	8	294
	W8	USA	NA	5	8	294
	W9	USA	NA	4	8	297
	WA–WG	USA (see W)				
	WH0	Northern Mariana Is	OC	27	64	35
	WH1	Baker Is	OC	31	61	349
	WH1	Howland Is	OC	31	61	349
	WH2	Guam	OC	27	64	36
	WH3	Johnston Is	OC	31	61	11
	WH4	Midway Is	OC	31	61	358
	WH5J	Jarvis Is	OC	31	62	339
	WH5	Palmyra Is	OC	31	61	339
	WH5K	Kingman Reef	OC	31	61	339
	WH6, 7	Hawaii	OC	31	61	338
	WH7K	Kure Is	OC	31	61	338
	WH8	American Samoa	OC	32	62	346
	WH9	Peale Is	OC	31	65	13
	WH9	Wake Is	OC	31	65	13
	WH9	Wilkes Is	OC	31	65	13
	WI–WK	USA (see W)				
	WL	Alaska	NA	1	1	348
	WL	Alaska	NA	1	2	348
	WM–WO	USA (see W)				
	WP1	Navassa Is	NA	8	11	270
	WP2	US Virgin Is	NA	8	11	262
	WP3, 4	Puerto Rico	NA	8	11	263
	WP5	Desecheo Is	NA	8	11	263
	WQ–WZ	USA (see W)				
XAA–XIZ	XE, XF	Mexico	NA	6	10	294
	XF4	Revilla Gigedo Is	NA	6	10	299
XJA–XOZ		Canada				
XPA–XPZ		Denmark				
XQA–XRZ	XQ	Chile				
XSA–XSZ		China				
XTA–XTZ	XT	Burkina Faso	AF	35	46	182
XUA–XUZ	XU	Kampuchea	AS	26	49	71
XVA–XVZ	XV	Vietnam				
XWA–XWZ	XW	Laos	AS	26	49	68
XXA–XXZ		Portugal				
	XX9	Macao	AS	24	44	58
XYA–XZZ	XY, XZ	Myanmar	AS	26	49	75
Y2A–Y9Z		Germany				
YAA–YAZ	YA	Afghanistan	AS	21	40	81
YBA–YHZ	YB–YH	Indonesia	OC	28	51	81
	YB–YD	Indonesia	OC	28	54	81
YIA–YIZ	YI	Iraq	AS	21	39	102
YJA–YJZ	YJ	Vanuatu	OC	32	56	20
YKA–YKZ	YK	Syria	AS	20	39	106
YLA–YLZ	YL	Latvia	EU	15	29	58
YMA–YMZ		Turkey				
YNA–YNZ	YN	Nicaragua	NA	7	11	274
YOA–YRZ	YO	Romania	EU	20	28	101
YSA–YSZ	YS	El Salvador	NA	7	11	278
YTA–YUZ	YT, YU	Yugoslavia	EU	15	28	113
YVA–YYZ		Venezuela				
	YV0	Aves Is	NA	8	11	259
	YV	Venezuela	SA	9	12	256
	YZ	Yugoslavia	EU	15	28	113
YZA–YZZ						
Z2A–Z2Z	Z2	Zimbabwe	AF	38	53	151
Z3A–Z3Z	Z3	Macedonia	EU	15	28	113
ZAA–ZAZ	ZA	Albania	EU	15	28	118
ZBA–ZJZ		UK				
	ZB	Gibraltar	EU	14	37	196
	ZC	Cyprus (UK bases)	AS	20	39	111
	ZD7	St Helena Is	AF	36	66	186
	ZD8	Ascension Is	AF	36	66	196
	ZD9	Gough Is	AF	38	66	188

ITU allocation	Prefix	Country	Continent	CQ	ITU	Deg
	ZD9	Tristan da Cunha Is	AF	38	66	188
	ZF	Cayman Is	NA	8	11	276
ZKA–ZMZ		New Zealand				
	ZK1	N Cook Is	OC	32	62, 63	331
	ZK1	S Cook Is	OC	32	62, 63	324
	ZK2	Niue Is	OC	32	62	342
	ZK3	Tokelau Is	OC	31	62	345
	ZL	New Zealand	OC	32	60	5–68
	ZL5	Antarctica				
	ZL7	Chatham Is	OC	32	60	40
	ZL8	Kermadec Is	OC	32	60	355
	ZL9	Auckland Is	OC	32	60	95
	ZL9	Campbell Is	OC	32	60	106
ZNA–ZOZ		UK				
ZPA–ZPZ	ZP	Paraguay	SA	11	14	230
ZQA–ZQZ		UK				
ZRA–ZUZ	ZR, ZS	South Africa	AF	38	57	160
	ZS7	Antarctica				
	ZS8	Marion Is	AF	38	57	154
	ZS8	Prince Edward Is	AF	38	57	154
ZVA–ZZZ		Brazil				
2AA–2ZZ		UK				
	2E	England	EU	14	27	–
	2D	Isle of Man	EU	14	27	321
	2I	Northern Ireland	EU	14	27	317
	2J	Jersey	EU	14	27	215
	2M	Scotland	EU	14	27	345
	2U	Alderney	EU	14	27	224
	2U	Guernsey	EU	14	27	224
	2U	Sark	EU	14	27	224
	2W	Wales	EU	14	27	285
3AA–3AZ	3A	Monaco	EU	14	27	143
3BA–3BZ		Mauritius				
	3B6	Agalega Is	AF	39	53	
	3B8	Mauritius	AF	39	53	125
	3B9	Rodriguez Is	AF	39	53	
3CA–3CZ	3C	Equatorial Guinea	AF	36	47	193
	3C0	Annobon	AF	36	52	173
3DA–3DM	3DA	Swaziland	AF	38	57	152
3DN–3DZ	3D2	Fiji	OC	32	56	3
3EA–3FZ		Panama				
3GA–3GZ		Chile				
3HA–3UZ		China				
3VA–3VZ	3V	Tunisia	AF	33	37	150
3WA–3WZ	3W	Vietnam	AS	26	49	70
3XA–3XZ	3X	Guinea	AF	35	46	195
3YA–3YZ		Norway				
	3Y	Antarctica (Norway)				
	3Y	Bouvet Is	AF	38	67	178
	3Y	Peter I Is				
3ZA–3ZZ		Poland				
4AA–4CZ		Mexico				
4DA–4IZ		Philippines				
4JA–4KZ	4J, 4K	Azerbaijan	AS	21	29	70
4LA–4LZ	4L	Georgia	AS	21	29	91
4MA–4MZ		Venezuela				
4NA–4OZ	4N	Yugoslavia	EU	15	28	113
4PA–4SZ	4S	Sri Lanka	AS	22	41	93
4TA–4TZ		Peru				
4UA–4UZ	4U1	United Nations				
4VA–4VZ		Haiti				
4XA–4XZ	4X	Israel	AS	20	39	113
4YA–4YZ		ICAO				
4ZA–4ZZ	4Z	Israel	AS	20	39	113
5AA–5AZ	5A	Libya	AF	34	38	143
5BA–5BZ	5B	Cyprus	AS	20	39	111
5CA–5GZ		Morocco				
5HA–5IZ	5H	Tanzania	AF	37	53	137
5JA–5KZ		Colombia				
5LA–5MZ		Liberia				
5NA–5OZ	5N	Nigeria	AF	35	46	171
5PA–5QZ		Denmark				
5RA–5SZ	5R	Madagascar	AF	39	53	137
5TA–5TZ	5T	Mauritania	AF	35	46	200
5UA–5UZ	5U	Niger	AF	35	46	163
5VA–5VZ	5V	Togo	AF	35	46	178
5WA–5WZ	5W	Western Samoa	OC	32	62	13
5XA–5XZ	5X	Uganda	AF	37	48	140
5YA–5ZZ	5Z	Kenya	AF	37	48	137
6AA–6BZ		Egypt				
6CA–6CZ		Syria				
6DA–6JZ		Mexico				
6KA–6NZ		Korea (RK)				
6OA–6OZ		Somalia				
6PA–6SZ		Pakistan				
6TA–6UZ		Sudan				
6VA–6WZ	6W	Senegal	AF	35	46	207
6XA–6XZ		Madagascar				
6YA–6YZ	6Y	Jamaica	NA	8	11	272
6ZA–6ZZ		Liberia				
7AA–7IZ		Indonesia				
7JA–7NZ	7J–7N	Japan	AS	25	45	35
7OA–7OZ	7O	Yemen	AF	21	39	119
	7O	Socotra Is	AF	37	48	120
7PA–7PZ	7P	Lesotho	AF	38	57	156
7QA–7QZ	7Q	Malawi	AF	37	53	145
7RA–7RZ		Algeria				
7SA–7SZ		Sweden				
7TA–7YZ	7X	Algeria	AF	33	37	175
7ZA–7ZZ	7Z	Saudi Arabia	AS	29	31	137
8AA–8IZ		Indonesia				
8JA–8NZ		Japan				
	8J1	Antarctica (Japan)	AN	39	74	
8OA–8OZ		Botswana				
8PA–8PZ	8P	Barbados	NA	8	11	254
8QA–8QZ	8Q	Maldives	AS	22	41	101
8RA–8RZ	8R	Guyana	SA	9	12	247
8SA–8SZ		Sweden				
8TA–8YZ		India				
8ZA–8ZZ		Saudi Arabia				
9AA–9AZ	9A	Croatia	EU	15	28	113
9BA–9DZ		Iran				
9EA–9FZ		Ethiopia				
9GA–9GZ	9G	Ghana	AF	35	46	188
9HA–9HZ	9H	Malta	EU	15	28	140
9IA–9JZ	9J	Zambia	AF	36	53	151
9KA–9KZ	9K	Kuwait	AS	21	39	103
9LA–9LZ	9L	Sierra Leone	AF	35	46	197
9MA–9MZ		Malaysia				
	9M2	West Malaysia	AS	28	54	78
	9M6	Sabah	AS	28	54	72
	9M8	Sarawak	AS	28	54	73
9NA–9NZ	9N	Nepal	AS	22	42	75
9OA–9TZ	9Q	Zaire	AF	36	52	160
9UA–9UZ	9U	Burundi	AF	36	52	145
9VA–9VZ	9V	Singapore	AS	28	54	78
9WA–9WZ		Malaysia				
9XA–9XZ	9X	Rwanda	AF	36	52	144
9YA–9ZZ	9Y	Trinidad, Tobago	SA	9	11	254

4 DXCC entities list

THIS list gives the ARRL DXCC entities (current at the time of going to press) arranged in alphabetical order. Sample prefixes are given to enable quick cross reference to Appendix 3 for zones and bearings; full callsign information will be found under the appropriate ITU allocation holder in Appendix 2.

The list also acts as a gazetteer for Appendix 1. The first character of the map reference is the map number, the second and third defining the square on the map. In the case of Map 1, the remaining characters (if shown) are locator squares. An

asterisk following the map reference indicates that the entity also appears in a map in the section for the ITU allocation holder in Appendix 2.

As readers will often be working for several different awards, the bands/modes have been left blank on the checklist columns.

Note: The official ARLL DXCC List contains much more information than given here. If you are applying for DXCC, you should obtain the official list, rules and form from ARRL or via www.arrl.org. However, some recent changes are detailed following the main table.

Country	Prefix	ITU alloc holder	Map	Worked/heard on										
Afghanistan	YA	Afghanistan	2Gb											
Agalega, St Brandon	3B6, 7	Mauritius	2Be,f											
Aland Is	OH0	Finland	IKP00*											
Alaska	KL7	USA	3Aa											
Albania	ZA	Albania	1											
Algeria	7T–7Y	Algeria	2Cc											
Amsterdam, St Paul Is	FT8Z	France	2Hg											
Andaman, Nicobar Is	VU	India	2Jd											
Andorra	C3	Andorra	1JN02											
Angola	D2, 3	Angola	2De											
Anguilla	VP2E	UK	6Gc											
Antarctica	CE9 etc	Various	5											
Antigua, Barbuda	V2	Antigua	6Gc											
Argentina	LO–LW	Argentina	3Hh*											
Armenia	EK	Armenia	—											
Aruba	P4	Aruba	6Fd											
Ascension Is	ZD8	UK	2Be											
Auckland, Campbell Is	ZL9	New Zealand	4Dg											
Austral Is	FO	France	4Ge											
Australia	VK	Australia	2Mg*											
Austria	OE	Austria	1*											
Aves Is	YV0	Venezuela	6Gc											
Azerbaijan	4J, 4K	Azerbaijan	—											
Azores	CU	Portugal	2Ab											
Bahamas	C6	Bahamas	6Da											
Bahrain	A9	Bahrain	2Fc											
Baker, Howland Is	KH1	USA	4Ec,d											
Balearic Is	EA6–EH6	Spain	1*											
Banaba Is (Ocean Is)	T33	Kiribati	4Dd											
Bangladesh	S2	Bangladesh	2Jc											

Country	Prefix	ITU alloc holder	Map	Worked/heard on								
Barbados	8P	Barbados	6Hd									
Belarus	EU	Belarus	1									
Belau	KC6	USA	4Bc									
Belgium	ON–OT	Belgium	1									
Belize	V3	Belize	6Bc									
Benin	TY	Benin	2Cd									
Bermuda	VP9	UK	3Gc									
Bhutan	A5	Bhutan	2Jc									
Bolivia	CP	Bolivia	3Gg*									
Bonaire, Curacao	PJ2, 4, 9	Netherlands Antilles	6									
Bosnia-Hercegovina	T9	Bosnia-Hercegovina	1									
Botswana	A2	Botswana	2Df									
Bouvet Is	3Y	Norway	2Ch									
Brazil	PP–PY	Brazil	3Hg*									
Brunei	V8	Brunei	2Kd									
Bulgaria	LZ	Bulgaria	1									
Burkina Faso	XT	Burkina Faso	2Bd									
Burundi	9U	Burundi	2Ee									
Cambodia	XU	Cambodia	2Jd									
Cameroon	TJ	Cameroon	2Cd									
Canada	VA–VE, VY	Canada	3Fb									
Canary Is	EA8–EH8	Spain	2Ac									
Cape Verde	D4	Cape Verde	2Ac									
Cayman Is	ZF	UK	6Cc									
Central African Rep	TL	Central African Rep	2Dd									
Ceuta, Melilla	EA9–EH9	Spain	1*									
Chad	TT	Chad	2Dc									
Chagos Is	VQ9	UK	2Ge									
Chatham Is	ZL7	New Zealand	4Ef									
Chile	CA–CE	Chile	3Gh*									
China	BY–BT	China	2									
Christmas Is	VK9X	Australia	2Ke*									
Clipperton Is	FO	France	3De									
Cocos Is	TI9	Costa Rica	3Fe									
Cocos-Keeling Is	VK9C	Australia	2Je									
Colombia	HJ–HK	Colombia	3Ge*									
Comoros	D6	Comoros	2Ee									
Congo	TN	Congo	2Ce									
Conway Reef	3D2	Fiji	—									
Corsica	TK	France	1									
Costa Rica	TE, TI	Costa Rica	6Ce									
Crete	SV9	Greece	1MV									
Croatia	9A	Croatia	1									
Crozet Is	FT8W	France	2Fh									
Cuba	CM, CO	Cuba	6Cb*									
Cyprus	5B	Cyprus	1									
Cyprus (UK sov bases)	ZC4	UK	1									
Czech Rep	OK–OL	Czech Rep	1									

Country	Prefix	ITU alloc holder	Map	Worked/heard on								
Dem Rep of Congo	9Q–9T	Dem Rep of Congo	2De									
Denmark	OZ	Denmark	1									
Desecheo Is	KP5	USA	6Fc									
Djibouti	J2	Djibouti	2Ed									
Dodecanese Is	SV5	Greece	1KM36*									
Dominica	J7	Dominica	6Gc									
Dominican Republic	HI	Dominican Republic	6Ec									
Easter Is	CE0	Chile	4Je									
Ecuador	HC–HD	Ecuador	3Ff									
Egypt	SU	Egypt	2Ec									
El Salvador	YS	El Salvador	3Fe									
England	G, M, 2E	UK	1									
Equatorial Guinea	3C	Equatorial Guinea	2Cd									
Eritrea	E3	Eritrea	2									
Estonia	ES	Estonia	1									
Ethiopia	ET	Ethiopia	2Ed									
European Russia	RA–RZ	Russia	—									
Falkland Is	VP8	UK	3Hj									
Faroe Is	OY	Denmark	1									
Fernando de Noronha	PP0–PY0	Brazil	*									
Fiji	3D2	Fiji	4De									
Finland	OF–OI	Finland	1*									
France	F	France	1									
Franz Josef Land	R1FJ	Russia	—									
French Guiana	FY	France	3He									
French Polynesia	FO	France	4Ge									
Gabon	TR	Gabon	2Ce									
Galapagos Is	HC8, HD8	Ecuador	3Ef									
Gambia	C5	Gambia	2Ad									
Georgia	4L	Georgia	—									
Germany	DA–DL	Germany	1									
Ghana	9G	Ghana	2Bd									
Gibraltar	ZB2	UK	1									
Glorioso Is	FR/G	France	2Fe									
Greece	SV–SZ	Greece	1*									
Greenland	OX	Denmark	2Ja									
Grenada	J3	Grenada	6Gd									
Guadeloupe	FG	France	6Gc									
Guam	KH2	USA	4Bc									
Guantanamo Bay	KG4	USA	6Db									
Guatemala	TD, TG	Guatemala	3Ee									
Guernsey	GU, MU, 2U	UK	1IN89									
Guinea	3X	Guinea	2Bd									
Guinea-Bissau	J5	Guinea-Bissau	2Bd									
Guyana	8R	Guyana	3He									
Haiti	HH	Haiti	6Ec									
Hawaiian Is	KH6, 7	USA	4Fb									
Heard Is	VK0	Australia	4Gh									

Country	Prefix	ITU alloc holder	Map	Worked/heard on								
Honduras	HQ–HR	Honduras	6Bc									
Hong Kong	VR2	China	2Kc									
Hungary	HA, HG	Hungary	1*									
Iceland	TF	Iceland	1*									
India	VU	India	2Hc									
Indonesia	YB–YH	Indonesia	2Ke*									
Iran	EP–EQ	Iran	2Fb									
Iraq	YI	Iraq	2Eb									
Irish Rep	EI–EJ	Irish Rep	1									
Isle of Man	GD, MD, 2D	UK	1IO74									
Israel	4X, 4Z	Israel	2Eb									
Italy	I	Italy	1*									
ITU HQ	4U1	ITU	1JN36									
Ivory Coast	TU2	Ivory Coast	2Bd									
Jamaica	6Y	Jamaica	6Dc									
Jan Mayen Is	JX	Norway	—									
Japan	JA–JS	Japan	2Mb*									
Jersey	GJ, MJ, 2J	UK	1IN89									
Johnston Is	KH3	USA	4Eb									
Jordan	JY	Jordan	2Eb									
Juan Fernandez	CE0	Chile	3Fh*									
Kaliningrad	UA2	Russia	1KO05									
Kazakhstan	UN–UQ	Kazakhstan	—									
Kenya	5Y, 5Z	Kenya	2Ed									
Kerguelen Is	FT8X	France	2Gh									
Kermadec Is	ZL8	New Zealand	4Ef									
Kingman Reef	KH5K	USA	4Fc									
Kiribati, Central	T31	Kiribati	4Ed									
Kiribati, East	T32	Kiribati	4Fd									
Kiribati, West	T30	Kiribati	4Dc, d									
Korea, North	P5	Korea (DPRK)	2Lb									
Korea, South	HL	Korea (RK)	2Lb									
Kure Is	KH7	USA	4Eb									
Kuwait	9K	Kuwait	2Fc									
Kyrgyzstan	EX	Kyrgyzstan	—									
Laccadive Is	VU	India	2Gd									
Laos	XW	Laos	2Jc									
Latvia	YL	Latvia	1									
Lebanon	OD	Lebanon	2Eb									
Lesotho	7P	Lesotho	2Dg									
Liberia	EL	Liberia	2Bd*									
Libya	5A	Libya	2Dc									
Liechtenstein	HB0	Switzerland	1JN47									
Lithuania	LY	Lithuania	1									
Lord Howe Is	VK9L	Australia	4Cf*									
Luxembourg	LX	Luxembourg	1									
Macao	XX9	Portugal	2Kc									
Macedonia	Z3	Macedonia	1									

Country	Prefix	ITU alloc holder	Map			Worked/heard on							
Macquarie Is	VK0	Australia	4Cg										
Madagascar Republic	5R, 5S	Madagascar Rep	2Ff										
Madeira Is	CT3	Portugal	2Ab										
Malawi	7Q	Malawi	2Ee										
Malaysia, East	9M6, 8	Malaysia	2Kd										
Malaysia, West	9M2, 4	Malaysia	2Jd										
Maldives	8Q	Maldives	2Gd										
Mali	TZ	Mali	2Bc										
Malpelo Is	HK0	Colombia	3Fe										
Malta	9H	Malta	1										
Malyj Vysotskij Is	R1MV	Russia	—										
Mariana Is	KH0	USA	4Bb										
Market Reef	OJ0, OH0M	Finland	1JP90*										
Marquesas Is	FO	France	4Gd										
Marshall Is	V7	USA	4Dc										
Martinique	FM	France	6Gd										
Mauritania	5T	Mauritania	2Bc										
Mauritius	3B8	Mauritius	2Ff										
Mayotte	FH	France	2Ee										
Mellish Reef	VK9M	Australia	4Ce										
Mexico	XA–XI	Mexico	3Ed										
Micronesia	V6	USA	4Cc										
Midway Is	KH4	USA	4Eb										
Minami Torishima	JD1	Japan	4Cb										
Moldova	ER	Moldova	1										
Monaco	3A	Monaco	1										
Mongolia	JT–JV	Mongolia	2Ja										
Montserrat	VP2M	UK	6Gc										
Morocco	CN	Morocco	2Bb										
Mount Athos	SV/A	Greece	1KN20										
Mozambique	C8, 9	Mozambique	2Ef										
Myanmar	XY, XZ	Myanmar	2Jc										
Namibia	V5	Namibia	2Df										
Nauru	C2	Nauru	4Dd										
Navassa Is	KP1	USA	6Ec										
Nepal	9N	Nepal	2Hc										
Netherlands	PA–PI	Netherlands	1										
New Caledonia	FK	France	4De										
New Zealand	ZL, ZM	New Zealand	4Df*										
Nicaragua	YN	Nicaragua	3Fe										
Niger	5U	Niger	2Cc										
Nigeria	5N	Nigeria	2Cd*										
Niue	ZK2	New Zealand	4Ee										
Norfolk Is	VK9N	Australia	4De*										
Northern Ireland	GI, MI, 2I	UK	1										
Norway	LA–LN	Norway	1										
Ogasawara Is	JD	Japan	4Bb*										
Oman	A4	Oman	2Fc										

Country	Prefix	ITU alloc holder	Map	Worked/heard on							
Pagalu Is	3C0	Equatorial Guinea	2Ce								
Pakistan	AP	Pakistan	2Gc								
Palestine	E4	Palestine	2Eb								
Palmyra, Jarvis Is	KH5	USA	4Fc,d								
Panama	HO, HP	Panama	3Fe								
Papua New Guinea	P2	Papua New Guinea	2Me								
Paraguay	ZP	Paraguay	3Hg*								
Peru	OA–OC	Peru	3Ff*								
Peter I Is	3Y	Norway	—								
Philippines	DU–DZ	Philippines	2Ld								
Pitcairn Is	VP6	UK	4He								
Poland	SN–SR	Poland	1*								
Portugal	CT	Portugal	1								
Pratas	BV9P										
Prince Edward, Marion Is	ZS8	South Africa	2Eh								
Puerto Rico	KP3, 4	USA	6Fc								
Qatar	A7	Qatar	2Fc								
Reunion Is	FR	France	2Ff								
Revilla Gigedo	XA4–XI4	Mexico	3Dd								
Rodrigues Is	3B9	Mauritius	2Ff								
Romania	YO–YR	Romania	1*								
Rotuma Is	3D2	Fiji	—								
Russia (Asiatic)	UA9, 0	Russia	—								
Russia (European)	UA	Russia	—								
Rwanda	9X5	Rwanda	2Ee								
Sable Is	CY0	Canada	3Gc								
Sahara, Western	S0	—	—								
St Helena	ZD7	UK	2Bf								
St Kitts, Nevis	V4	St Kitts, Nevis	6Gc								
St Lucia	J6	St Lucia	6Gd								
St Maarten, St Eustatius, Saba	PJ	Netherlands Antilles	6Gc								
St Martin	FJ, FS	France	6Gc								
St Paul Is	CY9	Canada	3Gb								
St Peter and Paul Rocks	PP0–PY0	Brazil	*								
St Pierre et Miquelon	FP	France	3Hb								
St Vincent	J8	St Vincent	6Gd								
Salvador	YS	Salvador	6Bd								
Samoa	5W	Samoa	4Ed								
Samoa, American	KH8	USA	4Ed								
San Andres, Providencia	HK0	Colombia	6Cd								
San Felix Is, San Ambrosio	CE0	Chile	3Fg								
San Marino	T7	San Marino	1								
São Tomé, Principe	S9	São Tomé, Principe	2Ce								
Sardinia	IS, IM	Italy	1*								
Saudi Arabia	HZ	Saudi Arabia	2Ec								
Scarborough Reef	BS7										
Scotland	GM, MM, 2M	UK	1								

Country	Prefix	ITU alloc holder	Map	Worked/heard on									
Senegal	6V–6W	Senegal	2Bd										
Seychelles	S7	Seychelles	2Fe										
Sierra Leone	9L	Sierra Leone	2Bd										
Singapore	9V	Singapore	2Jd										
Slovak Rep	OM	Slovak Rep	1*										
Slovenia	S5	Slovenia	1										
Solomon Is	H4	Solomon Is	4Cd										
Somalia	T5	Somalia	2Fd										
South Africa	ZR–ZU	South Africa	2Dg*										
South Georgia	VP8	UK	3Jj										
South Orkney Is	VP8	UK	5										
South Sandwich Is	VP8	UK	5										
South Shetland Is	various	various	—										
Sovereign Military Order of Malta	1A	—	1										
Spain	EA	Spain	1*										
Spratly Is	1S	—	2Kd										
Sri Lanka	4P–4S	Sri Lanka	2Hd										
Sudan	ST	Sudan	2Dc										
Suriname	PZ	Suriname	3He										
Svalbard	JW	Norway	—										
Swaziland	3DA	Swaziland	2Ef										
Sweden	SA–SM	Sweden	1*										
Switzerland	HB	Switzerland	1										
Syria	YK	Syria	2Eb										
Taiwan	BV	China	2Lc										
Tajikistan	EY	Tajikistan	—										
Tanzania	5H, 5I	Tanzania	2Ee										
Temotu Province	H40	Solomon Is	4Cd										
Thailand	E2, HS	Thailand	2Jc*										
Togo	5V	Togo	2Cd										
Tokelau Is	ZK3	New Zealand	4Ed										
Tonga	A35	Tonga	4Ee										
Trinidad and Tobago	9Y, 9Z	Trinidad and Tobago	6Gd										
Trinidade, Martim Vaz Is	PP0–PY0	Brazil	*										
Tristan Da Cunha, Gough Is	ZD9	UK	2Bg										
Tromelin Is	FR/T	France	2Ff										
Tunisia	3V	Tunisia	2Cb										
Turkey	TA–TC	Turkey	2Eb*										
Turkmenistan	EZ	Turkmenistan	—										
Turks, Caicos Is	VP5	UK	6Eb										
Tuvalu	T2	Tuvalu	4Dd										
Uganda	5X	Uganda	2Ed										
Ukraine	UR	Ukraine	1										
United Arab Emirates	A6X	United Arab Emirates	2Fc										
United Nations HQ	4U	United Nations	—										
Uruguay	CV–CX	Uruguay	3Hh*										
USA	K, W etc	USA	3										

Country	Prefix	ITU alloc holder	Map	Worked/heard on									
Uzbekistan	UJ–UM	Uzbekistan	—										
Vanuatu	YJ	Vanuatu	4De										
Vatican City	HV	Vatican City	1JN61										
Venezuela	YV, YY	Venezuela	3Ge*										
Vietnam	3W, XV	Vietnam	2Kd										
Virgin Is, British	VP2V	UK	6Gc										
Virgin Is, US	KP2	USA	6Gc										
Wake Is	KH9	USA	4Db										
Wales	GW, MW, 2W	UK	1										
Wallis, Futuna Is	FW	France	4Dd										
Willis Is	VK9W	Australia	4Ce*										
Yemen	7O	Yemen	2Fc										
Yugoslavia	YT–YU	Yugoslavia	1										
Zaïre	9O	Zaïre	2De										
Zambia	9I, 9J	Zambia	2De										
Zimbabwe	Z2	Zimbabwe	2Ef										

Recent changes

Prefix	Entity	Effective from
Added		
4J1, R1MV	Malyj Vysotskij Is	First operation
3D2	Rotuma Is	15 Nov 1945
3D2	Conway Reef	15 Nov 1945
T33	Banaba Is (Ocean Is)	15 Nov 1945
ZS9	Walvis Bay	1 Sep 1990
7O	Yemen	22 May 1990
ZS0, 1	Penguin Is	15 May 1945
9A	Croatia	26 Jun 1991
S5	Slovenia	26 Jun 1991
T9	Bosnia-Hercegovina	15 Oct 1991
Z3	Macedonia	8 Sep 1991
OK, OL	Czech Rep	1 Jan 1993
OM	Slovak Rep	1 Jan 1993
BV9P	Pratas	1 Jan 1994
BS7	Scarborough Reef	1 Jan 1995
P5	North Korea	14 May 1995
FO	Austral Is	1 Apr 1998
FO	Marquesas Is	1 Apr 1998
H40	Temotu Province	1 Apr 1998
E4	Palestine	1 Feb 1999
Reactivated		
E3	Eritrea	24 May 1991
Deleted		
Y2–9	German Democratic Rep	2 Oct 1990
4W	Yemen Arab Rep	21 May 1990
7O	PDR of Yemen	21 May 1990
—	Abu Ail	31 Mar 1991
OK-OM	Czechoslovakia	31 Dec 1992
ZS0, 1	Penguin Is	28 Feb 1994
ZS9	Walvis Bay	28 Feb 1994
ST0	Southern Sudan	31 Dec 1994

5 Amateur Service frequency allocations

THIS appendix shows all amateur allocations in the three ITU Regions. Where in a box of the table a band is indicated as allocated to more than one service, such services are listed in the following order:

1. Primary services, which are printed in **bold**.
2. Secondary services, which are printed in ordinary type.

The footnotes appearing after this table apply only to bands, or parts of bands, allocated to the Amateur Service. They comprise only a small proportion of the total number of footnotes which appear in the Radio Regulations, and not all the footnotes shown in the tables are listed. The footnote references which appear in the table below the allocated services apply to the whole of the allocation concerned. The footnote references which appear after the name of the service are applicable only to that particular service.

Region 1	Region 2	Region 3
1800–1810kHz **Radiolocation** [487] [485, 486]	1800–1850 **Amateur**	1800–2000 **Amateur, Fixed, Mobile** except aeronautical mobile, **Radionavigation,** Radiolocation
1810–1850 **Amateur** [490, 491, 492, 493]		
1850–2000 **Fixed, Mobile** except aeronautical mobile [484, 488, 495]	1850–2000 **Amateur, Fixed, Mobile** except aeronautical mobile, **Radiolocation Radionavigation** [494]	[489]
3500–3800 **Amateur** [510] **Fixed, Mobile** except aeronautical mobile [484]	3500–3750 **Amateur** [510] [509, 511]	3500–3900 **Amateur** [510]
3800–3900 **Fixed, Aeronautical Mobile** (OR), **Land Mobile**	3750–4000 **Amateur** [510] **Fixed, Mobile** except aeronautical mobile (R)	
3900–3950 **Aeronautical Mobile** (OR) [513]		3900–3950 **Aeronautical Mobile, Broadcasting**
3950–4000 **Fixed, Broadcasting**	[511, 512, 514, 515]	3950–4000 **Fixed, Broadcasting** [516]
7000–7100	**Amateur** [510], **Amateur-satellite** [526, 527]	
7100–7300 **Broadcasting**	7100–7300 **Amateur** [510] [528]	7100–7300 **Broadcasting**
10,100–10,150	**Fixed**, Amateur [510]	
14,000–14,250	**Amateur** [510], **Amateur-satellite**	
14,250–14,350	**Amateur** [510] [535]	

Region 1	Region 2	Region 3
18,068–18,168	**Amateur** [510], **Amateur-satellite** [537, 538]	
21,000–21,450	**Amateur** [510], **Amateur-satellite**	
24,890–24,990	**Amateur** [510], **Amateur-satellite** [542, 543]	
28–29.7MHz	**Amateur, Amateur-satellite**	
47–68 **Broadcasting** [553, 554, 555, 559, 561]	50–54 **Amateur** [556, 557, 558, 560]	
144–146	**Amateur** [510], **Amateur-satellite** [605, 606]	
146–149.9 **Fixed, Mobile** except aeronautical mobile (R) [608]	146–148 **Amateur** [607]	146–148 **Amateur, Fixed, Mobile** [607]
223–230 **Broadcasting**, Fixed Mobile [622, 628, 629, 631, 632, 633, 634, 635]	220–225 **Amateur, Fixed Mobile**, Radiolocation [607]	223–230 **Fixed, Mobile Broadcasting, Aeronautical Radionavigation,** Radiolocation [636, 637]
430–440 **Amateur, Radiolocation** [653, 654, 655, 656, 657 658, 659, 661, 662, 663, 664, 665]	430–440 **Radiolocation**, Amateur [653, 658, 659, 660, 663, 664]	
890–942 **Fixed, Mobile** except aeronautical mobile, **Broadcasting** [704], Radiolocation	902–928 **Fixed**, Amateur, **Mobile** except aeronautical mobile, Radiolocation [705, 707]	890–942 **Fixed, Mobile, Broadcasting** Radiolocation [706]

Region 1	Region 2	Region 3
1240–1260	**Radiolocation, Radionavigation-Satellite** [710] (space–to–Earth), Amateur [711, 712, 713, 714]	
1260–1300	**Radiolocation**, Amateur [664, 711, 712, 713, 714]	
2300–2450 **Fixed**, Amateur, **Mobile, Radiolocation** [664, 752]	2300–2450 **Fixed, Mobile, Radiolocation**, Amateur [664, 751, 752]	
3300–3400 **Radiolocation** [778, 779, 780]	3300–3400 **Radiolocation**, Amateur, Fixed, Mobile [778, 780]	3300–3400 **Radiolocation**, Amateur [778, 779]
3400–3600 **Fixed, Fixed-Satellite** (space-to-Earth), [781, 782, 785]	3400–3500 **Fixed, Fixed-Satellite** (space-to-Earth), Amateur, Mobile, Radiolocation [3736A] [664, 783]	
5650–5725	**Radiolocation**, Amateur, Space Research (deep space) [664, 801, 803, 804, 805]	
5725–5850 **Fixed–Satellite** (Earth-to-space), **Radiolocation,** Amateur [801, 803, 805, 800, 807, 808]	5725–5850 **Radiolocation,** Amateur [803, 805, 806, 808]	
5850–5925 **Fixed, Fixed-Satellite** (Earth-to-space), **Mobile**, Amateur, Radiolocation [806]	5850–5925 **Fixed, Fixed-Satellite** (Earth-to-space), **Mobile**, Amateur, Radiolocation [806]	5850–5925 **Fixed, Fixed-Satellite** (Earth-to-space), **Mobile**, Radiolocation [806]
10–10.45GHz **Fixed, Mobile, Radiolocation,** Amateur [828]	10–10.45 **Radiolocation,** Amateur [828, 829]	10–10.45 **Fixed, Mobile, Radiolocation,** Amateur [828]
10.45–10.5	**Radiolocation**, Amateur, Amateur-Satellite [830]	
24–24.05	**Amateur, Amateur–Satellite** [881]	
24.05–24.25	**Radiolocation**, Amateur, Earth Exploration–Satellite (active) [881]	
47–47.2	**Amateur, Amateur-Satellite**	
75.5–76	**Amateur, Amateur-Satellite**	
76–81	**Radiolocation**, Amateur, Amateur-Satellite [912]	
142–144	**Amateur, Amateur-Satellite**	
144–149	**Radiolocation**, Amateur, Amateur-Satellite [918]	
241–248	**Radiolocation**, Amateur, Amateur-Satellite [922]	
248–250	**Amateur, Amateur-Satellite**	

Footnotes

484 – Some countries of Region 1 use radiodetermination systems in the bands 1606.5–1625kHz, 1635–1800kHz, 1850–2160kHz, 2134–2300kHz, 2502–2850kHz and 3500–3800kHz. The establishment and operation of such systems are subject to agreement obtained under the procedure set forth in Article 14. The radiated mean power of these stations shall not exceed 50W.

485 – Additional allocation: in Angola, Bulgaria, Hungary, Mongolia, Nigeria, Poland, Germany, Chad, the Czech Republic and the CIS, the bands 1625–1635kHz, 1800–1810kHz and 2160–2170kHz are also allocated to the fixed and land mobile services on a primary basis subject agreement obtained under the procedure set forth in Article 14.

486 – In Region 1, in the bands 1625–1635kHz, 1800–1810kHz and 2160–2170kHz (except for the countries listed in No 485 and those listed in No 499 for the band 2160–2170kHz), existing stations in the fixed and mobile, except aeronautical mobile, services (and stations of the aeronautical mobile (OR) service in the band 2160–2170kHz) may continue to operate on a primary basis until satisfactory replacement assignments have been found and implemented in accordance with Resolution 38.

488 – In Germany, Denmark, Finland, Hungary, Ireland, Israel, Jordan, Malta, Norway, Poland, the United Kingdom, Sweden, Czech Republic and the CIS, administrations may allocate up to 200kHz to their amateur service in the bands 1715–1800kHz and 1850–2000kHz. However, when allocating the bands within this range to their amateur service, administrations shall, after prior consultation with administrations of neighbouring countries, take such steps as may be necessary to prevent harmful interference from their amateur service to the fixed and mobile services of other countries. The mean power of any amateur station shall not exceed 10W.

489 – In Region 3, the Loran system operates either on 1850kHz or 1950kHz, the bands occupied being 1825–1875kHz and 1925–1975kHz respectively. Other services to which the band 1800–2000kHz is allocated may use any frequency therein on condition that no harmful interference is caused to the Loran system operating on 1850 or 1950kHz.

490 – Alternative allocation: in Germany, Angola, Austria, Belgium, Bulgaria, Cameroon, Congo, Denmark, Egypt, Spain, Ethiopia, France, Greece, Italy, Lebanon, Luxembourg, Malawi, the Netherlands, Portugal, Syria, Somalia, Tanzania, Tunisia, Turkey and the CIS, the band 1810–1830kHz is allocated to the fixed and mobile, except aeronautical mobile, services on a primary basis.

491 – Additional allocations: in Saudi Arabia, Iraq, Israel, Libya, Poland, Roumania, Chad, the Czech Republic, Togo and Yugoslavia, the band 1810–1830kHz is also allocated to the fixed and mobile, except aeronautical mobile, services on a primary basis.

492 – In Region 1, the use of the band 1810–1850kHz by the amateur service is subject to the condition that satisfactory replacement assignments have been found and implemented in accordance with Resolution 38 for frequencies to all existing stations of the fixed and mobile, except aeronautical mobile, services operating in this band (except for the stations of the countries listed in Nos 490, 491 and 493). On completion of satisfactory transfer, the authorisation to use the band 1810–1830kHz by the amateur service in countries situated totally or partially north of 40°N shall be given only after consultation with the countries mentioned in Nos 490 and 491 to define the necessary steps to be taken to prevent harmful interference between amateur stations and stations of other services operating in accordance with Nos 490 and 491.

493 – Alternative allocations: in Burundi and Lesotho, the band 1810–1850kHz is also allocated to the fixed and mobile, except aeronautical mobile, services on a primary basis.

494 – Alternative allocation: in Argentina, Bolivia, Chile, Mexico, Paraguay, Peru, Uruguay and Venezuela, the band 1850–2000kHz is allocated to the fixed, mobile except aeronautical mobile, radiolocation and radionavigation services on a primary basis.

509 – Additional allocation: in Honduras, Mexico, Peru and Venezuela, the band 3500–3750kHz is also allocated to the fixed and mobile services on a primary basis.

510 – For the use of the bands allocated to the amateur service at 3.5MHz, 7.0MHz, 10.1MHz, 14.0MHz, 18.068MHz, 21.0MHz, 24.89MHz and 144MHz in the event of natural disasters, see Resolution 640.

511 – Additional allocation: in Brazil, the band 3700–4000kHz is also allocated to the radiolocation service on a primary basis.

512 – Alternative allocation: in Argentina, Bolivia, Chile, Ecuador, Paraguay. Peru and Uruguay, the band 3750–4000kHz is allocated to the fixed and mobile, except aeronautical mobile, services on a primary basis.

514 – Additional allocation: in Canada, the band 3950–4000kHz is also allocated to the broadcasting service on a primary basis. The power of broadcasting stations operating in this band shall not exceed that necessary for a national service within the frontier of this country and shall not cause harmful interference to other services operating in accordance with the Table.

515 – Additional allocation: in Greenland, the band 3950–4000kHz is also allocated to the broadcasting service on a primary basis. The power of the broadcasting stations operating in this band shall not exceed that necessary for a national service and shall in no case exceed 5kW.

526 – Additional allocation: in Angola, Iraq, Kenya, Rwanda, Somalia and Togo, the band 7000–7050kHz is also allocated to the fixed service on a primary basis.

527 – Alternative allocation: in Egypt, Ethiopia, Guinea, Libya, Madagascar, Malawi and Tanzania, the band 7000–7050kHz is allocated to the fixed service on a primary basis.

528 – The use of the band 7100–7300kHz in Region 2 by the amateur service shall not impose constraints on the broadcasting service intended for use within Region 1 and Region 3.

535 – Additional allocation: in Afghanistan, China, Ivory Coast, Iran and the CIS, the band 14,250–14,350kHz is also allocated to the fixed service on a primary basis. Stations of the fixed service shall not use a radiated power exceeding 24dBW.

537 – The band 18,068–18,168kHz is allocated to the fixed service on a primary basis subject to the procedure described in Resolution 8. The use of this band by the amateur and amateur-satellite services shall be subject to the completion of satisfactory transfer of all assignments to stations in the fixed service operating in this band and recorded in the Master Register, in accordance with the procedure described in Resolution 8.

538 – Additional allocation: in the CIS, the band 18,068–18,168kHz is also allocated to the fixed service on a primary basis for use within the boundary of the CIS, with a peak envelope power not exceeding 1kW.

542 – Additional allocation: in Kenya, the band 23,600–24,900kHz is also allocated to the meteorological aids service (radiosondes) on a primary basis.

543 – The band 24,890–24,990kHz is allocated to the fixed and land mobile services on a primary basis subject to the procedure described in Resolution 8. The use of this band by the amateur and amateur–satellite services shall be subject to the completion of the satisfactory transfer of all assignments to fixed and land mobile stations operating in this band and recorded in the Master Register, in accordance with the procedure described in Resolution 8.

556 – Alternative allocation: in New Zealand, the band 50–51MHz is allocated to the fixed, mobile and broadcasting services on a primary basis; the band 53–54MHz is allocated to the fixed and mobile services on a primary basis.

557 – Alternative allocation: in Afghanistan, Bangladesh, Brunei, India, Indonesia, Iran, Malaysia, Pakistan, Singapore and Thailand, the band 50–54MHz is allocated to the fixed, mobile and broadcasting services on a primary basis.

558 – Additional allocation: in Australia, China and the Democratic People's Republic of Korea, the band 50–54MHz is also allocated to the broadcasting service on a primary basis.

560 – Additional allocation: in New Zealand the band 51–53MHz is also allocated to the fixed and mobile services on a primary basis.

605 – Additional allocation: in Singapore, the band 144–145MHz is also allocated to the fixed and mobile services on a primary basis. Such use is limited to systems in operation on or before 1 January 1980, which in any case shall cease by 31 December 1995.

606 – Additional allocation: in China the band 144–146MHz is also allocated to the aeronautical mobile (OR) service on a secondary basis.

607 – Alternative allocation: in Afghanistan, Bangladesh, Cuba, Guyana and India, the band 146–148MHz is allocated to the fixed and mobile services on a primary basis.

629 – Additional allocation: in Oman, the United Kingdom and Turkey, the band 216–235MHz is also allocated to the radiolocation service on a secondary basis.

652 – Additional allocation: in Australia the United States, Jamaica and the Phillipines, the bands 420–430MHz and 220–450MHz are also allocated to the amateur service on a secondary basis.

653 – Additional allocation: in China, India, Germany, the United Kingdom and the CIS, the band 420–460MHz is also allocated to the aeronautical radionavigation service (radio altimeters) on a secondary basis.

654 – Different category of service: in France, the allocation of the band 430–434MHz to the amateur service is on a secondary basis (see No 424).

655 – Different category of service: in Denmark, Libya, Norway and Sweden, the allocation of the bands 430–432MHz and 438–440MHz to the radiolocation service is on a secondary basis (see No 424).

656 – Alternative allocation: in Denmark, Norway and Sweden, the bands 430–432MHz and 438–440MHz are allocated to the fixed and mobile, except aeronautical mobile, services on a primary basis.

657 – Additional allocation: in Finland, Libya and Yugoslavia, the bands 430–432MHz and 438–440MHz are also allocated to the fixed and mobile, except aeronautical mobile, services on a primary basis.

658 – Additional allocation: in Afghanistan, Algeria, Saudi Arabia, Bahrain, Bangladesh, Brunei, Burundi, Egypt, the United Arab Emirates, Ecuador, Spain, Ethiopia, Greece, Guinea, India, Indonesia, Iran, Iraq, Israel, Italy, Jordan, Kenya, Kuwait, Lebanon, Liechtenstein, Libya, Malaysia, Malta, Nigeria, Oman, Pakistan, the Philippines, Qatar, Syria, Singapore, Somalia, Switzerland, Tanzania, Thailand and Togo, the band 430–440MHz is also allocated to the fixed service on a primary basis and the bands 430–435MHz and 438–440MHz are also allocated to the mobile, except aeronautical mobile, service on a primary basis.

659 – Additional allocation: in Angola, Bulgaria, Cameroon, Congo, Gabon, Hungary, Mali, Mongolia, Niger, Poland, Germany, Roumania, Rwanda, Chad, the Czech Republic and the CIs, the band 430–440MHz is also allocated to the fixed service on a primary basis.

660 – Different category of service: in Argentina, Colombia, Costa Rica, Cuba, Honduras, Panama and Venezuela, the allocation of the band 430–440MHz to the amateur service is on a primary basis (see No 425).

661 – In Region 1, except in the countries mentioned in No 662, the band 433.05–434.79MHz (centre frequency 433.92MHz) is designated for industrial, scientific and medical (ISM) applications. The use of this frequency band for ISM applications shall be subject to special authorisation by the administration concerned, in agreement with other administrations whose radiocommunication services might be affected. In applying this provision, administrations shall have due regard to the latest relevant CCIR Recommendations.

662 – In Germany, Austria, Liechtenstein, Portugal, Switzerland and Yugoslavia, the band 433.05–434.79MHz (centre frequency 433.92MHz) is designated for industrial, scientific and medical (ISM) applications. Radiocommunication services of these countries operating within this band must accept harmful interference which may be caused by these applications. ISM equipment operating in this band is subject to the provisions of No 1815.

663 – Additional allocation: in Brazil, France and the French Overseas Departments in Region 2, and India, the band 433.75–434.25MHz is also allocated to the space operation service (Earth-to-space) on a primary basis until 1 January 1990, subject to agreement obtained under the procedure set forth in Article 14. After 1 January 1990, the band 433.75–434.25MHz will be allocated in the same countries to the same service on a secondary basis.

644 – In the bands 435–438MHz, 1260–1270MHz, 2400–2450MHz, 3400–3410MHz in Regions 2 and 3 only, 5650–5670MHz the amateur-satellite service may operate subject to not causing harmful interference to other services operating in accordance with the Table (see No 435). Administrations authorizing such use shall ensure that any harmful interference caused by emissions from a station in the amateur-satellite service is immediately eliminated in accordance with the provisions of No 2741. The use of the bands 1260–1270MHz and 5650–5670MHz by the amateur-satellite service is limited to the Earth-to-space direction.

665 – Additional allocation: in Austria, the band 438–440MHz is also allocated to the fixed and mobile, except aeronautical mobile, services on a primary basis.

666 – Additional allocation: in Canada, New Zealand and Papua New Guinea, the band 440–450MHz is also allocated to the amateur service on a secondary basis.

667 – Different category of service: in Canada, the allocation of the band 440–450MHz to the radiolocation service is on a primary basis (see No 425).

705 – Different category of service: in the United States, the allocation of the band 890–942MHz to the radiolocation service is on a primary basis (see No 14) and subject to agreement obtained under the procedure set forth in Article 425.

707 – In Region 2 the band 902–928MHz (centre frequency 915MHz) is designated for industrial, scientific and medical (ISM) applications. Radiocommunication services operating within this band must accept harmful interference which may be caused by these applications. ISM equipment opereting in this band is subject to the provisions of No 1815.

710 – Use of the radionavigation-satellite service in the band 1215–1260MHz shall be subject to the condition that no harmful interference is caused to the radionavigation service authorized under No 712.

711 – Additional allocation: in Afghanistan, Angola, Saudi Arabia, Bahrain, Bangladesh, Cameroon, China, the United Arab Emirates, Ethiopia, Guinea, Guyana, India, Indonesia, Iran, Iraq, Israel, Japan, Jordan, Kuwait, Lebanon, Libya, Malawi, Morocco, Mozambique, Nepal Nigeria, Oman, Pakistan, the Philippines, Qatar, Syria, Somalia, Sudan, Sri Lanka, Chad, Thailand, Togo and Yemen (PDR of), the band 1215–1300MHz is also allocated to the fixed and mobile services on a primary basis.

712 – Additional allocation: in Algeria, Germany, Austria, Bahrain, Belgium, Benin, Burundi, Cameroon, China, Denmark, the United Arab Emirates, France, Greece, India, Iran, Iraq, Kenya, Liechtenstein, Luxembourg, Mali, Mauritania, Norway, Oman, Pakistan, the Netherlands, Portugal, Qatar, Senegal, Somalia, Sudan, Sri Lanka, Sweden, Switzerland, Tanzania, Turkey and Yugoslavia, the band 1215–1300MHz is also allocated to the radionavigation service on a primary basis.

713 – In the bands 1215–1300MHz, 3100–3300MHz, 5250–5350MHz, 8550–8650MHz, 9500–9800MHz and 134–140GHz, radiolocation stations installed on spacecraft may also be employed for the earth exploration satellite and space research services on a secondary basis.

751 – In Australia, the United States and Papua New Guinea, the use of the band 2310–2390MHz by the aeronautical mobile service for telemetry has priority over other uses by the mobile services.

752 – The band 2400–2500MHz (centre frequency 2450MHz) is designated for industrial, scientific and medical (ISM) applications. Radio services operating within this band must accept harmful interference which may be caused by these applications. ISM equipment operating in this band is subject to the provisions of No 1815.

778 – In making assignments to stations of other services, administrations are urged to take all practicable steps to protect the spectral line observations of the radio astronomy service from harmful interference in the bands 3260–3267MHz, 3332–3339MHz, 3345.8–3352.5MHz and 4825–4835MHz. Emissions from space or airborne stations can be particularly serious sources of interference to the radio astronomy service (see Nos 343 and 344 and Article 36).

779 – Additional allocation: in Afghanistan, Saudi Arabia, Bahrain, Bangladesh, China, Congo, the United Arab Emirates, India, Indonesia, Iran, Iraq, Israel, Japan, Kuwait, Lebanon, Libya, Malaysia, Oman, Pakistan, Qatar, Syria, Singapore, Sri Lanka and Thailand, the band 3300–3400MHz is also allocated to the fixed and mobile services on a primary basis. The countries bordering the Mediterranean shall not claim protection for their fixed and mobile services from the radiolocation service.

780 – Additional allocation: in Bulgaria, Cuba, Hungary, Mongolia, Poland, Germany, Roumania, the Czech Republic and the CIS the band 3300–3400MHz is also allocated to the radionavigation service on a primary basis.

781 – Additional allocation: in Germany, Israel, Nigeria and the United Kingdom, the band 3400–3475MHz is also allocated to the amateur service on a secondary basis.

783 – Different category of service: in Indonesia, Japan, Pakistan and Thailand the allocation of the band 3400–3500MHz to the mobile, except aeronautical mobile, service is on a primary basis (see No 425).

785 – In Denmark, Norway and the United Kingdom, the fixed, radiolocation and fixed–satellite services operate on a basis of equality of rights in the band 3400–3600MHz. However, these administrations operating radiolocation systems in this band are urged to cease operations by 1985. After this date these administrations shall take all practicable steps to protect the fixed-satellite service and co-ordination requirements shall not be imposed on the fixed-satellite service.

801 – Additional allocation: in the United Kingdom, the band 5470–5850MHz is also allocated to the land mobile service on a secondary basis. The power limits specified in Nos 2502, 2505, 2506 and 2507 shall apply in the band 5725–5850MHz.

803 – Additional allocation: in Afghanistan, Saudi Arabia, Bahrain, Bangladesh, Cameroon, Central African Republic, China, Congo, Korea (Republic of), Egypt, the United Arab Emirates, Gabon, Guinea, India, Indonesia, Iran, Iraq, Israel Japan, Jordan, Kuwait, Lebanon, Libya, Madagascar, Malaysia, Malawi, Malta, Niger, Nigeria, Pakistan, the Philippines, Qatar, Syria, Singapore, Sri Lanka, Tanzania, Chad, Thailand and Yemen (PDR of) the band 5650–5850MHz is also allocated to the fixed and mobile services on a primary basis.

804 – Different category of service: in Bulgaria, Cuba, Hungary, Mongolia, Poland, Germany, the Czech Republic and the CIS, the allocation of the band 5670–5725MHz to the space research service is on a primary basis (see No 425).

805 – Additional allocation: in Bulgaria, Cuba, Hungary, Mongolia, Poland, Germany, the Czech Republic and the CIS, the band 5670–5850MHz is also allocated to the fixed service on a primary basis.

806 – The band 5725–5875MHz (centre frequency 5800MHz) is designated for industrial scientific and medical (ISM) applications. Radiocommunication services operating within this band must accept harmful interference which may be caused by these applications. ISM equipment operating in this band is subject to the provisions of No 1815.

807 – Additional allocation: in Germany, the band 5755–5850MHz is also allocated to the fixed service on a primary basis.

808 – The band 5830–5850MHz is also allocated to the amateur-satellite service (space-to-Earth) on a secondary basis.

828 – The band 9975–10,025MHz is also allocated to the meteorological satellite service on a secondary basis for use by weather radars.

829 – Additional allocation: in Costa Rica, Ecuador, Guatemala, and Honduras, the band 10–10.45GHz is also allocated to the fixed and mobile services on a primary basis.

830 – Additional allocation: in Germany, Angola, China, Ecuador, Spain, Japan, Kenya, Morocco, Nigeria, Sweden, Tanzania and Thailand, the band 10.45–10.5GHz is also allocated to the fixed and mobile services on a primary basis.

881 – The band 24–24.25GHz (centre frequency 24.125GHz) is designated for industrial, scientific and medical (ISM) applications. Radiocommunication services operating within this band must accept harmful interference which may be caused by these applications. ISM equipment operating in this band is subject to the provisions of No 1815.

912 – In the band 78–79GHz radars located on space stations may be operated on a primary basis in the earth exploration–satellite service and in the space research service.

915 – The band 119.98–120.02GHz is also allocated to the amateur service on a secondary basis.

918 – The bands 140.69–140.98GHz, 144.68–144.98GHz, 145.45–145.75GHz and 146.82–147.12GHz are also allocated to the radio astronomy service on a primary basis for spectral line observations. In making assignments to stations of other services to which the bands are allocated, administrations are urged to take all practicable steps to protect the radio astronomy service from harmful interference. Emissions from space or airborne stations can be particularly serious sources of interference to the radio astronomy service (see Nos 343 and 334 and Article 36).

922 – The band 244–246GHz (centre frequency 245GHz) is designated for industrial, scientific and medical (ISM) applications. The use of this frequency band for ISM applications shall be subject to special authorisation by the administration concerned in agreement with other administrations whose radiocommunication services might be affected. In applying this provision administrations shall have due regard to the latest CCIR Recommendations.

6 Foreign-language phone contacts

THOUSANDS of operators across the world have contacts with only 100 or so words in a second language. It gives them great satisfaction, as well as many contacts which would otherwise have been impossible.

A basic contact consists of 10 parts: (1) calling CQ, (2) acknowledge reply, (3) name, (4) QTH, (5) signal report, (6) equipment used, (7) weather, (8) QSL via bureau, (9) 'Thanks for contact', (10) 73. Nearly every contact comes over in that order and all 10 parts usually come over in three of four transmissions, grouping two or three parts together in each one.

Every language can be found on the 14MHz band. For familiarisation and pronunciation of any of these four languages, search around the band for someone calling CQ who is speaking clearly and slowly with a strong signal. Do not waste time with weak signals or trying to understand a ragchew!

After some listening practice, make out the memory page with your name, QTH, rig etc. There are eight types of weather to choose from and the numbers down the side will cover station reports and temperatures up to 30°C. The memory page is just a prompt so that you do not forget your words in the middle of a transmission and so that you are able to give the other operator the information required correctly.

When ready to start transmitting, you will find it easier to contact an English-speaking station. After the preliminaries are established, tell your contact that you wish to practice his language and change straight over, reading from the prepared memory page. Do not be put off if your contact continues in English – he may wish to practice as well!

When you advance to calling CQ it is advisable to tell the other station that you "only speak the language for the QSO". This should prevent questions coming across which cannot be understood; if they do, repeat the same statement and continue to the end of the contact.

While four foreign languages are given here, you are strongly advised not to attempt more than one at a time – probably the second language which you learned at school – as it is very easy to get the words mixed up, particularly when you start to rely on memory.

FRENCH

0 zero	Calling CQ, CQ, CQ, this is	CQ, CQ, CQ, (SAY KOO) appel général, appel
1 un, une		
2 deux	G . . . . calling CQ on	général, ici G . . . . qui lance appel
3 trois		
4 quatre	the . . . . . metre band	bande . . . . . mètres
5 cinq		
6 six	G . . . . standing by.	G . . . . pass a l'écoute.
7 sept		
8 huit		
9 neuf		
10 dix	F . . . . , G . . . . returning,	F . . . . , G . . . . qui revient
11 onze		
12 douze	The name here is	Le prénom ici est . . . .
13 treize		
14 quatorze	and the QTH	et le QTH (KOO TAY ASSH) . . . .
15 quinze		
16 seize	I spell	je vous épele . . . .
17 dix-sept		
18 dix-huit	Your report here is . . . . and	Votre rapport ici est . . . . . et . . . . . .
19 dix-neuf		
20 vingt	F . . . . , G . . . . over to you.	F . . . . , G . . . . à vous.
21 vingt et un		
22 vingt-deux		
23 vingt-trois		
24 vingt-quatre	F . . . . , G . . . . returning,	F . . . . , G . . . . de retour
25 vingt-cinque		
26 vingt-six	the rig here is	la condition de travail ici est . . . .
27 vingt-sept		
28 vingt-huit	with a . . . . antenna	avec une antenne . . . .
29 vingt-neuf		
30 trente	and a . . . . microphone	et un micro . . . .
40 quarante		
80 quatre-vingts	the WX here is	Le WX (DOUBLA VAY EKS) ici est . . . .

1. sunny	5. fog	1. ensoleillé	5. du brouillard
2. cloudy	6. warm	2. nuageux	6. chaud
3. raining	7. cold	3. il pleut	7. froid
4. windy	8. snowing	4. du vent	8. il neige

and the temperature C   et la température . . . . . . dégrées Centigrad

F . . . . , G . . . . back to you   F . . . . , G . . . . à vous

F . . . . , G . . . . returning for   F . . . . , G . . . . qui revient
the final   pour le final

I will send you my QSL card via the bureau,   Je vous enverrai ma carte QSL (KOO ES EL) via
     le bureau,

Thank you for the very good QSO,   Merci bien pour le très bon QSO (KOO ES AW)
73, OM,   Soixante-treize, cher OM

F . . . . , G . . . . closing,   F . . . . , G . . . . qui termine
cheerio.   maintenant, au revoir.

ADDITIONAL PHRASES WHICH MAY BE REQUIRED

I only speak French for the QSO.	Je parle français seulement pour le QSO.
Please speak slowly.	Parlez lentement, s'il vous plaît.
Please give me my report again.	Donnez-moi mon rapport encore une fois, s'il vous plaît.
Please give me your name again.	Donnez-moi votre prénom encore une fois, s'il vous plaît.
Please give me your QTH again.	Donnez-moi votre QTH encore une fois, s'il vous plaît.

GERMAN

0	null, zero
1	ein, eins
2	zwei, zwo
3	drei
4	vier
5	fünf
6	sechs
7	sieben
8	acht
9	neun
10	zehn
11	elf
12	zwolf
13	dreizehn
14	vierzehn
15	fünfzehn
16	sechszehn
17	siebzehn
18	achtzehn
19	neunzehn
20	zwanzig
21	ein und zwanzig
22	zwei und zwanzig
23	drei und zwanzig
24	vier und zwanzig
25	fünf und zwanzig
26	sechs und zwanzig
27	sieben und zwanzig
28	acht und zwanzig
29	neun und zwanzig
30	dreissig
40	vierzig
80	achtzig

CQ, CQ, CQ . . . . . metres

This is the English station G . . . . calling CQ

G . . . . going over to receive, stand by

D . . . . , G . . . . returning

My name is . . . .

and my QTH is . . . .

I spell . . . .

Your report here is . . . . and . . . . .

D . . . . , G . . . . back to you

---

D . . . . , G . . . . coming back

My station is . . . .

with a . . . . antenna

and a . . . . microphone

The weather here is . . . .

1. sunny   5. foggy
2. cloudy   6. warm
3. raining   7. cold
4. windy   8. snowing

and the temperature . . . . degrees C

back to you D . . . . , G . . . .

D . . . . G . . . . Thank you for the QSO

I will send you my QSL card via the bureau

Best wishes and good DX
73 until we meet again

D . . . . G . . . . cheerio

CQ, CQ, CQ (SEE KOO), ... meter Band

Hier ruft die englische Station G . . . . mit einem allgemeinen Anruf

G . . . . geht auf empfang, bitte kommen

D . . . . , G . . . . zurück

Mein Name ist . . . .

und mein QTH (KOO TAY HAA) ist . . . .

Ich buchstabiere . . . .

Ihr Rapport hier ist . . . . . und . . . . . .

D . . . . , G . . . . bitte kommen

---

D . . . . , G . . . . zurück

Meine Station ist . . . .

mit eine . . . . Antenne

und ein . . . . Mikrophon

Das Wetter hier ist . . . .

1. sunnig   5. nebelig
2. bewoelkt   6. warm
3. es regnet   7. kalt
4. windig   8. es schneit

und das Temperatur . . . . Grad

zurück zu Ihnen D . . . . , G . . . . bitte kommen

---

D . . . . G . . . . Danke schön für das QSO (KOO ES OH)

Ich werde meine QSL-Karte via Bureau senden

Die besten Grüsse und gut DX
drei und siebzig, auf wiederhören

D . . . . G . . . . Tschuess

ADDITIONAL PHRASES WHICH MAY BE REQUIRED

I only speak German for the QSO.	Ich spreche Deutsch nur für das QSO.
Please give me your callsign again.	Bitte, geben Sie mir Ihr Rufzeichen noch einmal.
Please give me your name again.	Bitte, geben Sie mir Ihr Name noch einmal.
Please give me your QTH again.	Bitte, geben Sie mir Ihr QTH (KOO TAY HAA) noch einmal.
I give you back the microphone.	Ich gebe Ihnen das Mikrophon wieder zurück.

ITALIAN

0	zero
1	uno
2	due
3	tre
4	quattro
5	cinque
6	sei
7	sette
8	otto
9	nove
10	dieci
11	undici
12	dodici
13	tredici
14	quattordici
15	quindici
16	sedici
17	diciasette
18	diciotto
19	diciannove
20	venti
21	ventuno
22	ventidue
23	ventitré
24	ventiquattro
25	venticinque
26	ventsei
27	ventisette
28	ventotto
29	ventinove
30	trenta
40	quaranta
80	ottanta

CQ . . . . . metres, this is the English station
G . . . . standing by

Chiamata generale . . . . . metri, questo è la stazione
inglese G . . . . e vi ascolto, avanti

---

I . . . . , G . . . . returning, many thanks for the call.

The name here is . . . .

and the QTH is . . . .

Your report is . . . . . and . . . . .

Mike back to you, I . . . . , G . . . .

I . . . . G . . . . ritornando
tante grazie per la risposta.

Il mio nome è . . . .

Il mio QTH (KOO TEE ACCA) è . . . .

Il vostro controllo è . . . . . e . . . . .

Vi ripasso il micro, I . . . . , G . . . .
avanti, cambio

I . . . . , G . . . . returning

The rig here is . . . .

with . . . . linear

The microphone is . . . .

My antenna is . . . .

The WX here is . . . .

1. sunny   5. foggy
2. cloudy   6. warm
3. raining   7. cold
4. windy   8. snowing

The temperature is . . . . . degrees

I . . . . , G . . . . over to you

I . . . . , G . . . . ritornando

Il mio apparechio è . . . .

con amplificatore . . . .

Il micro è . . . .

La mia antenna è . . . .

Il tempo qui è . . . .

1. bellissimo   5. nebbioso
2. nuvoloso   6. caldo
3. piovoso   7. freddo
4. fa vento   8. nevica

La temperatura è . . . . . gradi

Vi ripasso il micro, I . . . . , G . . . .
avanti, cambio

I . . . . , G . . . . returning for the final.

I will send my QSL card via the bureau.

Thank you for the excellent QSO, best wishes and good DX.

The mike to you for the final. Bye-bye.

I . . . . , G 73

I . . . . , G . . . . ritornando, per il finale.

Vi mandero la mia cartolina QSL (KOO ESS ELLE) via bureau (associatione).

Molti grazie per il QSO eccellente, tanti saluti e buon DX.

Il micro a voi caro amico per il finale. Ciao (CHOW).

I . . . . , G . . . .
Avanti, cambio, 73 (settanta tre)

ADDITIONAL PHRASES WHICH MAY BE REQUIRED

I only speak Italian for the QSO.   Parlo Italiano soltanto per il QSO.
Please speak slowly.   Prego, parlate lentamente.
I did not get your callsign.   Non ho capito il vostro nominativo.
What is my report?   Cos' è il mio controllo?

SPANISH

0 cero	
1 uno	
2 dos	
3 tres	
4 cuatro	
5 cinco	
6 seis	
7 siete	
8 ocho	
9 nueve	
10 diez	
11 once	
12 doce	
13 trece	
14 catorce	
15 quince	
16 diez y seis	
17 diez y siete	
18 diez y ocho	
19 diez nueve	
20 veinte	
21 veintiuno	
22 veintidos	
23 veintitres	
24 veinticuatro	
25 veinticinco	
26 veintiseis	
27 veintisiete	
28 veintiocho	
29 veintinueve	
30 treinta	
31 treinta y uno	
40 cuarenta	
80 ochenta	

Calling CQ CQ CQ, this is G . . . . calling CQ on . . . . . metres, G . . . . standing by.

Llamada general, G . . . . llamando, CQ . . . . . . metros, y G . . . . escuchando.

EA . . . . , from G . . . . returning,

EA . . . . , G . . . . retornando.

Thank you for returning my call.

Gracias por haber contestado a mi llamada.

My name is . . . . I spell . . . .

Mi nombre es . . . . como . . . .

My QTH is . . . . I spell . . . .

Mi QTH es . . . . como . . . .

Your report here is . . . . and . . . . .

Su control es . . . . . y . . . . . .

EA . . . . from G . . . . come in.

EA . . . . de G . . . . adelante

---

EA . . . . from G . . . . Thank you for your message.

EA . . . . de G . . . . Gracias por su mensaje.

My station is . . . .

Mis condiciones de trabajo . . . .

my antenna is dipole/multiband/beam.

mi antena es dipolo/multibanda/directional.

My microphone is crystal/dynamic.

Mi microfono es cristal/dinamico.

The weather here is

1. sunny	5. fog
2. cloudy	6. warm
8. raining	7. cold
4. windy	8. snowing

El tiempo aqui

1. hace sol	5. niebla
2. nublado	6. hace calor
3. lloviendo	7. hace frio
4. hace viento	8. nevando

The temperature is . . . . degrees.

La temperatura . . . . grados.

EA . . . . from G . . . . come in

EA . . . . de G . . . . adelante

---

EA . . . . from G . . . . for the final.

EA . . . . de G . . . . por el final.

I will send my QSL card via the bureau. Thank you for the QSO, 73, Cheerio.

Me enviaré la tarjeta de QSL via el bureau. Gracias por el QSO, 73 (setenta y tres).

EA . . . . de G . . . . adios!

ADDITIONAL PHRASES WHICH MAY BE REQUIRED

I only speak Spanish for the QSO.	Hablo el español solamente por el QSO.
Please speak slowly.	Por favor hable despacio.
Please repeat my report/your name/your QTH.	Por favor repete mi control/su nombre/su QTH.
Can you QSY . . . kHz higher/lower.	Puede Usted QSY . . . kHz alto/bajo.
I am sorry, I cannot copy, bad QRM.	Lo siento, no puedo copiar, mucho QRM.

7 Band plans

THE following pages give the IARU Region 1 band plans for all amateur bands from 1.8MHz to 47MHz, together with the UK interpretation. UK licence notes are also given so as to make this a comprehensive band-by-band reference.

Band plans are subject to slight changes from year to year – such changes are always described in *RadCom* as they happen and up-to-date plans are also published annually in the *RSGB Yearbook*.

HF band plan notes

1. The expression 'phone' includes all permitted forms of telephony.
2. If transmitting very close to a band edge, take care not to radiate outside of the band.
3. Before transmitting, all operators should check that the frequency is not already occupied. The normal advice is to use the phrase "Is this frequency in use?" on SSB or "QRL?" on CW.
4. Digimodes are defined as including: AMTOR, PACTOR, Clover, ASCII, RTTY (Baudot), AX25 packet, PSK and Feld Hell.
5. LSB is recommended on bands below 10MHz, and USB recommended on bands above 10MHz.

6. The IARU Region 1 HF band plans are designed to enable the best utilisation of the HF spectrum space available. They achieve this objective because the vast majority of licensed amateurs observe the voluntary recommendations. In some countries (eg the USA) licence regulations require that specific modes be confined to specific sections of each band.
7. The frequencies 14.230, 21.340 and 28.680MHz should be used as calling frequencies for SSTV and fax operators. After having established contact they should move to another free frequency within the telephony section of the band.

VHF band plan notes

1. The beacon and satellite services must be kept free of normal communication transmissions to prevent interference with these services.
2. The use of the FM mode within the SSB/CW section and CW and SSL in the FM-only sector is not recommended.
3. Repeater stations are primarily intended as an aid for mobile working and they are not intended to be used for DX communication. FM stations wishing to work DX should use the all-modes section, taking care to avoid frequencies allocated for specific purposes.

1.8MHz (160m)

LICENCE NOTES:

Amateur Service:	1.810 - 1.850MHz, Primary. Remainder secondary. *Available on the basis of non-interference to other services (inside or outside the UK)*
Satellite Service:	No allocation
Power limit:	1.810 - 1.850MHz: 26dBW PEP (Class A/B Licence 20dBW). Remainder 15dBW
Permitted modes:	Morse, telephony, RTTY, data, fax, SSTV

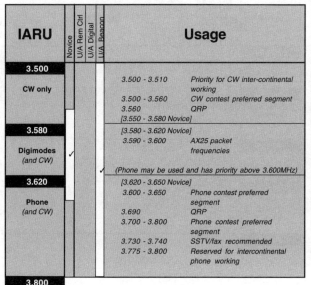

IARU	Novice	U/A Rem Ctrl	U/A Digital	U/A Beacon	Usage
1.810					
CW only					
1.838					
Digimodes *(and CW but excluding AX25 packet)*					RTTY (Baudot) is the preferred digital mode on this band Phone may be used above 1.840
1.842					
Phone *(and CW)*	✓			✓	1.843 QRP [1.950 - 2.000 Novice] 1.960 DF contest beacons (14dBW) 12.5kHz b/w max 1.970 Provisional Novice calling freq
2.000					Note: AX25 packet should not be used on the 1.8MHz band.

3.5MHz (80m)

LICENCE NOTES:

Amateur Service:	Primary, *Shared with other services*
Satellite Service:	No allocation
Power limit:	26dBW PEP (Class A/B Licence 20dBW)
Permitted modes:	Morse, telephony, RTTY, data, fax, SSTV
Unattended beacons:	Only for DF contests Sat & Sun only, 14dBW ERP PEP max

IARU	Novice	U/A Rem Ctrl	U/A Digital	U/A Beacon	Usage
3.500					
CW only					3.500 - 3.510 Priority for CW inter-continental working 3.500 - 3.560 CW contest preferred segment 3.560 QRP [3.550 - 3.580 Novice]
3.580					
Digimodes *(and CW)*	✓				[3.580 - 3.620 Novice] 3.590 - 3.600 AX25 packet frequencies
			✓		*(Phone may be used and has priority above 3.600MHz)*
3.620					
Phone *(and CW)*					[3.620 - 3.650 Novice] 3.600 - 3.650 Phone contest preferred segment 3.690 QRP 3.700 - 3.800 Phone contest preferred segment 3.730 - 3.740 SSTV/fax recommended 3.775 - 3.800 Reserved for intercontinental phone working
3.800					

7MHz (40m)

LICENCE NOTES:

Amateur Service:	Primary
Satellite Service:	Primary
Power limit:	26dBW PEP (Class A/B Licence 20dBW)
Permitted modes:	Morse, telephony, RTTY, data, fax, SSTV

IARU	Novice	U/A Rem Ctrl	U/A Digital	U/A Beacon	Usage
7.000					
CW only					7.030 QRP
7.035					
Digimodes *(and CW, SSTV, Fax but excluding AX25 packet)*					*(Phone may be used above 7.040)*
7.045					
Phone *(and CW)*					
7.100					Note: AX25 packet should not be used on the 7MHz band.

10MHz (30m)

LICENCE NOTES:

Amateur Service:	Secondary
Satellite Service:	No allocation
Power limit:	26dBW PEP (Class A/B Licence 20dBW)
Permitted modes:	Morse, telephony, RTTY, data, fax, SSTV

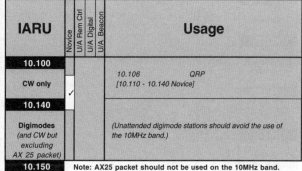

IARU	Novice	U/A Rem Ctrl	U/A Digital	U/A Beacon	Usage
10.100					
CW only	✓				10.106 QRP [10.110 - 10.140 Novice]
10.140					
Digimodes *(and CW but excluding AX 25 packet)*					*(Unattended digimode stations should avoid the use of the 10MHz band.)*
10.150					Note: AX25 packet should not be used on the 10MHz band.

10MHz Band Plan notes:

Note: The 10MHz band is allocated to the amateur service only on a secondary basis. Therefore IARU have agreed on a worldwide basis that only CW and digimodes, being narrow bandwidth modes, are to be used on this band. Likewise this band is not to be used for contests or news bulletins.

14MHz (20m)

LICENCE NOTES:

Amateur Service : Primary
Satellite Service : 14.000 - 14.250MHz: Primary
Power limit: 26dBW PEP (Class A/B Licence 20dBW)
Permitted modes: Morse, telephony, RTTY, data, fax, SSTV

IARU	Novice	U/A Rem Ctrl	U/A Digital	U/A Beacon	Usage
14.000					
CW only					14.060 — QRP 14.000 - 14.060 — CW only contest preferred segment
14.070					
Digimodes (and CW)					14.089 - 14.099 — No digimode mailbox or forwarding AX25 packet preferred frequencies
14.099					
Beacons only					14.099 - 14.101 — Reserved exclusively for beacons
14.101					
Digimodes (+ phone & CW)					14.101 - 14.112 — Digimode mailbox and forwarding AX25 packet preferred frequencies
14.112					
Phone (and CW)					14.125 - 14.300 — SSB only contest preferred segment 14.230 — SSTV/fax calling frequency 14.285 — QRP
14.350					

21MHz (15m)

LICENCE NOTES:

Amateur Service: Primary
Satellite Service : Primary
Power limit: 26dBW PEP (Class A/B Licence 20dBW)
Permitted modes: Morse, telephony, RTTY, data, fax, SSTV

IARU	Novice	U/A Rem Ctrl	U/A Digital	U/A Beacon	Usage
21.000					
CW only					21.060 — QRP [21.050 - 21.080 Novice]
21.080					
Digimodes (and CW)		✓			21.100 - 21.120 — AX25 packet preferred [21.080 - 21.149 Novice]
21.120					
CW only					
21.149					
Beacons only					21.149 - 21.151 — Beacons exclusive
21.151					
Phone (and CW)					21.285 — QRP 21.340 — SSTV/fax calling frequency
21.450					

18MHz (17m)

LICENCE NOTES:

Amateur Service: Primary
Satellite Service: Primary
Power limit: 26dBW PEP (Class A/B Licence 20dBW)
Permitted modes: Morse, telephony, RTTY, data, fax, SSTV

IARU	Novice	U/A Rem Ctrl	U/A Digital	U/A Beacon	Usage
18.068					
CW only					
18.100					
Digimodes (and CW)					
18.109					
Beacons only					18.109 - 18.111 — Exclusively beacons
18.111					
Phone (and CW)					
18.168					

24MHz (12m)

LICENCE NOTES:

Amateur Service: Primary
Satellite Service: Primary
Power limit: 26dBW PEP (Class A/B Licence 20dBW)
Permitted modes: Morse, telephony, RTTY, data, fax, SSTV

IARU	Novice	U/A Rem Ctrl	U/A Digital	U/A Beacon	Usage
24.890					
CW only					
24.920					
Digimodes (and CW)					
24.929					
Beacons only					24.929 - 24.931 — Beacons exclusive
24.931					
Phone (and CW)					
24.990					

28MHz (10m)

LICENCE NOTES:

Amateur Service:	Primary
Satellite Service:	Primary
Power limit:	26dBW PEP (Class A/B Licence 20dBW)
Permitted modes:	Morse, telephony, RTTY, data, fax, SSTV
Unattended beacons:	Only for DF contests (14dBW PEP max)

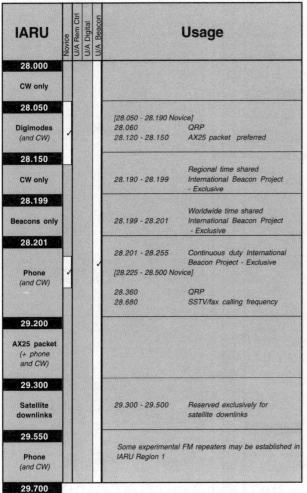

IARU	Novice	U/A Rem Ctrl	U/A Digital	U/A Beacon	Usage	
28.000						
CW only						
28.050						
Digimodes (and CW)		✓			[28.050 - 28.190 Novice]	
					28.060	QRP
					28.120 - 28.150	AX25 packet preferred
28.150						
CW only					28.190 - 28.199	Regional time shared International Beacon Project - Exclusive
28.199						
Beacons only					28.199 - 28.201	Worldwide time shared International Beacon Project - Exclusive
28.201						
				✓	28.201 - 28.255	Continuous duty International Beacon Project - Exclusive
Phone (and CW)		✓			[28.225 - 28.500 Novice]	
					28.360	QRP
					28.680	SSTV/fax calling frequency
29.200						
AX25 packet (+ phone and CW)						
29.300						
Satellite downlinks					29.300 - 29.500	Reserved exclusively for satellite downlinks
29.550						
Phone (and CW)					Some experimental FM repeaters may be established in IARU Region 1	
29.700						

50MHz (6m)

LICENCE NOTES:

Amateur Service:	50.0 - 51.0MHz, Primary; 51.0 - 52.0MHz, Secondary. *Available on the basis of non-interference to other services (inside or outside the UK).*
Satellite Service :	No allocation
Power limit:	50.0-51.0MHz, 26dBW PEP; 51.0-52.0MHz, 20dBW PEP
Permitted modes:	Morse, telephony, RTTY, data, fax, SSTV

IARU	Novice	U/A Rem Ctrl	U/A Digital	U/A Beacon	Usage	
50.000						
CW only					50.020 - 50.080	Beacons
					50.090	CW calling frequency
50.100						
					50.100 - 50.130	DX window - Note 1
					50.110	Intercontinental calling - Note 2
SSB and CW only		✓			50.150	SSB Centre of Activity
					50.185	Cross-band activity centre
					50.200	MS Reference frequency (CW & SSB)
50.500						
					50.500 - 50.700	Digital communications
All modes					50.510	SSTV
					50.550	Fax
					50.600	RTTY
51.000					50.710 - 50.910	FM repeater outputs
All modes		✓			51.210	Emergency comms. priority
					51.210 - 51.410	FM repeater inputs
51.410						
All modes					51.430 - 51.590	FM telephony - Note 3
					51.510	FM calling
					51.530	Note 4
51.830						
All modes					51.940 - 52.000	Emergency comms priority
52.000						

50MHz Band Plan notes:

1. Only to be used for QSOs between stations in different continents.
2. No QSOs on this frequency. Always QSY when working intercontinental DX.
3. 20kHz channel spacing. Channel centre frequencies start at 51.430MHz.
4. Used by GB2RS news and for slow Morse transmissions.

70MHz (4m)

LICENCE NOTES:

Amateur Service: Secondary. Available on the basis of non-interference to other services (inside or outside the UK).
Satellite Service: No allocation
Power limit: 22dBW PEP
Permitted modes: Morse, telephony, RTTY, data, fax, SSTV

IARU	Novice	U/A Rem Ctrl	U/A Digital	U/A Beacon	Usage	
70.000						
Beacons					70.030	Personal beacons
70.030						
SSB and CW only					70.150	Meteor scatter calling
					70.185	Cross-band activity centre
					70.200	SSB/CW calling
70.250						
All modes					70.260	AM/FM calling
70.300						
					70.3000	RTTY/fax calling/working
		✓	✓	✓	70.3125	Digital modes
					70.3250	Digital modes
					70.3375	Digital modes
					70.3500	Emergency comms priority
Channelised operation using 12.5 kHz channels					70.3625	Digital modes
					70.3750	Emergency comms priority
					70.3875	Digital modes
					70.4000	Emergency comms priority
					70.4125	Digital modes
					70.4250	FM simplex - used by GB2RS
					70.4375	Digital modes
					70.4500	FM calling
					70.4625	Digital modes
				✓	70.4875	Digital modes
70.500						

144MHz (2m)

LICENCE NOTES:

Amateur Service: Primary
Satellite Service: Primary
Power limit: 26dBW PEP
Permitted modes: Morse, telephony, RTTY, data, fax, SSTV
Unattended beacons: Only for DF Contests

IARU	Novice	U/A Rem Ctrl	U/A Digital	U/A Beacon	Usage	
144.000						
EME (SSB/CW)					144.000 - 144.035	Moonbounce (only)
144.035						
					144.050	CW calling frequency
CW only					144.100	MS CW ref frequency (Note 1)
					144.140 - 144.150	CW FAI/EME working
144.150	✓		✓	✓	144.150 - 144.160	SSB FAI/EME working
					144.175	Microwave talk-back (UK)
SSB and CW only					144.195 - 144.205	SSB random MS
					144.250	GB2RS and slow Morse
					144.260	Emerg comms priority
					144.300	SSB calling frequency
					144.390 - 144.400	SSB random MS
144.400						
Beacons					144.490	SAREX uplink
144.490						

Continued in next column

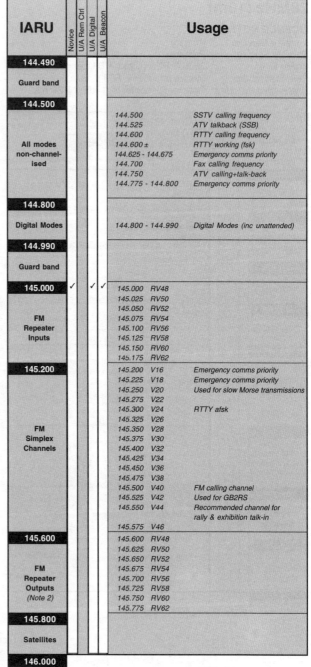

IARU	Novice	U/A Rem Ctrl	U/A Digital	U/A Beacon	Usage	
144.490						
Guard band						
144.500						
					144.500	SSTV calling frequency
					144.525	ATV talkback (SSB)
					144.600	RTTY calling frequency
All modes non-channel-ised					144.600 ±	RTTY working (fsk)
					144.625 - 144.675	Emergency comms priority
					144.700	Fax calling frequency
					144.750	ATV calling+talk-back
					144.775 - 144.800	Emergency comms priority
144.800						
Digital Modes					144.800 - 144.990	Digital Modes (inc unattended)
144.990						
Guard band						
145.000	✓		✓	✓	145.000	RV48	
					145.025	RV50	
					145.050	RV52	
					145.075	RV54	
FM Repeater Inputs					145.100	RV56	
					145.125	RV58	
					145.150	RV60	
					145.175	RV62	
145.200					145.200	V16	Emergency comms priority
					145.225	V18	Emergency comms priority
					145.250	V20	Used for slow Morse transmissions
					145.275	V22	
					145.300	V24	RTTY afsk
					145.325	V26	
					145.350	V28	
FM Simplex Channels					145.375	V30	
					145.400	V32	
					145.425	V34	
					145.450	V36	
					145.475	V38	
					145.500	V40	FM calling channel
					145.525	V42	Used for GB2RS
					145.550	V44	Recommended channel for rally & exhibition talk-in
					145.575	V46	
145.600					145.600	RV48	
					145.625	RV50	
					145.650	RV52	
					145.675	RV54	
FM Repeater Outputs (Note 2)					145.700	RV56	
					145.725	RV58	
					145.750	RV60	
					145.775	RV62	
145.800						
Satellites						
146.000						

144MHz Band Plan notes:

1. Meteor scatter operation can take place up to 26kHz higher than the reference frequency.
2. In the UK the frequency 145.790MHz is for point-to-point links, 16kHz maximum bandwidth.

430MHz (70cm)

LICENCE NOTES:

Amateur Service:	Secondary
Satellite Service:	435 - 438MHz, Secondary
Exclusion:	431 - 432 not available for use within 100 km radius of Charing Cross, London. (51° 30' 30"N, 00° 7' 24"W)
Power limit:	430 - 432MHz: 16dBW ERP PEP, 432 - 440MHz: 26dBW
Permitted modes:	Morse, telephony, RTTY, data, fax, SSTV, FSTV

IARU	Novice	U/A Rem Ctrl	U/A Digital	U/A Beacon	Usage		
430.000 All modes					430.000 - 430.810	Digital communications (Notes 6, 7)	
					430.600 - 430.800	Note 5	
430.810 Low power repeater i/p Note 1					430.810 - 430.990	Low power repeaters	
431.000 All modes Note 1					430.990 - 431.900	Digital communications (Note 6)	
432.000 CW only					432.000 - 432.025	Moonbounce	
					432.050	CW centre of activity	
432.150 SSB and CW only					432.200	SSB centre of activity	
					432.350	Microwave talk-back calling frequency (Europe)	
432.500 All modes non-chanelised		✓	✓		432.500 - 432.600	IARU Region 1 linear transponder outputs	
					432.600 - 432.800	IARU Region 1 linear transponder inputs	
					432.500	SSTV activity centre	
					432.600	RTTY (fsk) activity centre	
					432.625	Digital communications	
					432.650	Digital communications	
					432.675	Digital communications	
					432.700	Fax activity centre	
432.800 Beacons				✓	432.800 - 432.990	Beacons	
433.000 FM repeater outputs in UK only Note 1					433.000	RU240 (RB0)	
					433.025	RU242 (RB1)	
					433.050	RU244 (RB2)	
					433.075	RU246 (RB3)	
					433.100	RU248 (RB4)	
					433.125	RU250 (RB5)	
					433.150	RU252 (RB6)	
					433.175	RU254 (RB7)	
					433.200	RU256 (RB8)	
					433.225	RU258 (RB9)	
					433.250	RU260 (RB10)	
					433.275	RU262 (RB11)	
					433.300	RU264 (RB12)	
					433.325	RU266 (RB13)	
					433.350	RU268 (RB14)	
					433.375	RU270 (RB15)	
433.400							

Continued in next column

IARU	Novice	U/A Rem Ctrl	U/A Digital	U/A Beacon	Usage		
433.400 FM simplex channels			✓		433.400	U272 (SU16)	
					433.425	U274 (SU17)	
					433.450	U276 (SU18)	
					433.475	U278 (SU19)	
					433.500	U280 (SU20)	FM calling channel
					433.525	U282 (SU21)	
					433.550	U284 (SU22)	Recommended channel for rally and exhibition talk-in
					433.575	U286 (SU23)	
					433.600	U288 (SU24)	RTTY afsk
					433.625		Digital communications
					433.650		Digital communications
					433.675		Digital communications
					433.700		Notes 2, 3 and 5
					433.725		Notes 2 and 5
					433.750		Notes 2 and 5
					433.775		Notes 2 and 5
					433.800 - 434.250		Digital communications (Note 8)
434.600 FM repeater inputs (in UK only) - note 1; and fast scan television - note 4		✓			434.600	RU240 (RB0)	
					434.625	RU242 (RB1)	
					434.650	RU244 (RB2)	
					434.675	RU246 (RB3)	
					434.700	RU248 (RB4)	
					434.725	RU250 (RB5)	
					434.750	RU252 (RB6)	
					434.775	RU254 (RB7)	
					434.800	RU256 (RB8)	
					434.825	RU258 (RB9)	
					434.850	RU260 (RB10)	
					434.875	RU262 (RB11)	
					434.900	RU264 (RB12)	
					434.925	RU266 (RB13)	
					434.950	RU268 (RB14)	
					434.975	RU270 (RB15)	
435.000 Satellites and fast scan TV - note 4							
438.000 Fast scan TV					438.025 - 438.175	Note 5	
					438.200 - 439.425	Note 1	
438.425 Low power repeater o/p + fast scan TV					438.425 - 438.575	Low power repeaters	
438.575 Fast Scan TV					438.200 - 439.425	Note 1	
439.750 Packet radio					439.600 - 439.750	Digital communications (Note 6)	
					439.750 - 440.000	Digital communications (Note 6)	
440.000							

430MHz Band Plan notes:

1. In Switzerland, Germany and Austria, repeater inputs are 430.600 - 431.825MHz with 25kHz spacing, and outputs are 438.200 - 439.425MHz. In France and the Netherlands repeater outputs are 430.025 - 430.375MHz with 25kHz spacing and inputs at 431.625 - 431.975MHz. In other European countries repeater inputs are 433.000 - 433.375MHz with 25kHz spacing and outputs at 434.600 - 434.975MHz, ie the reverse of the UK allocation.
2. Emergency communications priority.
3. IARU Region 1 fax/AFSK.
4. Fast Scan Television carrier frequencies shall be chosen so as to avoid interference to other users, in particular the satellite service and repeater inputs. IARU Region 1 recommends that video carriers should be in the range 434.000 - 434.500MHz or 438.500 - 440.000MHz.
5. IARU Region 1 packet radio.
6. The DCC will recommend usage of this sub-band at a later date.
7. Users must accept interference from F/PA repeater output channels in 430.025 to 430.375MHz. Users with sites which allow propagation to other countries (notably F and PA) must survey the proposed frequency before use to ensure that they will not cause interference to users of repeaters in those countries.
8. Only vertical polarisation is permitted.

1.3GHz (23cm)

LICENCE NOTES:

Amateur Service: Secondary

Satellite Service: 1260 - 1270, Secondary *Earth to space only*

1296 - 1297, Secondary *Earth to space only*

Power limit: 26dBW PEP

Permitted modes: Morse, telephony, RTTY, data, fax, SSTV, FSTV

Unattended operation: Not permitted in Northern Ireland

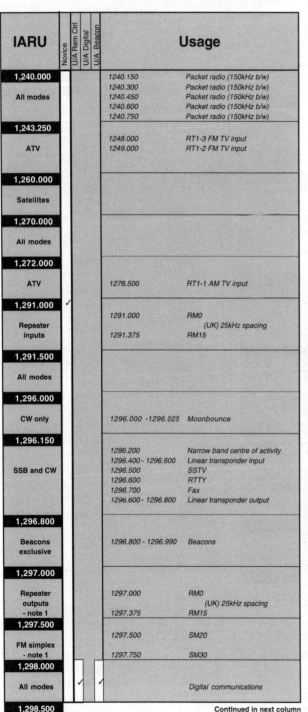

IARU	Novice	U/A Rem Ctrl	U/A Digital	U/A Beacon	Usage	
1,240.000 All modes					1240.150	Packet radio (150kHz b/w)
					1240.300	Packet radio (150kHz b/w)
					1240.450	Packet radio (150kHz b/w)
					1240.600	Packet radio (150kHz b/w)
					1240.750	Packet radio (150kHz b/w)
1,243.250 ATV					1248.000	RT1-3 FM TV input
					1249.000	RT1-2 FM TV input
1,260.000 Satellites						
1,270.000 All modes						
1,272.000 ATV					1276.500	RT1-1 AM TV input
1,291.000 Repeater inputs	✓				1291.000	RM0 (UK) 25kHz spacing
					1291.375	RM15
1,291.500 All modes						
1,296.000 CW only					1296.000 -1296.025	Moonbounce
1,296.150 SSB and CW					1296.200	Narrow band centre of activity
					1296.400 - 1296.600	Linear transponder input
					1296.500	SSTV
					1296.600	RTTY
					1296.700	Fax
					1296.600 - 1296.800	Linear transponder output
1,296.800 Beacons exclusive					1296.800 - 1296.990	Beacons
1,297.000 Repeater outputs - note 1					1297.000	RM0 (UK) 25kHz spacing
					1297.375	RM15
1,297.500 FM simplex - note 1					1297.500	SM20
					1297.750	SM30
1,298.000 All modes		✓	✓			Digital communications
1,298.500						

Continued in next column

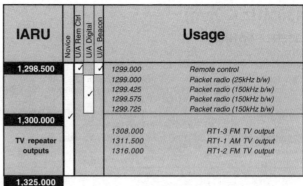

IARU	Novice	U/A Rem Ctrl	U/A Digital	U/A Beacon	Usage	
1,298.500		✓		✓	1299.000	Remote control
					1299.000	Packet radio (25kHz b/w)
			✓		1299.425	Packet radio (150kHz b/w)
					1299.575	Packet radio (150kHz b/w)
1,300.000	✓				1299.725	Packet radio (150kHz b/w)
TV repeater outputs					1308.000	RT1-3 FM TV output
					1311.500	RT1-1 AM TV output
					1316.000	RT1-2 FM TV output
1,325.000						

1.3GHz Band Plan notes:

1. Local traffic using narrow-band modes should operate between 1296.500 - 1296.800MHz during contests and band openings.

2. Stations in countries which do not have access to 1298 - 1300MHz (eg Italy) may also use the FM simplex segment for digital communications.

2.3GHz (13cm)

LICENCE NOTES:

Amateur Service: Secondary. *Users must accept interference from ISM users*

Satellite Service: 2400 - 2450, Secondary. *Users must accept interference from ISM users.*

Power limit: 26dBW PEP

Permitted modes: Morse, telephony, RTTY, data, fax, SSTV, FSTV

ISM = Industrial Scientific and Medical

IARU	Novice	U/A Rem Ctrl	U/A Digital	U/A Beacon	Usage	
2,310.000 Sub-regional (national band plans)					2310.000 - 2310.500	Repeater links
					2310.100	Packet radio (200kHz b/w)
					2310.300	Packet radio (200kHz b/w)
					2310.000 - 2310.500	Remote control
2,320.000 CW exclusive					2320.000 - 2320.025	Moonbounce
2,320.150 CW and SSB					2320.200	SSB centre of activity
2,320.800 Beacons exclusive	✓	✓	✓		2320.800 - 2320.990	Beacons
2,321.000 Simplex & repeaters (FM) - note 1						
2,322.000 All modes					2322.000 - 2355.000	ATV
					2355.100 - 2364.000	Repeater links
					2355.100	Packet radio (200kHz b/w)
					2355.300	Packet radio (200kHz b/w)
					2364.000	Packet radio (1MHz b/w)
					2365.000 - 2370.000	Repeaters
					2370.000 - 2390.000	ATV
					2390.000 - 2392.000	Moonbounce
2,400.000 Satellites						
2,450.000						

Notes continued in next column

2.3GHz Band Plan notes:

1. Stations in countries which do not have access to the All Modes section (2,322 - 2,390MHz), use the simplex and repeater segment 2,321 - 2,322MHz for data transmission
2. Stations in countries which do not have access to the narrow band segment 2,320 - 2,322 MHz, use alternative narrow band segments: 2,304 - 2,306MHz and 2,308 - 2,310MHz.

3.4GHz (9cm)

LICENCE NOTES:

Amateur Service:	Secondary
Satellite Service:	No allocation
Power limit:	26dBW PEP
Permitted modes:	Morse, telephony, RTTY, data, fax, SSTV, FSTV

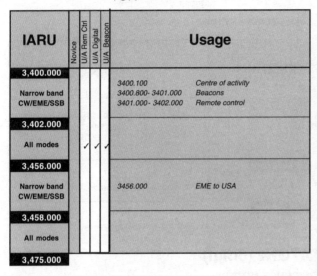

IARU	Novice	U/A Rem Ctrl	U/A Digital	U/A Beacon	Usage	
3,400.000						
Narrow band CW/EME/SSB					3400.100 3400.800- 3401.000 3401.000- 3402.000	Centre of activity Beacons Remote control
3,402.000						
All modes	✓	✓	✓			
3,456.000						
Narrow band CW/EME/SSB					3456.000	EME to USA
3,458.000						
All modes						
3,475.000						

Unattended (U/A) Operation

Frequencies on which unattended (U/A) operation is permitted by full licensees are shown in these band plans. Novice licensees can also operate their stations unattended but the frequencies and powers are different – please see the Novice licence for the details. Remember that unattended operation requires the prior consent of the local Radio Investigation Service before operation can begin, to enable close down arrangements to be made.

Unattended beacons are limited to 14dBW ERP max. Do not confuse this type of unattended beacon operation with the normal beacon sections of the bands (these are fully site cleared, have special licences and are co-ordinated on an international basis.

Unattended low power remote control is limited to –20dBW ERP and should not radiate outside the boundary of the premises from which you are operating.

Unattended digital operation is limited to 10dBW on the 50MHz band and 14dBW on the other bands where it is permitted.

5.7GHz (6cm)

LICENCE NOTES:

Amateur Service :	5,650 - 5,680, Secondary; 5,755 - 5,765 + 5820 - 5850: Secondary. Users must accept *interference from ISM users*
Satellite Service:	5,650 - 5,670 Secondary *Earth to Space only;* 5,830 - 5,850 Secondary *Users must accept interference from ISM users Space to Earth only*
Power limit:	26dBW PEP
Permitted modes:	Morse, telephony, RTTY, data, fax, SSTV, FSTV

ISM = Industrial, Scientific & Medical

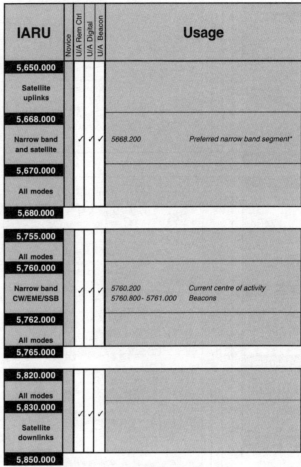

IARU	Novice	U/A Rem Ctrl	U/A Digital	U/A Beacon	Usage	
5,650.000						
Satellite uplinks						
5,668.000						
Narrow band and satellite	✓	✓	✓		5668.200	Preferred narrow band segment*
5,670.000						
All modes						
5,680.000						
5,755.000						
All modes						
5,760.000						
Narrow band CW/EME/SSB		✓	✓	✓	5760.200 5760.800- 5761.000	Current centre of activity Beacons
5,762.000						
All modes						
5,765.000						
5,820.000						
All modes						
5,830.000						
Satellite downlinks		✓	✓	✓		
5,850.000						

* IARU aim to move narrow band operation to this segment, but for the time being operation will continue in the 5760 - 5762 band.

10GHz (3cm)

LICENCE NOTES:

Amateur Service:	Secondary
Satellite Service:	10,450 - 10,475 Terrestrial and Space
	10,475 - 10,500 Space ONLY
Power limit:	26dBW PEP
Permitted modes:	Morse, telephony, RTTY, data, fax, SSTV, FSTV

IARU	Novice	U/A Rem Ctrl	U/A Digital	U/A Beacon		Usage
10,000.000						
All modes (ATV, data FM simplex, duplex and repeaters)	✓	✓	✓	✓	10,002.5 - 10,027.5	WB Transponders RMT 290/015 OUT
					10,027.5 - 10,052.5	WB Transponders RMT 315/040 OUT
					10,052.5 - 10,077.5	WB Transponders RMT 340/065 OUT
					10,080 - 10,090	Packet Links
					10,090 - 10,110	Wideband Beacons & Operating
					10,110 - 10,120	Speech Repeaters OUT
10,125.000						
10,225.000						
					10,227.5 - 10,252.5	WB Transponders RMT 240/425 OUT
					10,252.5 - 10,277.5	WB Simplex
					10,277.5 - 10,302.5	WB Transponders RMT 290/015 IN
					10,302.5 - 10,327.5	WB Transponders RMT 213/040 IN
					10,327.5 - 10,352.5	WB Transponders RMT 340/065 IN
					10,352.5 - 10,368	Wideband Modes
10,368.000	✓					
Preferred narrow band CW/EME/SSB Beacons					10,368 - 10,370	Narrowband modes
					10,368.1	Centre of Activity
					10,368.8 - 10,369	Beacons
10,370.000						
All modes					10,370 - 10,390	Wideband modes
					10,390 - 10,410	WB Beacons and Operating
					10,412.5 - 10,437.5	WB Transponders RMT240/425 (IN)
					10,440 - 10,450	Speech Repeaters IN
					[10,400 - 10,500 unattended operation]	
10,450.000						
All modes + satellites	✓	✓	✓		10,450 -10,452	Alternative Narrowband CW/EME/SSB - note 3
10,475.000						

10GHz Band Plan notes:

1. 10,400 is the preferred frequency for wideband beacons, but 10,100 is still used.
2. Wideband FM is preferred around 10,350 - 10,410 to encourage compatibility with narrowband systems; however, there is still activity around 10,050 - 10,150.
3. The current NB sub-band is at 10,368, however, a sub-band at 10,450 - 10,452 is an alternative.
4. Simplex TV operation could also take place on RMT inputs which are not used by local transponders.
5. Wideband transponder pairs are designated by input/output frequency. The pairings shown are recommended but occasionally variants may be needed suit local circumstances.

24GHz (12mm)

LICENCE NOTES:

Amateur Service:	24,000 - 24,050 Primary. *Users must accept interference from ISM users*; 24,050 - 24,150 Secondary, *May only be used with the written consent of the Secretary of State. Users must accept interference from ISM users*; 24,150 - 24,250 Secondary. *Users must accept interference from ISM users.*
Satellite Service:	24,000 - 24,050 Primary. *Users must accept interference from ISM users*
Power limit:	26dBW PEP
Permitted modes:	Morse, telephony, RTTY, data, fax, SSTV, FSTV

ISM = Industrial, Scientific & Medical

IARU	Novice	U/A Rem Ctrl	U/A Digital	U/A Beacon		Usage
24,000.000						
Satellites	✓	✓	✓		24,025	Preferred operating frequency wide band equipment
					24,048 - 24,050	Preferred narrow band operating*
24,050.000						
All modes					24,192 - 24,194	Narrow band op (UK)
24,250.000						

* Will eventually be used if and when allocation changes force this.

47GHz (6mm)

LICENCE NOTES:

Amateur Service:	Primary
Satellite Service:	Primary
Power limit:	26dBW PEP
Permitted modes:	Morse, telephony, RTTY, data, fax, SSTV, FSTV

IARU	Novice	U/A Rem Ctrl	U/A Digital	U/A Beacon		Usage
47,000.000		✓	✓	✓	47,088	Centre of narrow band activity
47,200.000						

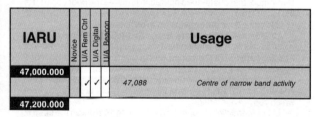

Note: The UK 76GHz band differs from the European band. The Eu centre of activity is 76.03GHz, just above the UK band.

Other amateur bands allocated in the UK are:

71.6 - 74.4kHz (Notice of Variation only. Available until 30 June 2000 only)
135.7 - 137.8kHz (Permanent beacons not recommended), and 75.5 - 76.0, 142.0 - 144.0,
248.0 - 250.0GHz.

8 Beacons

BEACONS are intended mainly as propagation indicators although, especially on the microwave bands, they may also serve as signal sources for alignment purposes. The table lists a selection, some of which may be heard regularly in the UK so that variation in strength gives an indication of conditions; others may be heard occasionally and the appearance of one can indicate exceptional propagation. For example, the 144MHz beacon GB3VHF can always be heard over much of the UK so, if its strength is above average, then there is a 'lift' on. If the 50MHz beacon in Namibia, not usually audible in the UK, appears then there is a path to Southern Africa. Conversely, the 28MHz beacon GB3RAL is of little interest to UK stations but can indicate to overseas operators the presence of propagation to the British Isles.

Some 25 years ago, in order to avoid interference between 144MHz beacons in Europe, the RSGB was asked by the International Amateur Radio Union (IARU) to co-ordinate their frequencies. The Society has done so since and has extended the service to other bands above 30MHz.

On HF, beacons are co-ordinated by IARU. The band 28.190–28.199MHz is used for regional networks, 28.200MHz is shared by the International Beacon Project stations and 28.201–28.225MHz is allocated to approved continous-cycle stations.

HF BEACONS

Freq (MHz)	Callsign	Nearest town	Locator	ERP (W)	Antenna	Beam direction	Mode	Status
1805	VO1NA	St John's		1		Omni		24
1817	ZS1J	Plettenberg Bay	KF15PF	1	1/2 Dip	E-W	A1	24
1840	OK0EK	Kromeriz	JN89QG	4	Vertical	Omni	A1	T Non Op
1845	OK0EV	Near Prague		100	PT			
3579	DK0WCY	Scheggerott	JO44VQ	30	Dipole		A1	0700-0800
3585	ZS1J	Plettenberg Bay	KF15PF	1	1/2 Dip		A1	24 1530-1700zz
3600.0	OK0EN	Kam. Zehrovice	JO70AC	0.15	Corner Dip	90/270	A1	24
5471.5	LN2A	Sveio	JO29PO	1kW	Vert Monop	Omni	A1/F1	ITU24+
5471.5	VL8IPS	Nr Darwin NT	PH57PJ	2kW	Vert Monop	Omni	A1/F1	ITU24+
7004	WQ4RP	Nr Raleigh NC		5mW			A1	?
7025	ZS1AGI	George Airport	KF16EA	1	1/2 Dipole	E-W	A1	24
7871.5	LN2A	Sveio	JO29PO	1kW	Vert Monop	Omni	A1/F1	ITU24+
7871.5	VL8IPS	Nr Darwin NT	PH57PJ	2kW	Vert Monop	Omni	A1/F1	ITU24+
10102.5	ZS1J	Plettenberg Bay	KF15PF	1	1/2 Dip		A1	24
10144.6	DK0WCY	Scheggerott	JO44VQ	30	Hor. Loop	Omni	A1	24zz
10408.5	LN2A	Sveio	JO29PO	1kW	Vert Monop	Omni	A1/F1	ITU24+
10408.5	VL8IPS	Nr Darwin NT	PH57PJ	2kW	Vert Monop	Omni	A1/F1	ITU24+
14100.0	4U1UN	UN NY	FN20AS	100-0.1	Vertical	Omni	A1	IBP cycle
14100.0	VE8AT	Eureka,Nunavut	EQ79AX	100-0.1	Vertical	Omni	A1	IBP cycle
14100.0	W6WX	Nr San Jose CA	CM87	100-0.1	Vertical	Omni	A1	IBP cycle
14100.0	KH6WO	Honolulu	BL11BK	100-0.1	Vertical	Omni	A1	IBP cycle
14100.0	ZL6B	Nr Masterton	RE77TW	100-0.1	Vertical	Omni	A1	IBP cycle
14100.0	VK6RBP	28k SE Perth	OF87BW	100-0.1	Vertical	Omni	A1	IBP cycle
14100.0	JA2IGY	Mt Asama	PM84JK	100-0.1	Vertical	Omni	A1	IBP cycle
14100.0	RR9O	Novosibirsk	NO14KX	100-0.1	Vertical	Omni	A1	IBP cycle
14100.0	VR2HK	Hong Kong	OL72CG					UC
14100.0	4S7B	Colombo	MJ96	100-0.1	Vertical	Omni	A1	IBP cycle
14100.0	ZS6DN	Pretoria	KF44DC	100-0.1	Vertical	Omni	A1	IBP cycle
14100.0	5Z4B	Kilifi	KI95	100-0.1	Vertical	Omni	A1	IBP cycle
14100.0	4X6TU	Tel Aviv	KM72JC	100-0.1	Vertical	Omni	A1	IBP cycle
14100.0	OH2B	Espoo	KP20KE	100-0.1	Vertical	Omni	A1	IBP cycle
14100.0	CS3B	Madeira	IM12	100-0.1	Vertical	Omni	A1	IBP cycle
14100.0	LU4AA	Buenos Aires	GF05	100-0.1	Vertical	Omni	A1	IBP cycle
14100.0	OA4B	Lima	FH17KV	100-0.1	Vertical	Omni	A1	IBP cycle
14100.0	YV5B	Caracas	FK60NL	100-0.1	Vertical	Omni	A1	IBP cycle
14100.2	VE9BEA	Crabbe Mtn NB	FN66	3	Vertical	Omni	A1	24
14396.5	LN2A	Sveio	JO29PO	1kW	Vert Monop	Omni	A1/F1	ITU24+
14396.5	VL8IPS	Nr Darwin NT	PH57PJ	2kW	Vert Monop	Omni	A1/F1	ITU24+
18068.1	IK6BAK	Montefelcino	JN63KR	12	Dipx2	Omni	A1	24
18068.5	HB9TC	Bellinzona	JN46ME	10	Dipole	N-S	A1	07-1700
18101	VE3RAT	Thornhill ONT	FN03GL	1	Vertical	Omni	A1	24
18102	I1M	Bordighera	JN33UT	10	5/8 Vert	Omni	A1	24
18110	DL0AGS	Kassel	JO41NL	5	GP	Omni	A1	24
18110.0	4U1UN	U. Nations NY	FN20AS	100-0.1	Vertical	Omni	A1	IBP cycle
18110.0	VE8AT	Eureka, Nunavut	EQ79AX	100-0.1	Vertical	Omni	A1	IBP cycle

Freq (MHz)	Callsign	Nearest town	Locator	ERP (W)	Antenna	Beam direction	Mode	Status
18110.0	KH6WO	Honolulu	BL11BK	100-0.1	Vertical	Omni	A1	IBP cycle
18110.0	ZL6B	Nr Masterton	RE77TW	100-0.1	Vertical	Omni	A1	IBP cycle
18110.0	VK6RBP	28k SE Perth	OF87BW	100-0.1	Vertical	Omni	A1	IBP cycle
18110.0	JA2IGY	Mt Asama	PM84JK	100-0.1	Vertical	Omni	A1	IBP cycle
18100.0	RR9O	Novosibirsk	NO14KX	100-0.1	Vertical	Omni	A1	IBP cycle
18100.0	VR2HK	Hong Kong	OL72CG	UC				
18110.0	4S7B	Colombo	MJ96	100-0.1	Vertical	Omni	A1	IBP cycle
18110.0	ZS6DN	Pretoria	KF44DC	100-0.1	Vertical	Omni	A1	IBP cycle
18110.0	5Z4B	Kilifi	KI95	100-0.1	Vertical	Omni	A1	IBP cycle
18110.0	4X6TU	Tel Aviv	KM72JC	100-0.1	Vertical	Omni	A1	IBP cycle
18110.0	OH2B	Espoo	KP20KE	100-0.1	Vertical	Omni	A1	IBP cycle
18110.0	CS3B	Madeira	IM12	100-0.1	Vertical	Omni	A1	IBP cycle
18110.0	OA4B	Lima	FH17KV	100-0.1	Vertical	Omni	A1	IBP cycle
18110.0	LU4AA	Buenos Aires	GF05	100-0.1	Vertical	Omni	A1	IBP cycle
18110.0	YV5B	Caracas	FK60NL	100-0.1	Vertical	Omni	A1	IBP cycle
20948.5	LN2A	Sveio	JO29PO	1kW	Vert Monop	Omni	A1/F1	ITU24+
20948.5	VL8IPS	Darwin NT	PH57PJ	2kW	Vert Monop	Omni	A1/F1	ITU24+
21150.0	4U1UN	UN New York	FM20AS	100-0.1	Vertical	Omni	A1	IBP cycle
21150.0	VE8AT	Eureka,Nunavut	EQ79AX	100-0.1	Vertical	Omni	A1	IBP cycle
21150.0	W6WX	Nr San Jose CA	CM87	100-0.1	Vertical	Omni	A1	IBP cycle
21150.0	KH6WO	Honolulu	BL11BK	100-0.1	Vertical	Omni	A1	IBP cycle
21150.0	ZL6B	Masterton	RE77TW	100-0.1	Vertical	Omni	A1	IBP cycle
21150.0	VK6RBP	28k SE Perth	OF87BW	100-0.1	Vertical	Omni	A1	IBP cycle
21150.0	JA2IGY	Mt Asama	PM84JK	100-0.1	Vertical	Omni	A1	IBP cycle
21150.0	RR9O	Novosibirsk	NO14KX	100-0.1	Vertical	Omni	A1	IBP cycle
21150.0	VR2HK	Hong Kong	OL72CG					UC
21150.0	4S7B	Colombo	MJ96	100-0.1	Vertical	Omni	A1	IBP cycle
21150.0	ZS6DN	Pretoria	KF44DC	100-0.1	Vertical	Omni	A1	IBP cycle
21150.0	5Z4B	Kilifi	KI95	100-0.1	Vertical	Omni	A1	IBP cycle
21150.0	4X6TU	Tel Aviv	KM72JC	100-0.1	Vertical	Omni	A1	IBP cycle
21150.0	OH2B	Espoo	KP20KE	100-0.1	Vertical	Omni	A1	IBP cycle
21150.0	CS3B	Madeira	IM12	100-0.1	Vertical	Omni	A1	IBP cycle
21150.0	LU4AA	Buenos Aires	GF05	100-0.1	Vertical	Omni	A1	IBP cycle
21150.0	OA4B	Lima	FH17KV	100-0.1	Vertical	Omni	A1	IBP cycle
21150.0	YV5B	Caracas	FK60NL	100-0.1	Vertical	Omni	A1	IBP cycle
21151.0	I1M	Bordighera	JN33UT	10	2 5/8 Vert	Omni	A1	24
24915.0	IK6BAK	Montefelcino	JN63KR	12	Dip x 2	Omni	A1	24
24930.0	4U1UN	UN NY	FN20AS	100-0.1	Vertical	Omni	A1	IBP cycle
24930.0	VE8AT	Eureka,Nunavut	EQ79AX	100-0.1	Vertical	Omni	A1	IBP cycle
24930.0	KH6WO	Honolulu	BL11BK	100-0.1	Vertical	Omni	A1	IBP cycle
24930.0	ZL6B	Nr Masterton	RE77TW	100-0.1	Vertical	Omni	A1	IBP cycle
24930.0	VK6RBP	28k SE Perth	OF87BW	100-0.1	Vertical	Omni	A1	IBP cycle
24930.0	JA2IGY	Mt Asama	PM84JK	100-0.1	Vertical	Omni	A1	IBP cycle
24930.0	RR9O	Novosibirsk	NO14KX	100-0.1	Vertical	Omni	A1	IBP cycle
24930.0	VR2HK	Hong Kong	OL72CG					UC
24930.0	4S7B	Colombo	MJ96	100-0.1	Vertical	Omni	A1	IBP cycle
24930.0	ZS6DN	Pretoria	KF44DC	100-0.1	Vertical	Omni	A1	IBP cycle
24930.0	5Z4B	Kilifi	KI95	100-0.1	Vertical	Omni	A1	IBP cycle
24930.0	4X6TU	Tel Aviv	KM72JC	100-0.1	Vertical	Omni	A1	IBP cycle
24930.0	OH2B	Espoo	KP20KE	100-0.1	Vertical	Omni	A1	IBP cycle
24930.0	CS3B	Madeira	IM12	100-0.1	Vertical	Omni	A1	IBP cycle
24930.0	LU4AA	Buenos Aires	GF05	100-0.1	Vertical	Beam	A1	IBP cycle
24930.0	OA4B	Lima	FH17KV	100-0.1	Vertical	Omni	A1	IBP cycle
24930.0	YV5B	Caracas	FK60NL	100-0.1	Vertical	Omni	A1	IBP cycle
24930.5	GB3...			100			A1	UC
24931	DK0HHH	Hamburg	JO53AM	10	Dip	Omni	A1	24
28124.7	KA5FYI	Austin TX	EM10DI	0.1	Vert Dip	Omni	A1	24
28175	VE3TEN	Ottawa	FN25	10	GP	Omni	A1	24
28178	PA3FCB/A		JO28BG	4				24?
28180.3	I1M	Bordighera	JN33UT	5/20	2x5/8 Vert	Omni	A1	Operational?
28182.0	SV3AQR	Amalias	KM07QS	4	GP	Omni	A1	Irregular
28186	ZS6PW	Pretoria	KG44DE	15	3 el Yagi	N	A1	24*
28188	JE7NYQ	Fukushima	QM07	50	Stack Dip	Omni		24
28190.2	LU4XS	Tierra/d/Fuego	FD65PA	5	Inv Vee	Omni	A1	24
28192	LU2DT		GF11FX				A1	24
28193.2	VE4ARM	Austin MB	EM09HW	5	GP	Omni	A1	24
28194	A47RB	Oman	LL93FO	10	AR-10	Omni	A1	24
28195.1	IY4M	Bologna	JN54QK	20	5/8 GP	Omni	A1	Non op
28196.0	VA2MGL	Saint-Agnes PQ	FN47UQ	2	Inv Vee	Omni	A1	24
28197.0	VE7MTY	Vancouver BC	CN89	5	Vertical	Omni	A1	24
28197.4	HB9TC	Bellinzona	JN46ME	4	Inv Vee		A1	07-1700
28199.3	LU1FHH	Santa Fe SF					A1	24
28200.0	4U1UN	UN New York	FM20AS	100-0.1	Vertical	Omni	A1	IBP cycle
28200.0	VE8AT	Eureka,Nunavut	EQ79AX	100-0.1	Vertical	Omni	A1	IBP cycle
28200.0	W6WX	Nr San Jose CA	CM87	100-0.1	Vertical	Omni	A1	IBP cycle
28200.0	KH6WO	Honolulu	BL11BK	100-0.1	Vertical	Omni	A1	IBP cycle
28200.0	ZL6B	Nr Masterton	RE77TW	100-0.1	Vertical	Omni	A1	IBP cycle
28200.0	VK6RBP	28k SE Perth	OF87BW	100-0.1	Vertical	Omni	A1	IBP cycle
28200.0	JA2IGY	Mt Asama	PM84JK	100-0.1	Vertical	Omni	A1	IBP cycle
28200.0	RR9O	Novosibirsk	NO14KX	100-0.1	Vertical	Omni	A1	IBP cycle
28200.0	VR2HK	Hong Kong	OL72CG					UC

Freq (MHz)	Callsign	Nearest town	Locator	ERP (W)	Antenna	Beam direction	Mode	Status
28200.0	4S7B	Colombo	MJ96	100-0.1	Vertical	Omni	A1	IBP cycle
28200.0	ZS6DN	Pretoria	KF44DC	100-0.1	Vertical	Omni	A1	IBP cycle
28200.0	5Z4B	Nr Mombasa	KI95	100-0.1	Vertical	Omni	A1	IBP cycle
28200.0	4X6TU	Tel Aviv	KM72JC	100-0.1	Vertical	Omni	A1	IBP cycle
28200.0	OH2B	Espoo	KP20KE	100-0.1	Vertical	Omni	A1	IBP cycle
28200.0	CS3B	Madeira	IM12	100-0.1	Vertical	Omni	A1	IBP cycle
28200.0	LU4AA	Buenos Aires	GF05	100-0.1	Vertical	Omni	A1	IBP cycle
28200.0	OA4B	Lima	FH17KV	100-0.1	Vertical	Omni	A1	IBP cycle
28200.0	YV5B	Caracas	FK60NL	100-0.1	Vertical	Omni	A1	IBP cycle
28202	ZS1J	Plettenberg Bay	KF15PF	5	1/2 Vert	Omni	A1	24*
28203.5	K6LLL	Laguna Beach CA	DM13CN	5	Vertical	Omni	A1	24
28204.7	S55ZRS	Mt Kum	JN76MC	10	1/4 GP	Omni	A1	24
28205.2	KB3BOE	Nr Ridgway PA	FN01PK	5	Dipole		A1	24
28206.1	VA3GRR	Brampton ON	FN03	1.75 ½	Vert	Omni	A1	24
28208.0	WN2A	Budd Lake NJ	FN20	3			A1	24
28210	N9YDZ	Brighton IL		1	Vertical	Omni	A1	24
28210	N7SCQ	Portola CA	CM99ST	5			A1	24
28211	KC4DPC	Wilmington NC	FM14BF	3	Indoor Dip		A1	24
28211.2	LA4TEN	Sotra I.	JP20LG	250	Vertical	Omni	A1	24
28212	W3AW	Springfield IL		1			A1	24
28213	PT7BCN	Fortaleza CE	HI06RF	5	GP	Omni	A1	24
28214	N3BUB	Kresgeville PA	FN20FV	5		Omni	A1	24
28215	KA9SZX	Champagne IL	EN50VD	1	CX 1000	Omni	A1	24
28215	GB3RAL	Nr Didcot	IO91IN	25	1/4 GP	Omni	F1	24
28216.0	N3FTI	Reading PA	FN20AH	5			A1	24
28218.5	W8MI	Mackinaw C. MI	EN75	0.5	Vertical	Omni	A1	24
28218.4	PT8AA	Rio Branco	AC FI60CA	5	GP	Omni	A1	Operational?
28219	K2KL	Monsey NY	FN21WC	3	Horiz Dip	Omni	A1	24
28219.9	KB9DJA	Mooresville IN	EM69RO	35	GP	Omni	A1	24
28220	5B4CY	Zyyi	KM64PR	26	GP	Omni	F1	24
28221.7	W6TOD	Ridgecrest CA	DM15	10	dipole		A1	24
28223	NC6DX	Nr Grass Valley	CM99	10	Vertical	Omni	A1	Operational?
28224	W6NW	Sunnyvale CA	CM87	10	Vertical	Omni	A1	Operational?
28225.2	KW7Y	Marysville Wa	CN88SD	4	Vertical	Omni	A1	24
28228	ZL3TEN	Rolliston	RE66	10	1/2 Vert	Omni	A1	24
28228	9A0TEN	Daruvar	JN85OO	50	Vertical	Omni	A1	Irregular
28228.3	WG8T	Fairmount WV	EN99WM	1	Dipole		A1	24
28230.4	PY3ARL	Port Alegre	GF49JW	5	GP@287m	Omni	A1	24
28232	W7JPI	Sonoita AZ	DM41QP	5	3 el Yagi	045	A1	24
28233.3	KD4EC	Jupiter FL	EL96WV	7	Vertical	Omni	A1	24
28233.4	N9RET	Riverside IL	EN61	5	5/8 vert	Omni	A1	24
28234.2	N1WRG	Wales MA	FN32VB	25			A1	24
28235.2	VE1CBZ	Fredericton NB	FN65	3	Vertical	Omni	A1	24
28236.1	N2VMF	Freehold NJ	FN20UF	5	Vertical	Omni	A1	24
28236	VE3GOP	Mississauga ON	FN03GD	5	3-el/whip		A1	24
28237.6	LA5TEN	Nr Oslo	JO59JV	10	5/8 GP	Omni	A1	24
28240	YO2X	Timisoara	KN05PS	3/.3/.1	Dipole		A1	Irregular?
28240	AB8Z	Parma OH	EN91DJ	5	5/8	Omni	A1	24
28240.0	VA3SBB	Thunder Bay ON	EN58	5	Vertical		A1	24?
28241.0	VE9MS	Fredericton NB	FN65	5	Loop		F1	24
28242.6	W2IK	Islip NY	FN30GR	25	Vertical	Omni	A1	24
28244.0	WA6APQ	Long Beach CA	DM13	30	Vertical	Omni	A1	24
28244.1	WA1RAJ	Hollis NH		1	1/4 vert	Omni	A1	24
28244.8	VE9BEA	Crabbe Mtn	NB FN66	3	Vertical	Omni	A1	24
28246	F5TMJ	Nr Toulouse	JN03SM	5	Horiz Loop	Omni	A1	TEST
28248	EA3JA	Barcelona	JN11BI				A1	Non op
28248.2	N1ME	Bangor ME	FN54PS	5			A1	24
28248.5	N7LT	Bozeman MT	DN45LQ	5/0.5/0.05	1/2GP	Omni	A1	24
28249.3	PY5ND	Curittba PR	GG54IN				A1	Irregular?
28249.7	PI7BQC	Haarlem	JO22HK	2			A1	24
28250.1	Z21ANB	Bulawayo	KG47	25	GP	Omni	F1	24
28250.0	WJ9Z	St Francis WI	EN62BX	15	Vertical	Omni	A1	24
28252.5	W6PC	Mt Woodson CA	DM13	10	Vee	Omni	A1	24
28253.5	VK3SIX	Wannon Falls	QF12WH	35	5el@32m		A1	17-0500
28254.7	W4STT	Hastings FL	EL99GQ	20	Vertical	Omni	A1	12-2400
28255	N0AR	St Paul MN	EN35KB	0.125	1/2 Vert	Omni	A1	24
28256	K5PF	Cary NC		8	A99		A1	24?
28256.5	VK3RMH	25kNEMelbourne	QF22JH	20/2	Vertical	Omni	A1	24
28257.5	DK0TEN	Konstanz	JN57NP	40	GP	Omni	F1	Non op
28259	F5KCF	Boussieres 59	JO10QD	6	3dbi Vert	Omni		24
28259.8	KA1NSV/4	Green Bay VA	FM07	25	Vert.Dip.	Omni	A1	24
28260	VK5WI	Nr Adelaide	PF95GD	10	GP	Omni	A1	24
28260	KF4FOF						A1	
28262	VK2RSY	Dural	QF56MH	25	1/2 Vert	Omni	A1	24
28263.0	VA3SRC	Toronto	FN03	5	Dipole		A1	?
28264	VK6RWA	Nr Perth	OF78WB	20	Vertical	Omni	A1	24
28264	JA5ZQM	Tokushima	PM74GA	10	1/2 GP	Omni	A1	24
28265.2	KK4XO	FtLauderale FL	EL96WE	10	H Loop@30'		A1	24
28266	VK6RTW	Albany	OF84	4	Vertical	Omni	F1	24
28266.1	LZ1TEN		KN12PO	1	GP	Omni	A1	24?
28268	OH9TEN	Pirttikoski	KP36OI	20	1/2 GP	Omni	A1	24

Freq (MHz)	Callsign	Nearest town	Locator	ERP (W)	Antenna	Beam direction	Mode	Status
28268	VK8VF	Darwin	PH57KP	40	Vertical	Omni	A1	24
28269.5	W3HH	Pittsburgh PA	EN90	8			A1	24
28270	VK4RTL	Townsville	QH30JS				F1	INT
28270.3	W8BEP	Sault S Marie MI	EN76				A1	24?
28274	KC2CZI						A1	?
28275	ZS1LA	Still Bay	KF05QK	20	3-el Yagi	N	F1	24
28277.6	DF0AAB	Kiel	JO54GH	10	GP	Omni	F1	24
28280	NO6J	1000 Oaks CA	DM04NF	5	Vertical	Omni	A1	Operational?
28280.4	K5AB	Austin TX	EM10DH	20	GP	Omni	A1	24
28282.5	N7GSU	McMinnville OR	CN85IF	0.5	1/2 Vert	Omni	A1	24
28282.5	KL7AQC	Fairbanks AK	BP64EU	1	GP	Omni	A1	24
28282.6	OK0EG	Hradec Kralove	JO70WE	10	GP	Omni	F1	24
28282.5	W0ERE	HighlandvilleMO	EM36IW	5	Vertical	Omni	A1	24
28284	KJ7AZ	Rawlins WY	DN61JS	5	Vertical	Omni	A1	24
28284.5	VP8ADE	Adelaide I.	FC52WK	8	Vertical	Omni	A1	24
28284.9	WL7IE/7	Olympia WA	CN87LC	5	Vertical	Omni	A1	Operational?
28285.0	N2JNT	Troy NY	FN32DR	1	GP	Omni	A1	24
28285	KB7EFZ	Portland OR	CN85	1	5/8 GP	Omni	A1	0, 15, 30, 45m
28285	KB7DQJ	Pt Orchard WA	CN87				A1	24?
28286	N5AQM	Chandler AZ	DM43AH	2	Vertical	Omni	A1	24
28286.6	WA8YWO	Nr Richwood WV	EM98	0.1	Slope dip		A1	24
28287.0	NQ2RP	Brockport NY	FN13AE	0.125	1/2 Vert	Omni	A1	24
28289.0	WJ5O	Cor. Christi TXEL17		2	Yagi	NE	A1	24
28290.3	N8NSY	Brighton MI	EN82CN	2	A99	Omni	A1	24
28290.4	SK5TEN	Strengnes	JO89KK	75	GP	Omni	A1	24
28290.5	WB4WOR	Sophia NC	FN05	3	Vertical	Omni	A1	24
28291.1	K9KXP	Collinsville IL	EM58AQ	5	Vertical	Omni	A1	24
28295.1	SK2TEN	Kristineberg	JP95HB	5	Vertical	Omni	A1	Non op
28295.8	W3VD	Laurel MD	FM19NE	10	Vert Dip	Omni	A1	24
28297	SK7TEN	Eksjo	JO77LS	10	Horiz Dip	E-W	A1	Non op
28297	ND0DX	Fargo ND		2	Dipole		A1	Operational?
28298.0	V73TEN	Roi Namur I	RJ39RJ					24?
28298.9	K4JDR	Raleigh NC					A1	24
28299.3	SK3TEN	Osterfernebo	JP80FD	50	2x6el Yagi	West	A1	Non op?
28301.0	PI7ETE	Amersfoort	JO22QD	0.5	Vertical	Omni	F1A	24?
28302.2	UA4NM		LO48UO					Irregular
28325	DF0THD		JN49HU				A2	Operational?

Notes

? Activity pattern uncertain

IBP International Beacon Project station. These beacons transmit for 10 seconds on each frequency in turn in the sequence shown on the right. Transmissions consist of callsign at 22WPM and 100W followed by four 1-second dashes at 100W, 10W, 1W and 0.1W. Equipment is TS-50, Cushcraft R-5 vertical and a Trimble Navigation GPS receiver to ensure sychronisation, with a control unit built by NCDXF.

DK0WCY
zz Normal transmission: DK0WCY beacon (x3) + 4 secs dash
During auroras: DK0WCY beacon (x3) aurora + short dashes or DK0WCY beacon (x3) strong aurora + 9 secs dash

At every full 5 minutes basic solar/geophysical information and forecast on CW updated aroung 0630UTC. 3.5MHz may not operate on contest weekends, and is one hour earlier in summer.

HF Beacons (1.8–30MHz):
Please notify errors/changes to Martin Harrison G3USF, HF Beacon Coordinator, Region 1 of the International Amateur Radio Union, (M.Harrison@pol.keele.ac.uk), 1 Church Fields, Keele, Staffs. ST5 5HP, England. Tel: (home) +44(0)1782 627396. Fax: (work) +44 (0) 1782 583592. A fully up-to-date version of the HF List is available on the Internet at www.keele.ac.uk/depts/por/28.htm.

Country	Callsign	Frequency (MHz)				
		14100	18110	21150	24930	28200
United Nations, NY	4U1UN	00.00	00.10	00.20	00.30	00.40
Northern Canada	VE8AT	00.10	00.20	00.30	00.40	00.50
USA (CA)	W6WX	00:20	-	00.40	-	01.00
Hawaii	KH6WO	00.30	-	00.50	-	01.10
New Zealand	ZL6B	00.40	00.50	01.00	01.10	01.20
West Australia	VK6RBP	00.50	01.00	01.10	01.20	01.30
Japan	JA2IGY	01.00	01.10	01.20	01.30	01.40
China	-	(01.10	01.20	01.30	01.40	01.50)
Siberia	-	(01.20	01.30	01.40	01.50	02.00)
Sri Lanka	4S7B	01.30	01.40	01.50	02.00	02.10
South Africa	ZS6DN	01.40	01.50	02.00	02.10	02.20
Kenya	5Z4B	01.50	02.00	02.10	02.20	02.30
Israel	4X6TU	02.00	02.10	02.20	02.30	02.40
Finland	OH2B	02:10	02.20	02.30	02.40	02.50
Madeira	CS3B	02.20	02.30	02.40	02.50	00.00
Argentina	LU4AA	02.30	02.40	02.50	00.00	00.10
Peru	OA4B	02.40	02.50	00.00	00.10	00.20
Venezuela	YV5B	02:50	00.00	00.10	00.20	00.30

W6WX and KH6WO are not currently licensed for 18 or 24MHz operation.

VHF/UHF/MICROWAVE BEACONS

Freq (MHz)	Callsign	Nearest town	Locator	ASL (m)	Antenna	Beam direction	ERP (W)	Info from	Notes
50	GB3BUX	Buxton, Derbys	IO93BF	457	Turnstile	Omni	20	G4IHO	
50.001	VE1SMU	Halifax NS	FN84		3 el Yagi	90°	40		
50.001	BV2FG	Taihoku	PL05		5/8 Vertical	Omni	3		QRT Sunday
50.003	7Q7SIX	Malawi	KH74						
50.004	IOJX	Rome	JN61HV						
50.004	PJ2SIX	Willemstad	FK52		4 x Horz dipole	Omni	22		
50.005	4N0SIX	Belgrade	KN04FU		Dipole	Omni	1		
50.006	V73SIX		RJ38		PAR Loop		10	V73AT	
50.008	VE8SIX	Inuvik NWT	CP38		Double Bay	0°/180°	80		
50.008	HI0VHF		FK58						
50.008	XE2HWB/B	La Paz Baja	DL44		6 el Yagi	0°	5		
50.0095	PY2SFY/B		GG77GA		5/8 Vert	Omni	5		
50.01	SV9SIX	Iraklio	KM25NH		Vertical dipole	Omni	30		
50.01	JA2IGY	Mie	PM84JK		5/8 G/Plane	Omni	10		
50.011	OK0EK	Kromeriz	JN89OF	300	2x Dipole	Omni	10-Jan	OK2PWM	Plan Q2/98
50.013	CU3URA	Terceira, Azores	HM68		5/8 Vertical		5		
50.014	S55ZRS	Mt. Kum	JN76MC	1219	Ground Plane	Omni	8	S57C	
50.0155	LU9EHF	Lincoln City	FF95		Dipole		15		
50.017	JA6YBR	Miyazaki	PM51		Turnstile		50		
50.018	V51VHF	Namibia	JG87		5/8 Vert	Omni	50		
50.019	CX1CCC	Montevideo	GF15		Ground Plane	Omni	5		
50.021	OZ7IGY	Tollose	JO55VO	92	Turnstile	Omni	20	OZ7IS	
50.0225	FR5SIX	Reunion Is	LG78	2896	Halo	Omni	2	F5QT	QRT
50.0225	XE1KK/B		EK09			Omni	20		
50.023	LX0SIX	Bourscheid	JN39AV	500	Horizontal Dipole	0°/180°	5	LX1JX	
50.023	SR5SIX	Wesola	KO02OF	130	Ground Plane	Omni	3	SP5TAT	
50.0235	ZP5AA	Asuncion	GG14		Vertical	Omni	5		
50.025	9H1SIX	Attard, Malta	JM75FV	75	Ground Plane	Omni	7	9H1ES	
50.025	OH1SIX	Ikaalinen	KP11QU	157	4 x Turnstile	Omni	40		
50.025	YV4AB	Valensia	FK50		Ringo		15		
50.027	CN8LI	Rabat	IM64		J Pole	Omni	8	CN8LI	
50.027	JA7ZMA		QM07		2 x Turnstile	Omni	50		
50.028	SR6SIX	Sztobno / Wolow	JO81HH		Ground Plane	Omni	10	SP6GZZ	
50.028	XE2UZL/B		DM10		2 Sq Loops		25		
50.029	SR8SIX	Sanok	KN19CN						
50.03	CT0WW	Portugal	IN61GE	400	H. Dipole	45°/225°	40		
50.032	CT0SIX		IM59						
50.032	JR0YEE	Niigata	PM97		Loop		2		
50.0325	ZD8VHF	Ascension Is	II22TB	723	5/8 Vertical	Omni	50		
50.036	VE4VHF		EN19		Vertical	Omni	35		
50.037	ES0SIX	Muhu Island	KO18PO	30	Hor. dipole	90°/270°	15	ES0CB	
50.037	JR6YAG	Okinawa	PL36		2 x 5/8 G Planes		10		
50.038	FP5XAB	St Pierre Miquelon Is	GN16		Dipole	Omni	15	FP5EK	
50.039	VO1ZA	St Johns	GN37		1/4 Wave Vert	Omni	10		
50.04	SV1SIX	Athens	KM17UX	130	Vertical Dipole	Omni	30		
50.04	ZL3SIX	Christchurch	RE66			45/225/135/315	70		
50.041	VE6EMU	Camrose	DO33		4el Yagi	22°	35		
50.042	GB3MCB	St Austell	IO70OJ	320	Dipole	90°/270°	40	G3YJX	
50.042	YB0ZZ	Jakarta	OI33		Ground Plane	Omni	15		
50.043	YO2S	Timisoara	KN05PS		Dipole		2	YO2IS	
50.044	VE6ARC		DO05		Ground Plane	Omni	25		
50.044	ZS6TWB/B	Haenertsburg	KG46XA		3 el Yagi	330°	20		
50.045	OX3VHF	Julianhaab	GP60XR	15	Ground Plane	Omni	20	OX3JUL	
50.046	VK8RAS		PG66		X Dipole		15		
50.047	TR0A		JJ40		5 el Yagi	0°	15		
50.047	4N1SIX	Belgrade	KN04OO		Vee	Omni	10		
50.047	JW7SIX	Svalbard	JQ78TF	25	4 el Yagi	190°	30	JW5NM	QRT
50.048	VE8BY	Iqaluit NT	FP53				30		
50.05	5W1WS	Western Samoa	AH46		5 el Yagi		10		
50.05		Reserved for IARU IBP							
50.05	ZS6DN/B	Pretoria	KG44DE		5 el Yagi	135°	100	ZS6DN	
50.051	LA7SIX	Senja	JP89MB	30	4 el Yagi	180°	30	LA5TFA	
50.052	Z21SIX	Zimbabwe	KH52NK		Ground Plane	Omni	8		
50.053	PI7SIX	Utrecht	JO22NC	40	Dipole	0°/180°	12	PA3FYM	
50.053	VK3SIX	Hamilton	QF12		Colinear		12		QRT ?
50.0535	KL7SIX	Alaska	BP51				20		Op Sep 21-Dec21 98
50.054	OZ6VHF	Oestervraa	JO57EI		Turnstile	Omni	25	OZ1IPU	
50.0555	V44K	St Kitts/Nevis	FK87		5/8 Vertical		3		
50.057	VK7RAE	Lonah	QE38		X Dipoles		20		
50.057	VK8VF	Darwin	PH57		1/4 Vertical		100		
50.058	VK4RGG	Nerang	QG62				6		
50.058	VE3UBL	Brougham	FN03		Turnstile		10		
50.059	PY2AA	Sao Paulo	GG66		Ground Plane		5		
50.059	JH0ZPI		PM96				10		
50.06	KA5FYI		EM10						
50.06	W5VAS		EM40		Squalo		50		
50.06	K4TQR/B		EM63		Dipole		3		
50.06	EA3VHF		JN11MV						
50.06	GB3RMK	Inverness	IO77UO	270	Dipole	0°/180°	10	GM3WOJ	

Freq (MHz)	Callsign	Nearest town	Locator	ASL (m)	Antenna	Beam direction	ERP (W)	Info from	Notes
50.061	KH6HME/B		BK29		Dipole		20		
50.061	KE7NS/B		DN31		Squalo		2		
50.061	W1VHF/b		FN41		Vertical		25		
50.061	WB0RMO		EN10		Squalo		50		
50.062	GB3NGI	Ballymena	IO65PA	240	Dipole	140°/320°	10	GI6ATZ	Plan Q2 99
50.062	W7HAH		DN28		Halo	Omni	25		
50.062	K8UK/b		EN82			Omni	2		
50.062	KA0NNO		EM24		Halo	Omni	8		
50.064	AA5ZD		EM12						
50.064	GB3LER	Lerwick	IP90JD	104	Dipole	0°/180°	30	GM4IPK	QRT 98
50.065	AB5L		EM13		Dipole		0.2		
50.065	W0IJR		DM79		2 x Ring Halo	Omni	20		
50.065	KG9AE		EM69		AR6	Omni	10		
50.065	KH6HI/b		BL01		Turnstile	Omni	15		
50.065	W3VD		FM19		Squalo		7		
50.065	W0MTK		DM59		4 V Dipoles		2		
50.0655	GB3IOJ	St Helier	IN89WE	115	Vertical	Omni	10	GJ4ICD	
50.066	W5OZI		DM90		Dipole		20		
50.066	VK6RPH		OF88		U Dipole		10		
50.066	WA1OJB		FN54		J Pole		30		
50.067	W3HH		EN90		Loop		10		
50.067	KQ4E		EM86		Halo		10		
50.067	W4RFR		EM66				2		
50.067	OH9SIX	Pirttikoski	KP36OI	192	2 x Turnstile	Omni	35	OH6DD	
50.068	W7US		DM42		4 el		50		
50.069	K6FV		CM87				100		
50.07	SK3SIX	Edsbyn	JP71XF	505	Hor X dipole	Omni	10	SM3EQY	
50.07	W2CAP/B		FN41		V Dipole		15		
50.07	W7WKR/B		CN87		Beam		10		
50.07	ZS1SES								
50.071	WB5LUA		EM12		Halo		1.5		
50.071	KA5BTP		EM40						
50.072	KS2T		FM29		Ground Plane		10		
50.072	WA4NTF/B		EM81						
50.072	W4IO		EM81		M2 Halo	Omni	1		
50.072	KW2T		FN13		Squalo		0.5		
50.073	WB4WTC/B		FM06		2 Loops		10		
50.073	WR7V/B		CN87		Halo		10		
50.073	ES6SIX	Voeru	KO37MT	85	Ground Plane	Omni	1	ES5MC	
50.073	NN7K		DM09		Ringo Ranger		1		
50.075	W6SKC/7		DM41		Halo		5		
50.075	VR2SIX		OL72		Ground Plane		7		
50.075	NL7XM/2		FN20				1		
50.076	KL7GLK/3		FM18			Omni	4		
50.077	VE3DRL								
50.077	N0LL		EM09		2 x Halo		21		
50.077	WB2CUS		EL98		Loop		1		
50.0775	VK4BRG		QG48		Turnstile		5		
50.078	OD5SIX	Lebanon	KM74WK		1/4 Vertical	Omni	7	OD5SB	
50.078	KE4SIX		EM83		Ringo Ranger	Omni	5		
50.079	JX7DFA	Jan Mayen Island	IQ50		5 el yagi	160°	40	LA7DFA	
50.079	TI2NA		EJ79		Dipole		20		
50.08	PP1CZ		GG99UQ		5 el Yagi	0°	3		
50.08	ZS1SIX		JF96		Halo		10		
50.082	CO2FRC		EL83		Dipole		2		
50.083	LZ1SIX		KN12						
50.086	VP2MO		FK86		6 el Yagi		10		
50.0875	VE9MS/B		FN65		2 H/Loops		40		
50.088	YU1SIX		KN03KN		Dipole		15		
50.089	VE2TWO	Radisson	FO13		Dipole		15		
50.095	PY5XX		GG54		Dipole		50		
50.162	IS0SIX	Sardinia	JM49NG		Dipole		1	IS0AGY	
50.306	VK6RBU		OF76		3 el Yagi	260°/80°	100/10		
50.315	FX4SIX	Neuville	JN06CQ	153	Turnstile	Omni	25	F5GTW	
50.325	F????	Cany-Barville	JN09HS	130	2 x Turnstile	Omni	5	F1EHX	Plan 5/99
50.48	JH8ZND/B		QN02		Ground Plane		10		
50.485	JH9YHP		PM86		X Dipole				
50.49	JG1ZGW	Tokyo	PM95		Dipole		10		
50.499	5B4CY	Zyghi, Cyprus	KM64PR	30	Ground Plane	Omni	20	5B4BBC	
50.521	SZ2DF		KM25		4 x 16 el	30°/330°	1000		
51.029	ZL2MHB	Hastings	RF80		1/2 Vertical	Omni			
52.345	VK4ABP	Longreach	QG26		1/4 Vertical	Omni	4		
52.42	VK2RSY	Dural	QF56		Turnstile	Omni	25		
52.45	VK5VF	Mt Lofty	PF95		Turnstile	Omni	10		
52.51	ZL2MHF	Mt Climie	RE78		Dipole		4		
70	GB3BUX	Buxton, Derbys	IO93BF	456	2 x Turnstile	Omni	20	G4IHO	
70.005	ZS5MTL		KG50IG			Omni	50		
70.01	GB3REB	Camberley	IO91OH	117	2 el Yagi	330°	28		QRT
70.014	S55ZRS	Mt. Kum	JN76MC	1219					Planned
70.015	ZR6FOR								Planned
70.02	GB3ANG	Dundee	IO86MN	370	3 el Yagi	160°	100	GM4ZUK	
70.025	GB3MCB	St Austell	IO70OJ	320	2 el Yagi	45°	40	G3YJX	

Freq (MHz)	Callsign	Nearest town	Locator	ASL (m)	Antenna	Beam direction	ERP (W)	Info from	Notes
70.03	UK Personal Beacons								
70.03	UK Personal Beacons								
70.05	Proposed for IARU IBP								
70.114	5B4CY	Zyghi, Cyprus	KM64PR	30	4 el Yagi	315°	15	5B4BBC	
70.13	EI4RF	Dublin	IO63WD	120	2 x 5 el Yagi	45°/135° seq	25	EI9GK	
144.282	W1RJA/B	Rhode Is	FN41CJ	140	5 el Yagi		500	W1RJA	QRT
144.3	VE1SMU/H	Nova Scotia	FN84CM		4 x 9 el Yagi	61°	4.8kW	VE1KG	
144.4	Transatlantic beacon								
144.402	EA8VHF	Grand Canary Is	IL28GC			Omni	10		
144.403	EI2WRB	Portlaw	IO62IG	248	5 el Yagi	95°	200	EI6GY	
144.404	EA1VHF	Curtis	IN53UG	100	5 el Yagi	45°	100	EA1DKV	
144.405	F5XAR	Lorient	IN87KW	165	9 ele Yagi	290°	400	F6ETI	Trans Atl.
144.407	GB3?	Planned UK Transatlantic Beacon							02/99 G4ASR
144.409	F5XSF	Lannion	IN88GS	145	9 el Yagi	90°	50	F6DBI	
144.41	DB0SI	Schwerin DOK V 14	JO53QP	90	Big wheel	Omni	10 TX	DL1SUZ	
144.41	ZS2VHF	Port Elizabeth	KF25UX		5 el Yagi	45°	160	ZS2FM	
144.411	I1G	La Spezia	JN44VC	745	4 el Yagi	22°	4	IK1LBW	
144.412	SK4MPI	Borlaenge	JP70NJ	520	4 x 6 el Yagi	45°/315°	1500	SM4HFI	
144.413	3A2B	Monaco	JN33RR	50	Yagi	90°	50	3A2LF	QRT ?
144.414	DB0JW	Wurselen DOK G 05	JO30DU	238	7 el Yagi	22°	50	DL9KAS	
144.415	I1M	Bordighera IM	JN33UT	300	Big wheel	Omni	20	IK1PCB	
144.416	PI7CIS	Delft	JO22DC	40	Omni	Omni	50	PA0CIS	
144.417	OH9VHF	Pirttikoski	KP36OI	310	10 dBd gain	200°	200	OH6DD	
144.418	ON4VHF	Louvain La Neuve	JO20HP	180	Clover leaf	Omni	15	ON7PC	
144.419	I2M	Cremona	JN55AD	46	Big wheel	Omni	10	IK2THZ	
144.42	DB0RTL	DOK P 60	JN48PL	480	Big wheel	Omni	15	DL8SDL	
144.422	DB0TAU	DOK F 11	JO40HG	326	4 x 4 el Yagi	Omni	15	DL3DC	
144.423	PI7FHY	Heerenveen	JO33WW	52	Vertical	Omni	10	PA3FHY	
144.424	IN3A	Trento	JN56NB	225	Ground plane	Omni	0.1	IN3IYD	
144.425	F5XAM	Blaringhem	JO10EQ	99	Big wheel	Omni	14	F6BPB	
144.426	EA6VHF	San Jose, Ibiza	JM08PV	150		Omni	20	EA6FB	
144.427	OK0EJ	Frydek-Mistek	JN99FN	1323	4 el Yagi	270°	0.3	OK2UWF	
144.427	PI7PRO	Nieuwegein	JO22NA	20	Halo	Omni	10	PI4VRZ	
144.428	DB0JT	Oberndorf DOK C 16	JN67JT	785	4 x Dipole	0°	30	DJ8QP	
144.429	IV3A	Cormons Go	JN65RW	130	2 x Turnstile	Omni	5	IV3HWT	
144.43	GB3VHF	Wrotham, Kent	JO01DH	268	2 x 3 el Yagi	315°	40	G8JNZ	
144.431	9A0BVH		JN85JO	489	V Dipole	Omni	1		QRT ?
144.432	9H1A	Malta	JM75FV	160	Turnstile	Omni	1.5	9H1BT	QRT ?
144.433	TF?								Planned
144.434	DB0LBV	DOK S 30	JO61EH	232	2 x Dipole	Omni	0.4 TX	DL1LWM	
144.435	HB9H	Locarno	JN46KE						Uncertain
144.435	SK2VHG	Svappavara	KP07MV	380	16 el Yagi	180°	800	SK2CP	
144.436	I3A		JN55						Plan 1/99
144.436	PI7NYV	Holtenburg	JO32EH	80	Halo	Omni	10	PI4NYV	QRT
144.437	LA1VHF	Oslo	JO59		Turnstile	Omni	12	LA4PE	QRT Q2 99
144.438	3A2B	Monaco	JN33RR	50					Planned
144.438	OK0EO	Olomouc	JN89QQ	602	Ring dipole	Omni	0.05	OK2VLX	Planned
144.439	SK3VHF	Oestersund	JP73HF	325	Horizontal Yagi	180°	500	SM3PXO	MS beacon
144.44	DL0UH	Melsungen DOK Z 25	JO41RD	385	V Dipole	Omni	1	DJ3KO	
144.441	LA4VHF	Bergen	JP20LG	30	2 x 8 el Yagi	0°	380	LA6LU	QRT
144.442	I4A	Bologna	JN54QK	300	4 x Dipole	Omni		IK4PNJ	
144.443	OH2VHF	Nummi	KP10VJ	76	9 el yagi	0°	150		
144.444	DB0KI	Bayreuth DOK Z42	JO50WC	1025	Dipole	Omni	2.5	DC9NL	
144.444	I5A	Lucca	JN53GW	1000	Big wheel	Omni	6	IW5BHY	On Test
144.445	GB3LER	Lerwick	IP90JD	108	2 x 6 el Yagi	45°/135°	500/500	GM4IPK	QRT 99
144.446	OK0EB	Ceske Budejovice	JN78DU	1084	3 x Dipole	Omni	70/7mW	OK1APG	
144.447	SK1VHF	Klintehamn	JO97CJ	65	2 x Cloverleaf	Omni	10		
144.448	HB9HB	Biel	JN37OE	1300	3 el Yagi	345°	120	HB9AMH	QRT
144.449	I0A	P.Mirteto RI	JN62IG	300	2 x Big wheel	Omni	10	IW0BCF	
144.45	DL0UB	Trebbin	JO62KK	120	4 x Dipole	Omni	10 TX	DL7ACG	
144.45	F5XAV	Remoulins	JN23GX	100	Halo	Omni	5	F5IHN	
144.451	LA7VHF	Senja	JP89MB	30	10 el Yagi	190°	150	LA5TFA	
144.452	OK0EC	As	JO60CF	778	3 el Yagi	90°	0.7	OK1VOW	
144.453	GB3ANG	Dundee	IO86MN	370	4 el Yagi	160°	20	GM4ZUK	QRT Q1 99
144.454	IS0A	Olbia SS	JN40QW	350	Turnstile	Omni	1	IW0UGR	
144.455	OH5ADB	Hamina	KP30NN	65	Dipole	135°/315°	0.1		
144.456	DB0GD	Rhoen DOK Z 62	JO50AL	930	Dipole	0°/180°	1 TX	DG6ZX	
144.457	SK2VHF	Vindeln	JP94TF	300	2 x 10 el Yagi	0°/225° 1.5	100		
144.458	F1XAT	Brive	JN15AO	913	Big wheel 6dB	Omni	25	F1HSU	
144.458	I0G	Foligno PG	JN63IB	1200	4 x dipole	Omni	10	IW0QIT	
144.459	LA5VHF	Bodo	JP77KI	260	2 x 6 el Quad	15°/180°	100	LA1UG	QRT
144.46	HG1BVA	Szentgotthard	JN86CW	370	Hybrid Quad	80°	40	HA1YA	
144.46	TF?								Plan ?
144.461	SK7VHF	Falsterbo	JO65KJ	25	2 x Cloverleaf	Omni	10		
144.462	I6A		JN72						Plan 1/99
144.463	LA2VHF	Melhus	JP53EG	710	10 el Yagi	15°	500	LA1BFA	
144.464	I7A	Bari	JN81EC	685	Big wheel	Omni	8	I7FNW	
144.465	DF0ANN	DOK B 25	JN59PL	630	V Dipole	Omni	0.3 TX	DL8ZX	
144.466	OZ4UHF	Osterlars Bornholm Is	JO75LD	130	Big wheel	Omni	10	OZ1HTB	
144.467	HB9RR	Zurich	JN47FI	871	4 x Dipole	Omni			
144.467	I8A	Reggio C.	JM78WD	1778	SqLo	Omni	8	I8GMP	
144.467	OK0ED	Frydek-Mistek	JN99DQ	290	2 x Dipole	Omni	0.1	OK2UWF	

Freq (MHz)	Callsign	Nearest town	Locator	ASL (m)	Antenna	Beam direction	ERP (W)	Info from	Notes
144.468	F1XAW	Beaune	JN26IX	561	Big wheel	Omni	10	F1RXC	
144.468	LA6VHF	Kirkenes	KP59AL	70	14 el Yagi	210°	300	LA4OO	
144.469	GB3MCB	St Austell	IO70OJ	320	3 el Yagi	45°	40	G3YJX	
144.469	IT9A	Alcamo TP	JM67LX	825	2 x Big wheel	Omni	10	IT9QPF	
144.47	OH2VAN	Vantaa	KP20			Omni			Planned
144.47	OK0EZ	Pardubice	JO70VB	250	vertical	Omni	2/0.5	OK1DXF	
144.471	OZ7IGY	Tollose	JO55VO	96	Big wheel	Omni	25	OZ7IS	
144.472	IT9G	Mondello PA	JM68QE	50				IT9BLB	Plan 1/99
144.472	TF?								Plan ?
144.473	OE3XAA	Hoher Lindkogel	JN88BA	834	Halo	Omni	0,2	OE1BKW	QRT
144.473	SK2VHH	Lycksele	JP94	300	Horizontal	22°	15000	Scientific	QRV summer
144.474	OK0EL	Benecko	JO70SQ	900	Dipole	Omni	0.004	OK1AIY	
144.475	DL0SG	DOK U 14	JN69KA	1024	4 x 4 el Yagi	Omni	5 TX	DJ4YJ	
144.475	LY2WN	Jonava	KO25GC		2 x Dipole	Omni	15	LY2WN	
144.475	YU1VHF	Pozarevac	KN04OO	200	2 x QQ	135°/337°	10	YU1AU	
144.476	F5XAL	Pic Neulos	JN12LL	1100	5 el Yagi	0°		F6HTJ	
144.477	DB0ABG	DOK U 01	JN59WI	522	Big Wheel	Omni	4 TX	DJ3TF	
144.478	LA3VHF	Mandal	JO38RA	30	16 el Yagi	180°	100	LA8AK	QRT
144.478	OM0MVA	Bratislava	JN88NE	440	Dipole	Omni	1		
144.478	S55ZRS	Mt.Kum	JN76MC	1219	Dipole	Omni	1	S57C	
144.479	SR5VHF	Wesola	KO02OF	130	Turnstile	Omni	0.75	SP5TAT	
144.479	IT9S	Acireale CT	JM77NO	800	2 x Loop	Omni	3	IW9AFI	
144.482	GB3NGI	Ballymena	IO65VB	528	2 x 4 el Yagi	45°/135°	120/120	GI6ATZ	
144.484	F?	Fréville	JN09JN	120	2 x turnstile	Omni	10	F1EHX	Plan 5/99
144.486	DL0PR	Garding DOK Z 69	JO44JH	75	4 x 6 el Yagi	0°/180°	200 TX	DL8LD	
144.49	DB0FAI	Langerringn DOK T01	JN58IC	590	16 el Yagi	305°	1000	DL5MCG	
144.825	OY6VHF	Faeroes (pl 144.402)	IP62OA	300	2 x 4 el Yagis	45°/135°	50		QRT Q2 99
144.922	ZS6TLB	Peitersburg	KG46RC		2 x 5 el Yagis	215°	10		QRT
144.955	YO2X		KN05OS		Turnstile		2	YO2IS	
432.128	S55ZNG	Trstelj	JN65UU	643	Horizontal Loop	Omni	0.1	S50M	
432.8	DB0GD	Rhoen	JO50AL	930	Dipole	0°/180°	1 TX	DG6ZX	
432.8	OE3XMB	Muckenkogel	JN77TX	1154			2	OE3FFC	
432.81	DB0ZW	DOK U 17	JN69EQ	825	Schlitz	Omni	1 TX	DC9RK	
432.82	LA8UHF	Tonsberg	JO59FB	30	8 el Yagi	180°	50	LA6LCA	
432.825	DB0ABG	DOK U 01	JN59WI	522	Big Wheel	Omni	1 TX	DJ3TF	QRT
432.83	F5XBA	Preaux	JN18KF	166	4 x HB9CV	Omni	10	F6HZA	
432.83	LA7UHF	Bergen	JP20LG	30	4 el Yagi	0°	200	LA6LU	
432.835	ES0UHF	Hiiumaa Island	KO18CW	105	Horizontal	Omni	50	ES0NW	
432.84	DB0KI	Bayreuth	JO50WC	925	Dipole	Omni	10	DC9NL	
432.84	OH6UHF	Uusikaarlepyy	KP13GM	55	3 x Big wheel	Omni	7	OH6UH	
432.845	DB0LBV	DOK S 30	JO61EH	234	Schlitz	Omni	2 TX	DL1LWM	
432.845	LA9UHF	Geilo	JP40CM	1000	2 x 13 el Yagi	33°	250	LA3SP	
432.847	9A0BUH		JN85JO	489	V dipole	Omni	1		
432.85	DL0UB	DOK Z 20	JO62KK	120	Malteser	Omni	10 TX	DL7ACG	
432.85	I5B	Vinci FI	JN53KN	300	2 x 10 el Yagi	0°/260°	2	I5WBE	
432.852	OH2UHF	Nummi	KP10VJ	76	2 x dipole	90°/270°	50		
432.855	LA5UHF	Bodo	JP66WX	1110	10 el Yagi	15°	100	LA1UG	QRT
432.855	SK3UHF	Nordingra	JP92FW	200	4 x Double quad	Omni	10	SM3AFT	
432.86	LA1UHF	Oslo	JO59	522	Mini wheel	Omni	10	LA4PE	QRT Q2 99
432.863	F5XAG	Lourdes	IN93WC	550	2 x 10 el	22°	40	F5HPQ	
432.865	LA6UHF	Kirkenes	KP59AL	70	15 el Yagi	210°	40	LA4OO	
432.87	EI2WRB	Portlaw	IO62IJ	248	5 el Yagi	95°	250	EI9GO	
432.87	LA2UHF	Melhus	JP53EG	710				LA1BFA	
432.873	PI7HVN	Heerenveen	JO22WW	50	Horizontal	Omni	0.5	PE1HUE	
432.875	DB0FAI	DOK T 01	JN58IC	610		Omni	10	DL5MCG	
432.875	OH7UHF	Kuopio	KP32TW	215	6 dBd	225°	15/1.5/.15		
432.875	SK2UHF	Vindeln	JP94WG	445	2 x 20 el coll	0°/225°	300	SK2AT	
432.88	LA3UHF	Mandal	JO38RA	12	15 el Yagi	180°	29	LA8AK	QRT
432.882	OE3XAA	Hoher Lindkogel	JN88BA	834			0,2	OE1BKW	QRT
432.885	OK0EP	Sumperk	JO80OC	1505	2 x 3 el Yagi	90°	6	OK1VPZ	
432.885	OY6UHF	Faroe Is	IP62OA	300	7 dB Group	135°	50	OY1A	QRT Q2 99
432.886	F5XAZ	St Savin	JN06KN	144	Big wheel	Omni	50	F5EAN	
432.888	OM0MUA	Bratislava	JN88NE	440	Dipole	Omni	1		
432.89	GB3SUT	Sutton Coldfield	IO92CO	270	2 x 8 el Yagi	0°/135°	10	G8XGG	
432.89	LA4UHF	Haugesund	JO29PJ	75	10 el Yagi	200°	50		
432.895	PI7YSS	Zutphen	JO32CD	45	Big wheel	Omni	4	PA0JAZ	
432.895	OZ4UHF	Bornholm Island	JO75KC	115	Clover leaf	Omni	30	OZ1HTB	QRT Q2/99
432.9	DB0YI	Hildesheim Z 35	JO42XC	480	Big wheel	Omni	3 TX	DL4AS	
432.9	ZS5UHF	Pietersburg	KG46RC		13 el Yagi	215°	10		QRT Q3/98
432.905	PI7QHN	Zandvoort	JO22KH	20	3 dB Gain	Omni	2	PA0QHN	
432.905	SK4UHF	Garphyttan	JO79LK	270	Horizontal	Omni	50	SM4RWI	
432.908	EA8UHF	Grand Canary Is	IL28GC			Omni	10		
432.91	GB3MLY	Emley Moor	IO93EO	600	6 el Yagi	150°	40	G3PYB	
432.918	EA6UHF	Ibiza Is	JM08PV			Omni	10	EA6FB	
432.918	FX3UHB	Locronan	IN78VC	285	Big wheel	Omni	15	F5MZN	QRT 99
432.92	DB0UBI	DOK N 59	JO42GE	125	8el Coll	45°	12	DD8QA	
432.92	SK7UHF	Taberg	JO77BQ	350	Big wheel	Omni	15	SM6DHW	QRT
432.925	DB0JG	Bocholt DOK N17	JO31GT	45	Clover Leaf	Omni	1 TX	DL3QP	
432.925	SK6UHF	Varberg	JO67EH	175	Clover Leaf	Omni	10	SM6ESG	
432.93	HG7BUA	Dobogoko	JN97KR	700	Slot	Omni	2	HG5ED	
432.93	OK0EA	Trutnov	JO70UP	1355	2 x 15 el Yagi	180°/270°	3	OK1AIY	*432.934
432.93	OZ7IGY	Tollose	JO55VO	93	Omni	Omni	30	OZ7IS	

Freq (MHz)	Callsign	Nearest town	Locator	ASL (m)	Antenna	Beam direction	ERP (W)	Info from	Notes
432.934	GB3BSL	Bristol	IO81QJ	252	4 x 3 el Yagi	90°	250	GW8AWM	
432.94	DL0UH	Melsungen DOK Z25	JO41RD	385	V-Dipole	Omni	1	DJ3KO	
432.94	SK7MHH	Faerjestaden	JO86GP	45	Horizontal	0°/Omni	300/30		
432.945	DB0LB	DOK P06	JN48NV	367	Corner dipole	0°/180°	0.2 TX	DK3PS	
432.945	DB0OS	Erndtebruck DOK N32	JO40CW	730	2 el Yagi	270°	0.3	DG6YW	
432.945	HG3BUA	Tubes	JN96CC	612	Slot	Omni	0.5 TX		
432.945	OH9UHF	Pirttikoski	KP36OI	307	9 dBd gain	200°	70	OH6DD	
432.947	HG6BUA	Kekes	KN07AU	1050	Slot	Omni	2	HG5ED	
432.95	DB0IH	Oberthal DOK Q 18	JN39ML	630	Big wheel	Omni	1	DC8DV	
432.95	S55ZRS	Mt Kum	JN76MC	1219	Slot Dipole	Omni	1	S50M	
432.95	SK1UHF	Klintehamn	JO97CJ	55	2 x Big wheel	Omni	10	SM1IUX	
432.955	OZ1UHF	Frederikshavn	JO57FJ	150	Big wheel	Omni	10	OZ9NT	
432.965	DF0ANN	Altdorf	JN59PL	630	Big wheel	Omni	1 TX	DL8ZX	
432.965	GB3LER	Lerwick	IP90JD	104	12 el Yagi	165°	675	GM4IPK	QRT
432.966	OK0EO	Olomouc	JN89QQ	602	Ring Dipole	Omni	0.05	OK2VLX	Planned
432.97	GB3MCB	St Austell	IO70OJ	320	4 el Yagi	45°	12	G3YJX	QRT
432.97	OK0EB	Ceske Budejovice	JN78DU	1084	Mini Wheel	Omni	0.03/0.16	OK1APG	
432.975	DB0JW	Aachen DOK G 05	JO30DU	238	2 x 11 el Yagi	45°	50	DL9KAS	
432.975	DL0SG	DOK U 14	JN69KA	1024	4 x 11 Yagi	Omni	5 TX	DJ4YJ	
432.975	HG1BUA	Hormann	JN87FI	700	Slot	Omni	0.5 TX		
432.98	GB3ANG	Dundee	IO86MN	370	9 el Yagi	170°	100	GM4ZUK	
432.98	OK0EC	As	JO60CF	778	10 el Yagi	90°	1	OK1VOW	
432.98	S55ZCE	Sv. Jungert	JN76OH	574	Ground plane (V)	Omni	0.07	S51KQ	
432.982	SR5UHF	Wesola	KO02OF	130	Turnstile	Omni	0.25	SP5TAT	
432.983	OZ2ALS	Sonderborg	JO45UB	28	4 x dipole	Omni	40	OZ9DT	
432.984	HB9F	Interlaken	JN36XN	3573	Corner reflector	0°	15	HB9MHS	
432.99	DB0VC	DOK Z 10	JO54IF	300	4 x DQ	Omni	10	DL8LAO	
432.99	ON4UHF	Brussels	JO20ET	180	Clover leaf	Omni	0.5	ON4LC	
432.995	DL0IGI	Mt Predigstuhl DOK Z 57	JN67KQ	1618	2 x DQ	315°	50	DJ1EI	
1296.063	S55ZNG	Trstelj	JN65UU	643	V-J Slot	Omni	0.1	S50M	
1296.38	S55ZRS	Kum	JN76MC	1219	Turnstile	Omni	1	S57C	
1296.739	FX6UHY	Strasbourg	JN38PJ	1070	Big wheel	Omni	4	F6BUF	
1296.8	DB0GD	DOK Z 62	JO50AL	930	Dipol	Omni	1 TX	DG6ZX	
1296.8	DB0HEG	DOK T 09	JN59GB	700	4 x Slot	Omni	0.5 TX	DL2QQ	
1296.8	OE3XMB	Muckenkogel	JN77TX	1154			0.1	OE3FFC	
1296.8	SK6UHI	Hallandsaas	JO66LJ	230	Big wheel	Omni	50	SM6IKY	
1296.805	DB0RIG	DOK P 17	JN48WQ	780	4 x Yagi Box	Omni	50	DG9SQ	
1296.81	DB0ZW	DOK U 17	JN69EQ	825	Slot	Omni	1 TX	DC9RK	
1296.81	GB3NWK	Orpington	JO01BI	180	15/15 Slot Yagi	293°	50	G8BJG	
1296.81	PI7DIJ	Dokkum	JO33AI	20	9 el Yagi	200°	1	PA3DIJ	
1296.812	FX6UHX	Petit Ballon	JN37NX	1278	4 el Yagi	135°	1	F1AHO	
1296.815	DB0VI	Saarbrucken DOK Z19	JN39MF	400	13 el Yagi		1 TX	DK1ME	
1296.82	DB0OT	Lathen DOK I 26	JO32QR	80	Big wheel	Omni	1 TX	DL1BFZ	
1296.82	LA8UHG	Oslo	JO59		14 el Yagi	160°	10	LA4PE	QRT Q3 99
1296.825	DB0ABG	DOK U 01	JN59WI	522	Slot	Omni	0.5 TX	DJ3TF	
1296.825	DB0HF	Wandsbek DOK E 27	JO53BO	65	Big wheel	Omni	0.3 TX	DK2NH	
1296.825	OE1XTB	Vienna	JN88EE	170	4 x dipole	Omni	10	OE1MOS	QRT
1296.83	GB3MHL	Martlesham	JO02PB	80	4 x 16 Slot wg	90°/270°	700	G4DDK	
1296.835	DB0AJ	DOK C 09	JN57	620	12 el Yagi	0°	50	DK2RV	QRT
1296.835	SK0UHG	Vaellingby	JO89WI	60	Horizontal	Omni	10		
1296.84	DB0KI	Bayreuth DOK Z 42	JO50WC	925	Slot	Omni	80	DC9NL	
1296.84	OH6SHF	Uusikaarlepyy	KP13GM		Dipole	Omni	8		
1296.845	DB0LBV	DOK S 30	JO61EH	234	4 x Slot	Omni	2 TX	DL1LWM	
1296.845	SR3SHF	Kalisz	JO91CQ					SP3JBI	
1296.847	FX1UHY	Faviers	JN18IR	160	Alford Slot	Omni	10	F6ACA	
1296.85	DL0UB	Berlin DOK Z 20	JO62KK	120	4 x Box	Omni	10 TX	DL7ACG	
1296.85	GB3FRS	Farnborough	IO91PH	120	Disc	Omni	3	G8ATK	
1296.854	DB0JO	Witten DOK Z 03	JO31SL	312	4 x 15 el Yagi	270°	350	DG8DCI	
1296.855	OZ3UHF		JO56CE	150	5 el Yagi	180°	6	OZ1GMP	QRT Q2 99
1296.855	SK3UHG	Nordingra	JP92FW	200		Omni	10		
1296.86	GB3MCB	St Austell	IO70OJ	300	15/15	45°	50	G3YJX	
1296.86	LA1UHG	Tonsburg	JO59FB	30	13 dB Horn	180°	60	LA6LCA	
1296.862	F1XAK		JN23	114		Omni		F1AAM	QRT 99
1296.865	DB0JK	Koln DOK Z 12	JO30LX	260	4 x 8 el Yagi	Omni	40	DK2KA	
1296.865	HB9WW	Neuchatel	JN37LA	1145	15 el loop	125°	30	HB9HLM	
1296.865	SK7MHG	Veberod	JO65SO	200		Omni	50		QRT
1296.87	DB0IBB	DOK N 49	JO32VG	200	4 x Slot	Omni	170	DB7QW	
1296.875	DB0FAI	DOK T 01	JN58IC	610		Omni	10	DL5MCG	
1296.875	FX3UHX	Landerneau	IN78UK	121	Quad	90°	1	F6CGJ	
1296.875	GB3USK	Bristol	IO81QJ	235	Slotted waveguide	90°	250	GW8AWM	
1296.88	LA3UHG	Fleckkeroy	JO38XB	5	2 x 15 el Yagi	180°	10	LA8AK	
1296.88	ON4SHF	Ellignies St Ann	JO10UN	130	Slotted	90°	10	ON5PX	
1296.883	DB0INN	DOK C 15	JN68GI	504	Schlitz	Omni	1 TX	DL3MBG	
1296.885	DB0TUD	DOK S07	JO61UA	260	Quad	Omni		DL4DTU	
1296.885	OE3XEA	Kaiserkogel	JN78SB	725			1	OE3EFS	QRT
1296.886	FX4UHY	Loudun	JN06BX	140	Alford Slot	Omni	25	F1AFJ	
1296.888	OM0MSA	Bratislava	JN88NE	440	Dipole	Omni	1		
1296.89	GB3DUN	Dunstable, Beds	IO91RV	263	Alford slot	Omni	2	G3ZFP	
1296.89	HG6BUB	Kekes	KN07AU	1050	Slot	Omni	1 TX	HG5ED	
1296.895	ON4RUG	Gent	JO11UB	95	Slotted	Omni	20	ON6UG	
1296.9	DB0AN	Muenster-Nienberger	JO31SX	100	Big wheel	Omni	1 TX	DF1QE	
1296.9	GB3IOW	Newport, IOW	IO90IO	250	Alford Slot	Omni	100	G3WXC	

Freq (MHz)	Callsign	Nearest town	Locator	ASL (m)	Antenna	Beam direction	ERP (W)	Info from	Notes
1296.9	OK0EA	Trutnov	JO70UP	1355	4 x 15 el Yagi	S/SW/W/NW	1.6	OK1AIY	
1296.902	LX0SHF	Walferdange	JN39BP	420	2 x Big wheel	Omni	3	LX1JX	
1296.905	DB0AD	DOK R14	JO40AQ	693	V dipole	Omni	1	DL7AJA	
1296.905	OH4SHF	Haukivuori	KP31OX	200	Alford Slot	Omni	15		
1296.905	SK4UHI	Garphyttan	JO79LK	270		Omni	10	SM4RWI	
1296.907	F5XAJ	Pic Neulos	JN12LL	1100	Slotted WG	Omni	100	F6HTJ	
1296.91	DB0UX	Karlsruhe DOK A 35	JN48FX	275	Big wheel	Omni	1	DK2DB	
1296.91	GB3CLE	Clee Hill, Salop	IO82RL	540	2 x 15/15 Yagi	0°/135°	20	G3UQH	
1296.915	DB0UBI	DOK N 59	JO42GE	165	Horn	45°	2.5	DD8QA	
1296.92	9A0BLB		JN83HG	778	Dipole		1		
1296.92	DB0VC	Lutjenberg DOK Z 10	JO54IF	300	2 x Big wheel	Omni	12	DL8LAO	
1296.92	SK7UHG	Taberg	JO77BQ	350	Big wheel	Omni	3	SM6DHW	QRT
1296.923	PI7QHN	Zandvoort	JO22KH	20	6 dB Gain	Omni	4	PA0QHN	
1296.925	DB0KME	DOK C 35	JN67HT	800	Vertical	Omni	1 TX	DL8MCG	
1296.925	SK6UHG	Hoenoe	JO57TQ	35	4 x Big wheel	Omni	10	SM6EAN	
1296.93	GB3MLE	Emley Moor	IO93EO	600	Corner Reflector	160°	50	G8AGN	
1296.93	OK0EL	Benecko	JO70SQ	1035	5 dB Horn	135°/270°	0.8	OK1AIY	
1296.93	OZ7IGY	Tollose	JO55VO	95	Big wheel	Omni	15	OZ7IS	
1296.935	DB0YI	Hucksheim DOK Z 35	JO42XC	480	Big wheel	Omni	3 TX	DL4AS	
1296.935	OH5SHF	Kuusankoski	KP30HV	145	Alford Slot	Omni	25		
1296.94	DL0UH	Melsungen DOK Z 25	JO41RD	385	V-Dipole	Omni	1	DJ3KO	
1296.94	SK7MHH	Farjestaden	JO86GP	45					
1296.945	DB0OS	Hitchembach DOK N32	JO40CW	730	6 el array	270°	1	DG6YW	
1296.945	HB9F	Bern	JN46SW	1015	Corner reflector	0°	15	HB9MHS	
1296.945	HG3BUB	Tubes	JN96CC	612	Slot	Omni	0.3 TX		
1296.945	OH9SHF	Pirttikoski	KP36OI	236	10 dBd	200°	30	OH6DD	
1296.948	FX4UHX	St Aignan	IN94UW	88	2 x Big wheel	Omni	50	F6CIS	
1296.95	DB0HG	DOK F11	JO40HG	300	Big wheel	Omni	3	DL3DC	
1296.95	OZ5UHF	Kobenhavn	JO65GQ	35	Collinear	Omni	1	OZ3TZ	
1296.955	OZ1UHF		JO57FJ	150	Big wheel	Omni	10	OZ9NT	
1296.96	HG7BUB	Dobogoko	JN97KR	700	Slot	Omni	0.5 TX		
1296.96	SK4UHG	Hagfors	JP60VA	440	2 x Helix	Omni	50	SM4DHN	QRT
1296.965	DF0ANN	Lauf DOK B 25	JN59PL	630	4 x DQ	Omni	0.5	DL8ZX	
1296.965	GB3ANG	Dundee	IO86MN	319	Slot Yagi	170°	40	GM4ZUK	
1296.975	DL0SG	DOK U 14	JN69KA	1024	4 x DQ	Omni	5 TX	DJ4YJ	
1296.975	HG1BUB	Hormann	JN87FI	700	Slot	Omni	2.5		
1296.975	OH3RNE	Tampere	KP11UM	247	Alford slot	Omni	35	rep/beac	
1296.975	ON4AZA	Antwerp	JO21EE	60	Clover leaf	Omni	1	ON1BPS	
1296.98	DB0JU	DOK L 04	JO31CV	150	Helical	Omni	2.4 TX	DF5EO	
1296.98	SK2UHG	Kristineberg	JP95HB	500	Horizontal	180°/Omni	500/80		
1296.983	OZ2ALS		JO45UB	28	2 x slot	Omni	8	OZ9DT	
1296.985	DB0AS	DOK C 29	JN67CR	1565	Dipolfeld	10°	0.5 TX	DL2AS	
1296.99	DB0FB	DOK Z 06	JN47AU	1495	8el Group	45°	5 TX	DJ3EN	
1296.99	GB3EDN	Edinburgh	IO85HW	117	2 x Corner Refl	45°/315°	25	GM8BJF	
1296.995	DB0WOS	DOK U 16	JN68ST	850	4 x DQ	Omni	5	DF8RU	
1297.01	DB0JW	DOK G 05	JO30DU	238	4 x 12 el Yagi	45°	70	DL9KAS	
1297.04	DB0LB	DOK P 06	JN48NV	367	Big wheel	Omni	0.3 TX	DK3PS	
2304.04	S55ZNG	Trstelj	JN65UU	643	V-J Slot	Omni	0.1	S50M	
2304.16	I3D		JN55		Slot	Omni	32	IW3FZQ	
2320.8	SK6MHI	Goeteborg	JO57XQ	135	Slotted WG	Omni	10	SM6EAN	
2320.805	SK0UHH	Taeby	JO99BM	90	Horizontal	Omni	25		
2320.81	DB0ZW	DOK U 17	JN69EQ	825	6 x Slot	Omni	1	DC9RK	
2320.815	DB0IH	Nohfelden DOK Q 18	JN39ML	630	Big wheel	Omni	5	DC8DV	
2320.82	DB0OT	Esterwegen DOK I 26	JO32QR	80	Big wheel	Omni	1 TX	DL1BFZ	
2320.825	DB0HF	Harksheide DOK E27	JO53BO	65	Big wheel	Omni	0.3 TX	DK2NH	
2320.825	OE1XTB	Vienna	JN88EE	170	4 x dipole	Omni	1	OE1MOS	QRT
2320.83	DB0JX	Willich DOK R21	JO31FF	115	Double helical	Omni	0.1 TX	DK4TJ	
2320.83	GB3MHS	Martlesham	JO02PB	85	Slotted WG	Omni	25	G4DDK	
2320.833	DB0FGB	DOK B 09	JO50WB	1150	Slot	Omni	12	DB8UY	
2320.838	F5XAC	Pic Neulos	JN12LL	1100	Slotted WG	Omni	20	F6HTJ	
2320.84	DB0KI	Bayreuth DOK Z42	JO50WC	925	Slot	Omni	40	DC9NL	
2320.845	DB0LBV	DOK S 30	JO61EH	234	DQ	135°/225°	1.5 TX	DL1LWM	
2320.845	SR3SHF	Kalisz	JO91CQ					SP3JBI	
2320.85	DB0GW	DOK L 01	JO31JK	80	2 x Helix	Omni	8	DL4JK	
2320.85	DL0UB	DOK Z 20	JO62KK	120	5 x Dipole	Omni	10 TX	DL7ACG	
2320.85	GB3NWK	Orpington	JO01BI	180	Alford Slot	Omni	5	G8BJG	
2320.855	DB0SHF	DOK Z 46	JN48XS	800	6 x Dipole	260°	0.2	DL1SBE	
2320.857	PI7GHG	Capelle	JO21CV	30	10 el Yagi	270°	30	PE1GHG	
2320.86	HG7BUC	Dobogoko	JN97KR	700	Slot	Omni	1 TX		
2320.86	LA1UHH	Tonsberg	JO59FB	30	13 dB Horn	180°	50	LA6LCA	
2320.862	F1XAH		JN23	114	Slotted WG	Omni	15	F1AAM	QRT 99
2320.865	PI7TGA	Nijmegen	JO21WU	75		135°/270°	50	PA0TGA	
2320.865	SK7MHG	Veberöd	JO65SO	200		Omni	50		QRT
2320.87	DB0IBB	DOK N 49	JO32VG	200	10 x Slot	Omni	4	DB7QW	
2320.88	DB0GO	DOK N 32	JO41ED	738	10 x Slot	Omni	50	DB1DI	
2320.88	DB0YI	Hildesheim DOK Z 35	JO42XC	480	Big Wheel	Omni	3 TX	DL4AS	
2320.88	LA3UHH	Flekkeroy	JO38XB	5	2 x 6 dB Horn	90°/180°	1	LA8AK	
2320.883	DB0INN	DOK C 15	JN68GI	504	Slot	Omni	1 TX	DL3MBG	
2320.885	DB0TUD	DOK S07	JO61UA	260	Slot	Omni		DL4DTU	
2320.885	PI7RMD		JO31AE		2 x Quad	180°	10	PE1KXH	
2320.89	GB3ANT	Norwich	JO02PP	75	Alford slot	Omni	5	G8VLL	
2320.895	HG3BUA	Tubes	JN96CC	612	Slot	Omni	1 TX		

Freq (MHz)	Callsign	Nearest town	Locator	ASL (m)	Antenna	Beam direction	ERP (W)	Info from	Notes
2320.9	DB0UX	Grotzingen DOK A 35	JN48FX	275	Big wheel	Omni	1	DK2DB	
2320.9	DB0JW	DOK G 05	JO30DU	238	6 el Array	45°	25	DL9KAS	
2320.902	LX0THF	Walferdange	JN39BP	420	Double quad	Omni	0.5	LX1JX	
2320.912	DL0UH	DOK Z 25	JO41RD	385	6 x Dipole	0°	2	DJ3K0	
2320.915	DB0UBI	DOK N 59	JO42GE	165	Collinear	45°	0.5	DD8QA	
2320.92	DB0VC	Albersdorf DOK Z 10	JO54IF	300	Big wheel	Omni	3	DL8LAO	
2320.92	PI7QHN	Zandvoort	JO22KH	20		Omni	0.2 TX	PA0QHN	
2320.925	GB3PYS	Newtown	IO82HL	436	Alford Slot	Omni	10	GW4NQJ	
2320.93	OK0EL	Benecko	JO70SQ	1035	5dB Horn	135°/270°	0.8	OK1AIY	
2320.93	OZ7IGY	Tollose	JO55VO	91	Alford slot	Omni	20	OZ7IS	
2320.935	PI7PLA	Zuidlaren	JO33IC	50		Omni	0.15 TX	PA0PLA	
2320.937	DB0JO	Kamp-Lintfort DOK Z03	JO31SL	312	Horn	270°	0.2 TX	DG8DCI	
2320.94	DB0DON	DOK T21	JN58KR	532	Slot	Omni	1	DL5MEL	
2320.94	SK7MHH	Farjestaden	JO86GP	45		270°	50		
2320.945	DB0OS	Hitchinbach DOK N 32	JO40CW	730	8 el array	270°	2	DG6YW	
2320.95	DB0KP	DOK P 09	JN47TS	435	Slot	Omni	0.1 TX	DL1GBQ	
2320.95	OZ9UHF		JO65HP	30	Slot	Omni	5	OZ2TG	
2320.955	GB3LES	Leicester	IO92IQ	220	Slot	160°	30	G3TQF	
2320.955	OZ1UHF		JO57FJ	150	Slot	Omni	8	OZ9NT	
2320.963	HG6BUC	Kekes	KN07AU	1050	Slot	Omni	1 TX		
2320.965	DF0ANN	Lauf DOK B 25	JN59PL	630	4 x D Q	Omni	5 TX	DL8ZX	
2320.967	DB0AS	Rosenheim DOK C 14	JN67CR	1560	28 el Yagi	337°	0.5 TX	DL2AS	
2320.975	DB0JL	DOK R 25	JO31MC	195	Slot	Omni	2	DF1EQ	
2320.975	HG1BUC	Hormann	JN87FI	700	Slot	Omni	1 TX		
2320.98	DB0JU	Doesburg DOK L 04	JO31CV	150	Helical	Omni	1 TX	DF5EO	
3400.018	PI7SHF	Schipol Airport	JO22JH	80	10 dB Slot	Omni	2 TX	PA0EZ	
3400.02	DB0AS	DOK C 29	JN67CR	1565	Double 8	10°	0.5 TX	DL2AS	
3400.025	DB0HF	DOK E 27	JO53BO	65		202°		DK2NH	
3400.04	DB0KI	Bayreuth DOK Z 42	JO50WC	925	Slot	Omni	50	DC9NL	
3400.05	DB0EZ	Kleve DOK L-IG	JO31BS	110	Slot	115°	0.I	DB9JC	
3400.05	DB0JL	DOK R 25	JO31MC	195	Helical	Omni	1	DF1EQ	
3400.17	PI7CKK	Groningen	JO33GE	55	10 dB Slot	Omni	5	PE1CKK	
3400.85	DB0GW	Duisburg DOK L 01	JO31JK	80	Double Helical	Omni	8	DL4JK	
3400.955	GB3LEF	Leicester	IO92IQ	222	Alford slot	135°	8	G3TQF	
3456.8	DB0KHT	DOK F 13	JO40FE	247	Horn	Omni	10	DJ1RV	
3456.83	DB0JX	DOK R 21	JO31FF	115	Helical	Omni	0.1 TX	DK4TJ	
3456.85	DL0UB	DOK Z 20	JO62KK	120	12 x Slot	Omni	10 TX	DL7ACG	
3456.85	DB0SHF	DOK Z 46	JN48XS	800	Horn	260°	0.5 TX	DL1SBE	
3456.883	DB0INN	DOK C 15	JN68GI	504	Slot	Omni	1 TX	DL3MBG	
3456.885	DB0TUD	DOK S07	JO61UA	260	Slot	Omni		DL4DTU	
3456.9	GB3OHM	S Birmingham	IO92AJ	171	16 Slot waveguide	Omni	8	G6KOA	
5760.03	OK0EL	Benecko	JO70SQ	1035	5dB Horn	135°/270°	0.08	OK1AIY	
5760.04	PI7EHG	Schipol Airport	JO22JH	80	Horizontal	Omni	2 TX	PA0EHG	
5760.04	OK0EA	Trutnov	JO70UP	1355	12 el Slot	180°/270°	0.5	OK1AIY	
5760.06	F1XAO	Plougonver	IN88HL	326	Slotted WG	Omni	10	F1LHC	
5760.07	DB0JL	DOK R 25	JO31MC	195	Slot	Omni	0.8	DF1EQ	
5760.08	DB0EZ	Kleve DOK L-IG	JO31BS	110	Slot	Omni	1	DB9JC	
5760.1	DB0AS	DOK C 29	JN67CR	1565	Double 8	10°	0.5 TX	DL2AS	
5760.8	DB0KHT	DOK F 13	JO40FE	247	Horn	Omni	0.5 TX	DJ1RV	
5760.8	SK6MHI		JO57XQ	135	Sectoral Horn	270°	5	SM6EAN	
5760.805	DB0RIG	DOK P 17	JN48WQ	780		Omni	15	DG9SQ	
5760.83	DB0JX	DOK R 21	JO31FF	115	Slot	Omni	0.08 TX	DK4TJ	
5760.83	F5XBE	Favières	JN18JS		Slot	Omni	2	F5HRY	
5760.833	DB0FGB	DOK B 09	JO50WB	1150	Slot	Omni	12	DB8UY	
5760.84	DB0KI	Bayreuth DOK Z 42	JO50WC	925	Slot	Omni	20	DC9NL	
5760.845	F1XBB	Orleans	JN07WV		Slot	Omni	2	F1JGP	
5760.85	DL0UB	DOK Z 20	JO62KK	120	12 x Slot	Omni	0.2 TX	DL7ACG	
5760.85	I3E	M.te PIZ (BL)	JN55WV	1400	Slot 10 dB	170°	1	I3EME	
5760.855	DB0SHF	DOK Z 46	JN48XS	800	Array	260°	0.4 TX	DL1SBE	
5760.86	DB0ARB	DOK U 02	JN69NC	1456	Slot	Omni	3	DJ4YJ	
5760.86	LA1SHF	Tonsberg	JO59FB	30	13 dB Horn	180°	25	LA6LCA	
5760.865	OE1XVB	Vienna Simmering	JN88EF	191	Slotted WG	Omni	4	OE1WRS	QRT
5760.883	DB0INN	DOK C 15	JN68GI	504	Slot	Omni	1 TX	DL3MBG	
5760.885	DB0TUD	DOK S07	JO61UA	260	Slot	Omni		DL4DTU	
5760.9	DB0CU	DOK A 28	JN48BI	970	Slot	Omni	5	DJ7FJ	
5760.9	HG6BSB	Kekes	KN07AU	1050	Slot	Omni	0.2 TX		
5760.93	OZ7IGY	Tollose	JO55VO	91	Slotted Waveguide	90°/270°	15	OZ7IS	
5760.95	OZ9UHF		JO65HP	30	Slotted Waveguide	Omni	50	OZ2TG	
5760.955	OZ1UHF		JO57FJ	150	Slotted Waveguide	Omni	8	OZ9NT	
5760.975	HG1BSA	Hormann	JN87FI	700	Slot	Omni	0.2 TX		
10100	GB3IOW	Newport, IOW	IO90IO	250	Slotted waveguide	Omni	1	G8MBU	
10120	GB3ALD	Alderney	IN89VR	90	Sectoral horn	30°	1		
10368.04	DB0EZ	Kleve DOK L-IG	JO31BS	110	Slot	115°	1	DB9JC	
10368.04	F5XBD	Favieres	JN18JS		Slot	Omni	4	F5HRY	
10368.037	PI7SHY	Eindhoven	JO21RK	80	21 dBi	315°	50	PA0SHY	
10368.05	F5XAY	Mont Alembre	JN24BW	1691	Slot / horn	Omni / 0°	03-Oct	F6DPH	
10368.05	LX0DU	Soleuvre	JN29XM	280	1.3m Dish	63°	20 kW		
10368.05	OK0EL	Benecko	JO70SQ	1035	Waveguide	135°/270°	0.05	OK1AIY	
10368.05	OZ9UHF		JO65HP	30	Slotted WG	Omni	3	OZ2TG	
10368.06	F1XAI	Orleans	JN07WT	160	Slotted WG	Omni	10	F1JGP	
10368.075	OK0EA	Trutnov	JO70UP	1355	12 el Slot	180°/270°	0.5	OK1AIY	
10368.09	PA0TGA	Nijmegen	JO21WU	75	16 dB	Omni	4	PA0TGA	

Freq (MHz)	Callsign	Nearest town	Locator	ASL (m)	Antenna	Beam direction	ERP (W)	Info from	Notes
10368.108	F1XAP	Plougonver	IN88HL	326	Slotted WG	Omni	5	F1LHC	
10368.12	DB0JL	DOK R 25	JO31MC	195	Slot	Omni	0.15	DF1EQ	
10368.142	ON4TNR	Namur	JO20KJ	250	17 dB Horn	292°	7	ON5VK	
10368.15	I3F	M.te PIZ (BL)	JN55WV	1400	Slot 10 dB	170°	1.5	I3EME	
10368.15	OE8XXQ	Dobratsch	JN76UO	2166	Horn	0°	1	OE8MI	
10368.175	DB0AS	DOK C 29	JN67CR	1565	Horn	10°	0.5 TX	DL2AS	
10368.205	PI7EHG	Schipol Airport	JO22JH	90	13 dBi Slot	Omni	30	PA0EHG	
10368.24	GB3SWH	Watford	IO91TP	187	Slotted waveguide	45°/225°	1	G4KUJ	
10368.27	DL0WY	Rosenheim DOK C29	JN67AQ	1838	10 dB Slot horn	45°/270°	0.1 TX	DJ8VY	
10368.755	F1XAE	Mt Ventoux	JN24PE	1910	Horn	270°	5	F1AAM	
10368.8	SK6MHI	Goteborg	JO57XQ	135	Slotted WG	Omni	5	SM6EAN	
10368.805	DB0XL	DOK E-IG	JO53HU	45	Slot	Omni	1	DK1KR	
10368.815	DB0MAX	DOK B 41	JN58SP	420				DL4MDQ	
10368.82	DB0KHT	DOK F 13	JO40FE	247	Horn	Omni	3	DJ1RV	
10368.825	DB0HRO	DOK V 09	JO64AD	185	Slot	Omni	0.2 TX	DL5CC	
10368.83	DB0JX	Wickrath DOK R 21	JO31FF	115	10 dB Slot	Omni	0.09 TX	DK4TJ	
10368.83	GB3MHX	Martlesham	JO02PB	80	12 Slot waveguide	Omni	1	G4DDK	
10368.833	DB0FGB	DOK B 09	JO50WB	1150	Slot	Omni	7	DB8UY	
10368.835	SK0SHG	Kista	JO89XJ	60	Horizontal	Omni	0.5	SM0KAK	
10368.84	DB0JO	Kamp-Lintfort DOK Z03	JO31SL	312	6 x Slot	Omni	1	DG8DCI	
10368.84	DB0KI	Bayreuth DOK Z 42	JO50WC	925	Slot	Omni	13	DC9NL	
10368.845	DB0SZB	DOK S 45	JO60JM	767	Slot	Omni	15	DG0YC	
10368.85	DB0GG	DOK P 24	JN48NS	400	Slot	Omni	0.05 TX	DL5AAP	
10368.85	DL0UB	DOK Z 20	JO62KK	120	12 x Slot	Omni	0.1 TX	DL7ACG	
10368.85	GB3SEE	Reigate	IO91VG	250	Slotted waveguide	Omni	3	G0OLX	
10368.855	DB0SHF	DOK Z 46	JN48XS	800	Horn	260°	0.1 TX	DL1SBE	
10368.86	DB0ARB	DOK U 02	JN69NC	1456	Slot	Omni	3	DJ4YJ	
10368.86	F1BDB	Nice	JN33OQ		Slot	Omni	1	F1BDB	
10368.86	F5XAD	Pic Neulos	JN12LL	1100	Slotted WG	0°	3	F6HTJ	
10368.86	LA1SHG	Tonsberg	JO59FB	30	13 dB Horn	180°	10	LA6LCA	
10368.865	DB0JK	Koln DOK Z 12	JO30LX	260	Slot	Omni	200	DK2KA	
10368.87	DB0IBB	DOK N 49	JO32VG	245	Slot	Omni	2	DB7QW	
10368.87	GB3KBQ	Taunton	IO80LW	167	Slotted waveguide	Omni	1	G4UVZ	
10368.87	HG3BSB	Tubes	JN97CC	612	Slot	Omni	0.2		
10368.87	OE8XGQ	Gerlitze	JN66WQ	1909	Slotted WG	Omni	1.5	OE8MI	
10368.875	OE5XBM	Breitenstein	JN78DJ	985	Slotted WG	Omni	10	OE5VRL	
10368.88	GB3CEM	Wolverhampton	IO82WO	165	Slotted waveguide	Omni	30	G4PBP	
10368.88	OE1XVB	Vienna, Simmering	JN88EF	185	Slotted WG	Omni	1.5	OE1WRS	
10368.883	DB0INN	DOK C 15	JN68GI	504	Slot	Omni	1 TX	DL3MBG	
10368.884	HB9G	Geneva	JN36BK	1600	Slotted waveguide	Omni	2	HB9PBD	
10368.885	DB0TUD	DOK S 07	JO61UA	285	Slot	Omni	5	DL4DTU	
10368.89	DB0KLX	DOK K 16	JN39VK	350	Slot	Omni	1 TX	DC2UG	
10368.895	DB0ECA	DOK C 08	JN57UV	705	Slot	Omni	10	DC8EC	
10368.9	DB0UX	DOK A 35	JN48FX	275	Slot	Omni	1	DK2DB	
10368.9	DB0CU	DOK A 28	JN48BI	970	Slot	Omni	5	DJ7FJ	
10368.9	GB3SCX	Swanage	IO90AP	200	Slotted waveguide	Omni	1	G4JNT	
10368.9	OZ5SHF		JO45WX	170	Slotted WG	Omni	4	OZ2OE	*10368.895
10368.91	DB0HEX	DOK Z 85	JO51HT	1341	Slot	Omni	8	DG0CBP	
10368.91	GB3RPE	Swansea	IO81AO	60	Slotted Waveguide	Omni	4	GW4ADL	
10368.915	OZ4SHF		JO65BV	22	Slotted WG	Omni	10	OZ1UM	*10368.907
10368.92	DB0VC	DOK Z 10	JO54IF	291	Slot	Omni	1	DL8LAO	
10368.92	OE2XBO	Haunsberg	JN67MW	740	Slotted WG	Beam	1.5	OE2HFO	
10368.925	F1XAU	Sombernon	JN27IH	516	Slot WG	Omni	1.5	F1MPE	
10368.925	OE3XMB	Muckenkogel	JN77TX	1154	Slotted WG	Omni	1.5	OE3FFC	QRT
10368.93	DB0HO	DOK Z 49	JN47QT	487	Slot	Omni	10	DF6TK	
10368.93	GB3MLE	Emley Moor	IO93EO	600	Sectoral horns	0°/180°	1	G8AGN	
10368.93	OZ7IGY	Tollose	JO55VO	92	WG Slot	Omni	0.4 TX	OZ7IS	
10368.94	DB0DON	DOK T21	JN58KR	532	Slot	Omni	1	DL5MEL	
10368.94	GB3CCX	Cheltenham	IO81XW	342	SlottedWG	Omni	3	G6AWT	
10368.945	HG7BSA	Dobogoko	JN97KR	700	Slot	Omni	0.2 TX		
10368.945	OE2XBN..	Sonnblick	JN67LA	3105	Slotted WG	Omni	12.5	OE1MCU	
10368.95	DB0FHR	DOK C 31	JN67BU	474	Slot	Omni	1	DL5MEA	
10368.95	ON4RUG	Ghent	JO11UB	95	Slotted	Omni	7	ON6UG	
10368.955	OZ1UHF		JO57FJ	150	Slotted WG	Omni	0.8	OZ9NT	
10368.955	GB3LEX	Leicester	IO92JP	220	Slotted WG	Omni	1	G3TQF	
10368.96	GB3CMS	Chelmsford	JO01GR	107	Slotted waveguide	Omni	3	G4GUJ	Temp QRT
10368.965	DF0ANN	DOK B 25	JN59PL	630	12 x Slot	Omni	0.2 TX	DL8ZX	
10368.975	HG1BSB	Hormann	JN87FI	700	Slot	Omni	0.2 TX		
10368.975	ON4KUL	Leuven	JO20IV	100	Slotted	Omni	5	ON7VQ	*10368.860
10368.975	OZ3SHF		JO45NL	58	Slotted WG	Omni	2	OZ1IN	
10368.977	HG6BSB	Kekes	KN07AU	1050	Slot	Omni	0.2		
10369	F1XAN	Bus St Remy	JN09TD	300	Slotted WG	Omni	1.5	F1PBZ	
24025	GB3IOW	Newport, IOW	IO90IO	250	Sectoral horn		8	G8IDZ	
24192	ON4RUG	Gent	JO11UB	95	Slotted	Omni	0.1	ON6UG	
24192.05	DB0KHT	DOK F 13	JO40FE		Horn	Omni	0.02 TX	DJ1RV	
24192.05	I3G	M.te PIZ (BL)	JN55WV	1400	Slot 8 dB	170°	0.25	I3EME	
24192.055	DB0JO	DOK Z 03	JO31SL	312	6 x Slot	Omni	0.6	DG8DCI	
24192.075	PI7EHG	Schipol Airport	JO22JH	90	30 cm Dish	266°	0.1 TX	PA0EHG	
24192.114	OK0EL	Benecko	JO70SQ	1035	Waveguide	135°/270°	20µW	OK1AIY	
24192.12	DB0JL	DOK R 25	JO31MC	195	Slot	Omni	0.01	DF1EQ	
24192.2	LX0DUF	Soleuvre	JN29XM	280	0.4m Dish	63°	1.2 kW		
24192.252	F1XAQ	Plougonver	IN88HL	326	Slotted WG	Omni	0.1	F1LHC	

Freq (MHz)	Callsign	Nearest town	Locator	ASL (m)	Antenna	Beam direction	ERP (W)	Info from	Notes
24192.405	DB0AS	DOK C 29	JN67CR	1565	Horn	10°	0.5 TX	DL2AS	
24192.8	SK6MHI	Goteborg	JO57XQ	135	2 x Sectoral Horn	225°/315°	1	SM6EAN	
24192.83	F5XAF	Paris	JN18DU		Parabola	90°	0.1	F5ORF	
24192.833	DB0FGB	DOK B 09	JO50WB	1150	Slot	Omni	0.6	DB8UY	
24192.84	DB0KI	Bayreuth DOK Z 42	JO50WC	925	Slot	0°	0.5	DC9NL	
24192.853	DL0WY	DOK C 29	JN67AQ	1838	Sectored horn	45°/270°	0.01	DJ8VY	
24192.86	DB0ARB	DOK U 02	JN69NC	1456	Parabola	225°		DJ4YJ	
24192.865	DB0JK	DOK Z 12	JO30LX	260	2 x H-Horn	Omni	1	DK2KA	
24192.875	DB0HW	DOK H46	JO51GT	1016	Slot	Omni	1	DL3AAS	
24192.875	OE5XBM	Breitenstein	JN78DJ	985	Slotted WG	Omni	0.5	OE5VRL	
24192.875	ON4AZC	Antwerp	JO21EE	60	Slotted	Omni		ON1BPS	
24192.885	DB0TUD	DOK S 07	JO61UA	260	Slot	Omni		DL4DTU	
24192.89	GB3DUN	Dunstable	IO91RV	260	Slotted WG	Omni	1	G3ZFP	
24192.895	DB0ECA	DOK C 08	JN57UV	705	Slot	0°		DC8EC	QRT
24192.9	DB0CU	DOK A 28	JN48BI	970	Horn	180°	5	DJ7FJ	
24192.91	DB0HEX	DOK Z 85	JO51HT	1341	Slot	Omni	8	DG0CBP	
24192.915	OZ4SHF		JO65BV	22	Slotted WG	Omni	10	OZ1UM	
24192.94	GB3AMU	Cardiff	IO81JN	266	Sectorial Horn	135°	1	GW3PPF	
24192.955	OZ1UHF		JO57FJ	150	Slotted WG	Omni	0.5	OZ9NT	
24192.975	ON4LVN	Leuven	JO20IV	120	Slotted Waveguide	Omni	0.5	ON4AOD	
47088.1	DB0AS	DOK C 29	JN67CR	1565	Horn	10°	0.5 TX	DL2AS	
47088.24	I3H	M.te PIZZOC (TV)	JN66EB	1570	Horn 25 dB	180°	1.2	I3OIB	
47088.833	DB0FGB	DOK B09	JO50WB	1150	Slot	Omni	0.2	DB8UY	
47088.853	DL0WY	DOK C29	JN67AQ	1830	Horn	0°/90°/270°	0.1	DJ8VY	
47088.865	DB0JK	DOK Z 12	JO30LX	260	2 x H-Horn	Omni	0.1 TX	DK2KA	
47088.875	OE5XBM	Hellmonsoedt	JN78DK	855	Slotted WG	Omni	0.25	OE5VRL	
47088.895	DB0ECA	DOK C 08	JN57UV	705	Horn	0°		DC8EC	
76032.833	DB0FGB	DOK B 09	JO50WB	1150	Slot	Omni	0.01	DB8UY	QRT
76032.895	DB0ECA	DOK C 08	JN57UV	705	Horn	0°		DC8EC	QRT

*   May break for QSOs

The data comes from the IARU Region 1 list, compiled by John Wilson, G3UUT of the RSGB VHF Committee. Thanks go to VHF/UHF/Microwave managers across Region 1, beacon keepers, beacon coordinators and VHF/UHF DXers too numerous to mention. The main Region 2 and 3 50MHz beacons are included for completeness; thanks to G4ICD, G3USF and the UK Six Metre Group for those. All inputs are welcome and should be sent to: John Wilson, G3UUT, QTHR. E-mail wilson@shelford.prestel.co.uk. The list is copyright and is reproduced with acknowledgement to IARU Region 1. A fully up to date version of the VHF/UHF list is available on the Internet at www.scit.wlv.ac.uk/vhfc/
.

9 Repeaters

THIS appendix gives a list of UK repeaters at the time of going to press (early 2000), including ATV and microwave repeaters, courtesy of the RSGB Repeater Management Committee website (members.aol.com/rmcweb/rep_list.htm).

For more information on the CTCSS system, please see Chapter 6.

More details on any repeater may be obtained by contacting the keeper listed in the table or by visiting the RMC web site.

Callsign	Channel	CTCSS	Location	Keeper
GB3AE	RF72 (R50-1)	F	Tenby	GW0WBQ
GB3EF	RF72 (R50-1)	H	Ipswich	G0VDE
GB3UM	RF74 (R50-3)	C	Leicester	M0BKH
GB3HF	RF76 (R50-5)		Hastings	G1DVU
GB3UK	RF77 (R50-6)	D	Winter Hill	G8NSS
GB3PX	RF78 (R50-7)	C	Hertfordshire	G4NBS
GB3SX	RF79 (R50-8)	G	Stoke on Trent	G8DZJ
GB3HX	RF80 (R50-9)	D	Huddersfield	G0PRF
GB3FX	RF81 (R50-10)	D	Surrey	G4EPX
GB3RR	RF82 (R50-11)	B	Nottingham	G4TSN
GB3WX	RF83 (R50-12)	C	Warminster	C
G3ZXX				
GB3AM	RF84 (R50-13)	C	Amersham	G0RDI
GB3PD	RF85 (R50-14)	B	Portsmouth	G4JXL
GB3BY	RF86 (R50-15)	A	Bewdley	G8EPR
GB3AS	RV48 (R0)	C		
GB3CF	RV48 (R0)	C	Leicester	G0ORY
GB3EL	RV48 (R0)	D	East London	G4RZZ
GB3LY	RV48 (R0)	H	Limavady	GI3USS
GB3MB	RV48 (R0)	D	Bury	G8NSS
GB3SR	RV48 (R0)	E	Brighton Centre	G8VEH
GB3SS	RV48 (R0)	A	Moray	GM7LSI
GB3WR	RV48 (R0)	F	Wells	G0MBX
GB3YC	RV48 (R0)		Driffield	G0OOI
GB3GD	RV50 (R1)	H	Snaefell	GD3LSF
GB3HG	RV50 (R1)	J	Northallerton	G0RHI
GB3KS	RV50 (R1)	G	Dover	G4HHX
GB3NB	RV50 (R1)	F	Norwich	G8VLL
GB3NG	RV50 (R1)	A	Fraserburgh	GM8LYS
GB3NW	RV50 (R1)	A	Worcester	G4IDF
GB3PA	RV50 (R1)		Renfrewshire	GM7OAW
GB3SC	RV50 (R1)	B	Central Bournemouth	G0API
GB3SI	RV50 (R1)	C	St Ives	G3NPB
GB3WL	RV50 (R1)	D	West London	G8SUG
GB3AY	RV52 (R2)	G	Ayrshire	GM3YKE
GB3BF	RV52 (R2)	C	Bedford	G1BWW
GB3EC	RV52 (R2)	A	Birmingham	G4YKE
GB3GJ	RV52 (R2)	C	St Hellier	GJ0NSG
GB3HS	RV52 (R2)	E	Hull	G7JZD
GB3MN	RV52 (R2)	D	Stockport	G8LZO
GB3OC	RV52 (R2)	C	Kirkwall	GM0HQG
GB3PO	RV52 (R2)	H	Ipswich	G8CPH
GB3SB	RV52 (R2)	J	Selkirk	GM0FTJ
GB3SL	RV52 (R2)	D	Crystal Palace, London	G4PEB
GB3TR	RV52 (R2)	F	Torquay	G8XST
GB3WH	RV52 (R2)	J	Swindon	G4LDL
GB3DW	RV53	J	Criccieth, N Wales	GW4KAZ
GB3BX	RV54 (R3)	A	Wolverhampton	G4JLI
GB3ES	RV54 (R3)	G	Hastings	G7LEL
GB3LD	RV54 (R3)	H	Dalton, Cumbria	G7MCE
GB3LG	RV54 (R3)	G	Lochgilphead	GM4WMM
GB3LU	RV54 (R3)	C	Lerwick	GM4SWU
GB3NA	RV54 (R3)	B	Barnsley	G4LUE

Callsign	Channel	CTCSS	Location	Keeper
GB3PE	RV54 (R3)	F	Peterborough	G1ARV
GB3PR	RV54 (R3)	F	Perth	GM8KPH
GB3RD	RV54 (R3)	J	Reading	G8DOR
GB3SA	RV54 (R3)	F	Swansea	GW6KQC
GB3WZ	RV54 (R3)	G	Wrexham, Clwyd	
GB3SQ	RV55	B	Salisbury	G3YWT
GB3AR	RV56 (R4)	H	Caernarfon	GW4KAZ
GB3BB	RV56 (R4)	G	Brecon	GW0ABT
GB3BT	RV56 (R4)	J	Berwick on Tweed	GM1JFF
GB3EV	RV56 (R4)	C	Cumbria	G0IYQ
GB3HH	RV56 (R4)	B	Buxton	G4IHO
GB3HI	RV56 (R4)	E	Oban	GM3RFA
GB3KN	RV56 (R4)	G	Maidstone	G3YCN
GB3VA	RV56 (R4)	J	Aylesbury	G6NB
GB3WD	RV56 (R4)	C	Plymouth	G6URM
GB3KY	RV57	F	Norfolk	G1HYU
GB3AG	RV58 (R5)	F	Forfar	GM1CMF
GB3BI	RV58 (R5)	A	Inverness	GM0JFK
GB3CG	RV58 (R5)		Gloucester	G6AWT
GB3DA	RV58 (R5)	H	Chelmsford	G4GUJ
GB3LM	RV58 (R5)	B	Lincoln	G8VGF
GB3NC	RV58 (R5)	C	St Austell	G3IGV
GB3NI	RV58 (R5)	H	Belfast	GI3USS
GB3SN	RV58 (R5)	B	Alton	G4EPX
GB3TP	RV58 (R5)	D	Keighley, Yorks	G7HEN
GB3TW	RV58 (R5)	J	Durham	G4GBF
GB3VT	RV58 (R5)	G	Stoke on Trent	G8DZJ
GB3ZA	RV59 (R5X)	J	Hereford	G0JWJ
GB3BC	RV60 (R6)	F	Newport, Gwent	GW8ERA
GB3CS	RV60 (R6)	G	Salsburgh	GM4COX
GB3MP	RV60 (R6)	H	Denbigh	G7OBW
GB3MX	RV60 (R6)	B	Mansfield	G0UYQ
GB3PI	RV60 (R6)	C	Royston	G4NBS
GB3WS	RV60 (R6)	E	Crawley	G4EFO
GB3NE	RV61 (R6X)	J	Berkshire	G8JIP
GB3DG	RV62 (R7)	G	Gatehouse of Fleet	GM4VIR
GB3FR	RV62 (R7)	B	Spilsby	G8LXI
GB3GN	RV62 (R7)	A	Aberdeen	GM8LYS
GB3IG	RV62 (R7)	E	Stornoway	GM4PTQ
GB3NL	RV62 (R7)	D	Enfield	G3TZZ
GB3PC	RV62 (R7)	B	Portsmouth	G4NAO
GB3PW	RV62 (R7)	G	Newtown, Powys	GW4NQJ
GB3RF	RV62 (R7)	D	Burnley	G4FSD
GB3TE	RV62 (R7)	G	Clacton-on-Sea	G7HJK
GB3WK	RV62 (R7)	A	Leamington	G6FEO
GB3WT	RV62 (R7)	H	Omagh	GI3NVW
GB3WW	RV62 (R7)	F	Cross Hands, Dyfed	GW6ZUS
GB3SF	RV63 (R7X)	D	Buxton	G4IHO
GB3BN	RU240 (RB0)	D	Bracknell	G4DDN
GB3CK	RU240 (RB0)	G	Ashford	G0GCQ

Callsign	Channel	CTCSS	Location	Keeper
GB3DT	RU240 (RB0)	B	Blandford Forum	G8BXQ
GB3EX	RU240 (RB0)	F	Exeter	G8UWE
GB3LL	RU240 (RB0)	H	Llandudno	GW8WFS
GB3MK	RU240 (RB0)	C	Milton Keynes	G4NJU
GB3NR	RU240 (RB0)	F	Norwich	G8VLL
GB3NT	RU240 (RB0)	J	Newcastle Upon Tyne	G4GBF
GB3NY	RU240 (RB0)	E	Scarborough	G4EEV
GB3PF	RU240 (RB0)	D	Blackburn	G4FSD
GB3PU	RU240 (RB0)	F	Perth	GM8KPH
GB3SO	RU240 (RB0)	B	Boston	G8LXI
GB3SV	RU240 (RB0)	H	Bishops Stortford	G1NOL
GB3US	RU240 (RB0)	G	Sheffield	G3RKL
GB3WN	RU240 (RB0)	A	Wolverhampton	G4OKE
GB3BA	RU242 (RB1)	A	Stonehaven	GM4NHI
GB3BV	RU242 (RB1)	D	Hemel Hempstead	G6NB
GB3DV	RU242 (RB1)	B	Doncaster	G4LUE
GB3EM	RU242 (RB1)		Waltham	G8WWJ
GB3HJ	RU242 (RB1)	J	Harrogate	G3XWH
GB3HO	RU242 (RB1)	E	Horsham	G7JRV
GB3MA	RU242 (RB1)	D	Bury	G8NSS
GB3WA	RU242 (RB1)		Warminster	G3ZXX
GB3AV	RU244 (RB2)	D	Aylesbury	G6NB
GB3CH	RU244 (RB2)	C	Liskeard	G1NSV
GB3CI	RU244 (RB2)	B	Corby	G8MLA
GB3EK	RU244 (RB2)	G	Margate	G4TKR
GB3FC	RU244 (RB2)	D	Blackpool	G6AOS
GB3HK	RU244 (RB2)	J	Selkirk	GM0FTJ
GB3LS	RU244 (RB2)	B	Lincoln	G8VGF
GB3LV	RU244 (RB2)	D	Enfield	G3KSW
GB3NN	RU244 (RB2)	F	Wells Next The Sea	G0FVF
GB3NX	RU244 (RB2)	E	Crawley	G0DSU
GB3OS	RU244 (RB2)	A	Stourbridge	G1PKZ
GB3PH	RU244 (RB2)	B	Portsmouth	G8PGF
GB3ST	RU244 (RB2)	G	Stoke on Trent	G8DZJ
GB3UL	RU244 (RB2)	H	Belfast	GI3USS
GB3YS	RU244 (RB2)	F	Yeovil	G0LHX
GB3CC	RU246 (RB3)	E	Chichester	G3UEQ
GB3ER	RU246 (RB3)	H	Chelmsford	G4GUJ
GB3HL	RU246 (RB3)	D	West London	G8SUG
GB3HU	RU246 (RB3)	E	Hull	G3TEU
GB3KA	RU246 (RB3)		Kilmarnock	GM3YKE
GB3KR	RU246 (RB3)		Kidderminster	G8TNU
GB3MD	RU246 (RB3)	B	Mansfield	G0UYQ
GB3NH	RU246 (RB3)	C	Northampton	G4IIO
GB3TD	RU246 (RB3)	J	Swindon	G4XUT
GB3VS	RU246 (RB3)	F	Taunton	G4UVZ
GB3GC	RU248 (RB4)	E	Goole	G0GLZ
GB3IH	RU248 (RB4)	H	Ipswich	G8CPH
GB3IW	RU248 (RB4)	B	Newport, IOW	G1VGM
GB3KL	RU248 (RB4)	F	Kings Lynn	G3ZCA
GB3LE	RU248 (RB4)	C	Leicester	G0ORY
GB3NK	RU248 (RB4)	G	Wrotham	G8JNZ
GB3OH	RU248 (RB4)	F	Bo'ness	GM6WQH
GB3SP	RU248 (RB4)	F	Pembroke	GW4VRO
GB3UB	RU248 (RB4)	J	Bath	G0LIB
GB3VE	RU248 (RB4)		Great Dunfell	G0IYQ
GB3EB	RU250 (RB5)	H	Brentwood	G6IFH
GB3GH	RU250 (RB5)	J	Cheltenham	G3LVP
GB3HY	RU250 (RB5)	E	Haywards Heath	G3XTH
GB3IM	RU250 (RB5)	H	Douglas	GD3LSF
GB3OV	RU250 (RB5)	F	Huntingdon	G8LRS
GB3WB	RU250 (RB5)	F	Weston Super Mare	G4SZM
GB3WJ	RU250 (RB5)	B	Scunthorpe	G3TMD
GB3BD	RU252 (RB6)	C	Ampthill, Beds	G1BWW
GB3BR	RU252 (RB6)	E	Brighton	G8VEH
GB3CR	RU252 (RB6)	H	Wrexham	G8UEK
GB3CW	RU252 (RB6)	G	Powys	GW4NQJ
GB3DI	RU252 (RB6)		Didcot	G8CUL
GB3FG	RU252 (RB6)			GM4TNP
GB3HA	RU252 (RB6)	E	Hornsea	G4YTV
GB3HC	RU252 (RB6)	J	Hereford	G4JSN
GB3LW	RU252 (RB6)	D	London	G7OMK
GB3ME	RU252 (RB6)	A	Rugby	G8DLX
GB3SK	RU252 (RB6)	G	Canterbury	G6DIK
GB3SY	RU252 (RB6)	B	Barnsley	G4LUE
GB3WG	RU252 (RB6)		Swansea	GW3VPL

Callsign	Channel	CTCSS	Location	Keeper
GB3BL	RU254 (RB7)	C	Bedford	G1BWW
GB3DE	RU254 (RB7)	H	Ipswich	G0YAP
GB3HZ	RU254 (RB7)		Amersham	G0RDI
GB3MF	RU254 (RB7)	G	Macclesfield	G0AMU
GB3MG	RU254 (RB7)	F	Bridgend	GW7NIS
GB3MS	RU254 (RB7)		Worcester	G4IDF
GB3NM	RU254 (RB7)	B	Nottingham	G2SP
GB3TS	RU254 (RB7)	J	Middlesborough	G8MBK
GB3WY	RU254 (RB7)	D	Halifax	G8NWK
GB3AN	RU256 (RB8)	H	Amlwch	GW6DOK
GB3CM	RU256 (RB8)	F	Carmarthen	GW0IVG
GB3EA	RU256 (RB8)	B	Eastleigh	G4MYS
GB3EH	RU256 (RB8)	A	Banbury	G4OHB
GB3LA	RU256 (RB8)	D	Leeds	G8ZXA
GB3PY	RU256 (RB8)	C	Cambridge	G4NBS
GB3TF	RU256 (RB8)	G	Telford	G3UKV
GB3BE	RU258 (RB9)	H	Bury St Edmunds	G8KMM
GB3CL	RU258 (RB9)	G	Clacton	G7HJK
GB3CV	RU258 (RB9)	A	Coventry	G3ZFR
GB3HD	RU258 (RB9)	D	Huddersfield	G1FYS
GB3SW	RU258 (RB9)	B	Salisbury	G4SXQ
GB3UO	RU258 (RB9)		Oswestry	G4UDE
GB3AW	RU260 (RB10)	B	Newbury	G8DOR
GB3BS	RU260 (RB10)	J	Bristol	G4SDR
GB3DD	RU260 (RB10)	F	Dundee	GM4UGF
GB3DY	RU260 (RB10)	B	Derby	G3ZYC
GB3LI	RU260 (RB10)	D	Liverpool	G3WIC
GB3LT	RU260 (RB10)	C	Luton	G6OUA
GB3ML	RU260 (RB10)	G	Airdrie	GM3SAN
GB3MW	RU260 (RB10)	A	Leamington Spa	G6FEO
GB3NS	RU260 (RB10)	D	Reigate	G0OLX
GB3PB	RU260 (RB10)	C	Peterborough	G1ARV
GB3WO	RU260 (RB10)	J	Witney	G4GUN
GB3AH	RU262 (RB11)		E Dereham	G8PON
GB3BK	RU262 (RB11)	J	Reading	G8DOR
GB3DC	RU262 (RB11)	J	Sunderland	G6LMR
GB3GR	RU262 (RB11)	B	Grantham	G4WFK
GB3GY	RU262 (RB11)	E	Grimsby	G1BRB
GB3HN	RU262 (RB11)	D		G4LOO
GB3HT	RU262 (RB11)	C	Hinckley	G4ALB
GB3LR	RU262 (RB11)	E	Newhaven	G7PUV
GB3RE	RU262 (RB11)		Maidstone	G4AKQ
GB3RH	RU262 (RB11)	F	Axminster, Devon	G6WWY
GB3SH	RU262 (RB11)	F	Honiton	G4MYS
GB3WP	RU262 (RB11)		Hyde	G6YRK
GB3ZI	RU262 (RB11)	G	Stafford	G1UDS
GB3EE	RU264 (RB12)	B	Chesterfield	G6SVZ
GB3FE	RU264 (RB12)	F	Dunfermline	GM4TNP
GB3GB	RU264 (RB12)	A	Great Barr	G8NDT
GB3GF	RU264 (RB12)	E	Guildford	G4EML
GB3HM	RU264 (RB12)	J	Boroughbridge	G0RHI
GB3MT	RU264 (RB12)	D	Bolton	G8NSS
GB3OX	RU264 (RB12)	J	Oxford	G4WXC
GB3PT	RU264 (RB12)	H	Royston, Herts	G4NBS
GB3CA	RU266 (RB13)	C	Carlisle	G0JGS
GB3CY	RU266 (RB13)	C	York	G4FUO
GB3DS	RU266 (RB13)	B	Worksop	G3XXN
GB3GU	RU266 (RB13)	C	St Peter Port, CI	GU4EON
GB3HW	RU266 (RB13)	H	Romford	G4GBW
GB3LC	RU266 (RB13)	B	Louth, Lincs	G4IPE
GB3SM	RU266 (RB13)	G	Leek	G8DZJ
GB3VH	RU266 (RB13)	D	Welwyn Garden City	G4THF
GB3XX	RU266 (RB13)	C	Daventry	G1ZJK
GB3CB	RU268 (RB14)	A	Birmingham	G8AMD
GB3CE	RU268 (RB14)	G	Colchester	G7BKU
GB3ED	RU268 (RB14)	F	Edinburgh	GM4GZW
GB3GL	RU268 (RB14)	G	Glasgow	GM3SAN
GB3HE	RU268 (RB14)	G	Hastings	G4FET
GB3HR	RU268 (RB14)	D	Harrow	G4KUJ
GB3LF	RU268 (RB14)	H	Lancaster	G8UHO
GB3MR	RU268 (RB14)	D	Stockport	G8LZO
GB3ND	RU268 (RB14)	F	Bideford	G4JKN
GB3SD	RU268 (RB14)	B	Weymouth	G1IKP
GB3TL	RU268 (RB14)	B	Spalding	G7JBA
GB3WF	RU268 (RB14)	D	Leeds	G8ZXA

Callsign	Channel	CTCSS	Location	Keeper
GB3YL	RU268 (RB14)	F	Lowestoft	G4TAD
GB3FN	RU270 (RB15)	D	Farnham	G4EPX
GB3HB	RU270 (RB15)	C	St Austell	G3IGV
GB3LH	RU270 (RB15)	G	Shrewsbury	G3UQH
GB3OM	RU270 (RB15)	H	Omagh	GI4SXV
GB3PP	RU270 (RB15)	D	Preston	G3SYA
GB3SG	RU270 (RB15)	F	Cardiff	GW7KWG
GB3SU	RU270 (RB15)	H	Sudbury	G8AAR
GB3SZ	RU270 (RB15)	B	Bournemouth	G0API
GB3TH	RU270 (RB15)	A	Tamworth	G4JBX
GB3WI	RU270 (RB15)	F	Wisbech	G4NPH
GB3WU	RU270 (RB15)	D	Wakefield	G0COA
GB3BH	RM0	D	Watford	G7LXP
GB3MC	RM0	D	Bolton	G8NSS
GB3NO	RM0	F	Norwich	G8VLL
GB3FM	RM2	100Hz	Farnham	G4EPX
GB3CP	RM3		Crawley, Sussex	
GB3PS	RM3	C	Royston	G4NBS
GB3SE	RM3	G	Stoke on Trent	G8DZJ
GB3CN	RM5	C	Northampton	G6NYH
GB3MM	RM6	A	Wolverhampton	G4OKE
GB3CO	RM8		Corby	G8MLA
GB3ZP	RM9		Essex	G4GUJ
GB3UY	RM13	J	York	G7AUP
GB3WC	RM15	D	Wakefield	G0COA
GB3EY	1308MHz		Hull	G8EQZ
GB3HV	1308MHz		High Wycombe	G8LES

Callsign	Channel	CTCSS	Location	Keeper
GB3GW	1310MHz		Criccieth	GW4KAZ
GB3KT	1310MHz		Kent	G8SUY
GB3VX	1310MHz		Eastbourne	G1IFV
GB3UT	1311.5MHz		Bath	G0LIB
GB3EN	1312MHz		Enfield	G4DVG
GB3AD	1316MHz		Stevenage	G0OVO
GB3AT	1316MHz		Winchester	G6HNJ
GB3DH	1316MHz		Derby	G8DKV
GB3GV	1316MHz	C	Leicestershire	G8OBP
GB3LO	1316MHz		Lowestoft	G4TAD
GB3PV	1316MHz		Cambridge	G4NBS
GB3RT	1316MHz		Leamington Spa	G1GPE
GB3TM	1316MHz		Amlwch	GW8PBX
GB3TN	1316MHz		Fakenham	G4WVU
GB3TT	1316MHz		Chesterfield	G1IOR
GB3VL	1316MHz		Lincoln	G7AVU
GB3VR	1316MHz		Brighton	G8KOE
GB3WV	1316MHz		Plymouth	G6URM
GB3YT	1316MHz		Bradford	G3TQA
GB3ZZ	1316MHz		Bristol	G6TVJ
GB3TV	1318.5MHz		Dunstable	G4ENB
GB3UD	1318.5MHz		Stoke on Trent	G0KBI
GB3XT	10GHz		Burton on Trent	G8OZP
GB3XG	10GHz		Bristol	G6TVJ
GB3XY	10GHz		Hull	G3RMX
GB3BG	10GHz		West Midlands	G6WJJ
GB3DJ	10GHz		Telford	G8VZT
GB3RV	10GHz		Brighton	G8KOE
GB3TG	10GHz		Bletchley	G3LMX

CTCSS REPEATER AREAS

Tone Area	CTCSS Tone (Hz)
A	67.1
B	71.9
C	77.0
D	82.5
E	88.5
F	94.8
G	103.5
H	110.9
J	118.8

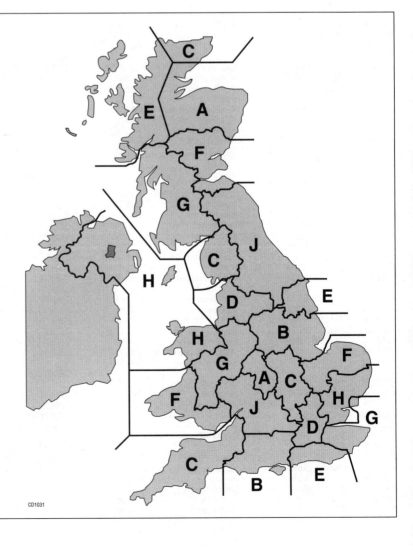

CD1031

144MHz REPEATERS

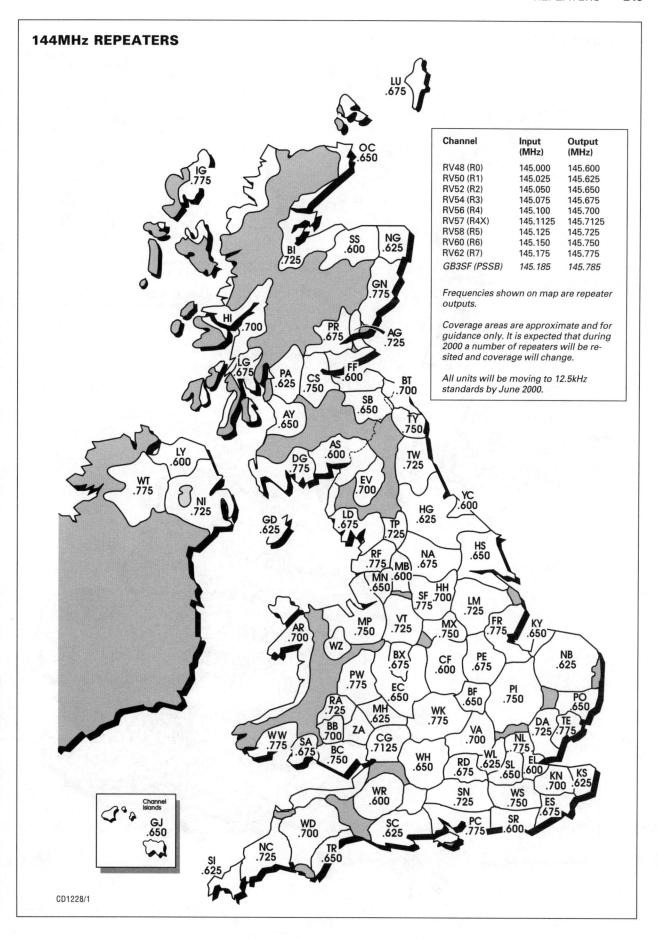

Channel	Input (MHz)	Output (MHz)
RV48 (R0)	145.000	145.600
RV50 (R1)	145.025	145.625
RV52 (R2)	145.050	145.650
RV54 (R3)	145.075	145.675
RV56 (R4)	145.100	145.700
RV57 (R4X)	145.1125	145.7125
RV58 (R5)	145.125	145.725
RV60 (R6)	145.150	145.750
RV62 (R7)	145.175	145.775
GB3SF (PSSB)	145.185	145.785

Frequencies shown on map are repeater outputs.

Coverage areas are approximate and for guidance only. It is expected that during 2000 a number of repeaters will be re-sited and coverage will change.

All units will be moving to 12.5kHz standards by June 2000.

Channel Islands

GJ .650

CD1228/1

432MHz REPEATERS

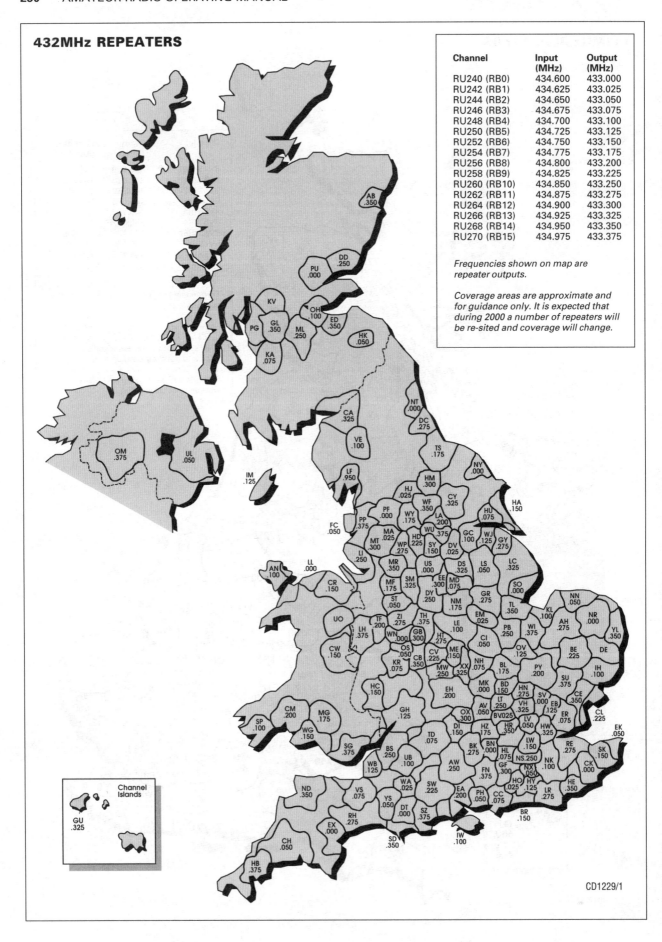

Channel	Input (MHz)	Output (MHz)
RU240 (RB0)	434.600	433.000
RU242 (RB1)	434.625	433.025
RU244 (RB2)	434.650	433.050
RU246 (RB3)	434.675	433.075
RU248 (RB4)	434.700	433.100
RU250 (RB5)	434.725	433.125
RU252 (RB6)	434.750	433.150
RU254 (RB7)	434.775	433.175
RU256 (RB8)	434.800	433.200
RU258 (RB9)	434.825	433.225
RU260 (RB10)	434.850	433.250
RU262 (RB11)	434.875	433.275
RU264 (RB12)	434.900	433.300
RU266 (RB13)	434.925	433.325
RU268 (RB14)	434.950	433.350
RU270 (RB15)	434.975	433.375

Frequencies shown on map are repeater outputs.

Coverage areas are approximate and for guidance only. It is expected that during 2000 a number of repeaters will be re-sited and coverage will change.

CD1229/1

50MHz REPEATERS

Frequencies shown on map are repeater outputs.

Channel	Input (MHz)	Output (MHz)
RF72 (R50-1)	51.220	50.720
RF74 (R50-3)	51.240	50.740
RF76 (R50-5)	51.260	50.760
RF77 (R50-6)	51.270	50.770
RF78 (R50-7)	51.280	50.780
RF79 (R50-8)	51.290	50.790
RF80 (R50-9)	51.300	50.800
RF81 (R50-10)	51.310	50.810
RF82 (R50-11)	51.320	50.820
RF83 (R50-12)	51.330	50.830
RF84 (R50-13)	51.340	50.840
RF85 (R50-14)	51.350	50.850

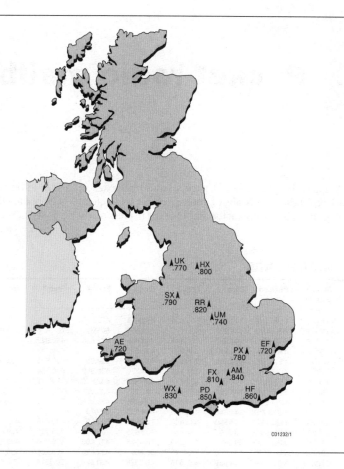

CD1232/1

MICROWAVE REPEATERS

While not being talked through, microwave repeaters revert to a beacon mode

23cm speech

Channel	Input (MHz)	Output (MHz)
RM0	1291.000	1297.000
RM2	1291.050	1297.050
RM3	1291.075	1297.075
RM5	1291.125	1297.125
RM6	1291.150	1297.150
RM9	1291.225	1297.225
RM12	1291.300	1297.300
RM15	1291.375	1297.375

There is also a range of 1.3 and 10GHz TV units as shown in the table.

= Speech

= TV unit

CD1231/1

10 Packet radio mailboxes

THIS appendix gives a list of packet radio mailboxes in the UK. These details are of course subject to frequent change, but are representative at the time of going to press (early 2000).

Up-to-date details of the mailboxes and their user groups can be obtained from the owners shown in the tables, or from the RSGB Data Communications Committee web site (linked from www.rsgb.org), from which these tables were taken.

AX25 MAILBOXES

Mailbox callsign	Owner callsign	Postcode	Freq 1	Freq 2	Freq 3	Freq 4	Freq 5	Freq 6	Freq 7	Freq 8	Freq 9	Freq10	Freq11
GB7ABB	GW0ENT	LL54 7YE	144.950	70.4875	432.675	430.650							
GB7ABN	GM1THS	AB1 4NG	144.950	432.675	430.650	50.650	70.4875	439.825	433.675				
GB7ADH	M0ADH	S19 6PA	144.950	70.3125	430.650	433.650	439.925						
GB7AHE	G7AHE	TA6 6JZ	144.9375	439.925	432.675								
GB7ATC	G4JDL	PE2 5SB	144.850	70.4875	432.650	439.850	430.675	433.650					
GB7AVM	G0DFP	OX9 4TL	433.650	144.950	70.4875	439.850	430.650						
GB7AYI	G7AYI	LE3 3LA	144.950	430.775	439.925	432.650	1240.750						
GB7AYR	GM0JQE	KA11 1PN	144.950	432.675	70.325	439.925							
GB7BAD	G4IRX	NG3 6BL	1240.750	144.850	430.625	433.675	70.3125	50.650					
GB7BAS	G0TJW	SS13 2HX	144.850	432.675	439.825	50.670							
GB7BAY	GW8FVI	LL52 0EF	144.850	433.650	70.4875	50.650	430.625	432.625	439.925				
GB7BED	G0BKN	MK41 9AL	50.670	144.850	432.675	70.4875	430.725	439.925					
GB7BEN	G8TTK	MK3 7BQ	50.670	144.950	70.4875	432.650	439.950	430.650					
GB7BEV	G8AEN	BL8 3DB	144.950	70.4875	50.650	432.675	1299.000	430.625	439.850				
GB7BHM	G1WXA	WS10 7RZ	144.850	433.675	432.625	430.675	70.3375	439.825	1240.700	1299.175	439.875		
GB7BLE	G0GTM	NE24 5TJ	432.675	144.950	439.975	430.675	70.3375	1299.000	433.675	50.690			
GB7BMR	G1GDB	CA5 2HZ	144.950	70.4875	432.675	439.875							
GB7BMT	G1GRB	BH8 9LT	144.9375	439.850	432.650	70.4875							
GB7BMX	G3YRH	NE12 6BT	144.950	70.4875	430.675	439.975	50.690						
GB7BNM	G4WPT	BH21 6QB	50.650	144.850	432.625	439.850	70.4875	1299.000					
GB7BOB	G0RLT	PR9 7AS	144.850	432.650	430.725	439.925	1299.000	70.3375	50.650				
GB7CAO	GM1AHC	KW14 8QB	144.850	432.675									
GB7CFB	G0CFB	IP19 0PY	50.650	144.950	432.675	70.4875	439.825	430.625					
GB7COS	G0MBA	CO16 8PT	144.950	433.675	439.850	430.675	1299.025	1240.950					
GB7COV	G3ZFR	CV7 8JJ	144.8625	430.625	432.650	1299.00	1240.150	70.4875	439.875	430.775	439.975	430.5875	439.7875
GB7CYM	G0PWO	YO2 3LF	144.850	432.675	1299.575	1240.450	70.3375	430.625	439.875				
GB7DAA	G3TIK	SG12 9JN	144.950	433.650	439.825								
GB7DAD	G0KLR	SK17 7EQ	144.850	433.650	439.975	70.3125	50.670	439.7625	430.5625				
GB7DBY	G8DBY	DE1 1RZ	144.950	432.675	70.4875	50.650							
GB7DDX	G0DDX	CB3 7XB	70.4875	144.950	1299.425	432.675	439.975						
GB7DEE	GW1SYG	*******	144.850	433.675	439.925								
GB7DEO	G0DEO	TW14 9JE	144.950	432.675	70.4875	50.570	439.850						
GB7DHI	G6DHI	M16 8PW	144.850	433.675	1299.00	439.875	144.975						
GB7DID	G1SSL	OX11 8HD	144.850	432.675	439.850	70.4875	430.725	439.675	430.475	1299.425	1240.150		
GB7DTX	G7DTX	LN1 3TJ	144.950	432.675	439.875	70.4875							
GB7EBN	G4MVN	BN20 8LT	144.950	70.4875	432.675	439.825							
GB7EDN	GM0ALS	EH12 7RZ	144.850	70.4875	432.625	439.850	430.625						
GB7ELS	G7ELS	LS15 0EZ	144.950	50.610	433.650	439.950	70.3125						
GB7ESX	G1NNB	CM8 2SZ	144.850	70.4875	432.675	50.650	433.650	439.825	430.650				
GB7EYM	G4HRM	YO13 9EY	50.650	144.850	1299.725	432.675	439.875						
GB7FBD	G7FBD	BS16 4JT	144.850	432.675	439.875	430.775							
GB7FEN	G7EBL	PE16 6PJ	144.9375	432.650	430.675	70.4875	50.690	439.825					
GB7FLG	G0FLG	LE6 4DF	50.670	1299.000	70.4875	432.675	439.850	430.675					
GB7FLY	G4FLY	RG1 7UG	144.950	432.675	439.850	430.650	50.650	1299.725					
GB7FUR	G3VUS	LA13 9AR	144.8625	433.675	1240.300	439.925	70.4875	50.650	430.625				
GB7GBY	G0MUY	DN32 0NP	70.4875	144.850	430.775	439.975							
GB7GCW	G0GCW	FY4 3HD	144.850	432.650	430.725	70.3375	50.650	439.850	1299.000				
GB7GFD	G4NBC	GU2 6TU	144.950	432.675	50.650	70.4875	1299.725	430.725	439.925				
GB7GLO	G4FPV	WR14 2BQ	144.950	70.3375	50.650	430.625	432.675	439.825	1299.575	1240.300			
GB7GSY	GU4WRP	GY1 1SF	144.850	432.650	430.775	439.925	70.3125						
GB7GUR	GU4YMV		50.650	144.950	432.675	70.4875	439.875	430.650					
GB7HFG	G0DEO	TW14 9JE	14.112	14.107									
GB7HSN	G1HSN	SE9 4AW	144.9375	433.650	70.4875	50.670	1299.425	430.725	439.875				
GB7HUL	G8UVQ	HU5 2AE	144.950	70.4875	50.650	439.925							
GB7HVU	M0CKO	BB6 7NL	144.950	432.625	430.650	70.4875	50.630	433.675					
GB7HXA	G4UXV	PE18 7JE	50.670	144.950	1299.425	432.675							
GB7IFI	G0IFI	TA6 4QJ	144.9375	432.675	439.925								
GB7INV	GM4JNB	IV3 6SD	144.850	432.675	430.650	439.850							

Mailbox callsign	Owner callsign	Postcode	Freq 1	Freq 2	Freq 3	Freq 4	Freq 5	Freq 6	Freq 7	Freq 8	Freq 9	Freq10	Freq11
GB7IOM	GD3YEO	IM2 2NP	144.950	70.3375	432.675	439.875	430.625	1240.450	1299.575				
GB7IPN	G8GCS	TQ12 5QS	144.950	432.675	439.875	70.3375	1299.425	430.650					
GB7IZR	G7IZR	WN5 7NB	144.525	430.625	1240.600	1299.000							
GB7JED	GM4UPX	TD8 6NP	50.650	144.850	70.4875	430.625	432.625	439.875	1299.725				
GB7JEM	G6UNC	HD5 0LW	50.650	144.950	430.675	432.675	70.4875	439.850					
GB7JSC	GM1VBE	G71 8AR	50.650	144.850	1299.425	439.925	433.650	432.650	70.4875				
GB7JYU	G4JYU	CT13 9JE	144.950	70.4875	430.625	433.650	439.975	1299.025	432.675				
GB7KAP	M1ADU	NG15 7NU	144.850	439.825	433.650	70.3125							
GB7KET	G7HIF	NN15 6TX	144.9375	430.625	432.650	439.825							
GB7KLY	G7HJT	BD22 7DX	144.950	50.650	432.675	1240.150	70.4875						
GB7LEN	G4LBJ	L31 5JU	70.4875	144.950	433.650	430.675	439.925	1299.125	1240.600				
GB7LWB	G4JLB	NN7 1AE	144.850	70.4875	50.650	433.675	1299.100	439.825	430.650				
GB7MAD	G7BTI	HR2 9NH	144.850	70.325	430.775	439.925	432.650						
GB7MAM	G1EQT	DE55 2DJ	144.950	432.650	50.650	439.875	430.650	70.4875					
GB7MAN	GD3YUM	IM1 2QD	144.950										
GB7MAX	G1DKI	WV6 7LX	144.950	432.650	439.875	1240.300	430.675						
GB7MRU	G4MRU	S30 5QB	144.950	432.675	50.570	439.925	430.5125	439.7125	70.3125	431.500			
GB7MSF	G4MSF	NE10 8EX	144.850	432.650	439.875	70.3375	430.625	1299.725	1240.300	50.650			
GB7MSF	G4MSF		14.093	14.097	14.103	14.107	29.250	3.593	3.597				
GB7MSW	G3MSW	AL5 5DW	432.675	144.850	1299.650	70.4875	50.670	439.850					
GB7MXM	G6SYW	IP14 5SN	70.4875	439.850									
GB7NEQ	G4OCO	TR9 6AY	144.850	432.675	70.4875	430.650							
GB7NLY	G8NLY	GU15 4DD	144.850	144.975	70.4875	430.650	433.650	439.850	1299.725	1240.150	432.650	50.670	
GB7NNA	G1NNA	CM8 2SZ	50.650	144.850	1299.000	70.4875	432.675						
GB7NND	G0WTK	S81 7DF	144.850	432.675	70.3375	430.775	439.975	50.630	1240.450	144.975	1299.725		
GB7NOS	GM0HBI		50.650	144.850	1299.000	432.675	430.650	439.850					
GB7NRC	G1HUL	LE67 9SN	430.775										
GB7NSY	G8ETC	KT19 9SF	144.850	70.4875	432.650	439.850	430.625						
GB7NXL	G0NXL	RM9 5LP	144.850	70.4875	433.675								
GB7OAR	G4OAR	L43 0TT	144.850	432.675	70.4875	1299.575	1240.150	439.925	430.775	432.650			
GB7ODM	G1VSJ	OL9 9TB	144.850	70.4875	1299.425	50.650	430.625	439.825	432.650	1240.150			
GB7OMN	G7OMN	PR8 6NF	144.975	433.650	1240.600	430.725							
GB7OPC	G0OPC	PE13 5BP	144.850	430.675	432.650	439.850	70.4875	433.650					
GB7OZV	G0WEI	WD1 6QG	144.950	50.650	432.675	1299.575	439.850	70.4875					
GB7PAB	G1RXR	PL4 7ER	144.850	432.675	439.875	70.4875							
GB7PEN	G6BSK	CA11 8LY	50.650	144.950	430.725	432.675	439.875	1299.425					
GB7PET	G1ARV	PE6 7TR	144.950	433.675	70.4875	432.675	439.850						
GB7PFD	G6HJP	GU31 4EE	144.8625	432.650	50.690	70.4875	430.650						
GB7PLM	G7DQC	PL5 1DS	144.9375	432.650	439.875								
GB7PLY	G3KFN	PL6 7AY	144.950	432.675	70.4875	50.650	1299.00	439.875					
GB7PMB	G3UQH	SY5 0HG	50.650	144.850	1299.725	439.900	70.3375	433.650	430.675				
GB7PZT	G8PZT	DY11 5EB	144.975	144.850	70.4875	433.650	430.625	430.725	439.825	439.850	432.675	430.700	439.900
GB7RDN	G0RDN	CT14 7EZ	144.950	70.4875	430.625	432.675	433.650	439.975	1299.025	50.670			
GB7RMS	G1UTF	TN29 0BU	144.950	70.4875	433.650	439.825							
GB7RQT	GW3RQT	SA70 7TS	144.950	430.675	432.675	70.4875							
GB7RUG	G1KQH	WS15 2XH	144.850	432.650	70.4875	430.775	50.650	433.675	439.825	430.5875	439.7875		
GB7SAM	G6HCI	ST5 3BS	144.950	432.675	439.825	430.675							
GB7SAN	GM3SAN	G69 7HW	50.650	144.950	70.4875	432.625	439.850	1299.575					
GB7SDN	G8VRI	SN3 5AG	144.850	432.675	1240.300								
GB7SDY	GM4PSX	KW17 2BJ	144.850	432.675	50.630	70.4875	29.250	29.230	28.130	28.140	21.080	3.605	14.093
GB7SEK	G7SEK	LE11 4TQ	144.850	432.675	439.850	70.3125							
GB7SFK	G3ZYP	IP12 2PL	50.650	430.650	432.650	439.825	1299.000						
GB7SIG	G3WGM	DT11 7SS	50.650	70.4875	28.080	28.081	21.080	21.081	14.075	14.076	14.077	10.145	10.146
GB7SIG_	G3WGM		7.038	7.039	7.040	3.580	3.588	3.589					
GB7SJP	G8SJP	HP7 9DZ	144.8625	70.3375	430.725	439.825	1240.450	1299.575	432.675	50.570	70.4875		
GB7SKG	G1ZCW	LN11 8QN	144.950	430.675	432.650	70.4875	439.975						
GB7SOL	G6URP	B93 9JD	144.9375	432.675	70.4875	430.675	439.825						
GB7SOU	G8IPG	SO40 7GF	144.950	439.850	432.650	430.625	70.3375						
GB7SQI	G1SQI	PL2 3RS	144.8625	432.650	439.875								
GB7SRL	GM4SRL	G76 7HG	144.950	432.650									
GB7SUF	GM4SUF	IV19 1LB	50.650	144.850	1299.000	432.675	430.650	439.850					
GB7SUN	G0OYN	PO2 8NH	144.850	432.675	439.850	50.650	430.725	1299.000	1240.150				
GB7SUT	G8AMD	B73 6PG	50.650	144.950	1299.000								
GB7SWN	GW8TVX	SA2 7UH	144.950	70.4875	432.675								
GB7SYP	G6TVA	S70 6RR	144.9375	433.700	50.670	70.3375	1299.575	439.850	430.675	1240.600	432.625	144.825	430.5125
GB7TAS	G4ZZZ	TR4 8QL	144.950	432.675	70.4875	50.650							
GB7TCM	G8ADH	WR8 0PQ	50.650	144.950	1240.600								
GB7TDG	G4VEL	IP24 3HQ	144.950	70.4875	432.675	430.675	439.825	433.675					
GB7TED	GI4AHP	BT27 5BF	50.650	144.850	430.625	432.675	439.825	70.325					
GB7TJF	G8TJF	TA2 8NA	144.950	432.650	439.925	430.625							
GB7TJZ	G6TJZ	BS17 1JW	144.950	433.650	432.650	70.4875	430.750	439.850	439.825	439.875			
GB7TUT	G4TUT	EN3 4QF	144.850	432.650	439.850	70.4875							
GB7ULV	G4BDE	LA12 9PF	50.650	144.850	432.675	439.925	1240.300						
GB7UWS	G1UWS	SE9 2PF	144.950	432.675	1299.425	430.700							
GB7VLS	G4VLS	NR1 2NX	50.65	144.950	432.675	70.4875	1299.425						
GB7VRB	G8DHE	BN14 7QW	144.950	432.675	1299.000	430.675	439.925	70.3375					
GB7WAM	G3SPX	WF3 2JG	144.8625	433.700	430.675	1240.600	70.3375	50.570					
GB7WAR	G0KPH	CV32 7QS	144.950	433.675	439.975	430.675	432.675						
GB7WIG	G6YLW	ME8 0NR	50.610	70.3125	144.950	430.625	439.825	433.650					
GB7WNM	G4RKN	PE33 0LB	144.850	430.675	432.675	70.4875	439.975						
GB7WRC	G7DMS	BB12 6NZ	50.630	144.850	432.675	439.925							
GB7WRG	G0COA	WF4 4TE	144.850	1299.000	50.650	432.650	70.4875	430.475	439.675	430.425	439.625		
GB7WRI	GI4WRI	BT41 3LT	50.650	144.950	432.675	430.625	439.825	70.4875					
GB7WSX	G8HQJ	RH11 8EQ	144.9375	432.650	439.850	430.625							

Mailbox callsign	Owner callsign	Postcode	Freq 1	Freq 2	Freq 3	Freq 4	Freq 5	Freq 6	Freq 7	Freq 8	Freq 9	Freq10	Freq11
GB7WXM	GW8HBP	LL14 2RL	144.950	432.675	439.900	70.4875	430.625						
GB7XDU	G4XDU	SL6 6DA	144.975	432.675	70.4875	50.670	70.4875	439.850					
GB7XJZ	G0SBV	SO45 2HP	144.950	430.625	432.650	70.3375	439.850						
GB7YEO	G7OGG	BA21 3UE	144.8625	433.650	439.925								
GB7YEW	GM3YEW	PH2 9LW	144.950	432.675	70.325	430.625	432.625						
GB7YUH	G3YUH	CT9 4DH	144.950	430.625	439.875	433.650	70.4875	432.675	1299.025				
GB7ZET	GM4IPK	ZE2 9JJ	144.950	433.650									
GB7ZZZ	G0JDR	BN8 5QD	144.9375	432.650	439.825	430.625							

AX25/IP MAILBOXES

Mailbox callsign	Owner callsign	Postcode	Freq 1	Freq 2	Freq 3	Freq 4	Freq 5	Freq 6	Freq 7	Freq 8	Freq 9	Freq10	Freq11
GB7CAM	G7AOR	TS12 3DN	144.925	432.675	70.4875	439.875							
GB7CIP	G4APL	CR3 5EL	432.625	433.625	50.650	70.3125	432.675	439.825					
GB7CRV	G6CRV	LA3 1NZ	144.8625	432.625	430.675	439.825	70.3125	1240.300					
GB7DON	G0UQQ	DN5 7RX	50.570	144.925	433.650	432.625	70.3125						
GB7FYL	G1MET	FY1 3RB	144.9375	432.675	70.3375	1299.000	1240.300	430.700	439.925	433.625			
GB7GBR	G4NQO	WC13 3AP	50.690	70.3125	1299.450	430.650	439.875						
GB7IPW	G4MDP	BN13 3NR	144.950	432.625	433.625	439.925	430.700	432.675	430.675				
GB7LPT	G7JJP	PE1 4SZ	144.8625	439.850	70.4875	432.625	433.625						
GB7LWS	G1TDM	BN7 1PS	430.625	439.875									
GB7MHD	GM0MHD	AB41 8LR	144.950	432.625	433.675	439.825	430.650						
GB7NET	GM7AOM	ML9 2DA	50.650	144.825	1299.425	439.925	433.625	432.625	70.4875				
GB7NHT	G0GSR	PL20 6SS	144.875	432.625	1240.150	70.4875	432.675	430.650					
GB7NOT	G1HTL	NG5 6QZ	50.690	144.875	433.625	439.825	1240.150	432.675	430.625	70.3375			
GB7NRY	G8NRY	HX3 7RD	144.925	432.625	70.4875	433.625	50.650	439.950					
GB7OSP	GW6HVA	LL30 2UU	144.9375	70.4875	50.650	432.675	430.725	1299.725	439.975				
GB7STU	G7JYF	DA11 7DE	144.925	432.625	439.950	430.725	433.625	439.675	430.475	430.700	70.3125		
GB7SXE	G1DVU	TN37 7PS	430.650	144.850	439.825	50.650	70.3125	1240.975	433.650	1299.00	432.675		
GB7ZPU	G1ZPU	SG19 2NE	50.670	144.9375	1299.000	70.4875	433.650	439.825					

TCP/IP MAILBOXES

Mailbox callsign	Owner callsign	Postcode	Freq 1	Freq 2	Freq 3	Freq 4	Freq 5	Freq 6	Freq 7	Freq 8	Freq 9	Freq10	Freq11
GB7BBC	G4HIP	London, W	144.925	70.4875	433.650	439.825	439.875						
GB7BIF	G4RWO	LA14 4ND	144.925	430.650									
GB7BIG	GD3UMW	IM4 2HE	144.975	70.4875	430.675	432.625	1240.450	1299.575					
GB7BIP	M1BSX	PO9 3JZ	144.875	433.625	432.625	70.3125	50.690	430.700					
GB7BVG	G3VQP	GU15 3NH	144.925	432.625	433.625	439.925	70.4875	430.725	1240.450	1299.425	144.825		
GB7DIP	M1ATV	DE55 7LA	144.925	432.625	70.3375	439.975	430.775	144.825	433.625	1299.575	1240.300		
GB7DIY	GU0FYR	GY1 1QR	144.875	433.625	439.875	430.650							
GB7DQY	G6DQY	SY4 2HY	70.3125	144.925	430.625	432.625	439.825						
GB7ECR	G0AHJ	SK10 4HY	144.925	433.625	430.675	439.850	432.625						
GB7EIP	G3XVV	CM8 3PH	144.925	432.625	70.4875	430.775	439.975	433.625	430.625	430.425	439.625	430.5125	439.7125
GB7EIY	GM8HBY	ML6 7DH	70.4875	144.925	432.625	1299.850							
GB7EMC	G8CZE	M43 6EF	144.875	432.625	439.850	433.625	430.650						
GB7ESC	G0WDA	GL4 6GD	144.875	50.630	439.850	430.700	432.625						
GB7FCR	M1CUK	FY1 4EQ	144.875	70.4875	430.700	433.625	1240.300						
GB7FDT	GI7FDT	BT63 5EQ	144.925	432.625	439.825	430.625	70.4875						
GB7FLK	M1CMN	CT19 5BZ	144.825	430.725	439.900	432.625	50.650	433.625					
GB7FLX	G7OCD	IP11 9NF	144.975	50.530	70.4875	432.625	1240.450	1299.575					
GB7HER	G3ZRK	DH7 6NX	144.925	430.675	439.975	70.4875	1299.000	432.625	50.690	70.3375			
GB7HIP	G7PEW	PO7 8HG	144.925	70.3125	432.625	439.850	430.725	1240.800	1299.000	433.625			
GB7IDD	G8NRB	LU6 1HB	144.875	430.675	439.900	70.3375	432.625						
GB7IHA	G1WKK	RG22 6LN	144.875	433.625	439.875								
GB7IHH	G0TCD	HP2 7QW	144.925	439.925	430.775	70.3375							
GB7IMK	G0TWN	MK3 7BQ	144.875	70.4875	433.625	439.900	430.625	430.675					
GB7IND	G7FCE	EN11 9QS	144.925	432.625	439.925	70.3375	430.650						
GB7INE	G8DZH	IG10 4RA	144.925	432.625	439.925								
GB7ING	G4XRT	N9 8HE	144.925	432.625	439.925								
GB7INH	G4YSJ	EN5 5LU	144.925	433.625	439.925								
GB7IPB	G4HJG	TW15 1EH	144.925	432.625	439.925	70.4875	1240.150	430.725					
GB7IPC	G4BIO	HP10 8AE	144.825	1240.150	432.625	430.675	439.950						
GB7IPD	G8ECJ	HP10 9LH	144.925	433.625	430.675	70.3375	439.825	439.950	439.875				
GB7IPE	G1LIG	SL6 9NF	144.875	433.625	430.725	70.3375							
GB7IPF	G4TNU	HP6 6RR	433.625	144.925	430.675	430.775	439.875	439.850					
GB7IPH	G4ZXI	TN27 9QQ	144.875	432.625	439.900	430.725	50.650	439.675	430.475				
GB7IPO	G4UDE	SY10 7RQ	144.925	70.3125	432.625	1299.425							
GB7IPS	G7WGK	SY1 4LG	144.925	439.825	432.625								
GB7IPT	G8JVM	TF4 3DD	144.925	432.625	430.625	70.325	439.975	430.5875	439.7875	439.900	430.700		
GB7IQB	G4ETG	SG2 8AN	144.925	439.925	70.4875	433.625							
GB7ITG	G6DZJ	HP23 4DS	144.925	439.900	433.625	430.625							
GB7KHW	G6KHW	MK45 3JP	144.875	50.670									
GB7KIP	G0CSF	TN15 8JU	433.625	432.625	430.725	439.675	430.475						
GB7LAG	G8LSN	RH9 8HB	144.925	433.625	432.625	1299.000							
GB7LGS	G0LGS	GL51 5HH	144.925	430.650	439.875	433.625	430.700	439.950					

Mailbox callsign	Owner callsign	Postcode	Freq 1	Freq 2	Freq 3	Freq 4	Freq 5	Freq 6	Freq 7	Freq 8	Freq 9	Freq10	Freq11
GB7MAI	G4EMC	ME20 6LY	144.8625	430.725	432.625	433.625	439.900	439.675	430.475				
GB7MAS	M0AGQ	BN1 6NL	144.925	70.3375	50.630	430.650	432.625	433.625	439.925				
GB7MBB	G6PHF	LA4 4UT	70.3375	144.925	432.625	430.650							
GB7MBR	M0AZM	LA4 5AA	430.675	439.825	1299.325	1240.950							
GB7MIP	G6KMQ	B91 3LL	144.925	430.675	432.625	433.625	439.825						
GB7NEW	G6RBP	RG14 2PN	144.925	433.625	432.625	430.700	439.925						
GB7NHV	G0TJH	BN9 9NH	144.875	432.625	430.625								
GB7NIP	G4MCF	TS6 0BS	144.925	439.875									
GB7NLW	G7PUN	WA12 0LN	144.925	70.325	433.625	439.975	430.675						
GB7NWI	G4TUP	PR8 5DA	50.650	144.925	430.675	70.4875	439.875	1240.950	1299.325	433.625			
GB7OIP	G6GZH	CB4 5NE	144.925	432.625	439.975	1299.425	70.3125	430.5125	439.7125				
GB7OQJ	G6OQJ	CM1 5SX	144.925	439.975	433.625	430.775							
GB7PAE	G0PAE	SS16 6SJ	144.925	439.975	433.625								
GB7PEJ	G0PEJ	SS5 5BP	144.925	433.625	439.975	430.775							
GB7PIT	G7PIT	LE65 2NN	144.925	70.4875	433.625								
GB7POZ	G7POZ	NR33 7SS	144.925	430.675	439.825								
GB7PVO	G0WWF	SK14 5QA	144.875	432.625	439.975	70.4875	1299.425	1240.150	430.675				
GB7RKN	G3RKN	CM20 3EF	144.925	439.925									
GB7RTW	G8SFR	TN2 4EY	144.925	432.625	430.725	439.675	430.475						
GB7SIP	G3ZIY	RH6 9NA	144.925	439.825									
GB7STW	G8STW	CM9 8XG	144.925	439.975	433.625	430.775							
GB7TON	G0DJC	TN10 3HG	144.925	432.625	430.725	439.675	430.475						
GB7TVG	G4WQS	SL4 5TD	144.925	70.3125	433.625	439.850							
GB7WFS	G0WFS	LE9 5BF	433.625	144.925	70.3125	439.925							
GB7WIP	G1FIP	CV34 5PB	144.925	439.975	430.675	432.625	1240.150	1299.000					
GB7WLR	G4EID	PR9 9TW	144.875	70.4875	430.675	432.625	1299.325	1240.950	439.850				
GB7WMD	G1WMD	GU14 9LE	144.925	432.675	1299.727								
GB7WPJ	G6WPJ	CO9 2HP	144.925	439.975	433.625	430.775							
GB7ZHR	G8ZHR	DA1 4RY	144.875	50.730	1299.725	432.625	430.725	439.900					

SATGATE MAILBOXES

Mailbox callsign	Owner callsign	Postcode	Freq 1	Freq 2	Freq 3	Freq 4	Freq 5	Freq 6	Freq 7	Freq 8	Freq 9	Freq10	Freq11
GB7LAN	G8TZJ	LA1 4LQ	145.975	145.900	432.675								
GB7LDI	G3LDI	NR14 8LQ	144.850	145.970	145.900	70.4875	432.675	439.825	50.650	430.625	439.825		
GB7SAT	G3WGM	DT11 7SS	50.650	70.4875	145.900	145.975	435.120						

DXCLUSTER MAILBOXES

Mailbox callsign	Owner callsign	Postcode	Freq 1	Freq 2	Freq 3	Freq 4	Freq 5	Freq 6	Freq 7	Freq 8	Freq 9	Freq10	Freq11
GB7ADX	GW4VEQ	LL65 3RD	144.900	70.325	432.675	50.650	439.850	1299.000	430.675				
GB7BAA	G8TIC	WR3 8TJ	144.900	432.650	70.3375								
GB7BDX	GM6YRZ	TD5 8LB	144.900	433.675	70.325	430.625	439.875						
GB7BPQ	G1MNB	********	144.900	430.675	50.650	70.325	433.675	432.625	1240.150	1299.975	439.900		
GB7CDX	G4BWP	IP28 8LQ	70.325	430.625	439.850	144.900							
GB7DJK	G1TLH	NR19 2ET	50.650	144.900	1299.000	432.675	70.325	430.625	439.825				
GB7DXA	G0RDI	HP7 9DZ	144.975	144.8875	1299.575	70.325	439.825	1240.450	50.530	430.650	432.675		
GB7DXB	G3VHB	WS13 8QZ	144.8875	70.325	439.900	433.675	1299.200						
GB7DXC	G0TYA	GL2 0SJ	144.900	1299.425	70.325	430.750	439.900						
GB7DXD	G0OFE	BH21 2NL	144.900	433.650	70.3250	439.850							
GB7DXG	GU0DXX	GY6 8XE	144.900	430.675	70.4875	433.675							
GB7DXH	G0UHK	HA1 3HT	144.900	144.8625	50.530	433.675	439.825	1299.950	1240.450	70.325			
GB7DXI	G4LJF	RG41 4AX	70.325	432.675	1299.950	144.900	430.725						
GB7DXK	M0BMD	TN37 7PS	144.900	1299.000	430.650	439.825	1240.975	50.650	70.3125	432.675			
GB7DXL	G6YIN	LS12 4HH	144.8625	433.675	70.325	430.625	439.825	439.875	144.975				
GB7DXM	G4PIQ	IP5 3RE	70.325	432.650	1299.500	439.825	430.625	1240.750					
GB7DXN	GM1THS	AB1 4NG	144.900	432.675	433.675	439.825							
GB7DXS	G3VKW	RH17 5AW	144.900	433.675	70.325	1299.875	430.650						
GB7DXW	G3PSM	SO40 3HY	144.900	70.325	50.610	430.625	433.675	1299.000	439.875				
GB7DXX	G3RTU	M7 2DH	144.900	430.625	439.875	439.975							
GB7DXY	G3RTU	M7 2DH	144.900	430.625	439.875	439.975							
GB7GDX	G0TYA	GL2 0SJ	144.900	1299.425	70.325	430.750							
GB7HDX	GM3WOJ	IV18 0PE	144.900	432.675	439.850	70.325							
GB7HFC	G0RDI		144.975	144.8875	1299.575	70.325	439.825	1240.450	50.530	430.650	432.675		
GB7MBC	G0VGS	LA4 6EQ	144.900	70.325	430.675	1240.950	1299.325	439.800					
GB7MDX	G4UJS	SY13 2PT	144.900	432.675	70.325	430.725	1240.950	439.825	1299.450				
GB7NDX	GM4WZY	DD9 6BT	70.3250	144.900	432.625	430.625	439.825						
GB7ODX	GM4PSX	KW17 2BA	144.900	432.650	70.4875	50.630	29.250	29.230	28.130	28.140	21.080	14.093	3.605
GB7PDX	G3ORH	EX17 6QY	432.625	433.650	439.875								
GB7SDX	GM0OPS	G46 7DP	144.900	70.3375	432.675	1299.575	439.975	430.775					
GB7TDX	G4NVR	NE65 9JN	144.900	433.675	439.975								
GB7TLH	G1TLH	NR19 2ET	50.650	144.900	1299	432.675	70.325	430.625	439.825				
GB7UDX	GI1ONL	BT28 2DW	144.900	432.675	70.3125	439.850	430.675	433.650					
GB7UJS	G4UJS	SY13 2PT	144.900	432.675	70.325	430.725	1240.950	439.825	1299.450				
GB7WDX	G7OPJ	BS22 9AQ	144.900	50.750	433.675	70.3125	430.650	439.925					
GB7YDX	G4DBN	DN14 7HE	50.690	144.900	433.675	70.325	1299.075	430.725	1240.300	439.875			

Index

A

Abbreviations, CW   22
Absorption, ionospheric   39
Accessibility of shack   9
Acoustics, shack   9
Allocations, frequency   3, 216
Amateur-Satellite Service   1
Amateur Service   1
　frequency allocations   216
Amplifiers, linear   34, 106
AMSAT   129
AMTOR   140
Antennas
　136kHz   49
　1.8MHz   50
　3.5MHz   52
　choosing   8
　contest   91, 106
　DX   33
　EMC aspects   15
　low-power operation   57
　mobile   114
　portable operation   122
　satellite   133, 136
　sense   109
　shipboard   120
　special-event   169
　VHF   16
ATV repeaters   161
Auroras   71
Awards
　HF   58
　VHF/UHF/microwave   82
Az-el rotator   79

B

Bands
　HF   49–55
　microwave   81
　VHF/UHF   66–69
Band plans   49, 133, 224
Beacon network   49, 233
Blackout   47
Breakthrough   15
Bulletin boards   148

C

Calling frequencies   65
Callsigns   1
　international callsign series
　　holders   179
　list   203
Capture effect   116
CEPT operation   4, 123, 124
Channels, VHF   116, 228
Check lists, contest   90
Clover   141
Comfort in shack   9
Contests
　HF   87
　VHF/UHF   96

Conversation   26
CQ call   18
CTCSS repeater access   117, 248
CW operating   19
　in a foreign language   24

D

Data communications   139
Demonstration stations   166
Direction finding contests   107
Ducts, atmospheric   80
Duplex operation   26
DX
　HF   33
　microwave   79
　VHF/UHF   64
DXCC entities list   208
DX cluster system   36, 149
DXpedition organising   124
DXNS Voicebank system   36

E

E-mail 'reflector' system   36
Earths   16
EMC   15
EME   78
Emissions, designation of   1
Emergency operation in the UK   28
Equipment, choosing   7

F

Facsimile   160
Fast-scan TV   160
Fax   160
Feeders   9, 64, 106
Foreign-country operation   4, 123, 124
Foreign-language contacts
　CW   24
　phone   27, 219
Foxhunting   110
Frequency allocations   3, 216

G

GTOR   141
Generators   103
Gin-pole   93, 102
Great-circle maps   35
Grey-line propagation   44, 57
Ground scatter   47

H

Harmonics   17
Hellscreiber   142
HF
　antennas   15
　awards   58
　band characteristics   39
　contests   87
　mobile whip antennas   114
　packet   141

I

IARU   4
　Monitoring System   5
Image techniques   156
Importation of amateur
　　equipment   124, 125
Insurance   15
Interference   15, 144
International Amateur Radio Union   4
Internet   36
Ionospheric absorption   40
ITU   1

L

Licensing
　CEPT   4, 123, 124
　reciprocal   123
Linear amplifiers   34, 106
Listening tips   35
Lists   38
Locator system   65
　map of Europe   173
Logkeeping   28, 89, 99
Low-power DXing   55

M

Maidenhead locator   65
Mailboxes, packet   148, 252
Maps
　continental and regional   173
　great-circle   35
Maritime mobile operation   119
Masts   92, 102, 122
Maximum usable frequency   39
Meteor-scatter propagation   73
Microphone technique   26
Microwaves   79
　award   84
MIR   137
Mobile operation   114
Moonbounce   78
Multipliers, contest   87

N

National telecommunication
　　administration   4
Net operation   28, 38
Northern lights   71

O

Operation abroad   4, 123, 124
Orbital data   131

P

Packet radio   142
　mailboxes   148, 252
　network   36, 73
　satellites   148
PacTOR   141
Permitted services   3

Phone operation
  foreign language   27, 219
  techniques and procedures   24
Phonetic alphabet   26
Planning permission   10
PMS   147
Portable operation, single
      operator   121
Power in the shack   9
Prefixes, callsign   4
Primary services   3
Privacy in the shack   9
Procedure signals, CW   22
Propagation
  HF   39, 91
  VHF/UHF   69
PSK31   141

Q
Q-code   21
QRP operation   55
QSL bureau   29
QSL cards   29

R
Radio Regulations   1
Receiver
  DF requirements   108, 111
  performance, HF   34, 51
  performance, VHF   104
  spurious signals   52

Reciprocal licensing   123
Repeaters   116, 246
RST code   22
RTTY   140
Rusty bolt effect   9

S
Safety precautions   12, 92, 101
Satellites, amateur   129
Secondary services   3
Security, shack   14
SID   47
Sites, VHF contest   97
Sked   18
Skip focussing   47
Slow-scan TV   156
Solar repeats   71
Solar variations   47
Solar wind   71
Special-event stations   165
Sporadic-E (Es) propagation   46, 70
Squares, locator   65
Station, main
  layout   11
  siting   8
Storms, magnetic   47
Sudden commencement geomagnetic
      disturbance   71
Suffixes, callsign   4
Sunspots   47
Super-refraction layer   80

SWLs   33

T
Tail-ending   57, 117
Talk-in stations   168
Telephony techniques and
      procedures   24
Television
  fast-scan   160
  slow-scan   156
Terminal node controller   143
TMOR reporting system   78
TNC   143
Tone access, repeater   117
Transceivers, HF   34
Transmitter checks and
      adjustments   15
Tropospheric propagation   69

U
Unilateral licensing   123

V
Voicebank system   36
VOX   26

W
World Administrative Radio
      Conference   1
WWV broadcasts   48